Globalization and Diversity

GEOGRAPHY OF A CHANGING WORLD

Les Rowntree
San Jose State University

Martin Lewis
Stanford University

Marie Price
George Washington University

William Wyckoff
Montana State University

PEARSON
Prentice
Hall

Upper Saddle River, NJ 07458

Library of Congress Cataloging-in-Publication Data

Globalization and diversity : geography of a changing world / Les Rowntree ... [et al.].
    p. cm.
  Includes index.
  ISBN 0-13-147739-0
   1. Economic geography. 2. Globalization. I. Rowntree, Lester

HF1025.G59 2005
330.9—dc22

2004044752

Executive Editor: Daniel Kaveney
Editor in Chief, Science: John Challice
Development Editor: Ginger Birkeland
Production Editor: Tim Flem/PublishWare
Vice President of Production and Manufacturing: David W. Riccardi
Executive Managing Editor: Kathleen Schiaparelli
Editor in Chief of Development: Carol Trueheart
Managing Editor, Science Media: Nicole Bush
Media Editor: Chris Rapp
Associate Editor: Amanda Griffith
Media Production Editors: John Cassar, Zach Hubert, and Tyler Suydam
Marketing Manager: Robin Farrar
Director of Creative Services: Paul Belfanti
Creative Director: Carole Anson
Art Director: Maureen Eide
Assistant to Art Director: Dina Curro
Page Composition: PublishWare
Manufacturing Manager: Trudy Pisciotti
Buyer: Alan Fischer
Managing Editor, AV Production & Management: Patricia Burns
AV Project Manager: Jess Einsig
Art Studio: Precision Graphics
Assistant Managing Editor, Science Supplements: Becca Richter
Supplements Production Editor: Christopher Sidor
Editorial Assistant: Margaret Ziegler
Production Assistant: Nancy Bauer
Director, Image Resource Center: Melinda Reo
Manager, Rights and Permissions: Zina Arabia
Interior Image Specialist: Beth Brenzel
Photo Coordinator: Debbie Latronica
Cover/Interior Designer: GRANNA design
Cover Photograph: © Galen Rowell / Mountain Light Photography, Inc.

© 2005 by Pearson Education, Inc.
Pearson Prentice Hall
Pearson Education, Inc.
Upper Saddle River, NJ 07458

Pearson Prentice Hall® is a trademark of Pearson Education, Inc.

Printed in the United States of America

10 9 8 7 6 5 4 3 2 1

ISBN 0-13-147739-0

Pearson Education Ltd., *London*
Pearson Education Australia Pty., Limited, *Sydney*
Pearson Education Singapore, Pte. Ltd.
Pearson Education North Asia Ltd., *Hong Kong*
Pearson Education Canada, Ltd., *Toronto*
Pearson Educación de Mexico, S.A. de C.V.
Pearson Education—Japan, *Tokyo*
Pearson Education Malaysia, Pte. Ltd.

Brief Contents

Contents

Preface

Places fascinate geographers. From an early age, we accumulate maps and dream about faraway corners of the world. Those of us lucky enough to make a living in geography also want to understand why the world works the way it does, how its unique regions have taken shape, and how those regions are increasingly interconnected. Those fundamental curiosities brought us together to create something new and different: an interpretation of world regional geography that was deeply appreciative of global diversity and that looked with a fresh and penetrating eye at the aspects of modern life that tie us all together.

The result has been our own odyssey of exploration, a pooling of expertise and enthusiasm that we hope can offer students a perspective that will stay with them long after they leave the classroom. As an introduction to the field of geography, this book provides a view of the world that will serve students well in the early twenty-first century. This view weds environment, people, and place; ponders how those relationships play out in particular regions; and assesses how powerful processes of globalization are reshaping those relationships in new and often unanticipated ways. *Globalization and Diversity* is the product of a shared vision that we believe demonstrates the essential and invaluable role that geography can play in all our lives.

Objective and Approach

Globalization and Diversity is an issues-oriented textbook for college and university world regional geography classes that explicitly recognizes the geographic changes accompanying today's rapid globalization. With this focus, we join the many who argue that globalization is the most fundamental reorganization of the planet's socioeconomic, cultural, and geopolitical structure since the Industrial Revolution. The explicit recognition of this premise provides the point of departure for this book. As geographers, we think it essential for students to understand two interactive tensions. First, they need to appreciate and critically ponder the consequences of converging environmental, cultural, political, and economic systems through forces of globalization. Second, they need to deepen their understanding of the creation and persistence of geographic diversity and difference. The interaction and tension between these opposing forces of homogenization and diversification forms a current running throughout the following chapters and is reflected in our title, *Globalization and Diversity.*

Globalization and Diversity is drawn from our longer and more comprehensive text, *Diversity Amid Globalization, Second Edition.* In this new book, we focus on the core topics that professors and students need in an introductory course about world regional geography, while simultaneously preserving the globalization approach that characterizes our larger text. *Globalization and Diversity* will be ideal for those professors and students who prefer a briefer approach in order to enhance classroom flexibility or to more easily supplement their lectures with outside materials.

Chapter Organization and Features

As are all other world regional geography textbooks, *Globalization and Diversity* is structured to explain and describe the major world regions of Asia, Africa, the Americas, and so on. These 12 regional chapters, however, depart somewhat from traditional world regional textbooks. Instead of filling them with descriptions of individual countries, we place most of that important material in readily accessible ancillaries, specifically, in the textbook website and in the instructor's manual. This leaves us free to develop five important thematic sections as the organizational basis for each regional chapter. We begin with "Environmental Geography," which discusses the physical geography of each region as well as current environmental issues. Next, we assess "Population and Settlement" geography, in which demography, land use, and settlement (including cities) are discussed. We also provide a section on "Cultural Coherence and Diversity," which examines the geography of language and religion, yet also explores current cultural tensions resulting from the interplay of globalization and diversity. The section on each region's "Geopolitical Framework" then treats the dynamic political geography of the region, including microregionalism, separatism, ethnic conflicts, global terrorism, and supranational organizations. Finally, we conclude each regional treatment with a section titled "Economic and Social Development," in which we analyze each region's economic framework as well as its social geography, including gender issues.

This regional treatment follows two substantive introductory chapters that provide the conceptual and theoretical framework of human and physical geography necessary to understand our dynamic world. In the first chapter, students are introduced to the notion of globalization and are asked to ponder the costs and benefits of the globalization process, a critical perspective that is becoming increasingly common and important to understand. Following this, the geographical foundation for each of the five thematic sections is examined. This discussion draws heavily on the major concepts fundamental to an introductory university geography course. The second chapter, "The Changing Global Environment," presents the essential concepts of global physical geography, including, climate, hydrology, and biogeography.

Globalization and Diversity offers a pedagogically unique cartography program. Seven of the maps in each chapter are constructed with the same theme and similar data and on the same base map so that readers can easily draw comparisons between different regions. Thus, in every regional chapter, readers will find an introductory regional place-name and feature map, a map of the physical geography of the region, a climate map, an environmental issues map, a population density map, a language geography map, and a map showing the geopolitical issues of the region. In addition, each regional chapter also presents other maps illustrating such major themes as ethnic tensions, social development, and economic activity.

In addition, six author field trips are included in the accompanying CD. These trips highlight localities that have been profoundly influenced through globalization. In selecting these places, the authors have drawn upon their own areas of research to illustrate many of the themes discussed in the regional chapters. We feel that this will be an exciting addition to the classroom that will help students visualize concepts and promote discussion.

Supplements: The Teaching and Learning Package

We have been pleased to work with Prentice Hall to produce an excellent supplements package that will enhance teaching and learning. Not only does this package contain the traditional supplements that students and professors have come to expect from authors and publishers, but it also folds together new digital technologies that complement and enhance the learning experience.

- **Instructor Resource Center on CD-ROM** (0-13-147832X): Everything instructors need where they want it. The Prentice Hall Instructor Resource Center helps make instructors more effective by saving them time and effort. All digital resources can be found in one, well-organized, easy-to-access place. The IRC on CD includes:
  - **Figures**: JPEGs of all illustrations and select photos from the text
  - **PowerPoint™ slides**: Pre-authored slides outline the concepts of each chapter with embedded art that can be used as is for lecture, or customized to fit instructors' lecture presentation needs
  - **TestGen**: The TestGen-EQ software, questions, and answers
  - **Electronic files of the Instructor's Manual and Test Item File**
- **Instructor's Resource Box** (0-13-147862-1): *Globalization and Diversity*'s resources are now organized the way you use them. For every chapter in *Globalization and Diversity*, you'll find a file in the Instructor's Resource Box containing:
  - **Transparencies**
  - **Printouts of the PowerPoint presentations**
  - **Printouts of graphics and animations** available on the Instructor Resource Center CD
  Included in the front of the box are text-wide resources:
  - **Instructor's Resource Center on CD-ROM**
  - **Instructor's Manual with Tests**
  - **TestGen EQ test generation software**
- **Transparencies** (0-13-147740-4): Includes more than 200 full-color images from the text, all enlarged for excellent classroom visibility. Every map in the book is included in the transparencies.
- *Globalization and Diversity WWW Site* (http://www.prenhall.com/rowntree): The Website, tied chapter-by-chapter to the text, provides students with three main resources for further study:
  - **On-line Study Guide:** The on-line study guide provides immediate feedback to study questions, access to current geographical issues, and links to interesting and relevant sites on the Web.
  - **Virtual Field Trips:** Integrated with the book's Website, each field trip gives students the on-the-ground feeling of visiting a place. The field trips include information about the history, places, people, and current events of locales in each world region.
  - **Country-by-Country Data:** Also integrated with the Web site, this material summarizes the vital statistics for each country within a region.
- **ABC Videos** (0-13-009774-8): Selected from the archives of ABC news, these timely and relevant video segments offer students insight into issues affecting the regions covered in the book.
- **Instructor's Manual** (0-13-147831-1): Includes Learning Objectives, Chapter Outlines, Activity Suggestions, Discussion Topics, Mapping Exercises, and Country-by-Country Profiles.
- **Test Item File** (0-13-147863-X): Includes more than 150 multiple-choice, true/false, and essay questions per chapter.
- **TestGen-EQ CD-ROM** (0-13-147833-8): TestGen-EQ is a computerized test generator that lets you view and edit test-bank questions, transfer questions to tests, and print the test in a variety of customized formats. Included in each package is the QuizMaster-EQ program that lets you administer tests on a computer network, record student scores, and print diagnostic reports. Both TestGen-EQ and QuizMaster-EQ are free to adopters of *Globalization and Diversity*.
- **On-Line Course Management Systems:** Prentice Hall offers content specific to *Globalization and Diversity*, ready for easy importation into the BlackBoard and WebCT course management system platforms. Each of these platforms lets you easily post your syllabus, communicate with students on-line or off-line, administer quizzes, and record student results and track their progress. Please call your local Prentice Hall representative for details.
- **Map Workbook to accompany** *Globalization and Diversity* (0-13-009777-2): This workbook, which can be used in conjunction with either the textbook or an atlas, features the base maps from the book, printed in black and white. The workbook contains a political and a physical base map for every region in the book, along with a list of key map items for the region at hand. The map workbook is available free when packaged with *Globalization and Diversity*. To order this package, use ISBN 0-13-163195-0.
- **Rand McNally Atlas of World Geography** (0-13-959339-X): Available packaged for free with *Globalization and Diversity*. To order this package use ISBN 0-13-149118-0.

Acknowledgments

We have many people to thank for their help in the conceptualization, writing, rewriting, and production of *Globalization and Diversity.* First, we'd like to thank the hundreds of students in our world regional geography classes who have inspired us with their energy, engagement, and curiosity; challenged us with their critical insights; and demanded a textbook that better meets their need to understand the diverse people and places of our dynamic world.

Next, we are deeply indebted to many professional geographers and educators for their assistance, advice, inspiration, encouragement, and constructive criticism as we labored through the different stages of this book. Among the many who provided invaluable comments on various drafts of *Globalization and Diversity* are:

Dan Arreola, Arizona State University
Bernard BakamaNume, Texas A&M University
Max Beavers, Samford University
James Bell, University of Colorado
William H. Berentsen, University of Connecticut
Kevin Blake, Kansas State University
Craig Campbell, Youngstown State University
Elizabeth Chacko, George Washington University
David B. Cole, University of Northern Colorado
Malcolm Comeaux, Arizona State University
Catherine Cooper, George Washington University
Jeremy Crampton, George Mason University
James Curtis, California State University, Long Beach
Jerome Dobson, University of Kansas
Caroline Doherty, Northern Arizona University
Vernon Domingo, Bridgewater State College
Jane Ehemann, Shippensburg University
Doug Fuller, George Washington University
Gary Gaile, University of Colorado
Steven Hoelscher, University of Texas, Austin
Eva Humbeck, Arizona State University
Richard H. Kesel, Louisiana State University
Rob Kremer, Front Range Community College
Robert C. Larson, Indiana State University
Alan A. Lew, Northern Arizona University
Catherine Lockwood, Chadron State College
Max Lu, Kansas State University
James Miller, Clemson University
Bob Mings, Arizona State University
Sherry D. Morea-Oakes, University of Colorado, Denver
Anne E. Mosher, Syracuse University
Tim Oakes, University of Colorado
Nancy Obermeyer, Indiana State University
Karl Offen, Oklahoma University
Jean Palmer-Moloney, Hartwick College
Bimal K. Paul, Kansas State University
Michael P. Peterson, University of Nebraska, Omaha
Richard Pillsbury, Georgia State University
David Rain, U.S. Census Bureau
Scott M. Robeson, Indiana State University
Susan C. Slowey, Blinn College
Philip W. Suckling, University of Northern Iowa
Curtis Thomson, University of Idaho
Suzanne Traub-Metlay, Front Range Community College
Nina Veregge, University of Colorado
Gerald R. Webster, University of Alabama
Emily Young, University of Arizona
Bin Zhon, Southern Illinois University at Edwardsville
Henry J. Zintambila, Illinois State University

In addition, we wish to thank the many publishing professionals who have been involved with this project. It has been a privilege to work with you. We thank Paul F. Corey, President of Prentice Hall's Engineering, Science, and Math division, for his early—and continued—support for this book project; Geosciences Editor and good friend Dan Kaveney, for his daily engagement, professional guidance, enduring patience, unyielding discipline, high standards, and warm companionship; Chris Rapp, Media Editor, for his creative and always stimulating additions and insights to our digital program; Amanda Griffith, Associate Editor, for gracefully taking care of the thousands of tasks necessary to this project; Marketing Manager Robin Farrar for her effective sales and promotion work, and for her tireless travel in pursuit of new adoptions; Editorial Assistant Margaret Ziegler, for gracefully meeting the incessant needs of four demanding authors who rarely remembered that she was also simultaneously doing the same for many other Geoscience authors; Developmental Editor Ginger Birkeland's assistance was invaluable as we worked to focus this brief treatment of world regional geography on core materials and as we clarified and simplified the narrative; Production Editor Tim Flem, for his daily miracles and steady hand on the tiller so that somehow thousands of pages of manuscript were turned into a finished book; copyeditor Bobbie Dempsey, for her eagle eyes and graceful hands; and photo researchers Linda Sykes and Jerry Marshall, for their creative solutions to indulging four geographers in our quest for outstanding pictures from strange parts of the world. To all of you, your professionalism is truly inspirational and very, very much appreciated.

Finally, the authors want to thank that special group of friends and family who were there when we needed you most—early in the morning and late at night; in foreign countries and familiar places; when we were on the verge of crying, yet needed to laugh; for your love, patience, companionship, inspiration, solace, understanding, and enthusiasm: Elizabeth Chacko, Meg Conkey, Rob Crandall, Allison Green, Karen Wigen, and Linda, Tom, and Katie Wyckoff. Words cannot thank you enough.

Les Rowntree

Martin Lewis

Marie Price

William Wyckoff

About the Authors

Les Rowntree teaches both Geography and Environmental Studies at San Jose State University in California, where he recently completed a term as the Chair of the interdisciplinary Department of Environmental Studies. As an environmental geographer, Dr. Rowntree's teaching and research interests focus on international environmental issues, the human dimensions of global change, biodiversity conservation, and human-caused landscape transformation. He sees world regional geography as a way to engage and inform students by giving them the conceptual tools needed to critically assess global issues. Dr. Rowntree has done research in Morocco, Mexico, Australia, and Europe, as well as in his native California. Current writing projects include a book on the natural history of California's Central Coast, along with textbooks in geography and environmental science.

Marie Price is an Associate Professor of Geography and International Affairs at George Washington University. A Latin American specialist, Marie has conducted research in Belize, Mexico, Venezuela, Cuba, and Bolivia. She has also traveled widely throughout Latin America and Sub-Saharan Africa. Her studies have explored human migration, natural resource use, environmental conservation, and regional development. Dr. Price brings to *Globalization and Diversity* a special interest in regions as dynamic spatial constructs that are shaped over time through both global and local forces. Her publications include articles in the *Annals of the Association of American Geographers, Geographical Review, Journal of Historical Geography, CLAG Yearbook, Studies in Comparative International Development, the Brookings Institution Survey Series,* and *Focus*.

Martin Lewis is a lecturer in International Affairs at Stanford University. He has conducted extensive research on environmental geography in the Philippines and on the intellectual history of global geography. His publications include *Wagering the Land: Ritual, Capital, and Environmental Degradation in the Cordillera of Northern Luzon, 1900–1986* (1992), and, with Karen Wigen, *The Myth of Continents: A Critique of Metageography* (1997). Dr. Lewis has travelled extensively in East, South, and Southeast Asia. His current research focuses on the geographical dimensions of globalization.

William Wyckoff is a geographer in the Department of Earth Sciences at Montana State University specializing in the cultural and historical geography of North America. He has written and co-edited several books on North American settlement geography, including *The Developer's Frontier: The Making of the Western New York Landscape* (1988), *The Mountainous West: Explorations in Historical Geography* (1995), and *Colorado: The Making of a Western American Landscape 1860–1940* (1999). In 1990 he received the Burlington Northern Corporation's Award for Outstanding Teaching. In addition, in 2003 he received Montana State's Cox Family Fund for Excellence Faculty Award for Teaching and Scholarship. A World Regional Geography instructor for 18 years, Dr. Wyckoff hopes that the fresh approach taken in *Globalization and Diversity* will more effectively highlight the tensions evident in the world today as global change impacts particular places and people in dramatic and often unpredictable ways.

Philosophy and Approach

FIGURE 4.9 • GUATEMALAN FARMER A peasant farmer works his small plot of subsistence and cash crops. Behind him rises one of the region's many volcanoes. The fertile volcanic soils, ample rainfall, and temperate climate of the Guatemala highlands have supported dense populations for centuries. *(Sean Sprague/Panos Pictures)*

Globalization and Diversity explicitly acknowledges the geographic changes accompanying today's rapid rate of globalization, emphasizing both the homogenizing and diversifying forces inherent to the globalization process. The text challenges students to make critical comparisons between the regions of the world and to understand more fully the interconnections that bind these regions together. Examples of topics used to accomplish this goal include:

- The rise of Islamic fundamentalism in Southwest Asia

- Aboriginal groups using high-technology tools to forge common political survival strategies

- The economic and political integration of the European Union, contrasted with micronationalism and factionalism in Europe

- Ethnic diversification in the face of strong participation in the global assembly line in Southeast Asia

- The globalization and localization of beer consumption and production in the United States and Canada

FIGURE 3.29 • GATED AMERICA Wealthier North Americans have had a tremendous impact on the region's cultural landscapes, and their influence far outweighs their relatively modest numbers. Many gated communities in North American suburbs and resorts display a desire for privacy, safety, and social exclusivity. *(James Patelli)*

THEMATIC STRUCTURE OF THE TEXT

To facilitate students' abilities to draw comparisons and contrasts between regions, each of the 12 regional chapters employs a thematic structure. This structure organizes each chapter into 5 thematic sections:

- Environmental Geography
- Population and Settlement
- Cultural Coherence and Diversity
- Geopolitical Framework
- Economic and Social Development

This structure allows a more penetrating treatment of themes and concepts than the traditional country-by-country approach and facilitates comparisons of specific topics between regions.

Focused, Brief Treatment of Core Topics

Globalization and Diversity provides a brief and focused treatment of world regional geography by highlighting the core topics professors need in an introductory course. Much of the text has been rewritten to enhance readability. This abbreviated treatment allows for more classroom flexibility, especially for those professors who prefer to use readings or activities in addition to the textbook.

FIGURE 1.7 • PROTESTS AGAINST GLOBALIZATION
Meetings of international groups such as the World Trade Organization, World Bank, and International Monetary Fund commonly draw large numbers of protesters against globalization. This demonstration took place at a Washington, D.C., meeting of the World Bank and IMF. *(Rob Crandall/Rob Crandall, Photographer)*

Cartography

Comparable, **standardized maps** in each chapter help students make comparisons between regions in cartographic form.

Seven of the maps in each chapter are organized around the same theme and designed to convey comparable material. These are:

- a **chapter-opening map** with countries and place names

- a **physical map**, showing landforms, hydrology, and tectonic boundaries

- a **climate map,** with climograph call-outs giving temperature and precipitation data for specific cities

- a **"transformation of Earth"** map with call-outs to environmental issues and solutions within the region

- a **population map** for the region

- a **map of regional languages**

- a **geopolitical map** with call-outs to current issues and tensions

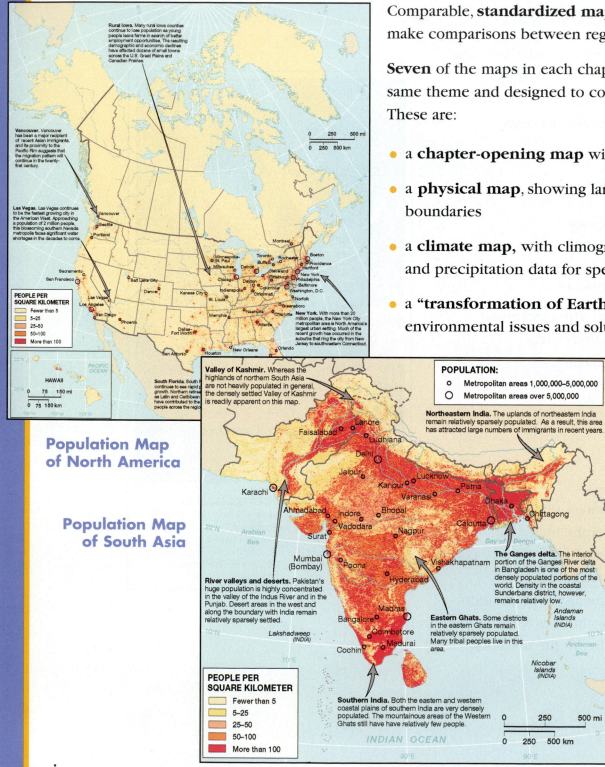

Population Map of North America

Population Map of South Asia

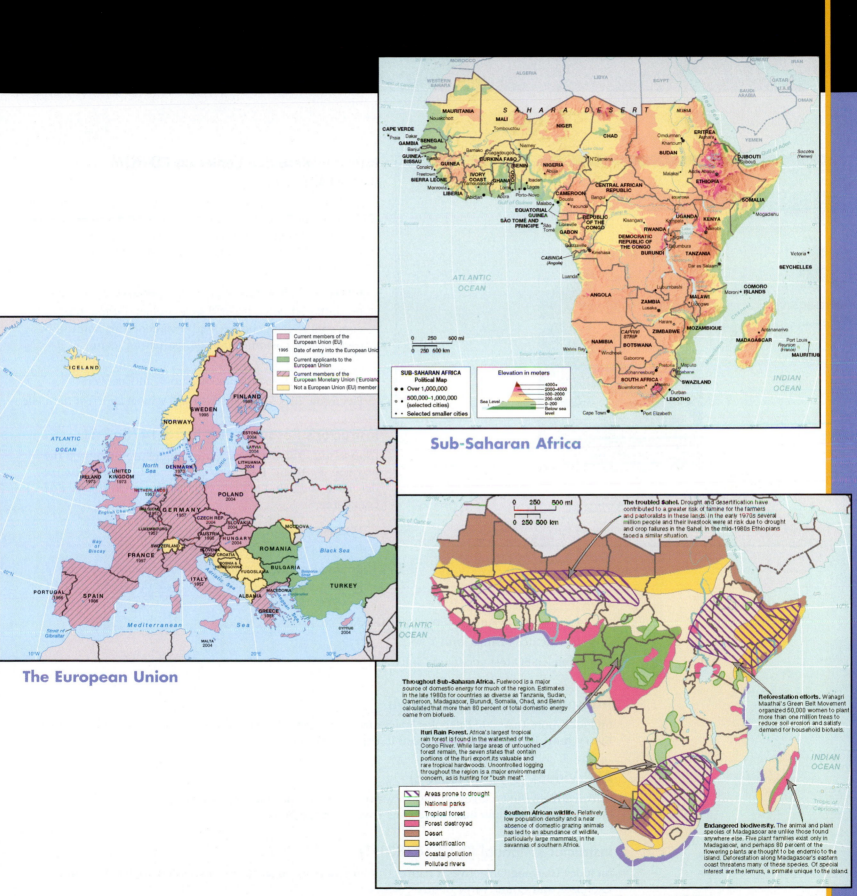

Sub-Saharan Africa

SUB-SAHARAN AFRICA
Political Map

Over 1,000,000
500,000–1,000,000
(selected cities)
Selected smaller cities

Elevation in meters

4000+
2000–4000
500–2000
200–500
0–200
Below sea
level
Sea Level

Current members of the
European Union (EU)
1995   Date of entry into the European Union
Current applicants to the
European Union
Current members of the
European Monetary Union ('Euroland')
Not a European Union (EU) member

The European Union

The troubled Sahel. Drought and desertification have
contributed to a greater risk of famine for the farmers
and pastoralists in these lands. In the early 1970s several
million people and their livestock were at risk due to drought
and crop failures in the Sahel. In the mid-1980s Ethiopians
faced a similar situation.

Reforestation efforts. Wangari
Maathai's Green Belt Movement
organized 50,000 women to plant
more than one million trees to
reduce soil erosion and satisfy
demand for household biofuels.

Throughout Sub-Saharan Africa. Fuelwood is a major
source of domestic energy for much of the region. Estimates
in the late 1980s for countries as diverse as Tanzania, Sudan,
Cameroon, Madagascar, Burundi, Somalia, Chad, and Benin
calculated that more than 80 percent of total domestic energy
came from biofuels.

Ituri Rain Forest. Africa's largest tropical
rain forest is found in the watershed of the
Congo River. While large areas of untouched
forest remain, the seven states that contain
portions of the Ituri export its valuable and
rare tropical hardwoods. Uncontrolled logging
throughout the region is a major environmental
concern, as is hunting for "bush meat".

Southern African wildlife. Relatively
low population density and a near
absence of domestic grazing animals
has led to an abundance of wildlife,
particularly large mammals, in the
savannas of southern Africa.

Endangered biodiversity. The animal and plant
species of Madagascar are unlike those found
anywhere else. Five plant families exist only in
Madagascar, and perhaps 80 percent of the
flowering plants are thought to be endemic to the
island. Deforestation along Madagascar's eastern
coast threatens many of these species. Of special
interest are the lemurs, a primate unique to the island.

Areas prone to drought
National parks
Tropical forest
Forest destroyed
Desert
Desertification
Coastal pollution
Polluted rivers

Environmental Issues in Sub-Saharan Africa

Aids for Testing, Teaching,

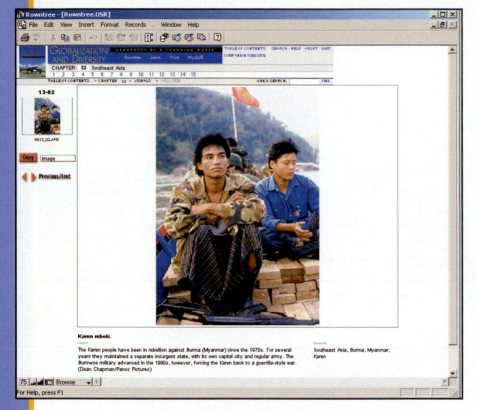

Instructor Resource Center on CD-ROM (0-13-147832X)

Everything instructors need where they want it. The Prentice Hall Instructor Resource Center helps make instructors more effective by saving them time and effort. All digital resources can be found in one, well-organized, easy-to-access place. The IRC on CD includes:

- JPEGs of all illustrations and select photos from the text
- Pre-authored PowerPoint™ slides, which outline the concepts of each chapter with embedded art and can be used as is or customized to fit instructors' lecture presentation needs
- The TestGen-EQ software, questions, and answers
- Electronic files of the Instructor's Manual and Test Item File

Instructor's Resource Box (0-13-147862-1)

Globalization and Diversity's resources are organized the way you use them. For every chapter in *Globalization and Diversity,* you'll find a file in the Instructor's Resource Box containing:

- Transparencies
- Printouts of PowerPoint presentations
- Printouts of graphics and animations available on the Instructor Resource Center CD. Included in the front of the box are text-wide resources:

- Instructor's Resource Center on CD-ROM
- Instructor's Manual with Tests
- TestGen-EQ test generation software

Transparencies (0-13-147740-4)

Includes more than 200 full-color images from the text, all enlarged for excellent classroom visibility. **Every map in the book** is included in the transparencies.

ABC Videos (0-13-009774-8)

Selected from the archives of ABC News, these timely and relevant video segments offer students insight into issues affecting the regions covered in the book.

Instructor's Manual (0-13-147831-1)

Includes Learning Objectives, Chapter Outlines, Activity Suggestions, Discussion Topics, Mapping Exercises, and Country-by-Country Profiles.

and Classroom Presentation

Test Item File (0-13-147863-X)

Includes more than 150 multiple choice, true/false, and essay questions per chapter.

TestGen-EQ CD-ROM (0-13-147833-8)

TestGen-EQ is a computerized test generator that lets you view and edit test-bank questions, transfer questions to tests, and print the test in a variety of customized formats. Included in each package is the QuizMaster-EQ program that lets you administer tests on a computer network, record student scores, and print diagnostic reports. Both TestGen-EQ and QuizMaster-EQ are free to adopters of *Globalization and Diversity*.

On-Line Course Management Systems

Prentice Hall offers content specific to *Globalization and Diversity*, ready for easy importation into the BlackBoard and WebCT course management system platforms. Each of these platforms allows you to easily post your syllabus, communicate with students on-line or off-line, administer quizzes, and record student results and track their progress. Please call your local Prentice Hall representative for details.

Optional Student Aids/Package Options

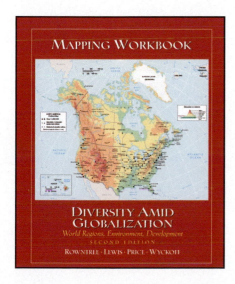

Mapping Workbook

Political and physical outline maps for each chapter, based on the maps in the text, ready for student use in identification exercises. Available at no extra charge when packaged with *Globalization and Diversity*. To order this package, use ISBN **0-13-163195-0**

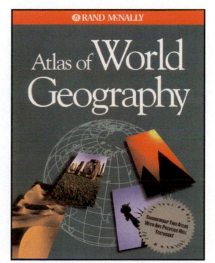

Rand McNally Atlas of World Geography

Available packaged FREE with *Globalization and Diversity*. To order this package, use ISBN **0-13-149118-0**.

CHAPTER ONE

Globalization and Diversity

1

FIGURE 1.1 World geography
Viewed from space, Earth's physical fabric is apparent. World regional geography studies both the physical environment as well as human activities.
(European Space Agency/Science Photo Library/Photo Researchers, Inc.)

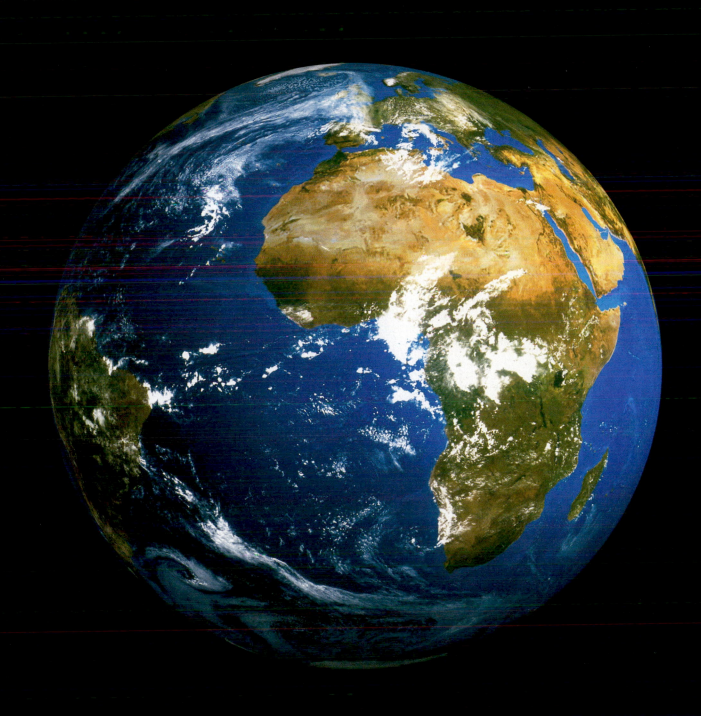

The most important challenges facing the world in the twenty-first century are associated with **globalization**, the growing interconnectedness of people and places through converging processes of economic, political, and cultural change. Once-distant regions are now increasingly linked together through commerce, communications, and travel. Many observers argue that globalization is the most fundamental reorganization of the planet's social and economic structures since the Industrial Revolution. While few overlook the widespread changes brought about by globalization, not everyone agrees on whether the benefits outweigh the costs.

Although economic activities may be the prime mover behind globalization, the consequences affect all aspects of land and life. Cultural patterns, political arrangements, and social development are all undergoing widespread change. Because natural resources like timber are now global commodities traded on world markets, the planet's physical environment is also affected by globalization. Often local ecosystems are altered by financial decisions made thousands of miles away. As a result, the cumulative effect of these far-ranging activities may have negative consequences for the world's climates, oceans, waterways, and forests (Figure 1.1).

These extensive global changes make understanding our world both a challenging and necessary task. Our future depends on understanding globalization in its varied expressions, because our lives are now deeply intertwined with these changes. Although studying globalization cuts across many academic disciplines, world regional geography is an effective starting point because of its focus on regions, environment,

FIGURE 1.2 • WORLD REGIONS   These regions are the basis for the 12 regional chapters in this book. Countries or areas within countries that are treated in more than one chapter are designated on the map with a striped pattern. For example, western China is discussed in both Chapter 10, Central Asia, and Chapter 11, East Asia. Also, 3 countries on the South American continent are discussed as part of the Caribbean region because of their close cultural similarities with the island region.

geopolitics, culture, and economic and social development. Thus, this book focuses on providing comprehensive knowledge of the world in which globalization is played out (Figure 1.2).

Diversity Amid Globalization: A Geography for the Twenty-first Century

This chapter introduces the framework for studying world regional geography by first examining the varied aspects of globalization in modern life. While the economic implications of globalization dominate most discussions, it is important to understand the cultural, geopolitical, environmental, and social expressions as well.

Converging Currents of Globalization

Most scholars agree that the most significant component of globalization is the economic reorganization of the world. Although different forms of a world economy have been in existence for centuries, a well-integrated global economy is primarily a product of the last several decades. The characteristics of this new world arrangement are

- global communication systems that link all regions on the planet instantaneously (Figure 1.3);
- global transportation systems capable of moving goods quickly by air, sea, and land;

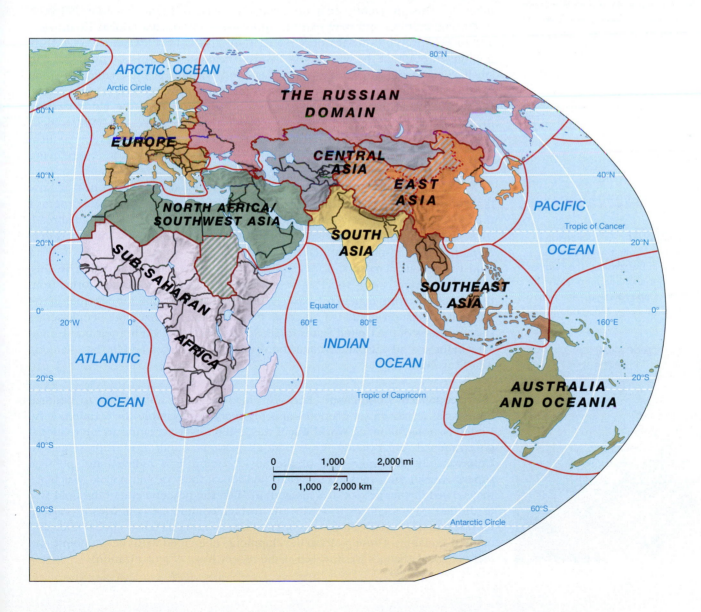

FIGURE 1.3 • GLOBAL CONNECTIONS The impacts of globalization, often through global TV, occur even in remote villages in developing countries. Here, in a small village in southwestern India, a rural family earns a few dollars a week by renting out viewing time on its globally linked television set. *(Rob Crandall/ Rob Crandall, Photographer)*

FIGURE 1.4 • HYBRID WORLD CULTURE Globalization is creating unique cultural expressions that often merge, or hybridize, the old with the new. Here, in Bangkok, a Thai woman eats lunch at a well-known hamburger chain under a poster for a uniquely Thai traditional dish, a tart papaya salad, which the hamburger chain is marketing as "Thai Spicy McSalad Shaker." *(AP/Wide World Photos)*

- transnational conglomerate corporate strategies that have created global corporations more economically powerful than many nation-states;
- new and more-flexible forms of monetary flow;
- international financial institutions that make possible 24-hour trading;
- global agreements that promote free trade;
- market economies that have replaced state-controlled economies;
- privatized firms and services, like water delivery, formerly operated by governments;
- an abundance of planetary goods and services that have arisen to fulfill consumer demand (real or imaginary); and, of course,
- an army of international workers, managers, and executives who give this powerful economic force a human dimension.

Global Consumer Culture Economic changes also trigger fundamental cultural change. Accompanying globalization is the spread of a global consumer culture that often erodes local diversity. This frequently sets up deep and serious social tensions between traditional cultures and new, external globalizing influences. Global TV, movies, and videos promote images of Western style and culture that are imitated by millions throughout the world. NBA T-shirts, sneakers, and caps are now found in small villages and large cities alike. Fast-food franchises are changing—some would say corrupting—traditional diets with the explosive growth of McDonald's, Burger King, and Kentucky Fried Chicken outlets in most of the world's cities, including Beijing, Moscow, Singapore, and Nairobi (Figure 1.4). While these cultural changes may seem harmless to North Americans because they are familiar, they illustrate the more profound cultural changes the world is experiencing through globalization.

It would be a mistake, however, to view cultural globalization as a one-way flow that spreads from the United States and other Western countries into all corners of the world. Actually, even when forms of American popular culture spread abroad, they are typically melded with local cultural traditions in a process known as *hybridization*. The resulting cultural "hybridities," such as world-beat music or Asian food, reverberate across the planet so that they are now found worldwide. To illustrate, U.S. culture is now much more strongly impacted by ideas and forms from the rest of the world than in the past (Figure 1.5). This is visible, for example, in the growing internationalization of American food, or in the fact that more people across large portions of the United States speak Spanish than English in their homes, and even in the spread of Japanese comic book culture—Pokémon and the like—among American children.

The Geopolitical Component Because globalization is not limited to national boundaries, it has a major geopolitical component. For example, the creation of the United Nations following World War II was a step toward creating an international governmental organization in which all nations could find representation. Unfortunately, the emergence of the Soviet Union as a military and political superpower in the 1950s led to a rigid division of the world into Cold War blocs that slowed further geopolitical integration. With the peaceful end of the Cold War in the late 1980s and early 1990s, the former communist countries of eastern Europe and the Soviet Union were opened almost immediately to global trade and cultural exchange. These political developments coincided with the economic and technological changes we now see as the recent wave of globalization.

Environmental Concerns Beyond geopolitics, the expansion of a globalized economy is also creating and intensifying environmental problems throughout the world. **Transnational firms** in particular disrupt local ecosystems in their search for natural resources and manufacturing sites. Landscapes and resources that were previously used by only small groups of local peoples are now thought of as global products to be sold and traded on the world marketplace. As a result, native peoples are often deprived of their traditional resource base. On a larger scale, economic globalization is aggravating worldwide environmental problems, such as climate change, air pollution, water pollution, and deforestation. And yet it is only through global cooperation, such as the United Nations treaties on biodiversity protection or the Kyoto protocol on global warming (discussed in Chapter 2), that these problems can be addressed.

Social Dimensions Globalization has a clear demographic dimension as well. Although international migration is nothing new, increasing numbers of people from all parts of the world are crossing national boundaries, often permanently. Migration from Latin America and Asia has drastically changed the demographic structure of the United States, just as migration from Africa and Asia has transformed western Europe. Countries such as Japan and South Korea that have long been perceived as ethnically homogeneous now have substantial immigrant populations. Even a number of relatively poor countries, such as Nigeria and the Ivory Coast in west Africa, encounter large numbers of immigrants coming from even poorer countries.

Finally, there is also a significant criminal element to contemporary globalization, including terrorism (discussed later in this chapter), drugs, pornography, and prostitution. Illegal narcotics, for example, are most definitely a global commodity (Figure 1.6). Some of the most remote parts of the world, such as the mountains of northern Burma, are thoroughly integrated into the global heroin trade with the production of opium. Many Caribbean countries have seen their economies become reoriented to drug smuggling and money laundering. Prostitution, pornography, and gambling (legal in some areas, illegal in others) have also emerged as highly profitable global businesses. Over the past decade, for example, parts of eastern Europe have become major sources of both pornography and prostitution, finding a profitable but morally questionable place in the new global economy.

FIGURE 1.5 • GLOBAL CULTURE IN THE UNITED STATES The hybridization of Vietnamese and American cultures dominates in the Eden Center shopping mall in Falls Church, Virginia, west of Washington, D.C. Often, transplanted foreign cultures attempt to create an idealized snapshot of their native culture. Here, the flag of the former South Vietnamese republic flies alongside the Stars and Stripes. *(Rob Crandall/Rob Crandall, Photographer)*

FIGURE 1.6 • THE GLOBAL DRUG TRADE The cultivation, processing, and transshipment of coca (cocaine), opium (heroin), and cannabis are global issues. The most important cultivation centers are Colombia, Mexico, Afghanistan, and northern Southeast Asia, whereas the major drug financing centers are mostly located in the Caribbean, the United States, and Europe. Additionally, Nigeria and Russia also have significant roles in the global transshipment of illegal drugs.

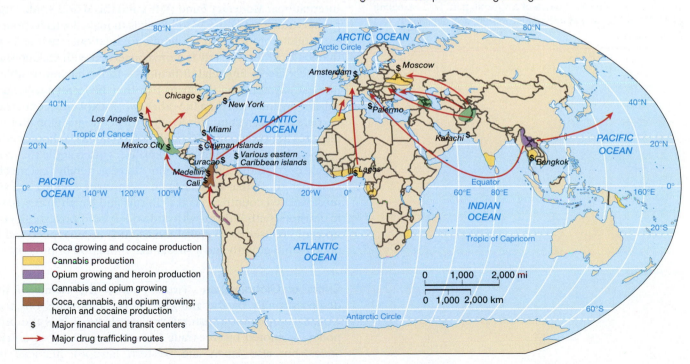

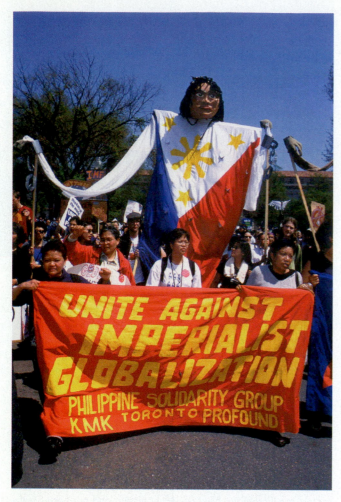

FIGURE 1.7 • PROTESTS AGAINST GLOBALIZATION
Meetings of international groups such as the World Trade Organization, World Bank, and International Monetary Fund commonly draw large numbers of protesters against globalization. This demonstration took place at a Washington, D.C., meeting of the World Bank and IMF. *(Rob Crandall/Rob Crandall, Photographer)*

Advocates and Critics of Globalization

Globalization, especially in its economic form, is one of the most controversial issues of the day (Figure 1.7). Supporters generally believe that it brings in greater economic efficiency that will eventually result in rising prosperity for the entire world. Critics think that it will largely benefit those who are already prosperous, leaving most of the world poorer than before while reducing cultural and ecological diversity. Economic globalization is generally applauded by corporate leaders and economists, and it has substantial support among the leaders of both the Republican and Democratic parties in the United States. But opposition to economic globalization is widespread in the labor and environmental movements and among many student groups.

The Pro-globalization Stance Advocates argue that globalization is a logical expression of modern international capitalism that will benefit all nations and all peoples by increasing global commerce and wealth. This new wealth, they say, will eventually trickle down to improve the conditions for even the poorest of peoples in all the world's different regions. Economic globalization can work such wonders, they contend, by increasing competition, allowing the flow of capital to poor areas, and encouraging the spread of beneficial new technologies and ideas. For example, if countries abolish trade barriers, inefficient local industries will become more efficient in order to compete with the new flood of imports. As a result, national productivity will increase. Every country and region of the world, moreover, ought to be able to concentrate on those activities for which each is best suited in the global economy. Thus the pro-globalizers argue that a more efficient world economy will result.

To the committed pro-globalizer, even the global spread of **sweatshops**—crude factories in which workers sew clothing, assemble sneakers, or perform similar labor-intensive tasks for extremely low wages—is to be applauded. People go to work in sweatshops, they say, because the alternatives in the local economy are even worse (Figure 1.8). Once countries achieve full employment, the argument goes, workers can gradually improve conditions and move into better-paying industries. The pro-globalizers in general strongly support the large multinational organizations that make possible the flow of goods and capital across international boundaries. Three such organizations are particularly important: the World Bank, the International Monetary Fund (IMF), and the World Trade Organization (WTO). The primary function of the World Bank is to make loans to poor countries so that they can build more modern economic foundations, such as highways or energy production facilities. The IMF is more concerned with making short-term loans to countries that are in financial difficulties—those having trouble, for example, making interest payments on the loans that they had previously taken. The WTO, a much smaller organization than the other two, is explicitly concerned with lowering trade barriers between countries to enhance economic globalization. It also tries to solve disputes between countries and trading blocks, although not always with success.

To support their claims, pro-globalizers often argue that countries that have been highly open to the global economy have generally had much more economic success than those that have isolated themselves, seeking self-sufficiency. The world's most isolated countries, such as Burma and North Korea, have often been economic disasters, with little growth and widespread poverty. In contrast, those that have opened themselves to global forces, such as Singapore and Thailand, have experienced rapid growth and a substantial reduction of poverty.

Critics of Globalization Virtually all of the claims of the pro-globalizers are denied strongly by the anti-globalizers. Opponents often begin by arguing that globalization is not a "natural" process. Instead, it is the product of an economic policy promoted by free-trade advocates, capitalist countries (mainly the United States, but also Japan and western Europe), financial interests, international investors, and multinational firms. These policies, they argue, have increased the differences between rich and poor in the world.

Since the globalization of the world economy appears to be creating greater inequality, the "trickle-down" model of developmental benefits for all people in all regions has yet to be demonstrated. In fact, the percentage of poor people in most world regions is actually increasing (Figure 1.9). On a global scale, the richest 20 percent of the world's people consume 86 percent of the world's resources, while the poorest 80 percent use only 14 percent of global resources. Opponents also argue that globalization promotes free-market, export-oriented economies at the expense of localized, sustainable activities. World forests, for example, are increasingly cut for export timber rather than serving local needs. As part of their economic structural adjustment package, the World Bank and the International Monetary Fund encourage developing countries to expand their resource exports so they will have more hard currency (often dollars) to make payments on their foreign debts. This strategy, however, usually leads to short-term exploitation of local resources.

Those who challenge globalization also worry that the entire system—with its immediate transfers of vast sums of money over nearly the entire world on a daily basis—is dangerously unstable. These international managers of capital tend to panic when they think their funds are at risk; and when they do so, the entire global financial system can quickly become destabilized, potentially leading to a crisis of global proportions. As vast sums of money flow into a developing country, they may create an inflated **"bubble economy"** that cannot be sustained.

Such a bubble economy emerged in Thailand and many other parts of Southeast Asia in the mid-1990s. So much money flooded into Thailand from global investors that property values in the capital city of Bangkok came to exceed those of most U.S.

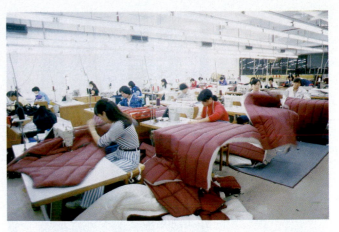

FIGURE 1.8 • GLOBAL SWEATSHOPS   One of the most debated aspects of economic globalization is the crude factories, or sweatshops, in which workers sew clothing, assemble sneakers, stitch together soccer balls, or perform similar labor-intensive tasks for low wages. While pro-globalizers and free-trade supporters argue that these sweatshops provide better opportunities than the local economy, critics of globalization point to long hours, low wages, uncertain employment, and unhealthy working conditions as the negative side of economic globalization. *(Macduff Everton/The Image Works)*

FIGURE 1.9 • GLOBAL ECONOMIC INEQUITY   One of the most troubling aspects of economic globalization is the tendency to increase income inequality. Put differently, the accusation that globalization has allowed the "rich to get richer while the poor stay poor" seems to be true despite the claim by pro-globalizers that everyone would benefit from greater global economic activity. This map shows the results of a United Nations study, "Inequality, Growth and Poverty in the Era of Liberalization and Globalization" (2001), that shows inequity increasing in 48 of the 73 countries studied. Further, though the UN study found inequity was unchanged in 16 countries, a high level of income inequity might still exist in these countries. For example, India and Brazil are mapped as "constant," yet have very high levels of inequity.

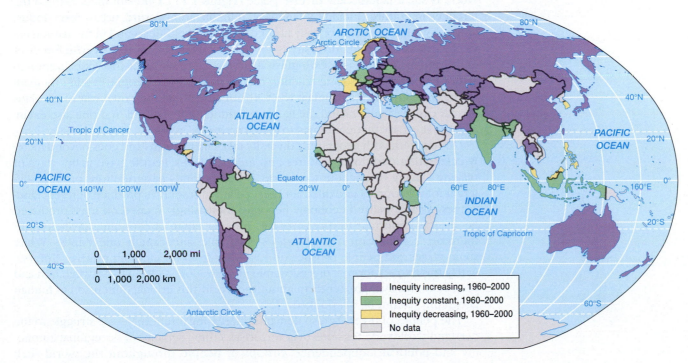

FIGURE 1.10 • ECONOMIC RECESSION IN THAILAND One of the unfortunate symbols of Thailand's 1997 economic financial crisis are more than 300 unfinished high-rise buildings that now litter Bangkok's downtown skyline. Before the economic collapse, Thailand's "bubble economy" led to widespread speculation in downtown property values and new buildings. When the bubble burst, bankrupt developers walked away from their high-rise office construction, leaving these abandoned steel skeletons. *(AFP/Corbis)*

cities. When the bubble burst in 1997, a severe depression crippled most of Southeast Asia (Figure 1.10). This depression still continues in Indonesia, the most populous country in the region. While this Asian crisis did not weaken the entire world economy, as many anti-globalizers feared it would, some evidence suggests that it very nearly did so.

A Middle Position? Not surprisingly, a number of experts argue that both the anti-globalization and the pro-globalization stances are exaggerated. Those in the middle ground tend to argue that economic globalization is indeed unavoidable. They point out that even the anti-globalization movement is made possible by the Internet and is, therefore, itself an expression of globalization. They further contend that globalization can be managed, at both the national and international levels, to reduce economic inequalities and protect the natural environment. Such scholars stress the need for strong yet efficient governments and international institutions (such as the UN, World Bank, and IMF), along with networks of watchdog environmental, labor, and human rights groups.

The debate about globalization is one of the most important and complicated issues of the day. While this book does not pretend to resolve the controversy, it does encourage readers to reflect on these critical points and apply them to different world regions and localities.

Diversity in a Globalizing World

The ever-increasing globalization of the world has led many observers to foresee a future world far more uniform than that of today. Optimists imagine a global culture uniting all humankind into a single community untroubled by war, ethnic conflict, or resource shortage—a global utopia of sorts. However, a more common view is that globalization causes different places, peoples, and environments to lose their distinctive character and become identical to their neighbors. While diversity may be the hardest thing for a society to live with, it may also be harmful to sacrifice. Nationality, ethnicity, cultural distinctiveness—all are the legitimate legacy of humanity. If this diversity is blurred, denied, or repressed through global uniformity, humanity loses one of its defining traits.

But even if globalization is generating a certain degree of homogenization, the world is still a fantastically diverse place (Figure 1.11). One still finds marked differences in culture (including language, religion, architecture, urban form, foods, and many other elements of daily life), economy, and politics—and in the natural environment as well. Such diversity is so vast that even the most powerful forces of globalization cannot easily extinguish it. Important to realize is that globalization itself often provokes a strong reaction on the part of local people, making them all the more determined to maintain what is distinctive about their own way of life. Thus it is far easier to understand the issues surrounding globalization if one also examines the diversity that continues to characterize the world.

Our concern with geographic diversity takes many forms and goes beyond merely celebrating traditional cultures and unique places. People have many ways of making a living throughout the world. Different kinds of agriculture produce different food resources, and different uses of natural resources produce different landscapes. As a globalized economy becomes increasingly fixated on mass-produced goods, it is important to recognize this broader range of economic diversity. Furthermore, a significant trait of today's economic landscape is unevenness: While some places and people prosper, others suffer from unrelenting poverty. Unfortunately, this also is a form of diversity amid globalization. As the gap between rich and poor becomes ever greater throughout the world, this critical dimension of human geography cannot be overlooked.

The politics of diversity also demands increasing attention as we struggle to understand worldwide tensions over terrorism, ethnic separateness, regional autonomy, and political independence. Groups of people throughout the world seek self-rule of territory they can call their own. Today most wars are fought *within* countries, not between them. Each conflict is unique, understandable only in the light of the specific cultural and political environments in which it occurs. But taken together, the conflicts also illustrate processes that are found across much of the globe. We must therefore seek a balance between describing the common features of the world's regions while also explaining the distinctiveness of places, environments, and landscapes.

Themes and Issues in World Regional Geography

Geography is the academic discipline that describes Earth and explains the patterns on its surface. Although this task involves different and often contrasting approaches and perspectives, the themes and concepts held in common give coherence to the field. A general background on global environmental geography is found in Chapter 2, "The Changing Global Environment." That chapter discusses some of the environmental elements that are important to human settlement.

Following the two introductory, thematic chapters, the book adopts a regional perspective, grouping all of Earth's countries into the framework of world regions. We begin with the region most familiar to most of our readers, North America, and then move to Latin America, the Caribbean, Africa, the Middle East, Europe, Russia, and the different regions of Asia, before concluding with Australia and Oceania. Each of the 12 regional chapters uses the same 5-part thematic structure: environmental geography; population and settlement; cultural coherence and diversity; geopolitical framework; and economic and social development. Let us now examine each of these major themes.

Population and Settlement: People on the Land

The human population is far larger than it ever was in the past, and there is considerable debate about whether continued growth will benefit or harm the world. There is concern not only with total numbers, but also with geographic patterns of human settlement across the planet. While some parts of the world are densely populated, other places remain almost empty (Figure 1.12).

Global population growth is one of the most troublesome issues facing the world. With more than 6 billion humans on Earth, we currently add another 86 million each year, or about 10,000 new people each hour. About 90 percent of this population growth takes place in the developing world, especially in Africa, Latin America, and South and East Asia. Because of rapid growth in developing countries, puzzling questions dominate discussions of all global issues. Can these countries absorb this rapidly increasing population and still achieve the necessary economic and social development to ensure well-being and stability for their total population? What role should the developed countries of North America, Europe, and East Asia play in those population issues now that their own populations have stabilized? While population is a complicated and controversial topic, several points help focus the issue:

1. Very different rates of population growth are found in various regions of the world. While some countries are growing rapidly, others have essentially no natural increase; instead, any population increase comes from in-migration. India is an example of the former, Europe of the latter. Other countries, such as those of the former Soviet Union, have both low birthrates and negligible rates of immigration. Unless circumstances change, those countries will soon see major population losses.

2. Population planning takes many forms, from the rigid one-child policies of China to the "more children, please" programs of western Europe. While

FIGURE 1.11 • LOCAL CULTURES This family in Ghana reminds us that although few places are beyond the reach of globalization, there are still many unique local landscapes, economies, and cultures. For example, most of the material culture in this photograph is local in origin and comes from long-standing cultural tradition. *(Caroline Penn/Panos Pictures)*

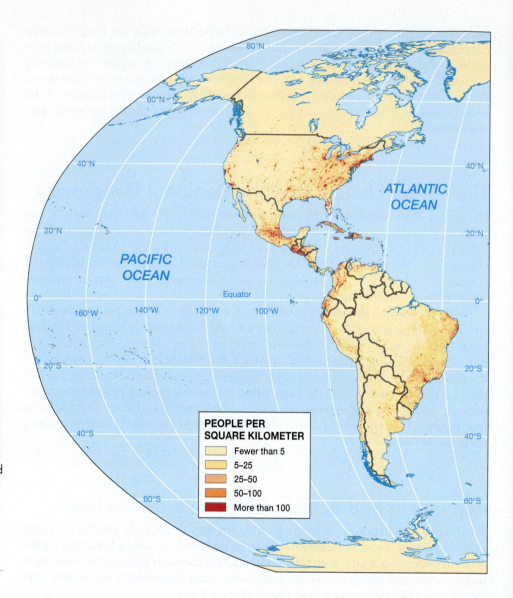

FIGURE 1.12 • WORLD POPULATION This world population map shows the differing densities of population in the regions of the world. East Asia stands out as the most populated region, with high densities in Japan, Korea, and eastern China. The second most populated region is South Asia, dominated by India, which is second only to China in population. In North Africa and Southwest Asia, population clusters are often linked to the availability of water for irrigated agriculture, as is apparent with the population cluster along the Nile River. Higher population densities in Europe, North America, and other countries are usually associated with large cities, their extensive suburbs, and nearby economic activities.

PEOPLE PER SQUARE KILOMETER

- Fewer than 5
- 5–25
- 25–50
- 50–100
- More than 100

FIGURE 1.13 • FAMILY PLANNING POLICIES Because unrestrained population growth may hinder economic improvement, many developing countries have put family planning policies in place. This poster in Vietnam urges smaller families. *(Chris Stowers/Panos Pictures)*

some programs attempt to achieve their goals through regulation, others rely on incentives and social cooperation (Figure 1.13).

3. Not all attention should be focused on natural growth since migration is increasingly a cause of population change in this globalized world. While most international migration is driven by a desire for a better life in the wealthier regions of the developed world, we should not lose sight of the millions of migrants who are refugees from civil unrest, political persecution, and environmental disasters.

4. The greatest migration in human history (at least in total numbers) is now going on as people move from rural to urban environments. If current rates continue, more than half the world's population will soon live in cities. Once again, it is the developing countries of Africa, Latin America, and Asia that are experiencing the most dramatic changes caused by urbanization.

Population Growth and Change

Familiarity with several population statistics is important for understanding population geographies of the countries in each world region. The most common population statistic is the **rate of natural increase** (abbreviated as RNI), which depicts the annual growth rate for a country or region as a percentage increase. This statistic is produced by subtracting a population's number of deaths in a given year from its

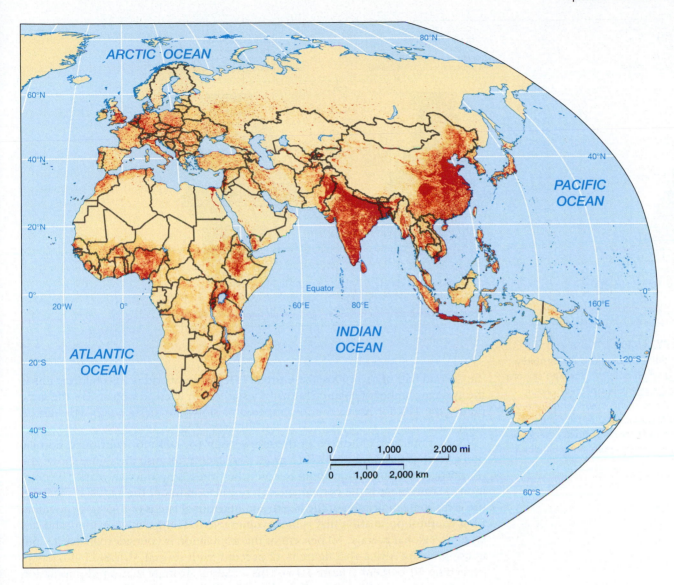

annual number of births: the result is the number of people added through natural increase. Gains or losses through migration are not considered in this figure. Instead of using raw numbers for a large population, demographers divide the total numbers of births or deaths by the total population, producing a figure per 1,000 of the population. This is referred to as the *crude birthrate* or the *crude death rate*. For example, current data for the world as a whole show a crude birthrate of 22 per 1,000 and a crude death rate of 9 per 1,000. Thus, the natural growth rate is 13 births per 1,000. This is expressed as the rate of natural increase simply by converting that figure to a percentage. As a result, the current RNI for the world is 1.3 percent per year. Because birthrates vary greatly between countries (and even regions of the world), rates of natural increase also vary greatly. In Africa, for example, many countries have crude rates of more than 40 births per 1,000 people, with several close to 50 per 1,000. Since their death rates are generally less than 20 per 1,000, we find RNIs around 3 percent per year, which are among the highest rates of natural growth found anywhere in the world.

   Though the crude birthrate gives some insight to current conditions in a country, demographers also use the **total fertility rate**, or TFR. The TFR is a rate that measures the fertility of an average group of women moving through their childbearing years. If women marry early and have many children over a long span of years, the TFR will be a high number. Conversely, if data show that women have few children, the number will be correspondingly low. According to population data collected in

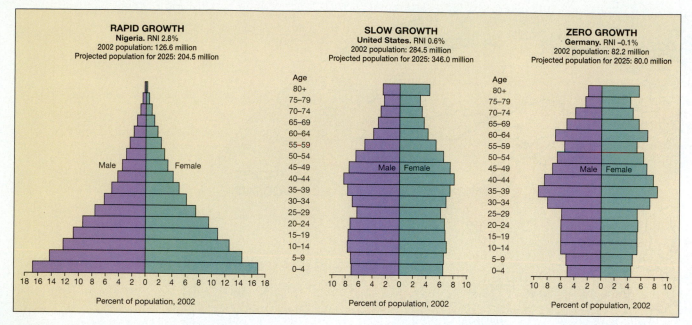

FIGURE 1.14 • POPULATION PYRAMIDS The term *population pyramid* comes from the form assumed by a rapidly growing country such as Nigeria when data for age and sex are plotted graphically as a percentage of total population. Rapid natural growth will probably continue in this country because of the high percentage of the population entering fertility years. In contrast, slow-growth and zero-growth countries have a very narrow base, which indicates relatively few people in the childbearing ages.

the second half of the 1990s, the current TFR for the world is 2.8 children; this is the average number of children borne by a statistically average woman. For the same period the TFR for Africa is 5.2, compared to slow-growth or no-growth Europe with 1.4.

One of the best indicators of the momentum (or lack) for continued population growth is the youthfulness of a given population, because this figure shows the proportion of a population about to enter the prime reproductive years. The common measurement is the percentage of a population under the age of 15. This figure gives some indication of the health and nutritional needs of a society providing for a youthful population that is most vulnerable to famine, malnutrition, and disease. As a global average, 30 percent of the population is younger than age 15, but in fast-growing Africa, that figure is 42 percent, with several African countries approaching 50 percent (Figure 1.14). This suggests strongly that rapid population growth in Africa will continue for at least another generation despite the tragedy of the AIDS epidemic. In contrast, Europe averages 18 percent and North America 21 percent.

At the other end of the age spectrum is the percentage of a population over the age of 65. This measure (also expressed as a percentage figure) suggests the need for social welfare services such as health care and social security, which are usually provided through taxation of those of working age. Japan and European countries, for example, have a relatively high proportion of people over age 65 and a very small percentage of young people approaching their reproductive years. In combination, these data raise concerns about whether these countries can continue to support social security and other services necessary for the elderly.

The Demographic Transition

Population growth rates change over time. For example, growth rates declined through history in Europe, North America, and Japan as these countries became increasingly industrialized and urbanized. From this historical data, demographers generated the **demographic transition** model, a four-stage model that conceptualizes changes in birthrates and death rates through time as a population urbanizes. However, while the demographic transition model seems valid for describing historical population change, there is considerable debate about whether it offers an accurate tool for predicting change in countries currently undergoing industrialization and urbanization (Figure 1.15).

In the demographic transition model, Stage 1 is characterized by both high birthrates and death rates, leading to a very slow rate of natural increase. Historically, this stage is associated with Europe's preindustrial period, which came before common public health measures such as sewage treatment, the scientific understanding of how disease spreads, or the most fundamental aspects of modern medicine. During this stage, living conditions and lack of medical knowledge combined to make death rates high and life expectancy short. Unfortunately, these conditions are still common in some parts of the world.

In Stage 2, death rates fall dramatically while birthrates remain high. This produces a rapid rise in the RNI. Again, both through history and in the present, this decrease in death rates is usually associated with the onset of public health measures and modern medicine. An assumption of the demographic transition model is that these services became increasingly available after some degree of economic development took place and, further, that they were increasingly accessible to the growing number of urban dwellers. However, even as death rates fall, it takes time for people to respond with lower birthrates.

In Stage 3, birthrates begin to decline as people become aware of the advantages of smaller families in an urban and industrial setting. Last, in Stage 4, a very low RNI results from a combination of low birthrates and very low death rates. As a result, there is very little natural population increase during this final phase.

For many professionals, the demographic transition model is a reasonable predictor of where and when we can expect population growth rates to slow in the developing world. However, critics do not believe the model can be universally applied in the twenty-first century because there are too many complicating variables and differences between countries. Minimally, though, the demographic transition concept is a useful tool for organizing information about the vast differences in birthrates and death rates that we find around the world (Figure 1.16).

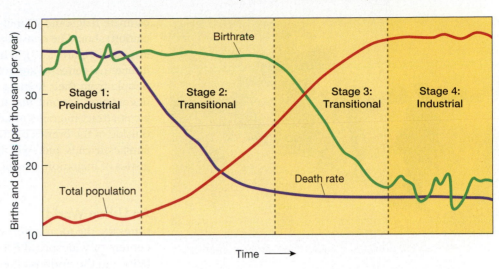

FIGURE 1.15 • DEMOGRAPHIC TRANSITION   As a country goes through industrialization, its population moves through the four stages in this diagram, referred to as the *demographic transition*. In the first stage, population growth is low because high birthrates are offset by high death rates. Rapid growth takes place in Stage 2 as death rates decline. Stage 3 is characterized by a decline in birthrates. The transition ends with low growth once again in Stage 4, resulting from a relative balance between low birthrates and death rates. *(From Knox and Marston, 2001,* Human Geography, *2nd ed., Upper Saddle River, NJ: Prentice Hall)*

Migration Patterns

Never before in human history have so many people been on the move and lived outside of their countries of birth. Today about 125 million, or 2 percent, of the total world population can be considered migrants of some sort. Estimates suggest that this figure may increase by some 4 million annually. Much of this international migration is directly linked to the new globalized economy. About half of the migrants live either in the developed world or in those developing countries with vibrant industrial, mining, or petroleum extraction economies. In the oil-rich countries of Kuwait and Saudi Arabia, for example, the labor force is primarily made up of foreign migrants. In terms of total numbers, fully a third of the world's migrants live in 7 industrial countries: Japan, Germany, France, Canada, the United States, Italy, and the United Kingdom. Since industrial countries usually have very low birthrates, immigration accounts for a large proportion of total population growth. Roughly one-third of the annual growth in the United States, for example, comes from in-migration.

But not all migrants move for economic reasons. War, discrimination, famine, and environmental destruction also cause people to flee to safer places. Accurate data on refugees are often difficult to obtain, but UN officials

FIGURE 1.16 • FERTILITY AND MORTALITY   Birthrates and death rates vary widely around the world. Fertility rates result from many variables, including state family-planning programs and the level of a woman's education. This family is in Bangladesh, a country that is working hard to reduce its birthrate through education programs. *(Trygve Bolstad/Panos Pictures)*

FIGURE 1.17 • REFUGEES About 27 million people, or 20 percent of the world's migrant population, are refugees from ethnic warfare. Most of these refugees are in Africa and western Asia. These women wait in line for water at a refugee camp in southern Sudan. *(Liz Gilbert/Corbis/Sygma)*

FIGURE 1.18 • GROWTH OF WORLD CITIES This map shows the largest cities in the world (more than 10 million), along with predicted growth by the year 2015. Note that the greatest population gains are expected in the large cities of the developing world, such as Lagos, Nigeria; Karachi, Pakistan; and Bombay (Mumbai), India. In contrast, large cities in the developed world are predicted to grow slowly over the next decades. Tokyo, for example, will add fewer than a million people.

estimate that some 27 million people (or about 20 percent of the total migrant population) should be considered refugees. More than half of these are in Africa and western Asia (Figure 1.17). While the causes of migration are often complicated, three interactive concepts are helpful in understanding this process. First, *push* forces drive people from their homelands. Examples are unemployment, war, or environmental degradation. Second, *pull* forces, such as better economic opportunity or health services, attract migrants to certain locations. These new destinations can be within or outside of the home country. Connecting the two are the informational *networks* of families, friends, and, sometimes, labor contractors who provide information on the logistics of migration, transportation details, or news of housing and jobs.

An Urban World

Cities are the focal points of the modern globalizing world. They are the fast-paced centers of widespread economic, political, and cultural change. Because of this vitality and the options cities offer to the poor and uprooted rural peoples, cities are also magnets for migration. The scale and rate of growth of some world cities is staggering (Figure 1.18): Estimates are that both Mexico City and São Paulo (Brazil), cities of more than 20 million, are adding nearly 10,000 new people each week. Urban planners predict that these two cities will actually double in size within the next 15 years. Assuming that predictions about the migration rate to cities are correct, the world is quickly approaching the point at which more than half of its people will be urban dwellers. Evidence for this comes from data on the urbanized population, which is that percentage of a country's population living in cities. Currently, 47 percent of the world's population lives in cities. Many demographers predict, however, that the world will be 60 percent urbanized by the year 2025.

For example, more than 75 percent of the populations of Europe, Japan, Australia, and the United States live in cities. Generally speaking, most countries with such high rates of urbanization are also highly industrialized, since manufacturing tends to locate around urban centers. In contrast, the urbanized rate for developing countries is usually less than 50 percent, with figures closer to 40 percent not uncommon. Roughly 30 percent of Sub-Saharan Africa's people, for example, live in cities. These data imply a low level of industrial activity. Urbanization figures also

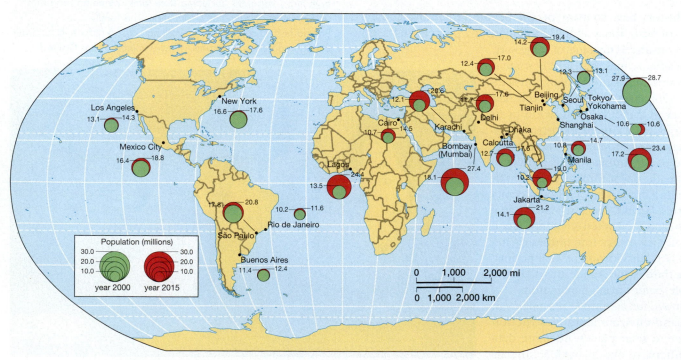

show where there is high potential for urban migration. If the urbanized population is relatively small, such as in Zimbabwe (Africa), where only 32 percent of the population lives in cities, the probability is great for high rates of urban migration during the next decades. Major changes in family structure, national policies, and market and manufacturing economics will likely occur as people move from the countryside to the cities.

Cultural Coherence and Diversity: The Geography of Tradition and Change

Culture can be thought of as the weaving that keeps the world's diverse social fabric together. However, a quick reading of the daily newspaper raises questions about whether the world is literally coming unraveled because of increased cultural conflict. With the recent rise of global communication systems (satellite TV, movies, video, etc.), stereotypic Western culture is spreading at a rapid pace. While many seek Western entertainment, others resist what they consider a new form of cultural imperialism. As a result, negative reaction results in the form of censorship and restrictions on Western film, TV, and music.

In response to the spread of a common world culture, many people across the globe are revisiting their traditional and historical identities as ethnic groups. This is not simply a matter of quaint customs and costumes; instead, it often leads to open conflict with neighboring groups that have different identities and self-interests. Consequently, over much of the world, we are seeing a new interest in ethnic and cultural politics. This is particularly pronounced in the former Soviet Union, where ethnic identities were suppressed under communism for most of the last century. But it is also encountered in other large nation-states—including those in western Europe—where new voices are demanding regional self-rule, separate territories, and control over local natural resources. Religious conflicts, too, seem to be on the increase. Some of this accompanies the return of religious choice to former Soviet lands, yet in other parts of the world tensions result as new forms of worship are introduced by missionaries. Religious fundamentalism is also on the rise as a reaction to globalizing culture, and this too causes cultural tensions.

The geography of cultural cohesion and tension, then, entails an examination of tradition and change, of the way language and religion are important to group belonging and identity. Additionally, this geography examines those complex and varied currents that underlie ethnic tensions throughout the world (Figure 1.19).

Culture in a Globalizing World

Because of the dynamic changes associated with globalization, traditional definitions of *culture* must be stretched somewhat to provide a conceptual framework for the twenty-first century. Though social scientists offer varied definitions of culture, there is enough agreement between these definitions to provide a starting point. **Culture** is learned, not innate, and it is shared, not individual. Further, culture is behavior held in common by a group of people, giving them what, for lack of a better term, could be called a "way of life."

Additionally, culture has both abstract dimensions, such as speech and religion, as well as material components like technology and housing. These varied expressions of culture are relevant to the study of world regional geography because they tell us much about the way people interact with their environment. Finally, especially given the widespread influences of globalization, it is best to think of culture as dynamic rather than static. That is, culture is a

FIGURE 1.19 • ETHNIC TENSIONS   Present day cultural change is often characterized by both convergence and similarity as people are brought together by globalized culture, or, alternatively, by factionalism and separatism, such as the religious strife shown here in India between Hindu and Muslim peoples. *(Robert Nickelsberg/Getty Images, Inc.)*

(a)

(b)

(c)

(d)

FIGURE 1.20 • FOLK, ETHNIC, POPULAR, AND GLOBAL CULTURE A broad range of cultural groups characterize the world's modern cultural geography, ranging from traditional folk cultures to the ever-present popular and global cultures. Representing global culture (a) is a Korean woman in downtown Washington, D.C., with laptop and cell phone, while the young couple (b) in Sofia, Bulgaria, express the universal style of popular culture. At the opposite end of the spectrum are the ethnic cultures of (c) Farifuna women in Honduras and (d) the Chonguinada in Peru. *(Rob Crandall/Rob Crandall, Photographer)*

changing process, not a fixed condition, for culture is constantly adapting to new circumstances. As a result, there are always tensions between the traditional elements of a culture and newer forces promoting change (Figure 1.20).

When Cultures Collide

Cultural change often takes place within the framework of international tensions. Sometimes a new cultural system will replace another, while at other times tradition will resist change. More commonly, however, a newer, hybrid form of culture results that is a mixture of two cultural traditions. Historically, colonialism was the most important force behind these cultural collisions; today, though, globalization in its varied forms is probably the major cause of cultural collision and change.

Cultural imperialism is the active promotion of one cultural system at the expense of another. Although there are still many expressions of cultural imperialism today, the most severe examples occurred in the colonial period when European cultures spread worldwide, sometimes overwhelming, eroding, and even replacing indigenous cultures. During this period, Spanish culture spread widely in Latin America; French culture diffused into parts of Africa; and British culture entered India. Of course, North America was also highly influenced by these three cultures. The effects of these cultural influences included the authorization of new languages, education systems, and administrative institutions. In addition, foreign dress styles, diets, gestures, and organizations were added to existing cultural systems. Many remnants of colonial culture are still evident today. In India the makeover was so complete among a few social groups that people are fond of saying, with only slight exaggeration, that "the last true Englishman will be an Indian." Today's cultural imperialism is seldom linked to a definitive colonizing force but more often comes as a fellow traveler with globalization. To illustrate, the spread of fast food restaurants or Hollywood movies throughout the world is more a search for new consumer markets on the part of U.S. corporations than a deliberate effort to spread North American culture into foreign countries.

The reaction against cultural imperialism, **cultural nationalism**, is the process of defending a cultural system against offensive cultural expressions while at the same time promoting local or national cultural values. Sometimes cultural nationalism takes the form of legislation or official censorship that simply prohibits the unwanted cultural traits. Examples of this form of cultural nationalism are common. France has long fought the Anglicization of its language by banning "Franglais," the use of English words such as "Le weekend," in official French. More recently, France has also sought to protect its national music and film industries by legislating that radio DJs play a certain percentage (40 percent at this writing) of French songs and artists each broadcast day. France is also erecting tax and tariff barriers to stave off Hollywood's products. Further, many Muslim countries censor Western cultural influences by restricting and censoring international TV, which they consider the source of many undesirable cultural influences. Most Asian countries, as well, are increasingly protective of their cultural values, and many are demanding changes to reduce the sexual content of MTV and other international TV networks.

As noted above, a common product of "cultural collision" is the blending of forces to form a new type of culture. This is called **cultural syncretism** or **hybridization** (Figure 1.21). To characterize India's culture as British, for example, would be to oversimplify and even exaggerate England's colonial influence even though Indians have adapted many British traits to their own unique circumstances. India's use of English has produced a unique form of "Indlish" that often confuses English-speaking visitors to South Asia. Nor should we forget how India has added many words to our own English vocabulary—khaki, pajamas, veranda, bungalow. In sum, both the Anglo and Indian cultures have been changed by the British colonial presence in South Asia through cultural hybridization.

FIGURE 1.21 • CULTURAL SYNCRETISM   When two cultures are brought together, a new culture is often created, as was the case in India as a result of the blending of British colonial culture and local South Asian influences. This photograph shows British colonial architecture and modern buses in Bombay (Mumbai), India, coupled with local cabs, scooters, and ox carts. *(Rob Crandall/Rob Crandall, Photographer)*

Language and Culture in Global Context

Language and culture are so closely tied that language is often the characteristic that best defines cultural groups. Furthermore, since language is the means of communication for a cultural group, it obviously brings together many other aspects of cultural identity, such as politics, religion, commerce, folkways, and customs. Thus language is fundamental to cultural cohesiveness (Figure 1.22).

Even individual languages often have very distinctive forms associated with specific regions. Such different forms of speech are called *dialects*. Although dialects of the same language have their own unique pronunciation and grammar (think of the distinctive differences, for example, between British, North American, and Australian English), they are usually mutually intelligible so that speakers of the mother language can understand each other despite different dialects. Additionally, when people from different cultural groups cannot communicate directly in their native languages, they often agree upon a third language to serve as a common tongue, or **lingua franca**. Swahili has long served that purpose in eastern Africa where many different languages are spoken. Other examples are that French was historically the lingua franca of international politics and diplomacy while today English is increasingly the common language of international business (Figure 1.23).

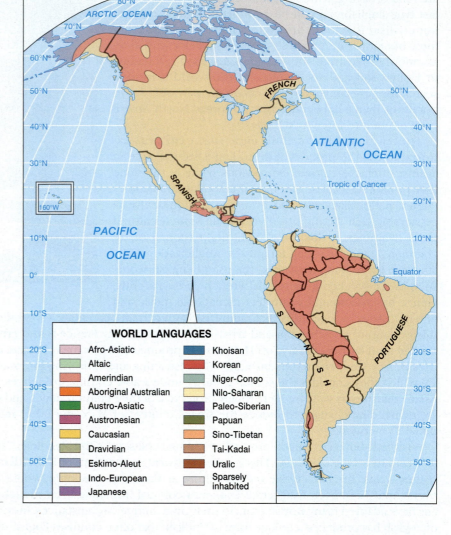

FIGURE 1.22 • WORLD LANGUAGES Most languages of the world belong to a handful of major language families. About 50 percent of the world's population speaks a language belonging to the Indo-European language family, which includes most European languages as well as major languages in South Asia . The next largest family is the Sino-Tibetan family, which includes languages spoken in China, the world's most populous country. (Adapted from Rubenstein, 2002, The Cultural Landscape: An Introduction to Human Geography, 7th ed., Upper Saddle River, NJ: Prentice Hall)

WORLD LANGUAGES

Afro-Asiatic		Khoisan	
Altaic		Korean	
Amerindian		Niger-Congo	
Aboriginal Australian		Nilo-Saharan	
Austro-Asiatic		Paleo-Siberian	
Austronesian		Papuan	
Caucasian		Sino-Tibetan	
Dravidian		Tai-Kadai	
Eskimo-Aleut		Uralic	
Indo-European		Sparsely inhabited	
Japanese			

FIGURE 1.23 • ENGLISH AS GLOBAL LANGUAGE A traffic sign in Saudi Arabia reminds us of how some familiar English phrases have become almost universal. In many ways, English has become a global *lingua franca* for diverse activities such as international business, aviation traffic, communications, popular music, and road signs. (Ray Ellis/Photo Researchers, Inc.)

A Geography of World Religions

Another extremely important and often defining trait of cultural groups is their religion (Figure 1.24). In fact, some would argue that much present-day ethnic tension is religious-based. Recent ethnic violence and unrest in far-flung places such as the Balkans, Afghanistan, and Indonesia illustrate the point. Additionally, widespread missionary activity, with its emphasis on convincing people to change religions, has spread specific forms of religion but has also aggravated religious tensions between ethnic groups. An example would be the spread of evangelical Protestantism into areas of Catholicism in Latin America.

Universalizing religions, such as Christianity, Islam, and Buddhism, attempt to appeal to all peoples regardless of location or culture; these religions usually have a proselytising, or missionary, program that actively seeks new converts. In contrast, **ethnic religions** remain identified closely with a specific ethnic, tribal, or national group. Judaism and Hinduism, for example, are usually regarded as ethnic religions because they normally do not actively seek new converts. Christianity is the world's largest in both areal extent and number of adherents, or believers. Although split into separate branches and churches, Christianity as a whole claims almost 2 billion adherents, or about a third of the world's population. The largest number of Christians can be found in Europe, Africa, Latin America, and North America. Islam, which has spread from its origins on the Arabian Peninsula as far east as Indonesia and the Philippines, has about 1.2 billion members. While not as severely

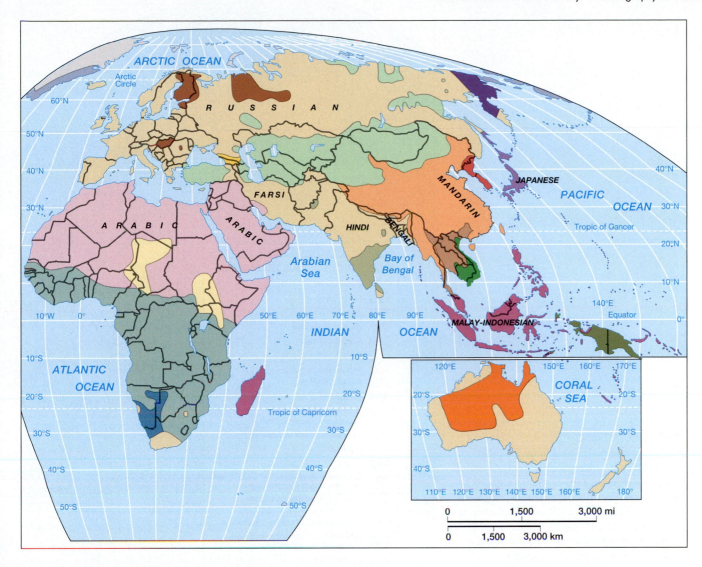

fragmented as Christianity, Islam is also split into separate groups. The major branches are Shiite Islam, which makes up about 11 percent of the total Islamic population and represents a majority in Iran and southern Iraq, and the more widespread Sunni Islam, which is found from the Arab-speaking lands of North Africa to Indonesia in Southeast Asia (Figure 1.25).

Hinduism, which is closely linked to India, has about 850 million adherents. Outsiders often regard Hinduism as polytheistic, since Hindus worship many deities or gods. Most Hindus argue, however, that all of their gods are merely representations of a single divine, cosmic unity. Historically, Hinduism is linked to the caste system with its segregation of peoples based upon ancestry and occupation. But because the caste system is now offensive to India's official democratic ideology, connections between religion and caste are now much less open than in the past. Buddhism, which had its roots in Hinduism 2,500 years ago as a reform movement, is widespread in Asia, extending from Sri Lanka to Japan and from Mongolia to Vietnam. It is difficult to accurately count the number of people practicing Buddhism because the religion exists alongside other faiths in many areas. However, estimates are that the total Buddhist population probably ranges between 350 million and 900 million people.

Judaism, the parent religion of Christianity, is also closely related to Islam. Although tensions are often high between Jews and Muslims because of the Israeli-Palestinian conflict, these two religions, along with Christianity, actually have

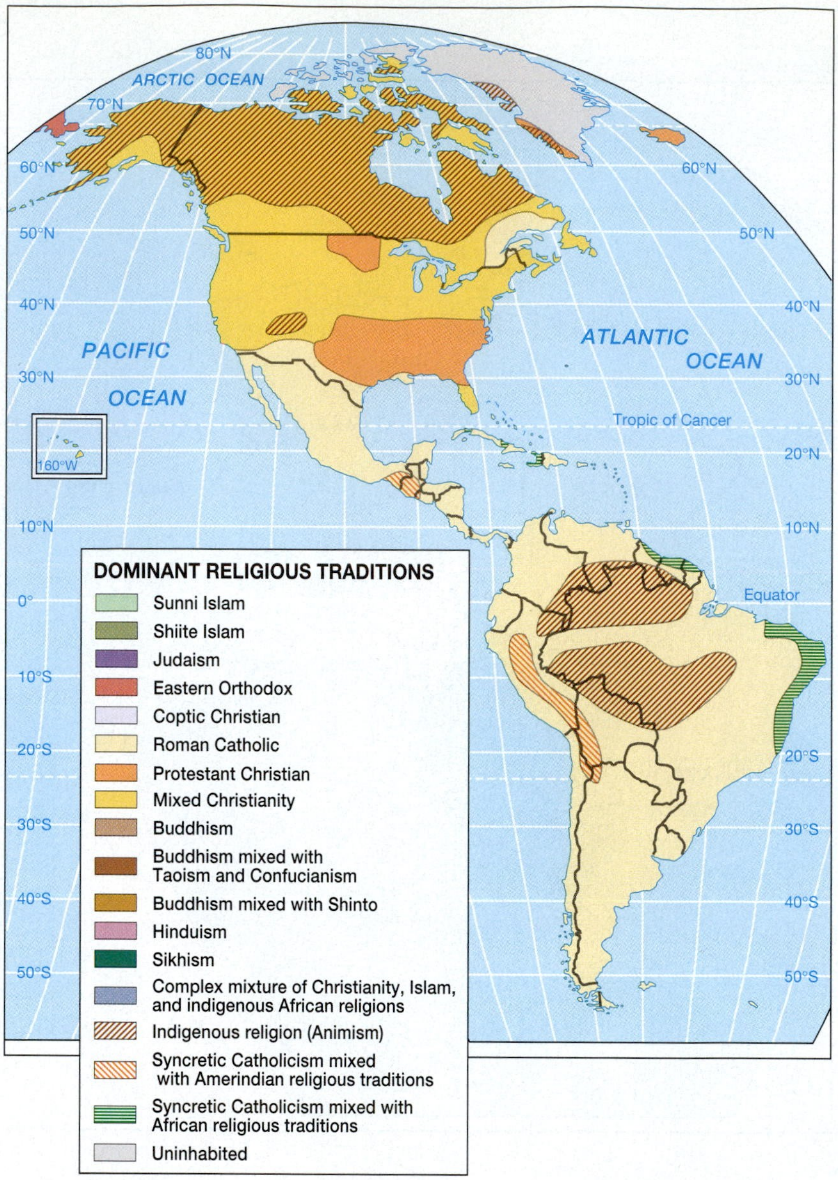

FIGURE 1.24 • MAJOR RELIGIOUS TRADITIONS This map shows the geographical pattern of major religious traditions found throughout the world. For most people, religious tradition is a major component of cultural and ethnic identity. While Christians of different sorts account for about 34 percent of the world's population, this religious tradition is highly fragmented. Within Christianity, there are about twice as many Roman Catholics as Protestants. Adherents to Islam make up about 20 percent of the world's population. The Sunni branch within Islam is three times larger than the Shiite. Hindus make up about 14 percent of the global population.

DOMINANT RELIGIOUS TRADITIONS

- Sunni Islam
- Shiite Islam
- Judaism
- Eastern Orthodox
- Coptic Christian
- Roman Catholic
- Protestant Christian
- Mixed Christianity
- Buddhism
- Buddhism mixed with Taoism and Confucianism
- Buddhism mixed with Shinto
- Hinduism
- Sikhism
- Complex mixture of Christianity, Islam, and indigenous African religions
- Indigenous religion (Animism)
- Syncretic Catholicism mixed with Amerindian religious traditions
- Syncretic Catholicism mixed with African religious traditions
- Uninhabited

FIGURE 1.25 • RELIGIOUS LANDSCAPES Minarets, from which Muslims are called to prayer, surround this mosque in Istanbul, Turkey, creating an instantly recognizable landscape linked to Islam. A similar landscape feature would be the towering steeple of a Roman Catholic cathedral, which would be associated with that religion regardless of location. *(Rob Crandall, Photographer)*

common historical and theological roots in the Hebrew prophets and leaders. Judaism now numbers about 18 million followers, having lost perhaps a third of its total population during the Nazis' systematic extermination of Jews during World War II.

Finally, in some parts of the world, religious practice has declined for different reasons, giving way to secularization, in which people consider themselves either nonreligious or outright atheistic. Though secularization is difficult to measure, social scientists estimate that about 1 billion people fit into this category worldwide. Perhaps the best example of secularization comes from the former communist lands of Russia and eastern Europe, where there was historically open hostility between government and church. Since the fall of Soviet communism in 1989, however, many of these countries have experienced a modest religious revival. In contrast, secularization has probably become more pronounced in recent years in western Europe. France now reportedly has more people attending Muslim mosques on Fridays than Christian churches on Sundays, even though this country's history and culture are closely tied to the Roman Catholic church. Japan and the other countries of East Asia are also noted for their high degree of secularization.

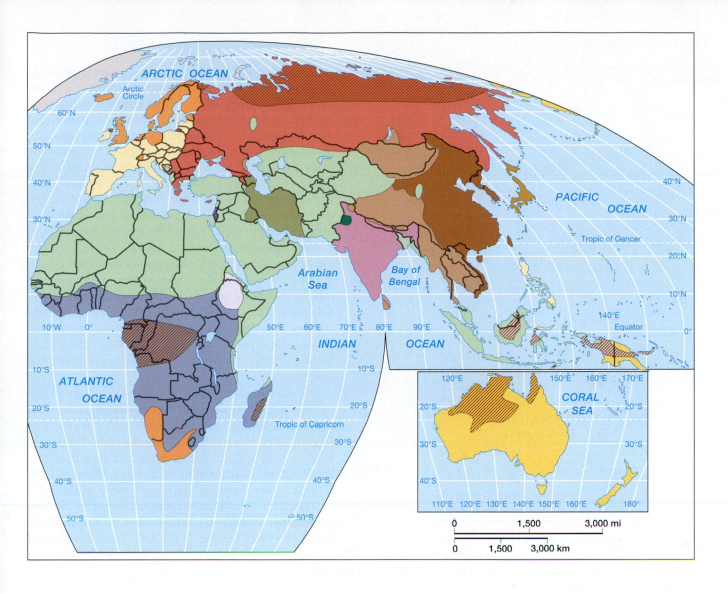

Geopolitical Framework: Fragmentation and Unity

Rapid and large-scale political change in various regions of the world has undoubtedly been one of the leading characteristics of the last several decades. The term *geopolitics* is used to describe and explain the close link between geography and political activity. More specifically, geopolitics focuses on the interaction between power, territory, and space at different scales, from the local to the global.

Nation-states

A conventional starting point for examining political geography is the concept of the nation-state. The hyphen links two different, yet related concepts: the term *state* describes a political unit with territorial boundaries recognized by other countries that is internally governed by an organizational structure. The second term *nation* refers to a large group of people who share numerous cultural elements, such as language, religion, tradition, or simple cultural identity, and, more importantly, view themselves as forming a single political community. The **nation-state**, then, is ideally a relatively homogeneous cultural group with its own fully independent political territory. Nation-states, however, do not always fit neatly into the boundaries of actual states. In fact, perfect nation-state overlap is relatively rare. While some large cultural groups may consider themselves to be nations lacking recognized, self-governed territory (such as the Kurdish people of Southwest Asia; Figure 1.26), many

FIGURE 1.26 • A NATION WITHOUT A STATE Not all nations or large cultural groups control their own political territories or states. For example, the Kurdish people of Southwest Asia traditionally occupy a large cultural territory that is currently in four different political states: Turkey, Iraq, Syria, and Iran. As a result of this political fragmentation, the Kurds are considered a minority in each of these four countries.

FIGURE 1.27 • END OF THE COLD WAR With the fall of Soviet communism, many nations rushed to remove the symbols of their former governments to forget the past and make room for a new future. In this photo taken in 1990, a monument to Lenin is toppled in Bucharest, Romania. *(Bi/Getty Images, Inc.)*

present-day politically independent states include diverse cultural and ethnic groups within their established boundaries. Often these ethnic groups seek self-rule. In Spain, for example, both the Catalans and the Basques think of themselves as forming separate, non-Spanish nations and thus seek political autonomy from the central government. In fact, cultural homogeneity within any political unit is actually rare. On a world scale, out of the more than 200 different countries that now make up the global geopolitical fabric, only several dozen would qualify as true nation-states.

Although the nation-state may be considered the ideal political model, multinational states containing different cultural and ethnic groups are much more numerous than homogeneous states with a unified culture. Because of this diversity, *micronationalism*, or group identity with the goal of self-rule within an existing nation-state, has been on the rise for the last half-century and remains a considerable source of geopolitical tension throughout the world.

For example, with the breakup of the Soviet Union in 1991 came opportunities for self-determination and independence in eastern Europe and Central Asia (Figure 1.27). As a result, there have been basic changes in economic, political, and even cultural arrangements. Furthermore, Soviet satellite states have different goals. While religious freedom helps drive national identities in some new Central Asian republics, eastern Europe is primarily concerned with economic and political links to western Europe. At the same time, Russia itself wavers perilously between different geopolitical and economic pathways (see Chapter 9, The Russian Domain).

Centrifugal and Centripetal Forces

While wide-ranging international conflicts remain a nagging concern of diplomats and governments, perhaps more common are strife and tension within—rather than between—nation-states. Civil unrest, tribal tensions, terrorism, and religious factionalism have created a new fabric of political tension. So widespread and problematic is this tension in some parts of the world that scholars increasingly describe the nation-state as a possibly outmoded political structure that may soon be replaced by tribal and ethnic micro-states or even warlord fiefdoms.

Cultural and political forces acting to weaken or divide an existing state are called **centrifugal forces**, since they pull away from the center. Many have already been mentioned: linguistic minority status, ethnic separatism, territorial autonomy, and disparities in income and well-being. Separatist tendencies in French-speaking Quebec (Canada) or the Basque region in Spain are good examples (Figure 1.28). Counteracting these destructive forces are those that promote political unity and reinforce the state structure. These are called **centripetal forces**, and they include a shared sense of history, a need for military security, an overarching economic structure, or simply the advantages that come from a larger unified political structure that builds and maintains the infrastructure of highways, airports, and schools.

The larger question in much of the world is whether centrifugal forces will increase in strength so that they overcome the centripetal forces, leading thereby to new, smaller independent units. Also important is whether this process takes place through violent struggle or peaceful means. Given the current widespread nature of these internal tensions, there is much evidence to suggest that separatist struggles will continue to dominate regional geopolitics for decades to come.

Global Terrorism

In many ways, the September 2001 terrorist attacks on the United States highlight the need to expand and redefine our thinking about globalization and geopolitics. Previously, on a global scale terrorist acts were usually connected to nationalist or regional geopolitical aspirations to achieve independence or autonomy. However, that was not the case with the attacks on the World Trade Center and the Pentagon. Unlike IRA terrorist bombings in Great Britain or those of the extremist wing of Basque nationalists in Spain, the September 2001 terrorism went beyond conventional geopolitics as a relatively small group of religious fanatics attacked the symbols of Western culture, power, and finance. These attacks were a reminder of the close and somewhat unpredictable links between politics, cultural identity, and economics in today's globalized world.

Many experts argue that global terrorism is both a product as well as an expression of globalization. Unlike earlier geopolitical conflicts, the geography of global terrorism is not defined by a war between well-established political states. Instead, the Al Qaeda terrorists appear to belong to a web of small, well-organized cells located in many different countries. These cells are linked together in a decentralized network that provides guidance, financing, and political autonomy. Further, Al Qaeda has used the tools of globalization to its advantage. For example, members communicate instantaneously via cell phones and the Internet. Transnational members travel between countries quickly and frequently. The network's activities are financed through a complicated array of holding companies and subsidiaries that traffic in a range of goods, such as honey, diamonds, and opium.

To recruit members and political support, the network feeds upon the unrest and inequalities (real and imagined) resulting from economic globalization, namely the differences between rich Western countries and poorer Muslim states. As a result, this form of global terrorism attacks targets that symbolize the modern global values and activities it opposes. Even though the September 2001 attacks focused terror on the United States, the casualties and economic damage were truly international. For example, citizens from more than 80 countries were killed in the World Trade Center tragedy alone. Additionally, the attacks may have pushed parts of the global economy from temporary economic slowdown to full-on recession. The devastating effects on U.S. and international airline companies as well as the loss of tourist revenue to Muslim countries such as Morocco and Egypt serve as examples.

The military and political responses to global terrorism also demand an expanded notion of geopolitics. In 2001, for example, in Afghanistan, the military muscle of a nation-state superpower—the United States—was directed not at another nation-state but rather at a confederation of religious extremists, the Taliban and Al Qaeda. Military strategists refer to this kind of fighting as **asymmetrical warfare**, a term that aptly describes the differences between a superpower's military technology and strategy and the lower-level technology and guerilla tactics used by Al Qaeda and the Taliban. Most superpower military strategists agree that the war on terrorism will be fought with political and battlefield asymmetry.

Further, because the global context for the war on terrorism is a complex mosaic of religious, cultural, political, and economic currents, there is debate in geopolitical circles about whether the United States can—or should—pursue its agenda as a member of a coalition of explicitly committed countries (Britain and Russia, for example), or whether it will be done essentially alone, or unilaterally. Many geopolitical thinkers posit that resolution of this coalition-unilateral issue will be decided by how the United States defines its goals for the war on terrorism. A global coalition seems likely if the United States sets a specific and limited goal such as the destruction of Al Qaeda. However, the broader goals discussed in Washington, D.C., namely the replacement of hostile governments in Iraq or North Korea or rebuilding nation-states in the Middle East, may have limited support from the world community.

FIGURE 1.28 • BASQUE SEPARATISM The demands of the Spanish Basques for independence are just one illustration of the ethnic separatism that is still common in contemporary Europe and, for that matter, the world. *(Alberto Arzoz/Panos Pictures)*

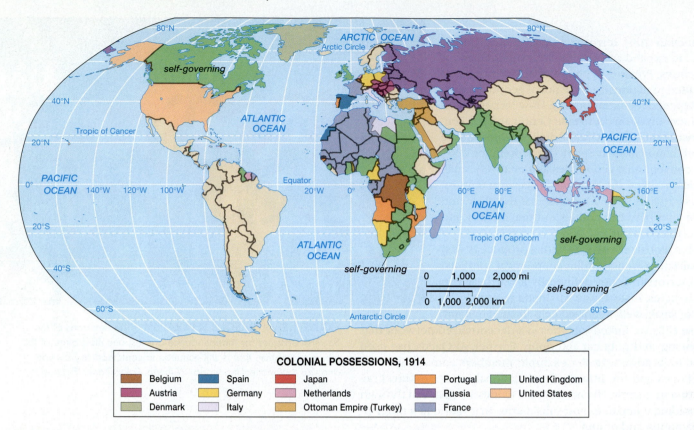

COLONIAL POSSESSIONS, 1914

Belgium	Spain
Austria	Germany
Denmark	Italy

Japan	Portugal
Netherlands	Russia
Ottoman Empire (Turkey)	France

United Kingdom
United States

FIGURE 1.29 • THE COLONIAL WORLD, 1914 This world map shows the extent of colonial power and territory just prior to World War I. At that time, most of Africa was under colonial control, as were Southwest Asia, South Asia, and Southeast Asia. Australia and Canada were very closely aligned with England. Also note that in Asia, Japan controlled colonial territory in Korea and northeastern China, which was known as Manchuria at that time.

Colonialism and Decolonialization

One of the major themes in world regional geography is the waxing and waning of European colonial power over much of the world. **Colonialism** refers to the formal establishment of rule over a foreign population. A colony has no independent standing in the world community, but instead is seen only as an appendage of the colonial power. Generally speaking, the main period of colonialization by European states was from 1500 through the mid-1900s, though even today a few colonies remain (Figure 1.29).

Decolonialization refers to the process of a colony's gaining (or regaining) control over its territory and establishing a separate, independent government. As was the case with the Revolutionary War in the United States, this process often begins as a violent struggle. As wars of independence became increasingly common in the mid-twentieth century, some colonial powers began working toward peaceful disengagement from their colonies. In the late 1950s and early 1960s, for example, Britain and France granted independence to most of their former African colonies, often after a period of warfare or civil unrest. This process was nearly completed in 1997 when Hong Kong was peacefully restored to China by the United Kingdom.

But decades and even centuries of colonial rule are not easily erased. As a result, the influence of colonialism is still found in the new nations' governments, education, agriculture, and economies. While some countries may enjoy special status with, and receive continued aid from, their former colonial masters, others remain disabled and disadvantaged because of a much-reduced resource base. Some scholars regard the continuing economic ties between certain imperial powers and their former colonies as a form of exploitative neo-colonialism in which the mother country is still dominant. On the other hand, some remaining colonies, such as the Dutch Antilles in the Caribbean, have made it clear that they have no wish for independence, finding both economic and political advantages in continued dependency. Because the consequences of colonialism differ greatly from place to place, the current expressions of this historical period are complex (Figure 1.30).

Economic and Social Development: The Geography of Wealth and Poverty

The pace of global economic development has accelerated dramatically in the last decade. As a result, we now talk about global assembly lines, transnational corporations, commodity chains, and international electronic offices. Few regions of the world are untouched by these changes. The primary question, however, is whether the positive changes of economic globalization outweigh the negative. While answers are uncertain, sometimes elusive, and often inconclusive, an important first step in world regional geography is to link economic change to social development.

Economic development is desirable because it generally brings increased prosperity to individuals, regions, and nation-states. By conventional thinking, this usually translates into social improvements such as better health care, improved education systems, and more enlightened labor practices. The operative assumption is that as the economy of an area develops, so will its social infrastructure consisting of schools, hospitals, and so on. One of the more troubling expressions of recent economic growth, however, has been the geographic unevenness of prosperity and social structure improvement. While some regions prosper, others stagnate or even fall behind. As a result, the gap between rich and poor regions has increased. This geographic unevenness in development, prosperity, and social infrastructure is a characteristic trait of the early twenty-first century. According to the World Bank, more than 1.3 billion people live on less than a dollar a day, with about 60 percent of them in Sub-Saharan Africa and South Asia.

Regional inequities are also problematic because of their interaction with political, environmental, and social concerns. Political instability and social conflict within a nation, for example, are often driven by the economic gap between a poor region and the country's rich, industrial core. In those less-favored areas, poverty, social tensions, and environmental degradation often drive civil unrest that ripples through the rest of the country. The World Bank's *World Development Reports* document these inequities. In China's Gansu province, for example, per capita income is almost 50 percent lower than the national average. While most of Gansu's population lives in poverty, many other regions of China prosper. Other examples of internal unevenness are common: Italy, Poland, Mexico, Kenya, Brazil, and India all have striking economic differences between rich and poor areas. Thus the theme of the unevenness of social and economic development is a major story line in the regional chapters of this book.

FIGURE 1.30 • COLONIAL VESTIGES IN VIETNAM The red star flag of communist Vietnam flies in front of the Hotel de Ville in Ho Chi Minh City (formerly Saigon). This juxtaposition of the contemporary government's symbol and a building of the French colonial period capture the process of decolonialization and independence. *(Catherine Karnow/Woodfin Camp & Associates)*

More- and Less-Developed Countries

Until recently, economic development was centered in North America, Japan, and Europe, while most of the rest of the world remained gripped in poverty. This uneven distribution of power led scholars to devise a **core-periphery model** of the world economy. According to this model, the United States and Canada, western Europe, and Japan constitute the global economic core of the North, while most of the areas to the south make up a less-developed global periphery. Although an oversimplification of sorts, this core-periphery dichotomy does contain some truth. All the G-7 countries—the exclusive club of the world's richest nations made up of the United States, Canada, France, England, Germany, Italy, and Japan—are located in the northern hemisphere core (for certain meetings, the G-7 includes Russia, at which time the group becomes the G-8). Another assumption of this core-periphery model is that the developed northern core achieved its wealth primarily by taking advantage of the southern periphery, either through historical colonial relationships or through more recent economic imperialism.

Today, for example, much is made of "North–South tensions," a phrase implying that the rich and powerful countries of the Northern Hemisphere are at odds with poor and less powerful countries to the south. Over the past several decades, however, the global economy has grown much more complicated. A few former colonies of the south, most notably Singapore, have become very wealthy. Meanwhile, the northern countries of the former Soviet Union have experienced an extremely rapid economic and political decline, and are therefore now sometimes placed within the zone of global poverty. Australia and New Zealand, moreover, never fit into the neat North–South division. For these reasons, many global experts conclude that the term *North–South* model is antiquated and should be avoided.

Another term, the *Third World*, is often used to refer to the developing world. This phrase suggests a low level of economic development, unstable political organizations, and a very basic social infrastructure. Originally, the term was a product of the Cold War between the Soviet Union and Western countries like the United States and Europe and was used to describe those countries that were independent and not allied with either the capitalist and democratic First World or communist Second World superpowers. Since the Second World of the Soviet Union and its eastern European allies no longer exists, the concept of a Third World must be questioned. Also significant is the fact that a number of so-called Third World countries, such as South Korea, Taiwan, and Singapore, have experienced such rapid growth over the past several decades that they are no longer underdeveloped. In this book, therefore, we avoid terms such as *Third World* and *First World* and instead use terms that capture the complex spectrum of economic and social development, namely *more-developed country* (MDC), and *less-developed country* (LDC). A world map of income per person shows the spatial patterns of rich, moderate, and poor countries (Figure 1.31).

Indicators of Economic Development

The terms *development* and *growth* are often used interchangeably when referring to international economic activities. There is, however, value in keeping them separate. A dictionary definition of *development* uses phrases such as "expanding or realizing potential; bringing gradually to a fuller or better state." Therefore, when we talk about economic development, we usually imply structural changes, such as a shift from agricultural to manufacturing activity. Changes in the uses of labor, capital, and technology often accompany such a shift. These changes are also assumed to bring improvements in standard of living, education, and even political organization.

Growth, in contrast, is simply the increase in size of a system. The agricultural or industrial output of a country may grow, as it has for India in the last decade,

FIGURE 1.31 • WORLD GNI PER CAPITA   Gross national income, or GNI (formerly referred to as GNP), is one of the best measures to show the geographic pattern of rich and poor countries. Whereas 75 percent of the world's population fall into the lower categories, only 10 percent are in the highest category. (Data from the World Bank Atlas, 2001)

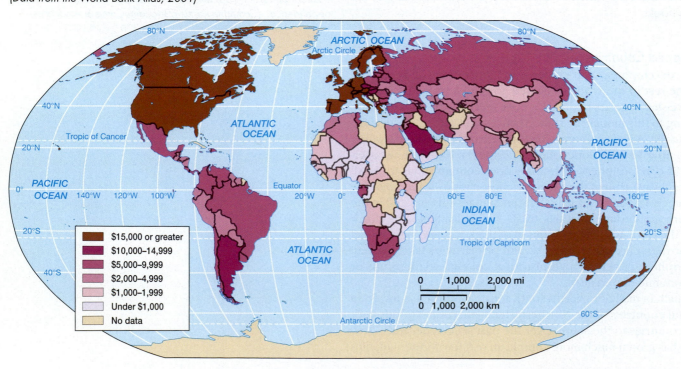

and this growth may—or may not—result in positive changes in the social structure. When something grows, it gets bigger; when it develops (sticking with the dictionary meaning, at least), it improves. Critics of the world economy are often heard to say that we need less growth and more development. This may be a realistic expectation.

Measuring Economic Wealth   The traditional measure for the size of a country's economy is the value of all final goods and services produced within its borders (gross domestic product, or GDP) plus the net income from abroad. These two factors constitute the gross national income, or GNI (formerly referred to as *gross national product*, or GNP). However, although widely used, GNI can be an incomplete and even misleading economic indicator because it ignores nonmarket economic activity, such as bartering or household work. It is also problematic because it does not take into account ecological degradation or depletion of natural resources, which may limit future economic growth. For example, if a country completely cuts its forest, which indeed might limit future economic growth, this act would nevertheless increase the GNI for that particular year. Further, GNI does not reveal the distribution of wealth within a country, the cost of living, or the social organizations supporting the economy. For instance, diverting educational funds to purchase military weapons might increase a country's GNI in the short run, but the economy would likely suffer in the future because of its less-educated population. For this reason, it is important to combine GNI data with other economic and social indicators.

A first step is to divide the GNI by the country's population, which produces a **GNI per capita** figure. For example, according to World Bank data, the annual GNI for the United States in 2002 was about $10.1 trillion. Dividing that figure by the population of around 288 million results in a GNI per capita of $35,060. Because of the ratio between population and GNI, similar figures will result for countries that have both a smaller GNI and a smaller population. Japan, for example, has a GNI about half that of the United States at $4.3 trillion. However, because that country also has a smaller population (127 million), its GNI per capita is slightly higher at $33,550.

An important qualification to these GNI per capita data is the concept of adjustment through **purchasing power parity (PPP)**, which gives us a comparable figure for a standard "market basket" of goods and services purchased with the local currency (Figure 1.32). Until recently, GNI data were usually based on the market exchange rate for a particular country's national currency as compared to the U.S. dollar. However, depending on the strength or weakness of that currency, the GNI data might be inflated or undervalued. For example, if the Japanese yen were to fall against the dollar overnight as a result of international traders speculating in the dollar, the official figure for Japan's GNI would correspondingly drop, despite the fact that Japan had experienced no real decline in economic output. Because of these problems, the PPP adjustment can provide a more accurate sense of the local cost of living, and these data can also be used to adjust GNI for currency differences. For example, when Japan's GNI per capita is adjusted for PPP, which takes out the inflationary factor, the figure is $25,170.

Lastly, the **economic growth rate** (the annual rate of expansion for GNI) can also be distorted by inflated currencies, but can be standardized by PPP. As an example, China's economic growth rate averaged nearly 9 percent per year during the 1980s and early 1990s based upon conventional GNI data. However, when adjusted for PPP-based currency, this annual growth rate is closer to 5 percent. Increasingly, world organizations are using PPP-based measures of economic activity for making international comparisons.

FIGURE 1.32 • VILLAGE FOOD MARKET   The concept of purchasing power parity (PPP), which is a common "market basket" of goods, is used to generate a sense of the true cost of living for different countries. This outdoor produce market is in Chichicastenango, Guatemala. *(Rob Crandall/Rob Crandall, Photographer)*

Indicators of Social Development

Although economic growth is a major component of development, there is equal interest in the conditions and quality of human life in this rapidly changing world. The standard assumption is that economic development and growth will spill over into the social infrastructure, thereby improving public health, working conditions, gender equality, and education. However, this assumption must be supported with objective data. For that reason, we include several measures of social development in the tables at the end of this book.

Life Expectancy One indicator of social development is *life expectancy*, the average length of life expected at birth for a typical male or female based upon national death statistics. A large number of social factors influence life expectancy, such as availability of health services, nutrition, frequency (or absence) of disease, accident rates, sanitation, and homicide rates. Additionally, life expectancy is affected by biological factors, which may explain why women usually live longer than men. When social and biological factors are combined, life expectancy data may provide important information into local conditions. For the few countries in which men live longer than women, for example, one can deduce that the social position of women remains quite low.

Life expectancy figures vary widely between countries in different world regions, but they have generally been improving over time. For the world as a whole, life expectancy in 1975 was 58 years, whereas today it is 67. Over the same period, Africa's figure has gone from 46 to 52, although in some African countries longevity has dropped recently because of the AIDS epidemic. In Russia, life expectancy has also fallen lately—especially for men—with the decline of economic conditions. Unfortunately, national life expectancy data often hide differences within a country, such as the wide disparity between rich and poor, differences between ethnic groups, or even discrepancies between specific regions. Within the United States, for example, the gap in life expectancy between African Americans and whites is 7 years, although this is an improvement over the 15-year gap recorded in 1900. Local-scale data also tell important stories. For example, partly because of the high homicide rates for African-American males in major urban areas such as New York City, Detroit, and Chicago, in these cities their calculated life expectancy is shorter than that of men in Bangladesh, one of the world's poorest countries.

Mortality and Illiteracy Rates *Under age 5 mortality*, which is the measure of the number of children who die per 1,000 persons, is another important indicator of social conditions. This figure provides information about conditions such as food availability, health services, and public sanitation. In the first 5 years of life, a child moves from the personal protection and nurturing of his or her mother into a larger social environment; these first 5 years are a time of high vulnerability to nutritional deprivation, infection, disease, accidents, and other human tragedies. In the tables, child mortality data are given for two points in time, 1980 and 1999, to indicate whether there has been an improvement in child mortality rates over that period.

Adult illiteracy rates provide two kinds of information relevant to the social development of a country. In general, high rates of illiteracy are usually products of low levels of national investment in education; given the importance of reading in this globalizing world, a high level of illiteracy can be a distinct liability for a country's future. Also, gender inequalities can also be evident in these data. In some countries, we see a striking difference between men and women, often with much higher rates of illiteracy for females than for males. This suggests that there are strong cultural, social, or economic limits on the education of girls and women. The consequences of low female literacy rates are many. For example, there is a demonstrated connection between acceptance of family planning and female education; usually a higher birthrate accompanies high female illiteracy. The role that women play in a country's economy is also closely linked to their levels of education and literacy (Figure 1.33).

Females in the Labor Force In many countries, women play a prominent role in economic life; thus, data on the percentage of *females in the labor force* are important (Figure 1.34). However, three points must qualify these data. First, much of women's work has traditionally been undervalued and hidden from the national census because it is considered part of the unpaid sector of household and agricultural work. Therefore, the long hours women spend gathering firewood, drawing water, cooking, tending gardens and fields, and raising children are not considered "employment" and therefore do not appear in the labor force statistics (men also perform work in the unpaid domestic sector, but usually to a lesser degree than women). Second, many women work in the "informal" economic sector, earning subsistence wages selling vegetables or handicrafts in village markets, cleaning houses, or doing other forms of temporary labor. National employment data rarely include such important forms of work. Third, even when employed in the formal economic sector, such as on a factory assembly line, women are usually paid less than men and their work is usually concentrated in jobs with little security or chance of advancement.

Even though women may make up the majority of workers in many developing countries, the quality and conditions of their work are reminders of the early and unpleasant days of European industrialization, when people worked long hours under dangerous conditions for low wages with little job security. While it took Europe almost a century to pass laws to improve these working conditions, it seems reasonable to expect that the economic, political, and social movements of the early twenty-first century might produce more humane working conditions at a faster pace.

FIGURE 1.33 • WOMEN AND LITERACY Gender inequalities are often found in national illiteracy rates. When there is a large difference between the percentage of males and females characterized as illiterate, often this can be explained by cultural, social, or economic limits placed on the education of girls and women. *(Ron Giling/Panos Pictures)*

FIGURE 1.34 • WOMEN IN THE WORKFORCE Female workers have become an important component of the modern globalized economy. However, women are often paid less than men and usually don't have the job security to prevent them from being the first to be fired or laid off as the global economy fluctuates. *(Nathan Benn/Woodfin Camp & Associates)*

CONCLUSION

As the world becomes increasingly interdependent, globalization is driving a fundamental reorganization of economies and cultures through trade agreements, supranational organizations, military alliances, and cultural exchanges. At one level, the world seems to be converging and becoming more homogeneous because of globalization. Nonetheless, there is still great diversity, for better or worse. While some areas of the world prosper, others stagnate or even decline; as global TV promotes a common world culture, small-group identity becomes increasingly important; as nation-states become world players through international agreements, separatist groups seek independence. This tension and interplay between globalization and diversity—and between the global, the regional, and the local—gives world regional geography its primary focus.

Our description and analysis of each world region are organized around five issue-oriented themes: environmental geography; population and settlement; cultural coherence and diversity; geopolitical framework; and economic and social development. Although the specific issues differ from region to region, certain general themes link the chapters. Environmental issues, be they at the global or local scale, affect all world regions. Examples include global climate change, water pollution, air pollution, forest destruction, and the protection of biodiversity.

In most regions of the developing world, population and settlement concerns revolve around four issues: rapid population growth, family planning (or its absence), migration to new centers of economic activity (both within and outside the region), and rapid urbanization (combined with the concern about whether cities can keep pace with the ever-increasing demand for jobs, housing, transportation, and public facilities).

A major theme in our treatment of global cultural geography is the tension and interplay between the forces of cultural homogenization and the countercurrents strengthening small-scale cultural and ethnic identity. Throughout the world, small groups are setting themselves apart from larger national cultures with renewed interest in ethnic traits, languages, religion, territory, and shared histories. This is not merely a matter of colorful folklore and revitalized local customs; in many parts of the world, cultural diversity is translated into politics with calls for regional autonomy or separatism.

The geopolitical issues of many of the world's regions are dominated by matters of global terrorism, ethnic strife and territorial disputes within nation-states, and border tensions between neighbors of different cultural traditions (Pakistan and India are examples). These modern geopolitical problems have their roots in cultural factionalism.

Finally, the theme of economic and social development is dominated by one issue: the increasing disparity between the rich and the poor, between countries and regions that already have wealth—and are getting even richer through globalization—and those that remain impoverished. Economic inequities are found on every scale from global to local, including pockets of poverty within the world's richest countries. On a global scale, the core of developed countries is surrounded by a periphery of less-developed nations aspiring to the same success; on the smaller scale of regions and individual nations, vibrant centers of economic development are linked to the globalized world, while more remote backwater areas stagnate. Often the same inequalities in social development, schools, health care, and working conditions accompany these differences in wealth.

Understanding this complex world is a challenging, yet necessary, task. Think of this book as a beginning rather than an end, as a way to gain skills in using the conceptual tools of geography to empower critical thinking about the complicated issues and themes of world regional geography as they interact with diversity amid globalization.

KEY TERMS

asymmetrical warfare (page 23)
bubble economy (page 7)
centrifugal forces (page 22)
centripetal forces (page 22)
colonialism (page 24)
core-periphery model (page 25)
cultural imperialism (page 16)

cultural nationalism (page 17)
cultural syncretism or
    hybridization (page 17)
culture (page 15)
decolonialization (page 24)
demographic transition (page 12)
economic growth rate (page 27)

ethnic religion (page 18)
globalization (page 2)
GNI per capita (page 27)
lingua franca (page 17)
nation-state (page 21)
purchasing power parity (PPP)
    (page 27)

rate of natural increase (page 10)
sweatshop (page 6)
total fertility rate (page 11)
transnational firms (page 5)
universalizing religion (page 18)

FIGURE 2.2 • TRASH ON MT. EVEREST   The human influence is found just about everywhere in the early twenty-first century. Even isolated Mt. Everest is littered with base camp garbage and discarded oxygen canisters at 17,000 feet. *(Mountain Light Photography, Inc.)*

The human influence is everywhere on Earth, from the highest mountains to the ocean depths; from dry deserts to lush tropical forests; from frozen Arctic ice caps to the cloudless atmosphere (Figure 2.1). Hundreds of spent oxygen canisters left by mountain climbers clutter the heights of Mt. Everest (Figure 2.2), and radioactive debris collects in deep offshore waters of the North Atlantic. While deserts in North America bloom with irrigated cotton, tropical forests in Brazil are laid waste by logging. Although many changes to the global environment are intentional and have improved the human condition, other environmental changes are accidental byproducts of human activities and are often harmful. Because of the importance of environmental issues, a study of the changing global environment is central to the study of world regional geography.

Environmental concerns are also closely linked with globalization and diversity. The destruction of tropical rain forests, for example, is a response to international demand for wood products and beef. Similarly, global warming through human-caused climate change is closely linked to patterns of world industry and commerce. With 6 billion people on Earth, there is a long list of the ways humans interact with—and change—the natural environment.

Global Climates: An Uncertain Forecast

Human settlement and food production are closely linked to patterns of local weather and climate. Where it is dry, such as in the arid parts of Southwest Asia, life and landscape differ considerably from the wet tropical areas of Southeast Asia. People in different parts of the world adapt to weather and climate in widely varying ways, depending on their culture, economy, and technology. For example, some desert areas of California are covered with high-value irrigated agriculture that produces vegetables for the global marketplace. In contrast, most of the world's arid regions support very little agriculture and thus barely participate in global commerce. Moreover, when drought hits a portion of the world (such as Russia's grain belt) the socioeconomic repercussions are felt throughout the world. In many ways, climate links us together in our globalized economy, providing opportunities for some, hardships for others, and challenges for all in the struggle to supply the world with food.

World Climate Regions

There are a number of atmospheric processes that interact to produce the world's weather and climate. Before going further, it is important to note the difference between these two terms. *Weather* is the short-term day-to-day (or even hourly) expression of atmospheric processes; our weather can be rainy, cloudy, sunny, hot, windy, calm, or stormy all within a short time period. Measures of this weather are taken at regular intervals each day, often hourly. Data are collected for temperature, pressure, precipitation, humidity, and so on. Over a period of time, statistical averages from these daily observations provide a picture of common conditions. From this long-term view, a sense of a regional *climate* is generated. Usually at least 30 years of daily weather data are required before climatologists and geographers construct a picture of an area's climate. In summary, weather data captures the short-term atmospheric processes, whereas climate is the long-term summation of average conditions.

Since weather observations have been taken for far longer than 30 years in most parts of the world, it is possible to use these data to generate a picture of the climate for thousands of places all over the globe. Boundaries are drawn around areas with similar average conditions to produce **climate regions**, which are then included in the larger classification of world climates. Knowing the climate type for a given part of the world conveys a clear sense of average rainfall and temperatures. As well, this information is valuable for inferences about human activities and settlement. If an area is categorized as desert, for example, then we conclude that rainfall is so limited that any agricultural activities require irrigation. Crops cannot be grown without supplemental watering, so people must turn to other strategies to obtain food.

In contrast, if we see an area characterized as tropical monsoon, we know there is enough rainfall for agriculture.

A standard system of climate types is used throughout this book, and each regional chapter contains a map showing the different climate regions in some detail. Figure 2.3 shows these climatic regions on a world scale. The regional climate maps also contain **climographs**, which are graphs of average high and low temperatures and precipitation for an entire year. In Figure 2.3 the climate of Cape Town, South Africa, is presented as an example of these graphs. Two lines for temperature data are presented on each climograph: The upper one plots average high temperatures for each month, while the lower shows average low temperatures. These monthly averages provide a good sense of what typical days might be like during different seasons. Besides temperatures, climographs also contain bar graphs depicting average monthly precipitation. Not only is the total amount of rain and snowfall important, but the seasonality of this precipitation provides valuable information related to agriculture and food production.

Global Warming

Human activities connected with economic development and industrialization are changing the world's climate in ways that may have significant consequences. More specifically, **anthropogenic**, or human-caused, pollutants into the lower atmosphere are increasing the natural greenhouse effect (discussed in detail below) so that worldwide global warming is taking place. This warming, in turn, may change rainfall patterns, increasing aridity in some areas; melt polar ice caps, causing higher sea levels; and possibly lead to a greater intensity in tropical storms. Moreover, because of climate change resulting from global warming, the world may experience a dramatic change in food production regions. While some prime agricultural areas such as the lower Midwest of North America may suffer because of increased dryness, the climate of other areas, such as the Russian steppes, may actually become better suited to agriculture. Though it is too early to tell if and when this will happen, these kinds of climate changes will have a dramatic effect on world food supplies.

Causes of Global Warming   The natural greenhouse effect provides us with a warm atmospheric envelope that supports human life. This envelope is formed as moisture and greenhouse gases (described below) in the lower atmosphere trap incoming and outgoing (or re-radiated) solar radiation. Although natural greenhouse gases vary over long periods of geologic time, they seem to have been relatively stable until recently. However, beginning 130 years ago with the Industrial Revolution in Europe and North America, the amount of these greenhouse gases have changed dramatically. This has primarily resulted from the burning of fossil fuels associated with industrialization. Four major greenhouse gases account for the bulk of this atmospheric change:

1. *Carbon dioxide* (CO_2) accounts for more than half of the human-generated greenhouse gases. The increase in atmospheric CO_2 is mainly a result of burning fossil fuels (coal and petroleum.). To illustrate this increase, in 1860 atmospheric CO_2 was measured at 280 parts per million (ppm). Today it is more than 370 ppm and is expected to exceed 450 ppm by 2050. Computer models predict that CO_2 could reach 970 ppm by the end of the century, which would be a 250 percent increase over pre-industrial levels.

2. *Chlorofluorocarbons (CFCs)*, which make up nearly 25 percent of the human-generated greenhouse gases, come mainly from aerosol sprays and refrigeration, including air conditioning. Although CFCs have been banned in North America and most of Europe, they are still increasing in the atmosphere at the rate of 4 percent each year. These gases reside in the atmosphere for a long time, perhaps as long as 100 years. As a result, their role in global warming is highly significant. Recent research has shown that a molecule of CFC absorbs a thousand times more infrared radiation from Earth than a molecule of CO_2.

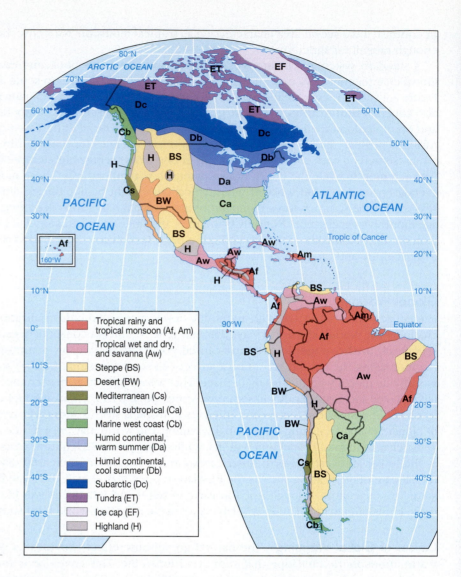

FIGURE 2.3 • WORLD CLIMATE REGIONS   A standard classification, called the *Köppen system* after the Austrian geographer who devised it in the early twentieth century, is used to describe the world's diverse climates. A combination of letters refers to the general climate type, along with its precipitation and temperature characteristics. More specifically, the A climates are tropical, the B climates are dry, the C climates are generally moderate and are found in the middle latitudes, and the D climates are associated with continental and high-latitude locations.

3. *Methane* (CH_4) has increased 151 percent since 1750 as a result of vegetation burning associated with rainforest clearing, anaerobic activity in flooded rice fields, byproducts of cattle and sheep digestion, and leakage of pipelines and refineries connected with natural gas production. Currently, CH_4 accounts for about 15 percent of anthropogenic greenhouse gases.

4. *Nitrous oxide* (N_2O) is responsible for just over 5 percent of human-caused greenhouse gases. It results primarily from the widespread usage of chemical fertilizers in modern agriculture.

Effects of Global Warming   The complexity of the global climate system leaves some uncertainty about how the world's climate may change as a result of human-caused greenhouse gases. Increasingly, though, the high-powered computer models used by scientists to improve our understanding of climate change are reaching agreement on the possible effects of global warming.

Unless countries of the world greatly reduce their emission of greenhouse gases in the next few years, computer models predict that average global temperatures will increase 2 to 4°F (1 to 2°C) by 2030. While this may not seem dramatic at first glance, it is about the same magnitude of change as the cooling that caused the Ice Age glaciers to cover much of Europe and North America 30,000 years ago. Further, this temperature increase is projected to double by 2100.

Such a change in climate could cause a shift in major agricultural areas. For example, the Wheat Belt in the United States might receive less rainfall and become

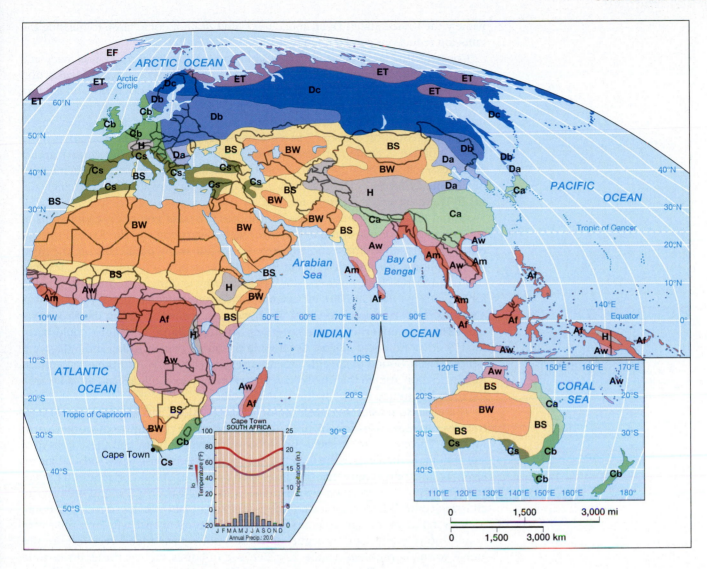

warmer and drier, endangering grain production as we know it today. While more northern countries, such as Canada and Russia, might experience a longer growing season because of global warming, the soils in these two areas are not nearly as fertile as in the United States. As a result, scientists are predicting a decrease in the world's grain production by 2030. Further, the southern areas of the United States and the Mediterranean region of Europe can expect a warmer and drier climate that will demand even more irrigation for crops.

Warmer global temperatures will cause rising sea levels as oceans warm and the melting of polar ice sheets occurs. Although forecasts differ on the predicted sea level rise—from several inches to almost 3 feet (1 meter)—even the smallest increase will endanger low-lying island nations throughout the world and coastal areas in Europe, Asia, and North America. Island nations in the Pacific and the Indian Ocean are particularly concerned, since these countries may be flooded out of existence.

Globalization and Climate Change: The International Debate on Limiting Greenhouse Gases   By the early 1990s, the United Nations recognized that climate change was "a common concern of humankind" and that all nations have a "common but differentiated responsibility" for fighting global warming. At the Rio de Janiero Earth summit in 1992, the first international agreement on global warming was prepared. Countries that signed the agreement were bound by international law to reduce their greenhouse gas emissions by agreed-upon target dates. Within a year, 167 countries had signed. Unfortunately, those countries emitting the most greenhouse gases, namely, the United States, Japan, India, and China, did not come close to meeting the emission reductions agreed upon in Rio.

Consequently, another attempt to develop an international treaty was made at a December 1997 meeting in Kyoto, Japan. The current plant for international action on global warming—the Kyoto Protocol—resulted from these discussions. In the Protocol, 38 industrialized countries agree to reduce their emissions of major greenhouse gases below 1990 levels. The United States, Japan, and Canada, for example, agreed to a reduction of about 7 percent below 1990 levels by the year 2012. If the Kyoto Protocol is ratified by countries emitting 55 percent of the world's greenhouse gases, it will become international law. However, since the original meeting in 1997, it has become clear that getting international approval will be extremely difficult. There are several reasons for this.

Within the United States, which contributes the most greenhouse emissions of any country in the world, the debate over global warming and the Kyoto Protocol remains politically charged and controversial. While many politicians see the need for greenhouse gas controls, others forcefully resist any action because they fear that controls will limit business, slow the economy, and increase the cost of living for U.S. citizens. Early in 2001, for example, while agreeing that global warming was a serious issue, President George W. Bush went on record opposing the Kyoto Protocol. He was concerned that restrictions on atmospheric emissions could harm the U.S. economy and that large developing countries, namely China and India, were not yet legally bound to specific greenhouse gas reductions. Additionally, the Bush Administration argued that the linkage between human activities and global warming must be studied further before action is taken.

There is also tension between the developed nations and less-developed countries over the Kyoto Protocol. Unrestricted emissions in the industrial world (North America, Europe, Japan) created the global warming problem, and these countries still emit more than half of the total human-caused greenhouse gases today. Many international experts argue that the developed countries should be required to take stringent steps to curb their own emissions, as well as subsidize and underwrite emission controls in developing countries. Understandably, the less-developed countries are reluctant to sign an emission control agreement that will limit their economic future when, as they argue, up to now they have added very little to the global warming problem (Figure 2.4). Because of issues such as these, at this time it appears that the Kyoto Protocol will not become international law. While the technology exists for controlling most greenhouse gas emissions, a strong commitment to reduce this pollution from countries like the United States and Russia appears to be lacking.

FIGURE 2.4 • BANGLADESH BRICK FACTORY Increasing industrialization in the developing world presents new threats to global environmental health. Often, developing countries lack pollution controls. As a result, environmental costs increase as industrial activity becomes more widespread. *(S. Noorani/Woodfin Camp & Associates)*

Human Impacts on Plants and Animals: The Globalization of Nature

One aspect of Earth's uniqueness compared to other planets in our solar system is the rich diversity of plants and animals that covers its continents. Geographers and biologists think of the cloak of vegetation as the "green glue" that binds together life, land, and atmosphere because it is both a product of and an influence on climate, geology, and hydrology. To better understand the vegetation of the world, geographers and ecologists use the concept of the **bioregion**, which is an assemblage of local plants and animals covering a large area such as a tropical forest or grassland. Figure 2.5 shows the global distribution of these bioregions.

Humans are very much a part of this interaction. Not only are we evolutionary products of a specific bioregion (most probably the tropical savanna of Africa), but human prehistory has been deeply connected with the domestication of certain plants and animals for crops and livestock. From this process of domestication has come agriculture. Further, humans have changed the natural pattern of plants and animals by plowing grasslands, burning woodlands, cutting forests, and hunting animals. The pace of such change has increased in recent decades, and these actions have led to an environmental crisis as forests are devastated, plants and animals exterminated, and watersheds denuded.

Many of these problems can be explained by the globalization of nature and of local ecologies. To illustrate, until the last half-century, tropical forests were primarily homes for small populations of native peoples who made modest demands on the environments for their sustenance and subsistence. Today, however, these same tropical forests are resources for multinational corporations as they clear-cut forests for international trade in wood products or search out plants and animals to meet the needs of far-removed populations. For example, Japanese lumber companies cut South American rain forests; German pharmaceutical corporations harvest medicinal plants in Africa; and international poachers kill North American bears in order to sell their gallbladders on the Asian black market as elixirs and sexual stimulants. The list of human impacts on nature is a long one. The following discussion focuses on such impacts in three of the major bioregions.

Tropical Forests and Savannas

Most tropical forests are found in climate zones near the equator with high average annual temperatures, long days of sunlight throughout the year, and heavy amounts of rainfall. This bioregion covers about 7 percent of the world's land area, which is roughly the size of the United States excluding Alaska. These forests are found primarily in Central and South America, Sub-Saharan Africa, Southeast Asia, Australia, and on many tropical Pacific islands. More than half of the known plant and animal species live in the tropical forest bioregion, making it the most diverse of all bioregions.

The dense tropical forest vegetation is usually arranged in three distinct levels that are adapted to decreasing amounts of sunlight closer to the forest floor. The tallest trees, around 200 feet (61 meters) high, receive open sunlight; the middle level (around 100 feet, or 31 meters) gets filtered sunlight; the third level is the forest floor, where plants can survive with very little direct sunlight (Figure 2.6). Even though organic material accumulates on the forest floor in the form of falling leaves, tropical forest soils tend to be very low in stored nutrients. The nutrients are stored instead in the living plants. As a result, tropical forest soils are not well suited for intensive agriculture.

Deforestation in the Tropics

Tropical forests are now being devastated at an unprecedented rate, creating a bioloigical crisis that tests our political, economic, and ethical systems. Although deforestation rates differ from region to region, each year an area of tropical forest about the size of Wisconsin or Pennsylvania is denuded. Geographically speaking, almost half of this activity is in the Amazon Basin of South America. However, the rate of deforestation appears to be occurring faster in Southeast Asia, where some estimates suggest logging takes place at rates three times faster than in the Amazon. If those estimates are accurate, Southeast Asia could be completely stripped of forests within 15 years. Behind the cause of this widespread cutting of tropical forests lies the recent globalization of commerce in international wood products. By all accounts, Japan was the first to globalize its timber business by reaching far beyond its national boundaries to purchase timber in North and South America and Southeast Asia. This was a strategy for meeting the increased demand for wood products in Japan while at the same time protecting its own rather limited forests for recreational use. Currently, about one-half of all tropical forest timber is used in Japan. Unfortunately, much of it is used for throwaway items such as chopsticks and newspapers.

Another factor contributing to the rapid destruction of tropical rain forests is the world's seemingly insatiable appetite for beef. Cattle species originally bred to survive the hot weather of India are now raised worldwide on grassland pastures created by cutting tropical forests. Unfortunately, because tropical forest soils are poor in nutrients, cattle ranching is not a sustainable activity in these new grassland areas. After a few years, soil nutrients become exhausted, making it necessary to move ranching activities into newly cleared forest land. As a result, more forest

FIGURE 2.5 • WORLD BIOREGIONS Despite human changes to global vegetation through commercial activities, there is still a recognizable pattern to the world's bioregions, ranging from tropical forests to arctic tundra. Each bioregion has its own group of ecosystems containing plants, animals, and insects. These different species constitute the biodiversity necessary for robust gene pools. Put differently, biodiversity can be thought of as the genetic library that keeps life going on Earth. *(Adapted from Clawson and Fisher, 2001, World Regional Geography, 7th ed., Upper Saddle River, NJ: Prentice Hall)*

Legend:
- Tropical forest
- Mediterranean woodland, shrub, and grassland
- Broadleaf or mixed broadleaf and coniferous forest
- Coniferous forest
- Tropical savanna, mixed grassland and woodland
- Middle-latitude prairie and steppe grassland
- Desert shrub
- Tundra
- Ice cap

FIGURE 2.6 • TROPICAL RAIN FOREST As fragile as they are diverse, tropical rain forest environments feature a complex, multilayered canopy of vegetation. Plants on the forest floor are well adapted to receiving very little direct sunlight. *(Gary Braasch/Woodfin Camp & Associates)*

is cut; more pasture created; more soil destroyed; and the process goes on at the expense of forest lands.

A third factor explaining tropical forest destruction is that these forest areas are often the last settlement frontiers for the rapidly growing population of the developing world. Brazil has used much of its interior rain forest for settlement in order to ease population pressure along its densely settled eastern coast, where rural lands are controlled by powerful landowners. Brazil had a choice: either address the troublesome issue of land reform and break up the coastal estates or open up the interior rain forests. It chose the path of least resistance, allowing settlers to clear and homestead the tropical forest lands of the interior. Many countries do the same, looking at the vast tracts of tropical rain forest as a safety valve of sorts that can be opened up for migration and settlement when land hunger becomes too great in other regions of the country.

Often these three processes work together. First, international logging companies are given timber concessions to log the forest. Following logging, deforested lands are then opened to settlement by migrants from other areas. At the same time, these deforested lands are also made available for cattle ranching by both domestic and foreign firms. When all three demands are combined—wood products, beef, and land hunger—the tropical rain forest and the plants, animals, and native tribal groups who live there are the losers.

Deserts and Grasslands

Poleward of the tropics (to both north and south, depending on the hemisphere) lie large areas of arid and semi-arid climates. Within these climate areas are found the world's deserts and grasslands. Fully a third of Earth's land area qualifies as true

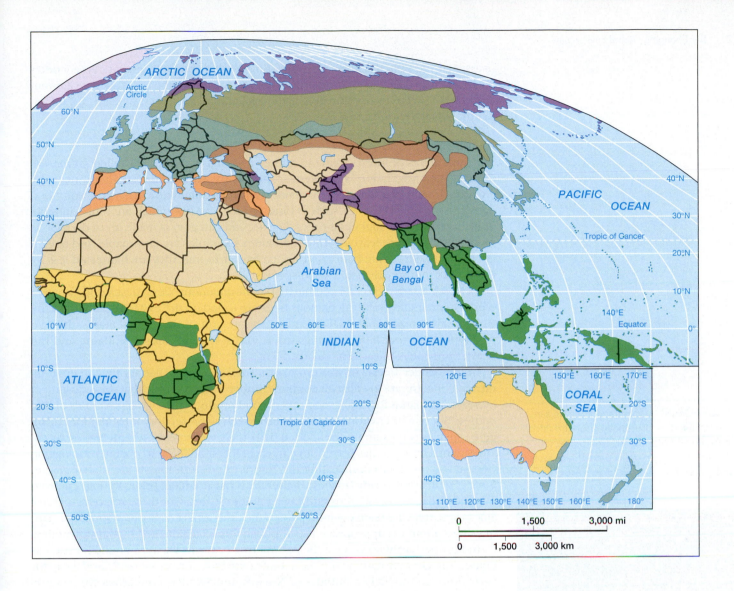

desert, which is commonly defined as places receiving less than 10 inches (25 centimeters) of rainfall per year. Semiarid grasslands often appear in those places receiving slightly more than 10 inches of rainfall. Here, grassy plants appear, forming a verdant cover during the wet season. In North America, the midsection of both Canada and the United States is covered by grassland known as **prairie**. In other parts of the world, such as Central Asia, Russia, and Southwest Asia, shorter, less dense grasslands form the **steppe**.

The boundary between desert and grassland has always changed naturally because of changes in the climate. During wet periods, grasslands might expand, only to contract once again during drier decades. The transition zone between the two can be a troublesome environment for humans, as the United States found out during the 1930s when the semiarid grasslands of the western prairie lands turned into the notorious "Dust Bowl." At that time, thousands of farmers saw their fields devastated by wind erosion and drought—a calamity that led to an out-migration of farmers from these once-productive farmlands.

Farming marginal lands may actually worsen the situation, leading to **desertification**, or a spread of desertlike conditions (Figure 2.7). This has happened on a large scale throughout the world—in Africa, Australia, and South Asia, to name just a few regions. In fact, in the last several decades an area estimated to be about the size of Brazil has become desertified through poor cropping practices, overgrazing, and the buildup of salt in soils from irrigation. In northern China an area the size of Denmark became desertified between 1950 and 1980 with the expansion of farming into marginal lands. Historically, this region averaged three sandstorms a year; today 25 such storms are common each year. Although some scientists say the case is somewhat

FIGURE 2.7 • DESERTIFICATION Climatic fluctuations and human misuse combine to produce desertification in different parts of the world. Usually this results when farming or grazing moves into semiarid grasslands. *(Mark Edwards/Still Pictures/Peter Arnold, Inc.)*

overstated, the United Nations recently estimated that about 60 percent of the world's rangelands are threatened by desertification. According to the UN, if such an amount of rangeland were to become desert, the agricultural livelihood of some 1.2 billion people would be threatened.

Temperate Forests

The large tracts of forests found in middle and high latitudes are called *temperate forests*. Their vegetation is different from the low-latitude forests found near the equator. In temperate forests, two major tree types dominate. One type is softwood conifers or evergreen trees, such as pine, spruce, and firs, which are found in higher elevations and at higher latitudes. The second category is made up of deciduous trees that drop their leaves during the winter. Examples are elm, maple, and beech. Because these trees are hardwood, hence harder to mill, they are harvested less by the timber industry than softwood species.

In North America, conifers dominate the mountainous West, Alaska, and Canada's western mountains, while deciduous trees are found on the eastern seaboard of the United States north to New England. In that region of the United States, the two tree types intermix before giving way to the softwood forests of Maine and the coastal provinces of eastern Canada.

In the conifer forests of western North America, timber harvesting and environmental concerns are in constant conflict. Timber interests argue that they must meet increased demands for lumber and other wood products through higher rates of tree cutting. But due to environmental concerns, such as protection of habitat for endangered species (for example, the spotted owl), the government—after lobbying by environmental groups—has placed large tracts of forest off-limits to commercial logging (Figure 2.8). Further complicating the future of western forests are global market forces. For example, many Japanese timber firms pay premium prices for logs cut from U.S. and Canadian forests, thus outbidding domestic firms for these scarce resources. Since these trees are often cut from public lands, this aspect of globalization raises hard questions about the appropriate use of public forests maintained by U.S. tax money.

In Europe, hardwoods were the natural tree cover in Western countries such as France and Great Britain before these once-extensive forests were cleared away for agriculture. In the higher latitudes of Norway and Sweden, coniferous species prevail in the remaining forests. These conifers also form extensive forests across Germany and eastern Europe, and through Russia into Siberia, creating an almost unbroken landscape of dense forests. This Siberian forest remains a resource that could become a major source of income for financially strapped Russia. Some argue that if the Siberian forests are put on the market for global trade, it will reduce logging pressure on North America's western forests, which, in turn, could make it easier to protect U.S. forests from cutting. This illustrates once again how the forces of globalization are interconnected with the fate of local ecosystems.

Food Resources: Environment, Diversity, Globalization

If the human population continues to grow at expected rates, food production must double by 2025 just to provide each person in the world with a minimum diet. Every minute of each day, 170 people are born who need food; during the same minute, about 10 acres of existing cropland are lost because of environmental problems such as soil erosion and desertification. Many experts argue that food scarcity will be the defining issue of the next several decades. Because of the close ties between world food problems, the environment, and globalization, we conclude this chapter by introducing the basic issues surrounding global food resources.

The Green Revolution

During the last 40 years of the twentieth century, the world's population doubled. Even more remarkably, during the same period global food production also doubled to keep pace with this population explosion. This increase in food production came

FIGURE 2.8 • CLEAR-CUT FOREST Commercial logging in Washington's Olympic Peninsula has drastically changed this landscape. Throughout the Pacific Northwest, environmental groups have successfully restricted logging to protect habitat for endangered species and recreation. *(Calvin Larsen/Photo Researchers, Inc.)*

primarily from the expansion of industrial agriculture into areas that previously produced subsistence crops through traditional means (Figures 2.9 and 2.10). It is because of these dramatic results that this recent transformation of global agriculture is called "revolutionary."

More specifically, since 1950 the increases in global food production have come from three interconnected processes that are known as the first stage of the **Green Revolution**. These three processes are

- the change from traditional mixed crops to monocrops, or single-crop fields, of genetically altered, high-yield rice, wheat, and corn seeds;
- intensive applications of water, fertilizers, and pesticides; and
- further increases in the intensity of agriculture by reduction in the fallow or field-resting time between seasonal crops.

Since the 1970s, a second stage of the Green Revolution has evolved. This emphasizes new types of fast-growing wheat and rice specifically bred for tropical and subtropical climates. When combined with irrigation, fertilizers, and pesticides, these new varieties allow farmers to grow two or even three crops a year on a piece of land that previously supported just one. Using these methods, India actually doubled its food production between 1970 and 1992.

However, many argue that these agricultural revolutions also carry high environmental and social costs. To illustrate, since the production of these crops draws heavily upon fossil fuels, there has been a 400 percent increase in the agricultural use of petroleum during the last several decades. As a result, Green Revolution agriculture now consumes almost 10 percent of the world's annual oil output. The environmental costs of the Green Revolution also include damage to habitat and wildlife from diversion of natural rivers and streams to agriculture; pollution of rivers and water sources by pesticides and chemical fertilizers applied in heavy amounts to fields; and increased regional air pollution from factories and chemical plants that produce these agricultural chemicals.

There is also evidence that the Green Revolution carries some social costs. This results from the fact that the financial costs to farmers participating in the Green Revolution are higher than with traditional farming. Money is needed for hybrid seeds, fertilizers, pesticides, and even new machinery. For people with high social standing and good credit, the rewards can be great. But for those without access to loans or family support, the costs can be high. Why is that? Usually traditional farmers cannot compete against those farmers raising Green Revolution crops in the regional marketplace. As a result, traditional farmers often struggle at a poverty level while Green Revolution farmers prosper. In some wheat-growing areas of India, inequality between the well-off Green Revolution farmers and poor, traditional farmers has become a major source of economic, social, and political tension. What was once an area where everyone shared a common plight has now become a society of rich and poor. These social costs can be accepted only because of the pressing need for food in this rapidly growing country.

Problems and Projections

Even though agriculture has been able to keep up with population growth in the last decades, few experts are sure that this can happen again in the twenty-first century. While fuller discussion of food and agriculture issues is found in the regional chapters of this book, four key points offer a starting point for understanding the issues.

1. While overall food production remains an important global issue, it is in fact local and regional problems that often keep people from obtaining food. Many

FIGURE 2.9 • INDUSTRIAL FARMING IN IOWA The harvesting of corn in Iowa reveals the effect of technology on modern agriculture. In such settings, large capital investments in machinery, fuel, and agrochemicals produce high yields while keeping labor costs low. *(Andy Levin/Photo Researchers, Inc.)*

FIGURE 2.10 • SUBSISTENCE FARMING IN BURKINA FASO These West African women are harvesting sorghum. Such traditional agricultural operations remain crucial for feeding people in the less-developed world. While this kind of agriculture requires low inputs of technology, it is often quite labor-intensive. *(Mark Edwards/Still Pictures/Peter Arnold, Inc.)*

experts believe that it is the widespread poverty and civil unrest at local levels that keeps people from growing, buying, or receiving adequate food supplies. If the total global agricultural output were somehow shared equally among all people in the world, each person would have approximately four times his or her basic daily needs. Given this view, some say that there is already enough food to feed the world's population. The real issue, they argue, is distribution of food and people's ability to buy adequate food for a healthy life.

2. Political problems are usually more responsible for food shortages and famines than are natural events, such as drought and flooding. Food aid goes to political friends and allies, while enemies go without. During the Cold War, both the Soviet Union and the United States provided food aid to their ideological allies, but never to those of the opposite camp. While the Cold War is over, we still see food being used as a political weapon to support allies and deprive opposition groups of sustenance. This occurs frequently in refugee camps in Pakistan, Africa, and the Balkans, for example.

3. Globalization is causing dietary preferences to change worldwide, and the implications of this change could be widespread as people add more meat to their diet. Currently two-thirds of the world population is primarily vegetarian, eating only small portions of meat because it is so expensive. Today, though, because of recent economic booms in some developing countries, an increasing number of people are now eating meat (Figure 2.11). In many cases this change in diet comes from changing cultural tastes and values as people are exposed to new products through economic and cultural globalization. There are, however, limits on the world's ability to supply meat. Some food experts say that the global food production system could sustain only half the current world population if everyone ate the standard meat-rich diet of North America, Europe, and Japan.

4. Most food supply experts agree that the two world regions most likely to face food shortages are Africa and South Asia. Until 1970 Africa was self-sufficient in food, but since that time there have been serious disruptions in the food supply system. The causes for these problems in Africa are twofold—rapid population increase and civil disruption from tribal warfare. As a result, one of every four people faces food shortages in Sub-Saharan Africa. More detail on these issues is found in Chapter 6. South Asia's future is also of concern. The United Nations predicts that by the year 2010 almost 200 million people in South Asia will suffer from chronic undernourishment. Further discussion of these problems is found in Chapter 12.

There is some good news in this otherwise bleak picture concerning food production. Because population growth rates are generally declining in the industrializing areas of East Asia and food production there is still increasing, the UN predicts that the percentage of undernourished people in that region will actually drop by almost 5 percent in the next 10 years. Similarly, gains against hunger are being made in Latin America because of lower population growth and higher agricultural production.

FIGURE 2.11 • CATTLE RANCHING IN BRAZIL Growing demands for meat are changing the world's landscapes. Here, cattle ranching in former rain forest areas of western Brazil shows how global market forces create different—and possibly harmful—environmental conditions. *(Martin Wendler/Peter Arnold, Inc.)*

CONCLUSION

Environmental geography is basic to the study of world regional geography because of its close links to issues of globalization and diversity. Global, regional, and local environmental change, which is a focus for environmental geography, results from a number of different processes. Sometimes these processes are natural, such as the unpredictable climatic fluctuations that produce a droughts or floods. Increasingly, however, global environmental change is being driven by human activities. While some environmental changes, such as the logging of an Alaskan rain forest, are expected by-products of world and national economic activity, others are unanticipated. For example, decades ago only a few scientists anticipated the consequences of global climate change resulting from fossil fuel consumption. Today, however, the problems of global warming are accepted by most scientists.

Globalization is both a help and a hindrance to world environmental problems. From a positive perspective, some would argue that the world's nation-states are increasingly willing to sign international agreements to solve environmental problems. Examples are treaties on whaling, ocean pollution, fisheries, and the protection of wildlife species. Because of these agreements, much progress has been made in some areas of environmental protection.

A conflicting view argues that globalization has aggravated global environmental problems. This is the case, say the critics, because of the environmental damage resulting from super-heated global economic activity that exploits the widest possible range of international resources, regardless of the consequences. Additionally, since unrestricted world free trade is part of globalization, trade agreements often conflict with national and local environmental protection. Demonstrations and protests at World Trade Organization (WTO) meetings around the globe underscore that concern. The conflict between globalization and the environment is an important theme discussed in the chapters that follow.

KEY TERMS

anthropogenic (page 33)

bioregion (page 36)

climate region (page 32)

climograph (page 33)

desertification (page 39)

Green Revolution (page 41)

prairie (page 39)

steppe (page 39)

North America

FIGURE 3.1 North America
North America plays a key role in globalization. The region also contains one of the world's most highly urbanized and culturally diverse populations. With 323 million people and extensive economic development, North America is also one of the largest consumers of natural resources on the planet.

SETTING THE BOUNDARIES

The United States and Canada are commonly referred to as "North America," but that regional terminology can sometimes be confusing. In some geography textbooks the realm is called "Anglo America" because of its close connections with Britain and its Anglo-Saxon cultural traditions. The increasingly visible cultural diversity of the realm, however, has discouraged the widespread use of the term in more recent years. While more culturally neutral, the term "North America" also has its problems. As a physical feature, the North American continent commonly includes Mexico, Central America, and, often, the Caribbean. Culturally, however, the United States–Mexico border seems a better dividing line. However, the growing Hispanic presence in the southwest United States, as well as ever-closer economic links across the border, make even that regional division problematic. While the future may find Mexico even more closely tied to its northern neighbors, our coverage of the "North American" realm concentrates on Canada and the United States, two of the world's largest and wealthiest nation-states.

North America includes the United States and Canada, a culturally diverse and resource-rich region that has seen tremendous human modification of its landscape and extraordinary economic development over the past two centuries (Figure 3.1; see also "Setting the Boundaries"). The result is one of the world's wealthiest regions, where two highly urbanized and mobile countries are associated with the processes of globalization and the highest rates of resource consumption on Earth. Indeed, the realm exemplifies a postindustrial economy in which human geographies are shaped by modern technology, by innovative financial and information services, and by a popular culture that dominates both North America and the world beyond.

Globalization has fundamentally reshaped North America. A walk down any busy street in Toronto, Tucson, or Toledo reveals how international products, foods, culture, and economic connections shape the everyday scene (Figure 3.2). From farmers to computer workers, most North Americans have jobs that are either directly or indirectly linked to the global economy. Large foreign-born populations in each nation also provide direct links to every part of the world. Tourism brings in millions of additional foreign visitors and billions of dollars that are spent everywhere from Las Vegas to Disney World. In more subtle ways, North Americans see globalization in their everyday lives. They consume ethnic foods, tune in to international sporting events on television, enjoy the sounds of salsa and Senegalese music, and surf the Internet from one continent to the next.

North American capital, culture, and power also influence the rest of the world. By any measure of global investment and trade, the region plays a dominant role that far outweighs its population of 323 million residents. North American automobiles, consumer goods, information technology, and investment capital circle the globe. In addition, North American foods and popular culture are quickly reaching all parts of the globe. A new McDonald's restaurant opens every few hours somewhere on the planet. North American music, cinema, and fashion have also spread rapidly around the world.

Globalization has also brought political risk and uncertainty to the region as the sudden terrorist attacks on the World Trade Center and Pentagon demonstrated in September 2001 (Figure 3.3). The attacks were a tragic reminder that the global reach of U.S. power and influence makes the country a highly vulnerable target. In addition, its traditions of freedom and openness offer opportunities for terrorism because it allows for the easy movement of people into and within the country. It is clear that the United States stands at the center of the globalization process, and it should not be a surprise that its largest metropolitan area, New York City (20.2 million people), was the primary objective of the 2001 attacks.

North of the United States, Canada is the other political unit within the region. While slightly larger in area than the United States [3.83 million square miles (9.97 million square kilometers) versus 3.68 million square miles (9.36 million square kilometers)], Canada's population is only about 10 percent that of the United States.

Widespread abundance and affluence characterize North America. The region is extraordinarily rich in natural resources, such as navigable waterways, good farmland,

FIGURE 3.2 • TORONTO'S CULTURAL LANDSCAPE
Toronto's varied ethnic population is celebrated during the Caribana Parade through the city. Powerful forces of globalization have reshaped the cultural and economic geographies of dozens of North American cities. *(Canadian Tourism Commission)*

FIGURE 3.3 • ATTACK ON THE PENTAGON Symbol of U.S. political power, the Pentagon was forever changed on September 11, 2001, as terrorists crashed a commercial airliner into the heart of the nation's military control center. The act demonstrated that even the world's greatest superpower was vulnerable to acts of terrorist violence and was a sobering reminder that globalization often comes with a price. *(Rob Crandall/Rob Crandall, Photographer)*

fossil fuels, and industrial metals. Combine that good fortune with the arrival of European colonizers and a growing pace of technological innovation, and the results are reflected everywhere on the modern scene. Indeed, contemporary North America displays both the bounty and the price of the development process. On one hand, the realm shares the benefits of modern agriculture, globally competitive industries, excellent transport and communications infrastructure, and two of the most highly urbanized societies in the world. The cost of development, however, has been high: Native populations were all but eliminated by European settlers, forests were logged, grasslands converted into farms, valuable soils eroded, numerous species threatened with extinction, great rivers diverted, and natural resources often wasted. Today, although home to only about 6 percent of the world's population, the region consumes 30 percent of the world's energy and produces carbon dioxide emissions at a per person rate almost 10 times that of Asia.

Nevertheless, economic growth has vastly improved the standard of living for many North Americans, who enjoy high rates of consumption and varied urban amenities that are the envy of the less-developed world. Satellite dishes, sushi, and shopping malls are within easy reach of most North American residents. Amid this material abundance, however, there are continuing differences in income and in the quality of life. Poor rural and inner-city populations still struggle to match the affluence of their wealthier neighbors. These patterns of poverty have been slow to disappear, even given the unprecedented economic growth of the second half of the twentieth century.

North America's unique cultural character also defines the realm. The cultural characteristics that hold this region together include a common process of colonization, a heritage of Anglo dominance, and a shared set of civic beliefs in representative democracy and individual freedom. But the history of the region has also juxtaposed Native Americans, Europeans, Africans, and Asians in fresh ways, and the results are two societies unlike any other. Adding to the mix is a popular culture that today exerts a powerful homogenizing influence on North American society.

Environmental Geography: A Threatened Land of Plenty

North America's physical geography is incredibly diverse (Figure 3.4). This region's natural environment has also been extensively modified by humans over the past 400 years. California's largest lake, the Salton Sea, illustrates the nature of these changes. The 35-mile-long lake, located in the southeast corner of the state, resulted from a simple accident in 1905. An irrigation diversion project on the nearby Colorado River failed, and for 18 months the river's entire drainage flooded into the low-lying Salton Basin. The huge inland "sea" transformed the environment and the economy of the entire region, with consequences that can still be seen a century later. A great variety of birds were drawn to the area, and in 1930 a large wildlife refuge was created to protect them. More than 350 types of birds now visit the refuge. Fish were also introduced into the lake, and by the 1950s a large sportfishing industry had attracted thousands of tourists to the desert setting. Nearby irrigated agriculture in the Imperial Valley also benefited because water from farm operations seeped into the lake instead of waterlogging the fields. Problems have also surfaced. The Salton Sea is getting larger (from agricultural overflows), saltier (from high rates of evaporation, leaving dissolved salts behind), and dirtier (agricultural and urban pollution drain into the area). Periodically, large die-offs of bird and fish populations (from diseases, toxic salts, and algae blooms) have harmed the regional environment. The lake has also raised international tensions between the United States and Mexico. A major borderlands issue in the area involves which country should contain the northward flow of Mexicali's (a city in nearby Mexico) industrial waste into the Salton Sea. Indeed, the story of the Salton Sea is a splendid example of how quickly people can transform the environment and how those sudden changes can have far-reaching and long-lasting consequences.

Seattle. Residents of the Pacific Northwest enjoy the amenities of their mountain scenery, but many Cascade peaks, such as Washington's Mt. Rainier, are still-active volcanoes that threaten nearby metropolitan areas.

Lower St. Lawrence River. Entryway to a continent, the St. Lawrence and Great Lakes system offers an extraordinary natural pathway deep into the heart of North America.

Iowa. Focus of one of the world's great food-producing regions, Iowa and much of North America's interior benefit from good soils and growing conditions for many crops and livestock products.

San Francisco and Los Angeles. California's two largest cities share one of the world's great earthquake hazards. The San Andreas Fault as well as countless smaller rifts pass through these large metropolitan areas and pose great seismic risks in the twenty-first century.

Gulf Coast and Florida. Much of the southeastern United States faces the annual summer and autumn threat from tropical hurricanes. Although they bring benefits of needed moisture, their winds and stormy seas periodically threaten cities such as New Orleans and Miami.

Salton Sea. California's largest lake was accidentally created in 1905 during an irrigation project along the Colorado River. A century later, the lake has extensively altered both the area's environment and economy.

Mississippi River. The Mississippi River and other major North American waterways often flood in the spring and summer as melting winter snows combine with heavy precipitation.

A Diverse Physical Setting

The North American landscape is dominated by vast interior lowlands bordered by more mountainous topography in the western portion of the region (see Figure 3.4). In the eastern United States, extensive coastal plains stretch from southern New York to Texas and include a sizable portion of the lower Mississippi Valley. The Atlantic coastline is complex, and is made up of drowned river valleys, bays, swamps, and low barrier islands (Figure 3.5). The nearby Piedmont, which is the transition zone between nearly flat lowlands and steep mountain slopes, consists of rolling hills and low

FIGURE 3.4 • PHYSICAL GEOGRAPHY OF NORTH AMERICA
North America's diverse physical setting includes both Arctic tundra and tropical forests. Stretching from northern Canada to the Hawaiian Islands, the region is characterized by varied climates, vegetation, and landforms.

FIGURE 3.5 • SATELLITE IMAGE OF THE CHESAPEAKE BAY
This view of the Middle Atlantic Coast reveals the complex shore-line of the Chesapeake Bay (lower center). The coastal area is characterized by drowned river valleys, barrier islands, and sandy beaches. The Piedmont zone and Appalachian Highlands appear to the northwest. *(Earth Satellite Corporation/Science Photo Library/Photo Researchers, Inc.)*

FIGURE 3.6 • LANDFORMS OF THE COLORADO PLATEAU
Monument Valley's colorful sedimentary rocks are part of the unique regional character of the Colorado Plateau. The region's aridity and lack of vegetation help to highlight the dramatic colors of these buttes and mesas, which are common landforms across Arizona and southern Utah. *(Rob Crandall/Rob Crandall, Photographer)*

mountains that are much older and less easily eroded than the lowlands. West and north of the Piedmont are the Appalachian Highlands, an internally complex zone of higher and rougher country reaching altitudes from 3,000 to 6,000 feet (915 to 1,829 meters). Far to the southwest, Missouri's Ozark Mountains and the Ouachita Plateau of northern Arkansas resemble portions of the southern Appalachians. Much of the North American interior is a vast lowland extending east–west from the Ohio River valley to the Great Plains, and north–south from west central Canada to the coastal lowlands near the Gulf of Mexico. Glacial forces, particularly north of the Ohio and Missouri rivers, have actively carved and reshaped the landscapes of this lowland zone.

In the West, mountain-building (including large earthquakes and volcanic eruptions), alpine glaciation, and erosion produce a regional topography quite unlike that of eastern North America. The Rocky Mountains reach more than 10,000 feet (3,048 meters) in height and stretch from Alaska's Brooks Range to northern New Mexico's Sangre de Cristo Mountains. West of the Rockies, the Colorado Plateau is characterized by highly colorful sedimentary rock eroded into spectacular buttes and mesas (Figure 3.6). Nevada's sparsely settled basin and range country features north–south-trending mountain ranges alternating with structural basins with no outlet to the sea. North America's western border is marked by the mountainous and rain-drenched coasts of southeast Alaska and British Columbia; the Coast Ranges of Washington, Oregon, and California; the lowlands of the Puget Sound (Washington), Willamette Valley (Oregon), and Central Valley (California); and the complex uplifts of the Cascade Range and Sierra Nevada.

Patterns of Climate and Vegetation
North America's climates and vegetation are highly diverse, mainly as a response to the region's size, latitudinal range, and varied terrain (Figure 3.7). Much of North America south of the Great Lakes is characterized by a long growing season, 30 to 60 inches (76.2 to 152.4 centimeters) of precipitation annually, and a deciduous broadleaf forest (later cut down and replaced by crops). From the Great Lakes north, the coniferous evergreen or **boreal forest** dominates the continental interior. Near Hudson Bay and across harsher northern tracts, trees give way to **tundra**, a mixture of low shrubs, grasses, and flowering herbs that grow briefly in the short growing seasons of the high latitudes. Drier continental climates found from west Texas to Alberta feature large seasonal ranges in temperature and unpredictable precipitation that averages between 10 and 30 inches (25.4 and 76.2 centimeters) annually. The soils of much of this region are fertile, and originally supported **prairie** vegetation dominated by tall grasslands in the East and by short grasses and scrub vegetation in the West. Western North American climates and vegetation are greatly complicated by the region's many mountain ranges. The Rocky Mountains and the intermontane interior experience the typical seasonal variations of the middle latitudes, but patterns of climate and vegetation are greatly modified by the effects of topography. Farther west, marine west coast climates dominate north of San Francisco, while a dry summer Mediterranean climate occurs across central and southern California.

The Costs of Human Modification
The story of North America's environment becomes more complex when we consider how human modifications have transformed the region's soils and vegetative cover, water resources, and atmosphere (Figure 3.8). The speed and magnitude of that transformation are remarkable, largely taking place over the past 200 years and leaving few areas of North America free from human influence.

Transforming Soils and Vegetation The arrival of Europeans to the North American continent impacted the region's flora and fauna as countless new species were introduced, including wheat, cattle, and horses. As the number of settlers increased, forest cover was removed from millions of acres. Grasslands were plowed under and replaced with grain and forage crops not native to the region. Widespread soil erosion was increased by unsustainable cropping and ranching practices, and many areas of the Great Plains and South suffered lasting damage.

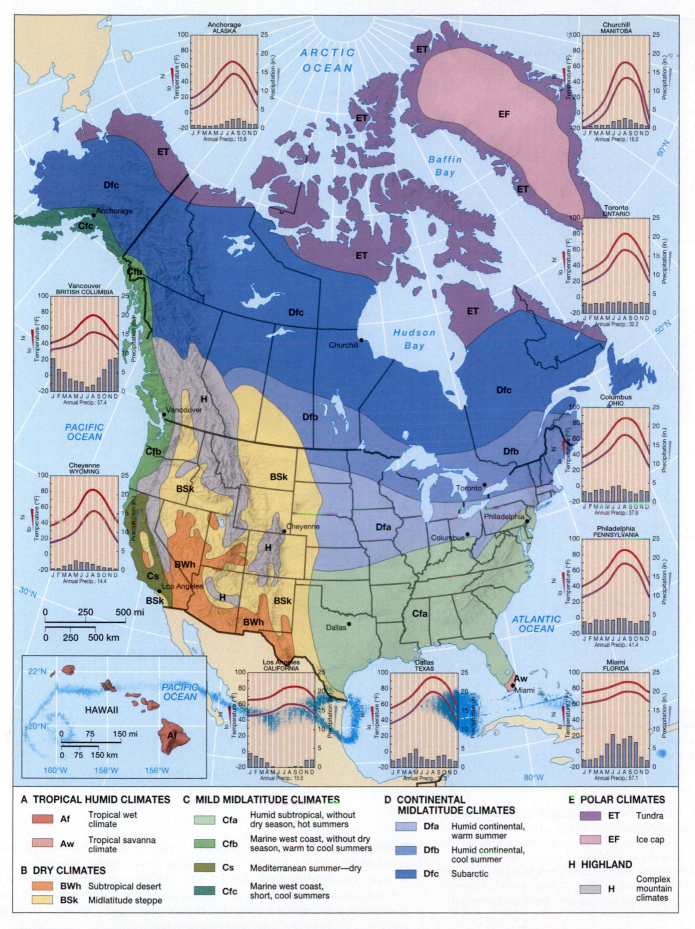

A TROPICAL HUMID CLIMATES

	Af	Tropical wet climate
	Aw	Tropical savanna climate

B DRY CLIMATES

	BWh	Subtropical desert
	BSk	Midlatitude steppe

C MILD MIDLATITUDE CLIMATES

	Cfa	Humid subtropical, without dry season, hot summers
	Cfb	Marine west coast, without dry season, warm to cool summers
	Cs	Mediterranean summer—dry
	Cfc	Marine west coast, short, cool summers

D CONTINENTAL MIDLATITUDE CLIMATES

	Dfa	Humid continental, warm summer
	Dfb	Humid continental, cool summer
	Dfc	Subarctic

E POLAR CLIMATES

	ET	Tundra
	EF	Ice cap

H HIGHLAND

	H	Complex mountain climates

FIGURE 3.7 • CLIMATE MAP OF NORTH AMERICA North American climates include everything from tropical savanna (Aw) to tundra (ET) environments. Most of the region's best farmland and densest settlements lie in the mild (C) or continental (D) midlatitude climate zones.

49

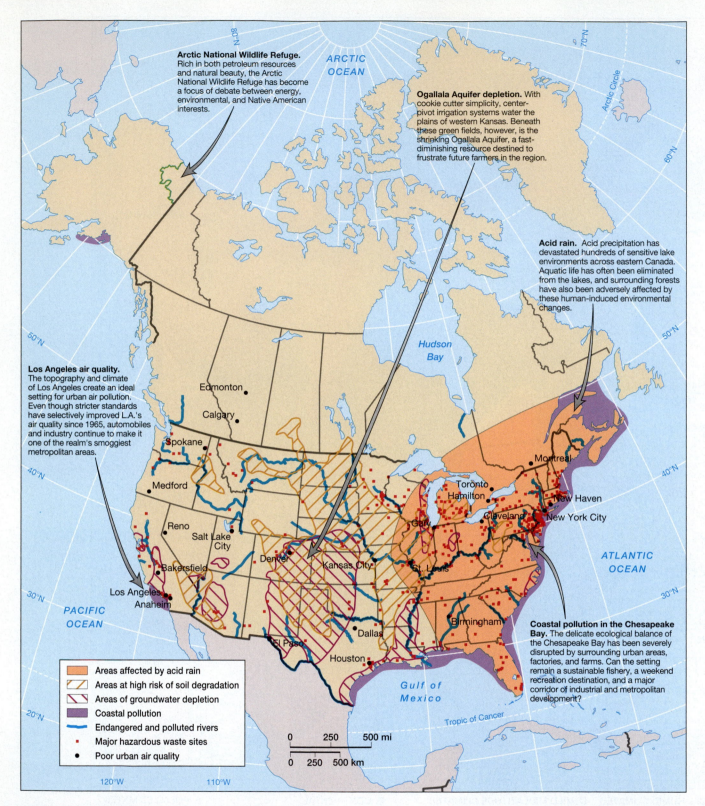

Arctic National Wildlife Refuge. Rich in both petroleum resources and natural beauty, the Arctic National Wildlife Refuge has become a focus of debate between energy, environmental, and Native American interests.

Ogallala Aquifer depletion. With cookie cutter simplicity, center-pivot irrigation systems water the plains of western Kansas. Beneath these green fields, however, is the shrinking Ogallala Aquifer, a fast-diminishing resource destined to frustrate future farmers in the region.

Acid rain. Acid precipitation has devastated hundreds of sensitive lake environments across eastern Canada. Aquatic life has often been eliminated from the lakes, and surrounding forests have also been adversely affected by these human-induced environmental changes.

Los Angeles air quality. The topography and climate of Los Angeles create an ideal setting for urban air pollution. Even though stricter standards have selectively improved L.A.'s air quality since 1965, automobiles and industry continue to make it one of the realm's smoggiest metropolitan areas.

Coastal pollution in the Chesapeake Bay. The delicate ecological balance of the Chesapeake Bay has been severely disrupted by surrounding urban areas, factories, and farms. Can the setting remain a sustainable fishery, a weekend recreation destination, and a major corridor of industrial and metropolitan development?

Legend:
- Areas affected by acid rain
- Areas at high risk of soil degradation
- Areas of groundwater depletion
- Coastal pollution
- Endangered and polluted rivers
- Major hazardous waste sites
- Poor urban air quality

FIGURE 3.8 • ENVIRONMENTAL ISSUES IN NORTH AMERICA Many environmental issues threaten North America. Acid rain damage is widespread in regions downwind from industrial source areas. Elsewhere, widespread air and water pollution present health dangers and economic costs to residents of the region. Since 1970, however, both Americans and Canadians have become increasingly responsive to these environmental challenges.

Managing Water North Americans consume huge amounts of water. While conservation efforts and better technology have slightly reduced per capita rates of water use over the past 25 years, city dwellers still use an average of more than 170 gallons daily. Approximately 45 percent of the water used in the United States is employed in manufacturing and energy production, 40 percent in agriculture, and the remainder for home and business use.

Many places in North America are threatened by water shortages. Metropolitan areas such as New York City struggle with outdated municipal water supply systems. Beneath the Great Plains, the waters of the Ogallala Aquifer are also being depleted.

The largest in North America, this aquifer is a huge reserve of precious groundwater created during the last Ice Age, and today it irrigates about 20 percent of all U.S. cropland. Center-pivot irrigation systems are steadily tapping into the supply; water tables across much of the region have fallen more than 100 feet (30 meters) in the past 50 years; and the the costs of pumping are rising steadily. Farther west, California's complex system of water management is a reminder of that state's ever-growing demands (Figure 3.9).

Water quality is also a major issue. North Americans are exposed to water pollution every day, and even environmental laws and guidelines, such as the U.S. Clean Water Act or Canada's Green Plan, cannot eliminate the problem. Mining operations and industrial users such as chemical, paper, and steel plants generate toxic waste-water or metals that enter surface and groundwater supplies. Metropolitan areas also produce vast amounts of raw sewage that annually costs billions to repurify. America's most toxic places include the petroleum-rich Texas and Louisiana Gulf coasts, the older industrial centers of the Northeast and Midwest, and nuclear fuel and chemical warfare storage areas such as Hanford, Washington.

Altering the Atmosphere   North Americans modify the very air they breathe; in doing so, they change local and regional climates as well as the chemical composition of the atmosphere. The development associated with urban settings often produces nighttime temperatures some 9 to 14°F (5 to 8°C) warmer than nearby rural areas. Air pollution affects large numbers of plants, animals, and people in many settings. At the local level, industries, utilities, and automobiles contribute carbon monoxide, sulfur, nitrogen oxides, hydrocarbons, and particulates to the urban atmosphere. While some of the region's worst offenders are U.S. cities such as Houston and Los Angeles, Canadian cities such as Toronto, Hamilton, and Edmonton also experience significant problems of air quality.

On a broader scale, North America is plagued by **acid rain**, industrially produced sulfur dioxide and nitrogen oxides in the atmosphere that damage forests, poison lakes, and kill fish. Many acid rain producers are located in the Midwest and southern Ontario, where industrial plants, power-generating facilities, and motor vehicles contribute atmospheric pollution. Prevailing winds transport the pollutants and deposit damaging acid rain and snow across the Ohio Valley, Appalachia, the northeastern United States, and eastern Canada.

The Price of Affluence
Globalization has brought many benefits to North America, but with the accompanying urbanization, industrialization, and heightened consumption, the realm is also paying an environmental price for its wealth. Energy consumption within the region, for example, remains extremely high, imposing a growing list of environmental and economic costs both within North America and beyond. For all its economic growth and material affluence, the twentieth century in North America will also be remembered for its toxic waste dumps, frequently unbreathable air, and wildlands lost to development. Still, many environmental initiatives in the United States and Canada have addressed local and regional problems. For example, the improved water quality of the Great Lakes over the past 30 years is an achievement to which both nations contributed and that benefits both. Tougher air quality standards have also reduced certain types of emissions in many North American cities.

Population and Settlement: Reshaping a Continental Landscape

The North American landscape is the product of four centuries of extraordinary human change. During that period, Europeans, Africans, and Asians arrived in the region, disrupted Native American peoples, and created new patterns of human settlement. Today, more than 323 million people live in the region, and they are some of the world's most affluent and highly mobile populations.

FIGURE 3.9 • CALIFORNIA AQUEDUCT   Thirsty Californians have extensively modified the geography of water in their state. Large aqueducts have greatly altered the distribution of this valuable resource, promoting both urban and agricultural growth within western North America. *(Alexander Lowry/Photo Researchers, Inc.)*

Modern Spatial and Demographic Patterns

Large metropolitan areas (including both central cities and suburbs) dominate North America's population geography, producing very uneven patterns of settlement across the region (Figure 3.10). The largest number of cities and the densest collection of rural settlements are found south of central Canada and east of the Great Plains. Within this broad region, Canada's "Main Street" corridor contains most of that nation's urban population, led by the cities of Toronto (4.2 million) and Montreal (3.3 million). **Megalopolis**, the largest settlement cluster in the United States, includes Washington, D.C. (4.7 million), Baltimore (2.5 million), Philadelphia (6 million), New York City (20.2 million), and Boston (3.3 million). Beyond these two national core areas, other sprawling urban centers cluster around the southern Great Lakes, in various parts of the South, and along the Pacific Coast.

Occupying the Land

Europeans began occupying North America about 400 years ago. They were not settling an empty land. North America was populated for at least 12,000 years by peoples as culturally diverse as the Europeans who came to conquer them. Native Americans were broadly distributed across the realm and adapted to its many natural environments. Cultural geographers estimate Native American populations in A.D. 1500 at 3.2 million for the continental United States and another 1.2 million for Canada, Alaska, Hawaii, and Greenland. In many areas, European diseases and disruptions reduced these Native American populations by more than 90 percent as contacts increased.

The first stage of a dramatic new settlement geography began with a series of European colonies, mostly within the coastal regions of eastern North America (Figure 3.11). Established between 1600 and 1750, these regionally distinct societies were anchored on the north by the French settlement of the St. Lawrence Valley and extended south along the Atlantic Coast, including separate English colonies. Scattered developments along the Gulf Coast and in the Southwest also appeared before 1750.

The second stage in the Europeanization of the North American landscape took place between 1750 and 1850, and it was highlighted by settlement of much of the better agricultural land within the eastern half of the continent. Following the American Revolution (1776) and a series of Indian conflicts, pioneers surged across the Appalachians. They found much of the Interior Lowlands region almost ideal for agricultural settlement. Much of southern Ontario, or Upper Canada, was also opened to widespread development after 1791.

The third stage in North America's settlement expansion picked up speed after 1850 and continued until just after 1910. During this period, most of the region's remaining agricultural lands were settled by a mix of native-born and immigrant farmers. Farmers were challenged and sometimes defeated by drought, mountainous terrain, and short northern growing seasons. In the American West, settlers were attracted by opportunities in California, the Oregon country, Mormon Utah, and the Great Plains. In Canada, thousands occupied southern portions of Manitoba, Saskatchewan, and Alberta. Gold and silver discoveries led to initial development in areas such as Colorado, Montana, and British Columbia's Fraser Valley.

Incredibly, in a mere 160 years, much of the North American landscape was occupied as expanding populations sought new land to settle and as the global economy demanded resources to fuel its growth. It was one of the largest and most rapid human transformations of the landscape in the history of the human population. This European-led advance forever reshaped North America in its own image, and in the process it also changed the larger globe in lasting ways by creating a "New World" destined to reshape the Old.

North Americans on the Move

From the mythic days of Davy Crockett and Daniel Boone to the twentieth-century sojourns of John Steinbeck and Jack Kerouac, North Americans have been on the move. Indeed, almost one in every five Americans moves annually, suggesting that residents of the region are quite willing to change addresses in order to improve their income or their quality of life. Several trends dominate the picture.

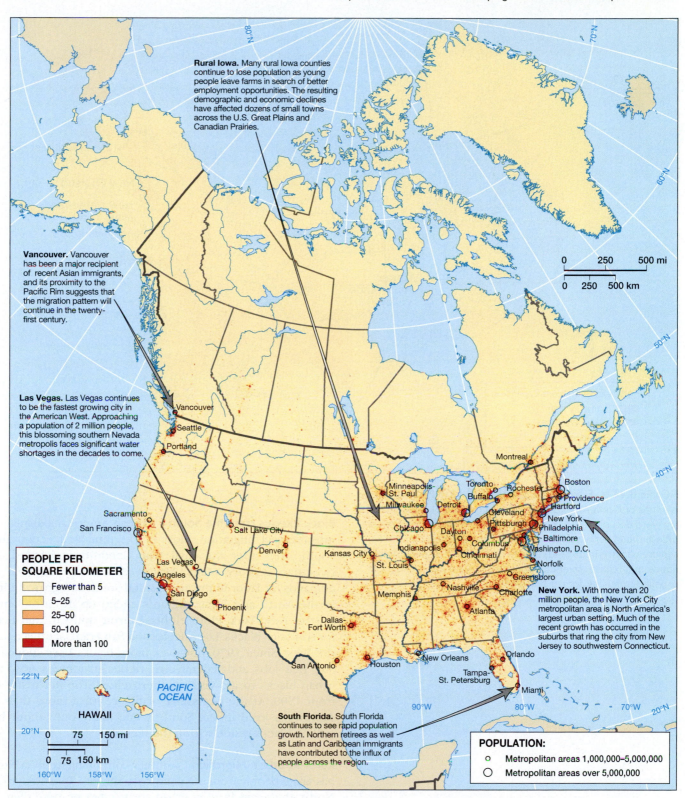

Rural Iowa. Many rural Iowa counties continue to lose population as young people leave farms in search of better employment opportunities. The resulting demographic and economic declines have affected dozens of small towns across the U.S. Great Plains and Canadian Prairies.

Vancouver. Vancouver has been a major recipient of recent Asian immigrants, and its proximity to the Pacific Rim suggests that the migration pattern will continue in the twenty-first century.

Las Vegas. Las Vegas continues to be the fastest growing city in the American West. Approaching a population of 2 million people, this blossoming southern Nevada metropolis faces significant water shortages in the decades to come.

New York. With more than 20 million people, the New York City metropolitan area is North America's largest urban setting. Much of the recent growth has occurred in the suburbs that ring the city from New Jersey to southwestern Connecticut.

South Florida. South Florida continues to see rapid population growth. Northern retirees as well as Latin and Caribbean immigrants have contributed to the influx of people across the region.

PEOPLE PER SQUARE KILOMETER
- Fewer than 5
- 5–25
- 25–50
- 50–100
- More than 100

POPULATION:
- ○ Metropolitan areas 1,000,000–5,000,000
- ◯ Metropolitan areas over 5,000,000

HAWAII

PACIFIC OCEAN

FIGURE 3.10 • POPULATION MAP OF NORTH AMERICA
North America's population is strongly centered on large cities with more thinly settled areas in between. Notable concentrations are found on the eastern seaboard between Boston and Washington, D.C., along the shores of the Great Lakes, and across the Sunbelt from Florida to California.

Westward-moving Populations   The most persistent regional migration trend in North America has been the tendency for people to move west. Indeed, the dominant thrust in the past two centuries (see Figure 3.11) has been to follow the setting sun, and many North Americans continue that pattern to the present. By 1990, more than half of the population of the United States lived west of the Mississippi River, a dramatic shift from colonial times.

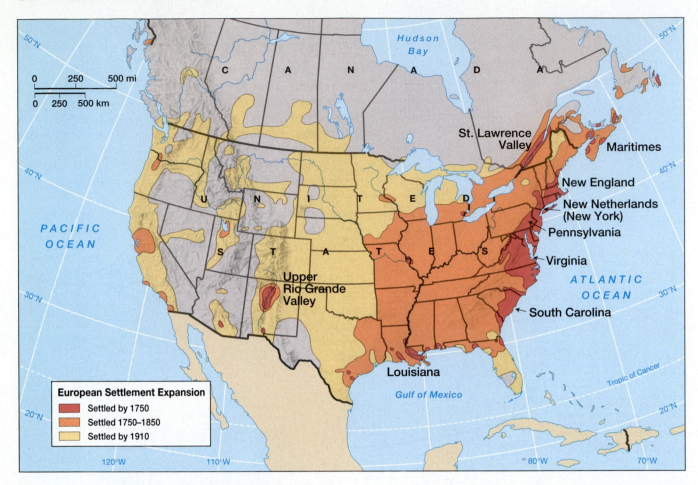

FIGURE 3.11 • EUROPEAN SETTLEMENT EXPANSION Large portions of North America's East Coast and the St. Lawrence Valley were occupied by Europeans before 1750. The most remarkable surge of settlement, however, occurred during the next century as vast areas of land were opened to European dominance and Native American populations were exterminated or expelled from their former homelands.

Black Exodus from the South African Americans displayed different patterns of interregional migration. Most blacks remained economically tied to the rural South, even after the Civil War. Conditions changed, however, in the early twentieth century. Many African Americans migrated because of declining demands for labor in the agricultural South and growing industrial opportunities in the North and West. Migrants ended up in cities where jobs were located. Boston, New York, Philadelphia, Detroit, and Chicago became key destinations for southern blacks. Los Angeles and San Francisco/Oakland drew many blacks to the West. Since 1970, however, more blacks have moved from North to South. Sun Belt jobs and federal civil rights guarantees now attract many northern urban blacks to growing southern cities. The net result is still a major change from 1900: At the beginning of the century, more than 90 percent of African Americans lived in the South, while today only about half of the nation's 35 million blacks reside within the region.

Rural-to-Urban Migration Another continuing trend in North American migration has taken people from the country to the city. Two centuries ago, only 5 percent of North Americans lived in urban areas (cities of more than 2,500 people), whereas today more than 75 percent of the North American population is urban. Shifting economic opportunities account for much of the transformation: As mechanization on the farm reduced the demand for labor, many young people left for new employment opportunities in the city.

Growth of the Sun Belt South Particularly after 1970, southern states from the Carolinas to Texas grew much more rapidly than states in the Northeast and Midwest. During the 1990s, Georgia, Florida, Texas, and North Carolina each grew by more than 20 percent. The South's expanding economy, modest living costs, adoption of air conditioning, attractive recreational opportunities, and appeal to snow-weary retirees all contributed to its growth. Atlanta's bustling metropolitan area (3.8 million) is now larger than that of Boston (3.3 million) (Figure 3.12).

The Counterurbanization Trend   During the 1970s, certain nonmetropolitan areas in North America saw significant population gains, including many rural settings that had previously lost population. Selectively, that pattern of **counterurbanization**, in which people leave large cities and move to smaller towns and rural areas, continues today. Some participants in counterurbanization are retirees, but a large number are younger, so-called *lifestyle migrants*. They find employment in affordable smaller cities and rural settings that are pleasant to live in and often removed from the perceived problems of urban America. In fact, recent nonmetropolitan population growth has exceeded metropolitan growth in most western states. Other smaller communities outside the West, such as Mason City, Iowa; Mankato, Minnesota; and Traverse City, Michigan, also are seen as desirable destinations for migrants interested in leaving behind the problems of metropolitan life.

Settlement Geographies: The Decentralized Metropolis

Settlement landscapes of North American cities are characterized by **urban decentralization**, in which metropolitan areas sprawl in all directions and suburbs take on many of the characteristics of traditional downtowns. Although both Canadian and U.S. cities have experienced decentralization, the impact has been particularly profound in the United States, where inner-city problems, poor public transportation, widespread automobile ownership, and fewer regional-scale planning initiatives have encouraged middle-class urban residents to move beyond the central city.

Historical Evolution of the City in the United States   Changing transportation technologies decisively shaped the evolution of the city in the United States (Figure 3.13). The pedestrian/horsecar city (pre-1888) was compact, essentially limiting urban growth to a 3- or 4-mile-diameter ring around downtown conveniently accessible by foot or horse-powered trolley cars. The invention of the electric trolley in 1888 expanded the urbanized landscape farther into new "streetcar suburbs," often 5 or 10 miles from the city center. Indeed, the electric streetcar city (1888–1920) offered commuters affordable mass transit at the dizzying speed of 15 or 20 miles an hour. A star-shaped urban pattern resulted, with growth extending outward along and near the streetcar lines.

The biggest technological revolution came after 1920 with the widespread adoption of the automobile. The automobile city (1920–1945) promoted the growth of suburbs beyond the reach of the streetcar and added even more distant settlement in the surrounding countryside. Essentially, it allowed many middle-income residents, particularly whites, to leave the central city in favor of low-density suburbs that had fewer ethnic minorities. Following World War II, growth in the outer city (1945 to the present) promoted more decentralized settlement along commuter routes as built-up areas appeared 40 to 60 miles from downtown.

Urban decentralization also reconfigured land-use patterns in the city, producing metropolitan areas today that are strikingly different from their counterparts of the early twentieth century. In the city of the early twentieth century, idealized in the **concentric zone model**, urban land uses were neatly organized in rings around a highly focused central business district (CBD) that contained much of the city's retailing and office functions. Residential districts beyond the CBD were added as the city expanded, with higher-income groups seeking more desirable locations on the outside edge of the urbanized area.

Today's **urban realms model** recognizes these new suburbs characterized by a mix of peripheral retailing, industrial parks, office complexes, and entertainment facilities. These areas of activity, often called "edge cities," have fewer functional connections with the central city than they have with other suburban centers. For most residents of an edge city, jobs, friends, and entertainment are located in other

FIGURE 3.12 • DOWNTOWN ATLANTA, GEORGIA   Sunbelt cities such as Atlanta have been transformed by rapid job creation. Healthy growth in office space, specialty retailing, and entertainment districts has fueled downtown Atlanta's expansion and reshaped the look of the central city skyline. *(Bill Bachmann/Photo Researchers, Inc.)*

FIGURE 3.13 • GROWTH OF THE AMERICAN CITY   Many U.S. cities became increasingly decentralized as they moved through the pedestrian/horsecar, electric streetcar, automobile, and freeway eras. Each era left a distinctive mark on metropolitan America, including the recent growth of edge cities on the urban periphery.

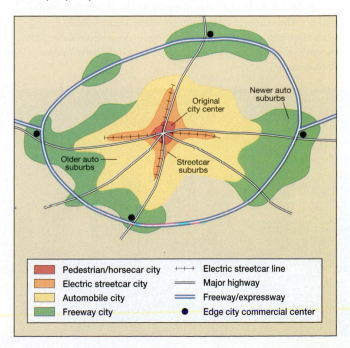

Newer auto suburbs

Original city center

Older auto suburbs

Streetcar suburbs

▢ Pedestrian/horsecar city	+++	Electric streetcar line
▢ Electric streetcar city	══	Major highway
▢ Automobile city	══	Freeway/expressway
▢ Freeway city	●	Edge city commercial center

FIGURE 3.14 • TYSONS CORNER, VIRGINIA North America's edge-city landscape is nicely illustrated by Tysons Corner, Virginia. Far from a traditional metropolitan downtown, this sprawling complex of suburban offices and commercial activities reveals how and where many North Americans will live their lives in the twenty-first century. *(Rob Crandall/Rob Crandall, Photographer)*

FIGURE 3.15 • BALTIMORE'S HARBORPLACE The renovation of Baltimore Harbor is a good example of inner-city revitalization. The elaborate development offers visitors food, entertainment, and specialty shopping in an upscale setting that serves a diverse urban clientele. *(Rob Crandall/Rob Crandall, Photographer)*

surbubs, rather than in the old downtown. Tysons Corner, Virginia, located west of Washington, D.C., is an excellent example of the edge-city landscape on the expanding periphery of a North American metropolis (Figure 3.14).

The Consequences of Sprawl As suburbanization increased in the 1960s and 1970s, many inner cities, especially in the Northeast and Midwest, suffered absolute losses in population, increased levels of crime and social disruption, and a shrinking tax base. Poverty rates average almost three times those of nearby suburbs. Unemployment rates remain above the national average. Central cities in the United States also remain places of racial tension, the product of decades of discrimination, segregation, and poverty.

Even with these challenges, inner-city landscapes are also enjoying selective improvement. Referred to as **gentrification**, the process involves the displacement of lower-income residents of central-city neighborhoods with higher-income residents, the improvement of deteriorated inner-city landscapes, and the construction of new shopping complexes, entertainment attractions, or convention centers in selected downtown locations. The older buildings of the central city are also a draw. They serve as specialty shops and restaurants for urban residents and offer residential opportunities for wealthier singles who wish to live near downtown. Seattle's Pioneer Square, Toronto's Yorkville district, and Baltimore's Harborplace offer examples showing how such new public and private investments shape the central city (Figure 3.15).

The suburbs are also changing. Construction of new corporate office centers, fashion malls, and industrial facilities has created true "suburban downtowns" in suburbs that are no longer simply bedroom communities for central-city workers. Indeed, such localities have been growing players in the continent's globalization process. For example, many of North America's key internationally connected corporate offices (IBM, Microsoft), industrial facilities (Boeing, Sun Microsystems, Oracle), and entertainment complexes (Disneyland, Walt Disney World, and the Las Vegas Strip) are now in such settings, and they are intimately tied to global information, technology, capital, and migration flows.

Settlement Geographies: Rural North America

Rural North American landscapes trace their origins to early European settlement. Over time, these immigrants from Europe showed a clear preference for a dispersed rural settlement pattern as they created new farms on the North American landscape. In portions of the United States settled after 1785, the federal government surveyed and sold much of the rural landscape. Surveys were organized around the simple, rectangular pattern of the federal government's township-and-range survey system, which offered a convenient method of dividing and selling the public domain in six-mile-square townships (Figure 3.16). Canada developed a similar system of regular surveys that stamped much of southern Ontario and the western provinces with a strikingly rectilinear character.

Commercial farming and technological changes further transformed the settlement landscape. Railroads opened corridors of development, provided access to markets for commercial crops, and helped to establish towns. By 1900, several transcontinental lines spanned North America, radically transforming the farm economy and the pace of rural life. After 1920, however, even greater change accompanied the arrival of the automobile, farm mechanization, and better rural road networks. The need for farm labor declined with mechanization, and many smaller market centers became unnecessary as farmers equipped with automobiles and trucks could travel farther and faster to larger, more diverse towns.

Today, many areas of rural North America face population declines as they adjust to the changing conditions of modern agriculture. Both U.S. and Canadian farm populations fell by more than two-thirds during the last half of the twentieth century. Typically, a fewer number of farms (but larger in acreage) dot the modern rural scene, and many young people leave the land to obtain employment elsewhere. The population drain in settings such as rural Iowa, southern Saskatchewan, and eastern Montana also affects towns within these regions. The visual record of abandonment offers a painful reminder of the economic and social adjustments that come from population losses. Weed-choked driveways, empty farmhouses, roofless barns, and the empty marquees of small-town movie houses tell the story more powerfully than any census or government report.

Other rural settings show signs of growth. Some places begin to experience the effects of expanding edge cities. Other growing rural settings lie beyond direct metropolitan influence but are seeing new populations who seek amenity-rich environments removed from city pressures. These trends are shaping the settlement landscape from British Columbia's Vancouver Island to Michigan's Upper Peninsula. Newly subdivided land, numerous real estate offices, and the presence of espresso bars are all signs of growth in such surroundings.

FIGURE 3.16 • IOWA SETTLEMENT PATTERNS   The regular rectangular look of this Iowa town and the nearby rural setting are common cultural landscape features across North America. In the United States, the township-and-range survey system stamped such predictable patterns across vast portions of the North American interior. *(Craig Aurness/Corbis)*

Cultural Coherence and Diversity: Shifting Patterns of Pluralism

North America's cultural geography exerts global influence. At the same time, it is internally diverse. On one hand, history and technology have produced a contemporary North American cultural force that is second to none in the world. Yet North America is also a collection of different peoples who retain part of their traditional cultural identities. In fact, North Americans celebrate their varied roots and acknowledge the region's multicultural character.

The Roots of a Cultural Identity

Powerful historical forces formed a common dominant culture within North America. While both the United States (1776) and Canada (1867) became independent from Great Britain, the two countries remained closely tied to their Anglo roots. Key Anglo legal and social institutions solidified the common set of core values that many North Americans shared with Britain and, eventually, with one another. Traditional Anglo beliefs emphasized representative government, separation of church and state, liberal individualism, privacy, pragmatism, and social mobility. From those shared foundations, particularly within the United States, consumer culture blossomed after 1920, producing a common set of experiences oriented around convenience, consumption, and the mass media.

But North America's cultural unity coexists with pluralism, the persistence and assertion of distinctive cultural identities. Closely related is the concept of **ethnicity**, in which a group of people with a common background and history identify with one another, often as a minority group within a larger society. For Canada, the early and enduring French colonization of Quebec complicates its modern cultural geography. Canadians face the challenge of creating a truly bicultural society where issues of language and political representation are central concerns. Within the United States, given its unique immigration history, a greater diversity of ethnic groups exists, and differences in cultural geography are often found at both local and regional scales.

Peopling North America

North America is a region of immigrants. Quite literally, global-scale migrations to the region created its unique cultural geography. Decisively displacing Native Americans in most portions of the realm, immigrant populations created a new cultural geography of ethnic groups, languages, and religions. Early migrants had considerable cultural influence, even though their numbers were very small. Over time, immigrant groups and their changing destinations produced a varied cultural geography across North America. Also varying between groups was the pace and degree of **cultural assimilation**, the process in which immigrants were absorbed by the larger host society.

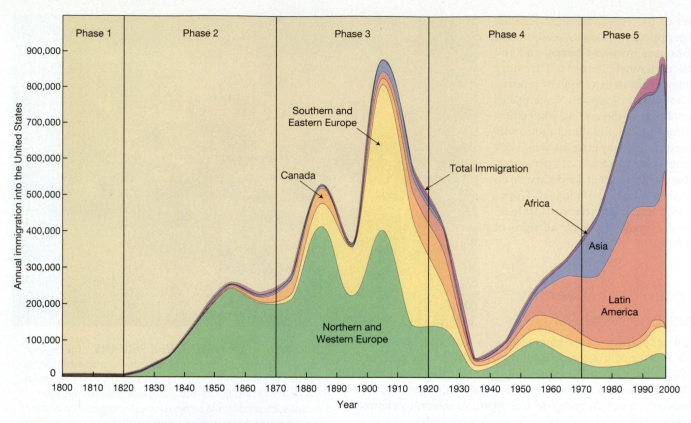

Phase 1 | Phase 2 | Phase 3 | Phase 4 | Phase 5

Southern and Eastern Europe

Canada

Total Immigration

Africa

Asia

Latin America

Northern and Western Europe

FIGURE 3.17 • UNITED STATES IMMIGRATION, BY YEAR AND GROUP Annual immigration rates peaked around 1900, declined in the early twentieth century, and then surged again, particularly since 1970. The source areas of these migrants have also shifted. Note the decreased role Europeans currently play versus the growing importance of Asians and Latin Americans. *(Modified from Rubenstein, 2002,* An Introduction to Human Geography, *Upper Saddle River, NJ: Prentice Hall)*

Migration to the United States In the United States, variations in the number and source regions of migrants produced five distinctive chapters in the country's history (Figure 3.17). In Phase 1 (prior to 1820), English and African influences dominated. Slaves, mostly from West Africa, contributed additional cultural influences in the South. Northwest Europe served as the main source region of immigrants between 1820 and 1870 (Phase 2). The emphasis, however, shifted away from English migrants. Instead, Irish and Germans dominated the flow and provided more cultural variety.

As Figure 3.17 shows, immigration reached a much higher peak around 1900, when almost 1 million foreigners entered the United States *annually*. During Phase 3 (1870–1920), the majority of immigrants were southern and eastern Europeans. Political strife and poor economies in Europe existed during this period. News of available land and expanding industrialization in the United States offered an escape from difficult conditions that caused many Europeans to move. Between 1880 and 1920, large numbers of Scandinavians settled the northern interior. By 1910, almost 14 percent of the nation was foreign-born. Very few of these immigrants, however, targeted the job-poor U.S. South, creating a cultural divergence that still exists.

Between 1920 and 1970 (Phase 4), more immigrants came from neighboring Canada and Latin America, but overall totals fell sharply, a function of more restrictive federal immigration policies (the Quota Act of 1921 and the National Origins Act of 1924), the Great Depression, and the disruption caused by World War II. Since 1970 (Phase 5), the number of immigrants has again increased, with total numbers matching those of the early twentieth century (see Figure 3.17). The current surge was made possible by economic and political instability abroad, a growing U.S. economy, and a loosening of immigration laws (the Immigration Acts of 1965 and 1990, and the Immigration Reform and Control Act of 1986). Most migrants since 1970 came from Latin America or Asia.

Today, about one-fourth of U.S. immigrants arrive from Mexico, and Mexicans make up more than 60 percent of the nation's Hispanic population. Although most Mexican migrants live in California or Texas, they are increasingly moving to other areas such as Wisconsin, Georgia, and Arkansas. The cultural consequences of these Hispanic migrations are important: Today the United States is the fifth largest Spanish-speaking nation on Earth.

In percentage terms, Asian newcomers are the fastest-growing immigrant group, and they account for almost 4 percent of the population. Asian migrants often move to large cities. In fact, more than 40 percent of the nation's Asian population lives in the Los Angeles (1.7 million), San Francisco (1.2 million), and New York City (1.2 million) metropolitan areas. Filipinos, Vietnamese, Chinese, and Indians dominate the flow, but Korean and Japanese communities are also well established, particularly in the West Coast states and Hawaii.

The future cultural geography of the United States will be dramatically redefined by these recent immigration patterns. The increasing diversity of the country is an expression of the globalization process and the powerful pull of North America's economy and political stability. By 2070, Asians may total more than 10 percent of the U.S. population, and almost one American in three will be Hispanic. Indeed, it is likely that the U.S. non-Hispanic white population will achieve minority status by that date.

The Canadian Pattern   The peopling of Canada is similar to the U.S. story. Early French arrivals concentrated in the St. Lawrence Valley. After 1765 many migrants came from Britain, Ireland, and the United States. Canada then experienced the same surge and reorientation in migration flows seen in the United States around 1900. Between 1900 and 1920, more than 3 million foreigners ventured to Canada, an immigration rate far higher than for the United States, given Canada's much smaller population. Eastern Europeans, Italians, Ukrainians, and Russians were the most important nationalities in these later movements. Today, more than 60 percent of Canada's recent immigrants are Asians. As in the United States, more liberal immigration laws since the 1960s encouraged movement to Canada, and its 16 percent foreign-born population is among the highest in the developed world.

Culture and Place in North America

Cultural and ethnic identity are often strongly tied to place. North America's cultural diversity is expressed geographically in two ways. First, people with similar backgrounds congregate near one another and derive meaning from the territories they occupy together. Second, these distinctive cultures leave their mark on the everday scene: The landscape is filled with the artifacts, habits, language, and values of different groups. Boston's Italian North End simply looks and smells different than nearby Chinatown.

Persisting Cultural Homelands   French-Canadian Quebec is an excellent example of a cultural homeland: It is a culturally distinctive nucleus of settlement in a well-defined geographical area, and its ethnicity has survived over time, stamping the cultural landscape with an enduring personality (Figure 3.18). More than 80 percent of the population of Quebec speaks French, and language remains the "cultural glue" that holds the homeland together. Indeed, policies adopted after 1976 strengthened the French language within the province by requiring French instruction in the schools and national bilingual programming by the Canadian Broadcasting Corporation (CBC). The group is also unified by its minority status within a larger Anglo-dominated Canada (only about 25 percent of Canadians are French). Ironically, many Quebeçois feel that the greatest cultural threat may come not from Anglo-Canadians but rather from recent immigrants to the province. Southern Europeans or Asians in Montreal, for example, show little desire to learn French, preferring instead to put their children in English-speaking private schools.

Another well-defined cultural homeland is the Hispanic Borderlands (see Figure 3.18). It is similar in size to French-Canadian Quebec, significantly larger in total population, but more diffuse in its cultural and political expression. Historical roots of the homeland are deep, extending back to the sixteenth century, when Spaniards opened the region to the European world. A rich legacy of Spanish place names, earth-toned Catholic churches, and traditional Hispanic settlements dot the rolling highlands of northern New Mexico and southern Colorado. From California to Texas, other historical sites, place names, missions, and presidios (military forts) also reflect the Hispanic legacy.

Unlike Quebec, however, large twentieth-century migrations from Latin America brought an entirely new wave of Hispanic settlement to the Southwest. More than 32 million Hispanics now live in the United States, with more than half in California, Texas, New Mexico, and Arizona combined. Indeed, by 2015, Hispanics will likely outnumber non-Hispanic whites in California. Hispanics have created a distinctive Borderlands culture that mixes many elements of Latin and North America. Cities such as San Antonio and Los Angeles play leading roles in revealing the Hispanic presence within the Southwest. Regionally distinctive Latin foods and music add internal cultural variety to the region.

African Americans also retain a cultural homeland, but it has become less important because of out-migration (see Figure 3.18). Reaching from coastal Virginia and North Carolina to east Texas, a zone of enduring African-American population remains a legacy of the Cotton South, when a vast majority of U.S. blacks resided as slaves within the region. Dozens of rural counties in the region still have large black majorities. More broadly, the South is home to many black folk traditions, including such music as black spirituals and the blues, which have now become popular far beyond their rural origins.

A second rural homeland in the South is Acadiana, a zone of persisting Cajun culture in southwestern Louisiana (see Figure 3.18). This homeland was created in the eighteenth century when French settlers were expelled from eastern Canada and relocated to Louisiana. Nationally popularized today through their food and music, the Cajuns still have a long-lasting attachment to the bayous and swamps of southern Louisiana.

FIGURE 3.18 • SELECTED CULTURAL REGIONS OF NORTH AMERICA From northern Canada's Nunavut to the Southwest's Hispanic Borderlands, different North American cultural groups strongly identify with traditional local and regional homelands. *(Modified from Jordan, Domosh, and Rowntree, 1998, The Human Mosaic, Upper Saddle River, NJ: Prentice Hall)*

Native American populations are also strongly tied to their homelands. Almost 4 million Indians, Inuits, and Aleuts live in North America, and they claim allegiance to more than 1,100 tribal bands. Particularly in the American West and the Canadian and Alaskan North, native peoples control sizable reservations, including the 16-million-acre (6.5-million-hectare) Navajo Reservation in the Southwest, as well as self-governing Nunavut in the Canadian North (see Figure 3.18). Although these homelands preserve traditional ties to the land, they have also been settings for pervasive poverty and increasing cultural tensions (Figure 3.19). Within the United States, many Native American groups have taken advantage of the special legal status of their reservations and have built gambling casinos and tourist facilities that bring in much-needed capital, but also challenge traditional lifeways.

A Mosaic of Ethnic Neighborhoods North America's cultural mosaic is also characterized by smaller-scale ethnic signatures that shape both rural and urban landscapes. When much of the agricultural interior was settled, immigrants often established close-knit communities. Among others, German, Scandinavian, Slavic, Dutch, and Finnish neighborhoods took shape, held together by common origins, languages, and religions. Although many of these ties weakened over time, rural landscapes of Wisconsin, Minnesota, the Dakotas, and the Canadian prairies still display these cultural imprints. Folk architecture, distinctive settlement patterns, ethnic place names, and the simple elegance of rural churches survive as signatures of cultural diversity upon the visible scene of rural North America.

Ethnic neighborhoods are also a part of the urban landscape and reflect both global-scale and internal North American migration patterns. The ethnic geography of Los Angeles is an example of both economic and cultural forces at work (Figure 3.20). Since most of its economic expansion took place during the twentieth century, the city's ethnic patterns reflect the movements of more recent migrants. African-American communities on the city's south side (Compton and Inglewood) represent the legacy of black population movements out of the South. Hispanic (East Los Angeles) and Asian (Alhambra and Monterey Park) neighborhoods are a reminder that almost 40 percent of the city's population is foreign-born.

Particularly in the United States, ethnic concentrations of nonwhite populations increased in many cities during the twentieth century as whites left for the perceived safety of the suburbs. In terms of central-city population, African Americans make up 75 percent of Detroit and more than 60 percent of Atlanta, while Los Angeles is now more than 40 percent Hispanic. Often these central-city ethnic neighborhoods are further isolated economically from most urban residents by high levels of poverty and unemployment. Ethnic concentrations are also growing in the suburbs of some U.S. cities: Southern California's Monterey Park has been called the "first suburban Chinatown," and growing numbers of middle-class African Americans and Hispanics shape suburban neighborhoods in metropolitan settings such as Atlanta and San Antonio.

FIGURE 3.19 • NATIVE AMERICAN POVERTY   Navajo youngsters enjoy a game of basketball on the reservation. Poor housing, low incomes, and persistent unemployment plague many Native American settings across the rural West. *(Jim Noelker/The Image Works)*

FIGURE 3.20 • ETHNICITY IN LOS ANGELES   Economic opportunities have attracted a wide variety of migrants to North American cities, producing an ethnic mixture of distinctive neighborhoods and communities. In Los Angeles, several cycles of economic expansion have attracted a diverse collection of residents from around the globe. *(Reprinted from Rubenstein, 2002, An Introduction to Human Geography, Upper Saddle River, NJ: Prentice Hall)*

Patterns of North American Religion

Reflecting its early settlement history, Protestantism is dominant within the United States, accounting for about 60 percent of the population. Within Canada, almost 40 percent of the population is Protestant, with the United Church of Canada and the Anglican Church claiming the largest numbers of adherents. French-Canadian Quebec is a stronghold of Catholic tradition and makes Canada's population (44 percent) distinctly more Catholic than that of the United States (25 percent). Still, major regional concentrations of American Catholics persist, particularly in the urban Northeast and the Hispanic Southwest. Eastern Orthodox Christians are clustered in the urban Northeast, where many Greek, Russian, and Serbian Orthodox communities were established between 1890 and 1920. The telltale domes of Ukrainian Orthodox churches still dot the Canadian prairies of Alberta, Saskatchewan, and Manitoba. More than 7 million Jews live in North America, concentrated in East and West Coast cities. In the United States, the rapidly growing Nation of Islam (Black Muslims) also has a strong urban presence, reflecting its appeal to many poor African Americans. In addition, a large number of Asians have fueled the recent growth of North American Islamic (6 million), Buddhist (1 million), and Hindu (1 million) populations.

The Globalization of American Culture

Simply put, North America's cultural geography is becoming more global at the same time that global cultures are becoming more North American (influenced particularly by the United States). As a dominant player in globalization, the United States is without doubt changing in the process, but to what extent? Similarly, how is the globalization of American culture transforming the larger world?

North Americans: Living Globally   More than ever before, North Americans in their everyday lives are exposed to people from beyond the region. In the United States, about 70,000 foreigners arrive daily, mostly as tourists. Annually, more than 22 million tourists visit the country, while 4 million business visitors spend time within its borders. At U.S. colleges and universities, 500,000 international students add variety to campus life. Canada experiences similar global influences, with the U.S. presence particularly dominant.

Globalization presents cultural challenges for North Americans. In the United States, one key issue revolves around the English language, which some have described as the key "social glue" that holds the nation together. Since 1980, the continuing flood of non-English-speaking immigrants into the country has sharpened the debate over the role English should play in American culture. Many in the United States argue that English should be the country's only officially recognized language in order to force immigrants to learn it and thus speed their entry into the host culture. On the other hand, immigrant groups suggest that they need to maintain their traditional languages both to function within their ethnic communities and to preserve their cultural heritage.

North Americans are going global in other ways. By 2000, more than half the region's people were Internet users, thus opening the door for far-reaching journeys in cyberspace. The popularity of ethnic restaurants has peppered the realm with a bewildering variety of Cuban, Ethiopian, Basque, and Pakistani eateries. Chinese and Italian foods dominate the taste buds of all North Americans, with Mexican food a rapidly growing choice in the United States. The growing demand for foreign beverages mirrors the pattern; imported beer sales in the United States are growing 14 percent annually, with the top producers selling 159 million cases during 1997 (Figure 3.21). Americans also have increased their consumption of foreign red wines and have rapidly Europeanized their coffee-drinking habits. In fashion, Gucci, Armani, and Benetton are household words for millions who keep their eyes on European styles. While British pop music has been an accepted part of North American culture for four decades, the beat of German techno bands, Gaelic instrumentals, and Latin rhythms also have become a part of daily life within the

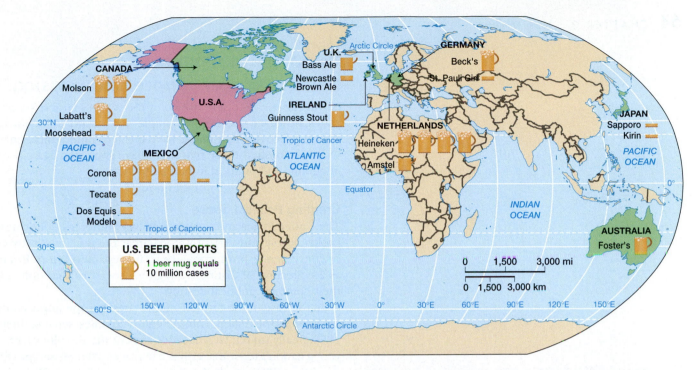

FIGURE 3.21 • ANNUAL BEER IMPORTS TO THE UNITED
STATES Whether they are aware of it or not, North Americans
are increasingly eating and drinking globally. Rising beer imports,
including many more expensive foreign brands, exemplify the pattern. The nation's beer drinkers know no bounds to their thirsts,
preferring a wide variety of Asian, Australian, European, and
Latin American producers. (Data from Forbes, October 5, 1998)

realm. Indeed, from acupuncture and massage therapy to soccer and New Age religions, North Americans are borrowing, adapting, and absorbing the larger world around them.

The Global Diffusion of U.S. Culture

In a similar fashion, the lives of billions of people beyond the region have been forever changed by U.S. culture. Rapid improvements in global transportation and information technologies, many of them engineered in the United States, brought the world more surely under the region's dominance. Perhaps most critical was the marriage between growing global demands for consumer goods and the rise of the U.S. multinational corporation, which was superbly structured to meet and cultivate those needs. While the results are not a simple Americanization of traditional cultures or a single, synthesized global culture shaped in the U.S. image, millions of people, particularly the young, are strongly attracted by that image's emphasis on individualism, wealth, freedom, youth, and mobility.

Global information flows illustrate the country's cultural influence on the world. Companies such as Time Warner and Walt Disney increasingly dominate the media world. The emergence of English as the *de facto* global language has obviously been an advantage to a variety of U.S. media outlets. The United States dominates television; as television sets, cable networks, and satellite dishes spread, so do U.S. sitcoms, CNN, and MTV. Movie screens tell a similar story. In western Europe, the United States claims 70 percent of the film market, and more than half of Japan's movie entertainment is dubbed fresh from Hollywood.

The United States shapes the popular cultural landscape of every corner of the globe. Global corporate advertising, distribution networks, and mass consumption bring Big Macs to Moscow, golf courses to Thai jungles, Mickey and Minnie Mouse to Tokyo, and Avon cosmetics to millions of beauty-conscious Chinese (Figure 3.22). Western-style business suits have become the professional uniform of choice, while T-shirts and jeans offer standardized global comfort on days away from work. In the built landscape, central city skylines become identical to one another, and one airport hotel looks the same as another eight time zones away.

The country's global cultural influence has not gone unchallenged. Canadian government agencies routinely chastise their radio, television, and film industries for letting in too much U.S. cultural influence. The French have also been critical of U.S. dominance in such media as the Internet. Public subsidies to France's Centre National de la Cinematographie are designed to foster French-language filmmaking for a national audience deluged with English-language productions.

FIGURE 3.22 • TOKYO DISNEYLAND While some traditional
cultures resist American influences, many people around the
world have embraced globalization, especially when it is
brought by popular characters such as Mickey and Minnie
Mouse. Tokyo Disneyland is one of a growing number of Magic
Kingdoms around the planet. (AP/Wide World Photos)

Geopolitical Framework: Patterns of Dominance and Division

North America is home to two of the world's largest states. The creation of these states, however, was neither simple nor preordained, but rather the result of historical processes that might have created quite a different North American map. Once established, these two states have coexisted in a close relationship of mutual economic and political interdependence.

Creating Political Space

The United States and Canada have very different political roots. The United States broke cleanly and violently from Great Britain. Canada, by contrast, was a country of convenience, born from a peaceful separation from Britain and then assembled as a collection of distinctive regional societies that only gradually acknowledged their common political destiny.

In the case of the future United States, Europe imposed its own political boundaries across the region. The 13 English colonies, sensing their common destiny after 1750, finally united two decades later in the Revolutionary War. By the 1790s, the young nation's territorial claims had reached the Mississippi River. Soon the Louisiana Purchase (1803) nearly doubled the national domain. By midcentury, Texas had been annexed, treaties with Britain had secured the Pacific Northwest, and an aggressive war with Mexico had captured much of the Southwest. The acquisition of Alaska (1867) and Hawaii (1898) rounded out what became the 50 states.

Canada was created under quite different circumstances. After the American Revolution, England's remaining territories in the region were controlled by administrators in British North America. In 1867 the British North America Act united the provinces of Ontario, Quebec, Nova Scotia, and New Brunswick in an independent Canadian Confederation. Within a decade, the Northwest Territories (1870), Manitoba (1870), British Columbia (1871), and Prince Edward Island (1873) joined Canada, and the continental dimensions of the country took shape. Soon, the Yukon Territory (1898) separated from the Northwest Territories; Alberta and Saskatchewan gained provincial status (1905); and Manitoba, Ontario, and Quebec were enlarged (1912) north to Hudson Bay. Newfoundland finally joined in 1949. The addition of Nunavut Territory (1999) represents the latest change in Canada's political geography.

Continental Neighbors

Geopolitical relationships between Canada and the United States have always been close: Their common 5,525-mile (8,900-kilometer) boundary requires both nations to pay close attention to one another. During the twentieth century, the two countries have lived largely in political harmony with one another. In 1909, the Boundary Waters Treaty created the International Joint Commission, an early step in the common regulation of cross-boundary issues involving water resources, transportation, and environmental quality. The St. Lawrence Seaway (1959) opened the Great Lakes region to better global trade connections. The two nations also joined in cleaning up Great Lakes pollution and in making plans to reduce acid rain in eastern North America.

It has been in the area of trade relations that the close political ties between these neighbors have mattered most. The United States receives 80 percent of Canada's exports and supplies more than two-thirds of its imports. Conversely, Canada is the United States' most important trading partner, accounting for roughly 20 percent of its exports and imports. One landmark agreement reached in 1989 was the signing of the bilateral (two-way) Free Trade Agreement (FTA). Five years later, the larger **North American Free Trade Agreement (NAFTA)** extended the alliance to Mexico. Paralleling the success of the European Union (EU), NAFTA has forged the world's largest trading bloc, including more than 400 million consumers and a huge free-trade zone that stretches from beyond the Arctic Circle to Latin America.

Political conflicts still divide the two countries. Environmental issues produce cross-border tensions, especially when environmental degradation in one nation affects the other. For example, Montana's North Flathead River flows out of British Columbia, where Canadian logging and mining operations periodically threaten fisheries and recreational lands south of the border. Agricultural and natural resource competition also causes occasional controversy between the two neighbors. For example, problems recently developed when Canadian wheat and potato growers were accused of dumping their products into U.S. markets, thus depressing prices and profits for U.S. farmers. Similar issues have arisen in the logging industry. American timber companies argue that Canadian companies are given special economic support by Canadian provincial governments. In the 1990s "salmon wars" flared along the Pacific Coast, where U.S. and Canadian fishermen compete for fish claimed by both sides. By 2000 conditions improved when the two countries signed an agreement on fishing rights, but other border disputes still remain (Figure 3.23).

FIGURE 3.23 • GEOPOLITICAL ISSUES IN NORTH AMERICA
Although Canada and the United States share a long and peaceful border, many political issues still divide the two countries. In addition, internal political conflicts, particularly in bicultural Canada, cause tensions. Given its global status and the openness of its society, the United States also remains vulnerable to acts of global terrorism.

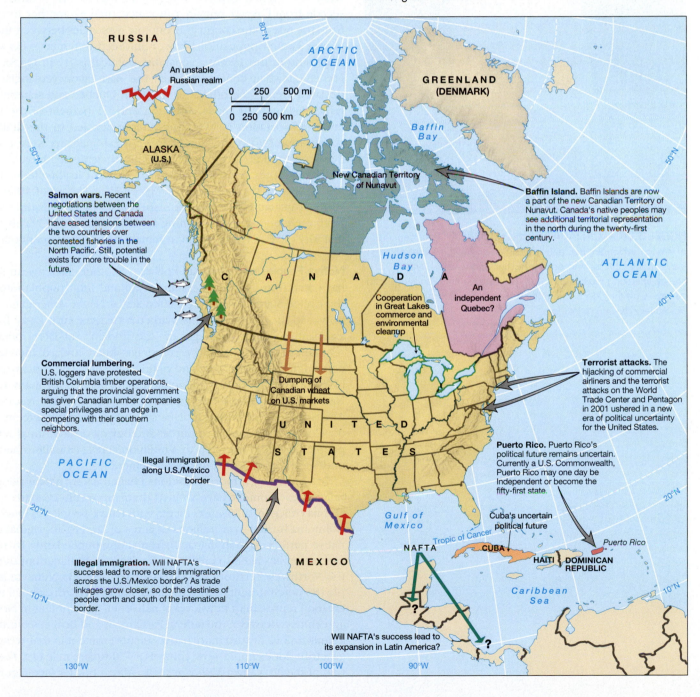

FIGURE 3.24 • CANADA'S FRENCH LEGACY Four centuries of French settlement in Canada, particularly within Quebec, leave many cultural and political issues unanswered. Most French-speaking Canadians want to preserve their language and culture. Can they do it within the present political framework? (R. Sidney/The Image Works)

The Legacy of Federalism

The United States and Canada are **federal states** in that both nations allocate considerable political power to units of government beneath the national level. Other nations, such as France, have traditionally been **unitary states**, in which power is centralized at the national level. Federalism leaves many political decisions to local and regional governments and often allows distinctive cultural and political groups to be recognized within a country. The U.S. Constitution (1787) limited centralized authority, giving all unspecified powers to the states or the people. In contrast, the Canadian Constitution (1867), which created a federal state under a parliamentary system, reserved most powers to central authorities. Ironically, the evolution of the United States produced an increasingly powerful central government, while Canada's geopolitical balance of power shifted toward more provincial autonomy and a relatively weak national government.

Quebec's Challenge One of the greatest challenges facing Canada is determining what role French Quebec will play in the country's future (Figure 3.24). Will it secede from Canada, remain a province, or seek another form of political self rule from the rest of the country? Economic disparities between the wealthier Anglo and poorer French populations have reinforced differences between the two groups. Beginning in the 1960s, a separatist political party in Quebec increasingly voiced French-Canadian concerns. When the party won provincial elections in 1976, it declared French the official language of Quebec and scheduled a popular vote on remaining within Canada. Although only 41 percent of the voting public were in favor of separation in 1980, the referendum led to the drafting of a new federal constitution two years later. Many French-Canadians were not impressed with the new constitution, however, fearing that their distinctive cultural identity might be threatened. They refused to sign the document, leading federal lawmakers in Ottawa to propose another solution, the Meech Lake Accord of 1990. The Accord included stronger guarantees of Quebec's special status as a "distinct society." Quebec again held a referendum on separating from Canada in 1995. This time, only a razor-thin majority (50.6 percent) voted to remain within the larger country.

Native Peoples and National Politics Another challenge to federal political power has come from North American Indian and Inuit populations, in both Canada and the United States. Within the United States, Native Americans asserted their political power in the 1960s and this marked a decisive turn away from earlier policies of assimilation. Since passage of the Indian Self-Determination and Education Assistance Act of 1975, the trend has been toward increased Native American control of their economic and political destiny. The Indian Gaming Regulatory Act (1988) offered potential economic independence for many tribes as it allowed for Native Americans to set up casino operations on many reservations. In the western American interior, where Indians control roughly 20 percent of the land, tribes are developing their natural resources and buying up additional acreage. In Alaska, native peoples gained ownership to 44 million acres (18 million hectares) of land in 1971 under the Alaska Native Claims Settlement Act.

In Canada even more ambitious challenges by native peoples to a weaker centralized government have yielded dramatic results. Canada established the Native Claims Office in 1975. Agreements with native peoples in Quebec, Yukon, and British Columbia turned over millions of acres of land to aboriginal control and increased native participation in managing remaining public lands. By far, the most ambitious agreement has been to create the territory of Nunavut out of the eastern portion of the Northwest Territories in 1999 (see Figure 3.23). Nunavut is home to 25,000 people (85 percent Inuit) and is the largest territorial/provincial unit within Canada. Its creation represents a new level of native self-government in North America. Recently, agreements between the federal Parliament and British Columbia tribes have begun a similar move toward more native self-government in that western province. The Nisga'a tribe, for example, now controls a 770-square-mile (1,992-square-kilometer) portion of the Nass River Valley near the Alaska border (see Figure 3.18).

A Global Reach

The geopolitical reach of the United States, in particular, has taken its influence far beyond the borders of the region. The Monroe Doctrine (1824) asserted that U.S. interests extended throughout the Western Hemisphere and went beyond national boundaries. In the Pacific, the United States claimed the Philippines as a prize of the Spanish-American War (1898), and the further annexation of Hawaii (1898) contributed to the country's twentieth-century dominance of the region. In Central America and the Caribbean, the growing role of the American military between 1898 and 1916 shaped politics in Cuba, Puerto Rico, Panama, Nicaragua, Haiti, Mexico, and elsewhere.

World War II and its aftermath forever redefined the role of the United States in world affairs. Victorious on both Atlantic and Pacific fronts, postwar America emerged from the conflict as the world's dominant political power. The United States also developed multinational political and military agreements, such as those establishing the North Atlantic Treaty Organization (NATO) and the Organization of American States (OAS). Violent conflicts in Korea (1950–1953) and Vietnam (1961–1973) pitted U.S. political interests against communist attempts to extend their control beyond the Soviet Union and China. Tensions also ran high in Europe as the Berlin Wall crisis (1961) and nuclear weapons deployments by NATO- and Soviet-backed forces brought the world closer to another global war. The Cuban missile crisis (1962) reminded Americans that traditional political boundaries provided little defense in a world uneasily brought closer together by technologies of potential mass destruction.

Even as the Cold War gradually faded during the late 1980s, the global political reach of the United States continued to expand. Policies in Central America favored governments friendly to the United States. President Carter's successful Middle East Peace Treaty between Israel and Egypt (1979) guaranteed a continuing diplomatic and military presence in the eastern Mediterranean. In the late 1990s, Serbian aggression within Kosovo prompted an American and NATO-led intervention, which included major air attacks on the Serbian capital of Belgrade (1999) and a peacekeeping presence (with the United Nations) in the disputed area of Kosovo. The recent war in Iraq (2003) offered yet another example of America's global political presence.

Economic and Social Development: Geographies of Abundance and Affluence

North America possesses the world's most powerful economy and its wealthiest population. Its 323 million people consume huge quantities of global resources, but also produce some of the world's most sought-after manufactured goods and services. The region's human capital—the skills and diversity of its population—has enabled North Americans to achieve high levels of economic development.

An Abundant Resource Base

North America is blessed with varied natural resources, which have provided diverse raw materials for development. Indeed, the direct extraction of natural resources still makes up 3 percent of the U.S. economy and more than 6 percent of the Canadian economy. Some of these North American resources are then exported to global markets, while other raw materials are imported to the region.

Opportunities for Agriculture North Americans have created one of the most efficient food-producing systems in the world and agriculture remains a dominant land use across much of the region (Figure 3.25). Farmers practice highly commercialized, mechanized, and specialized agriculture. The system emphasizes the importance of efficient transportation, global markets, and large capital investments in farm machinery. Today, agriculture employs only a small percentage of the labor force in both the United States (2.6 percent) and Canada (3.7 percent). At the same time, changes in farm ownership have sharply reduced the number of operating units while average farm sizes have steadily risen. Agriculture remains a

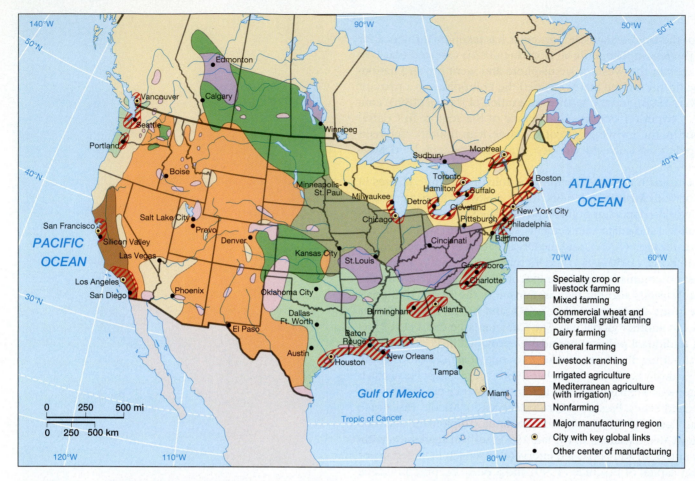

FIGURE 3.25 • MAJOR ECONOMIC ACTIVITIES OF NORTH AMERICA Varied environmental settings, settlement histories, and economic conditions have produced the modern map of North American economic activities. An increasing number of major metropolitan areas are playing increasingly visible roles in the global economy. *(Modified from Clawson and Fisher, 1998, World Regional Geography, Upper Saddle River, NJ: Prentice Hall)*

critically important part of the North American economy, producing between $200 billion and $300 billion annually in sales.

The geography of North American farming represents the combined impacts of 1) diverse environments, 2) varied continental and global markets for food, and 3) historical patterns of settlement and agricultural evolution. In the Northeast, dairy operations and truck farms take advantage of their nearness to major cities in Megalopolis and southern Canada. Corn and soybeans dominate the Midwest and western Ontario, where a tradition of mixed farming combines the growing of feed grains with the production and fattening of livestock. To the south, only remnants of the old Cotton Belt remain, largely replaced by varied subtropical specialty crops; poultry, catfish, and livestock production; and commercial logging (Figure 3.26). Farther west, extensive, highly mechanized commercial grain-growing operations stretch from Kansas to Saskatchewan and Alberta. Depending on surface and groundwater resources, irrigated agriculture across western North America also offers opportunities for farming. Indeed, California's agricultural output, nourished by large agribusiness operations in the irrigated Central Valley, accounts for more than 10 percent of the nation's farm economy.

Industrial Raw Materials North Americans produce and consume huge quantities of other natural resources. While the region possesses a variety of energy and metals resources, the size and diversity of the North American economy have led to the need to import additional raw materials. The U.S. produces about 12 percent of the world's oil but consumes more than 25 percent. As a result, the United States imports more than half of its oil. Within the realm, the major areas of oil and gas production are the Gulf Coast, the Central Interior, Alaska's North Slope, and Central Canada (especially Alberta). The most abundant fossil fuel in the United States is coal, but its relative importance in the overall energy economy declined in the twentieth century as industrial technologies changed and environmental concerns grew. Still, the country's 400-year supply of coal reserves (23 percent of the world's total) will be an important energy resource for the future. North America also remains a major producer of metals resources, although global competition, rising extraction costs, and environmental concerns present challenges for this sector of the economy.

Creating a Continental Economy

The timing of European settlement in North America was critical in its rapid economic transformation. The region's abundant resources came under the control of Europeans possessing new technologies that reshaped the landscape and reorganized its economy. By the nineteenth century, North Americans actively contributed to those technological changes. In addition, new natural resources were developed in the interior and new immigrant populations arrived in large numbers. In the twentieth century, although natural resources remained important, industrial innovations and more jobs in the service sector added to the economic base and extended the country's global reach.

Connectivity and Economic Growth Dramatic improvements in North America's transportation and communication systems laid the foundation for urbanization, industrialization, and the commercialization of agriculture. Indeed, the region's economic success was a function of its **connectivity**, or how well its different locations became linked with one another through improved transportation and communications networks. Those links greatly facilitated the potential for interaction between locations and dramatically reduced the cost of moving people, products, and information over long distances.

Tremendous technological breakthroughs revolutionized North America's economic geography between 1830 and 1920. By 1860 more than 30,000 miles (48,387 kilometers) of railroad track had been laid in the United States, and the network grew to more than 250,000 miles (403,226 kilometers) by 1910. Farmers in the Midwest and Plains found ready markets for their products in cities hundreds of miles away. Industrialists collected raw materials from faraway places, processed them, and shipped manufactured goods to their final destinations. The telegraph brought similar changes to information: Long-distance messages flowed across eastern North America by the late 1840s, and 20 years later undersea cables linked the realm to Europe, another milestone in the process of globalization.

North America's transportation and communications systems were modernized further after 1920. Automobiles, mechanized farm equipment, paved highways, commercial air links, national radio broadcasts, and dependable transcontinental telephone service reduced the cost of distance across the region. Perhaps most importantly, the region has taken the lead in the global information age, integrating computer, satellite, and Internet technologies in a web of connections that assists the flow of knowledge both within the realm and beyond.

The Sectoral Transformation Changes in employment structure signaled North America's economic modernization just as surely as its increasingly interconnected society. The **sectoral transformation** refers to the evolution of a nation's labor force from one dependent on the *primary* sector (natural resource extraction) to one with more employment in the *secondary* (manufacturing or industrial), *tertiary* (services), and *quaternary* (information processing) sectors. For example, with agricultural mechanization, lower demands for primary-sector workers are replaced by new opportunities in the growing industrial sector. In the twentieth century, new services (trade, retailing) and information-based activities (education, data processing, research) created other employment opportunities. Both the United States and Canada have seen these dramatic changes in employment. Today, the tertiary and quaternary sectors employ more than 70 percent of the labor force in both countries.

FIGURE 3.26 • SPECIALTY AGRICULTURE IN THE AMERICAN SOUTH Catfish ponds near Greenville, Mississippi, offer a new look to the South's rural landscape. Today, the old Cotton Belt has largely been replaced by specialty operations engaged in subtropical cropping, commercial forestry, and poultry, catfish, and livestock production. *(C. C. Lockwood/DRK Photo)*

FIGURE 3.27 • GULF COAST PETROLEUM REFINING
Petroleum-related manufacturing has transformed many Gulf Coast settings. Much of Houston's twentieth-century growth has been fueled by the dramatic expansion of oil-related industries. The port of Houston remains a major center of North America's refining and petrochemical operations. *(Walter Frerck/Odyssey Productions)*

FIGURE 3.28 • SILICON VALLEY The high-technology industrial landscape of California's Silicon Valley differs from the look of traditional manufacturing centers. Here similar industries form complex links, benefiting from their proximity to one another and to nearby universities such as Stanford and Berkeley. *(George Hall/Woodfin Camp & Associates)*

Regional Economic Patterns The locations of North America's industries show important regional patterns. **Location factors** are the varied influences that explain *why* an economic activity is located where it is. Many influences, both within and beyond the realm, shape patterns of economic activity. Patterns of industrial location illustrate the concept (see Figure 3.25). The historical manufacturing core includes Megalopolis (Boston, New York, Philadelphia, and Baltimore), southern Ontario (Toronto and Hamilton), and the industrial Midwest. The region's proximity to *natural resources* (farmland, coal, and iron ore); increasing *connectivity* (canals and railroad networks, highways, air traffic hubs, and telecommunications centers); a ready supply of *productive labor*; and a growing national, then global, *market demand* for its industrial goods encouraged continued *capital investment* within the core. Traditionally, the core has dominated in the production of steel, automobiles, machine tools, and agricultural equipment and played a critical role in producer services such as banking and insurance.

In the last half of the twentieth century industrial and service-sector growth shifted to the South and West. Cities of the South's Piedmont manufacturing belt (Greensboro to Birmingham) grew after 1960, partly because lower labor costs and Sun Belt amenities attracted new investment. The Gulf Coast industrial region is strongly tied to nearby fossil fuels that provide raw materials for its many energy-refining and petrochemical industries (Figure 3.27). The varied West Coast industrial region stretches from Vancouver, British Columbia, to San Diego, California (and beyond into northern Mexico), and it demonstrates the increasing importance of Pacific Basin trade. Large aerospace operations in the West also suggest the role of *government spending* as a location factor. Silicon Valley is now North America's leading region of manufacturing exports. Its proximity to Stanford, Berkeley, and other universities demonstrates the importance of *access to innovation and research* for many fast-changing high-technology industries. Silicon Valley's location also shows the advantages of *agglomeration economies*, in which many companies with similar and often integrated manufacturing operations locate near one another (Figure 3.28). Many smaller industrial centers also thrive throughout North America. Places such as Provo, Utah, and Austin, Texas, that specialize in high-technology industries demonstrate the growing role of *lifestyle amenities* in shaping industrial location decisions, both for entrepreneurs and for the skilled workers who need to be attracted to such opportunities.

North America and the Global Economy

Together with Europe and Japan, North America plays a key role in the global economy. When the economy is thriving, the realm benefits from global economic growth, but in periods of international instability, globalization means that the region is more vulnerable to economic downturns. Increasingly, North American workers are directly tied to export markets in Latin America, the rise and fall of Asian imports, or the pattern of global investments in U.S. stock and bond markets. No community remains untouched by these links, and the consequences will shape the lives of every North American in the twenty-first century.

The region is home to a growing number of truly "global cities" that serve as key connecting points and decision-making centers in the world economy (see Figure 3.25). While New York City is the largest, smaller metropolitan areas such as Miami, Toronto, Chicago, and Vancouver are emerging as critical urban players on the global stage.

The United States, with Canada's firm support, also played a formative role in creating much of this new global economy and in shaping many of its key institutions. In 1944 allied nations met at Bretton Woods, New Hampshire, to discuss economic affairs. Under U.S. leadership, the group set up the International Monetary Fund (IMF) and the World Bank and gave these global organizations the responsibility for defending the world's monetary system. The United States also was the driving force for the creation (1948) of the General Agreement on Tariffs and Trade (GATT). Renamed the World Trade Organization (WTO) in 1995, its 140 member states are dedicated to reducing global barriers to trade. In addition, the United States and Canada participate in the **Group of Seven (G-7)**, a collection of powerful countries (including Japan, Germany, Great Britain, France, Italy, and sometimes Russia) that regularly meets for discussions on key global economic and political issues.

Patterns of Trade   Global trade patterns show North America's importance in both the sale and purchase of goods and services within the international economy. Both countries import diverse products from many global sources. Dominated overwhelmingly by trade with the United States, Canada imports large quantities of manufactured parts, vehicles, computers, and foodstuffs. Imports to the United States continue to grow. Canada, Japan, Mexico, China, and world oil exporters supply the United States with a diversity of raw materials, low-cost consumer goods, and high-quality vehicles and electronics products.

Outgoing trade flows suggest the items that North Americans produce most cheaply and efficiently. Canada's exports include large quantities of raw materials (grain, energy, metals, and wood products), but manufactured goods are becoming increasingly important, particularly in its trade with the United States. The United States also enjoys many profitable global economic ties and its exports reveal strong links to other portions of the more developed world. By 2000, international sales of U.S. software and entertainment products totaled more than $60 billion annually, more than any other industry. Sales of automobiles, aircraft, computer and telecommunications equipment, financial and tourism services, and food products also contributed to the nation's flow of exports. The destinations for these products and services spanned the globe, with Canada, Japan, Mexico, and western Europe serving as major purchasers.

Patterns of Investment in North America   Patterns of capital investment and corporate power place the North American realm at the center of global money flows and economic influence. Given its relative stability, the region attracts huge inflows of foreign capital, both as investments in North American stocks and bonds and as foreign direct investment (FDI) by international companies. For Canada, the wealth and proximity of the United States have meant that 80 percent of foreign-owned corporations in the country are based in the United States.

Doing Business Globally   The growing impact of U.S. investments in foreign stock markets and the dominance of U.S.-controlled multinational companies suggest how U.S. capital is transforming the way business is done throughout the world. Since 1980, aging U.S. baby boomers have poured billions of pension fund and investment dollars into Japanese, European, and "emerging" stock markets. U.S. investments in foreign countries are also made directly by multinational corporations based in the United States. Coca-Cola, GE, Intel, and Procter & Gamble are household words from Bangladesh to Bolivia. The economic influence of these corporations exceeds that of many nation-states. The success of these multinational companies is increasingly related to their participation in foreign markets and their growing use of foreign labor.

Persisting Social Issues

Profound economic and social problems shape the human geography of North America. Even with its continental wealth, great differences persist between rich and poor. High median household incomes in the United States and Canada do not reveal the differences in wealth within the two countries. Broader measures of social well-being suggest that the poorer people in these countries have less access to quality education and health care. In addition, both nations face enduring issues related to gender inequity and social and economic problems related to aging populations. One consequence of globalization is that many of these economic and social challenges are increasingly shaped by changes outside the region. Poverty and a loss of jobs in the rural U.S. South may be related to low Asian wage rates, and a viral outbreak in Hong Kong might be only a plane flight away from suburban Vancouver.

FIGURE 3.29 • GATED AMERICA Wealthier North Americans have had a tremendous impact on the region's cultural landscapes, and their influence far outweighs their relatively modest numbers. Many gated communities in North American suburbs and resorts display a desire for privacy, safety, and social exclusivity. *(James Patelli)*

FIGURE 3.30 • INNER-CITY NEIGHBORHOOD Many poor inner-city neighborhoods, such as this community in New York City, suffer from substandard housing and aging infrastructure. Shared ethnic bonds often supply the social support necessary to ease the challenges of high unemployment and troubled family life. *(Erica Lananer/Black Star)*

Wealth and Poverty The North American landscape displays contrasting scenes of wealth and poverty. Elite suburbs, gated neighborhoods, upscale shopping malls, and posh resorts are all expressions of private and exclusive landscape settings that characterize wealthier North American communities (Figure 3.29). On the other hand, substandard housing, abandoned property, aging infrastructure, and unemployed workers are visual reminders of the gap between rich and poor within the realm (Figure 3.30). Specifically, within the United States, black household incomes remain only 64 percent of the national average, while Hispanic incomes fare slightly better at 72 percent of the national average.

The distribution of wealth varies widely across the region. Many of the wealthiest communities in the United States are suburbs on the edge of large metropolitan areas. Resort and retirement communities are homes for the rich as well: Palm Beach, Florida, and Aspen, Colorado, have some of the nation's most desirable real estate and costliest housing. The Northeast and West remain America's richest regions, although the South is now within 8 percent of the national norm. In Canada, Ontario and British Columbia are the country's wealthiest provinces, with Vancouver's high house prices rivaling those of San Francisco.

Poverty levels have declined in both countries since 1980, but poor populations are still grouped into a variety of geographical settings. By 2000, poverty rates in the United States had fallen to about 12 percent of the population; a similar measure of low-income Canadians declined to just below 20 percent. The problems of the rural poor remain major regional social issues in the Canadian Maritimes, Appalachia, the Deep South, and the Southwest. Most poor people in the United States, however, live in central city locations, and links between ethnicity and poverty are strong in these communities. Nationally, 24 percent of the country's African Americans and 23 percent of its Hispanic populations live below the poverty line, and the great majority reside in poorer central city communities. In the context of the information economy, there is also a pronounced **digital divide** in North America that suggests that the region's poor and underprivileged groups have significantly less access to Internet communications than the wealthy. Blacks, for example, are far less likely to be linked to the Web, and, within the United States, households making more than $75,000 annually are 20 times more likely to use the Internet than Americans living below the poverty level.

Twenty-first Century Challenges Measures of social well-being in North America compare favorably with those of most other world regions. Still, many economic and social challenges exist for this region. As workers, North Americans compete globally for jobs in a fast-changing world economy. American companies and workers are scrambling to adjust to the new uncertainties in the global economy.

Since World War II, both nations have seen great changes in the role that women play in society, but the "gender gap" is yet to be closed when it comes to salary issues, working conditions, and political power. Women widely participate in the workforce in both countries and are as educated as men, but they still earn only about 75 cents for every dollar that men earn. Women also head the vast majority of poorer single-parent families in the United States; more than 40 percent of these women are unwed mothers. Canadian women, particularly single mothers who work full-time, are also greatly disadvantaged, averaging only about 65 percent of the salaries of Canadian men.

Health care and aging are also key concerns within a region full of graying baby boomers. A recent report on aging in the United States predicted that 20 percent of the nation's population will be older than 65 by 2050 and that the most elderly (age 85+) are the fastest-growing part of the population. The geographical consequences of aging are already abundantly clear. Whole sections of the United States—from Florida to southern Arizona—have become increasingly oriented around retirement (Figure 3.31). Communities cater to seniors with special assisted-living arrangements, health-care facilities, and recreational opportunities.

Although Canada and the United States have different health-care delivery systems, they share many of the health-related benefits and costs of living within the realm. As long average lifespans and low childhood mortality rates suggest, the two countries benefit from modern health-care systems that offer the latest in high technology. Still, North Americans voice concern over rising health-care costs and uneven access to premium care. While U.S. residents spend more than 14 percent of GDP on health care (Canadians spend slightly less), there are 45 million Americans without any health-care insurance, 30 percent more than in 1990.

The rising incidence of chronic diseases associated with aging (heart disease, cancer, and stroke are the three leading causes of death) will continue to pressure both health-care systems. Another cost has been the care and treatment of the realm's 800,000 AIDS victims. The disease will hit poorer black and Hispanic populations particularly hard, where rates of new infection are still on the rise.

FIGURE 3.31 • TOMORROW'S BABY BOOM LANDSCAPE?
This Arizona retirement community, its circular shape boldly stamped on the desert landscape, suggests the twenty-first-century destination for many aging baby boomers. As North Americans grow older, entire subregions may be devoted to age-dependent demands for housing, recreation, and health services. (Cont/Frank Fournier/Woodfin Camp & Associates)

CONCLUSION

North Americans have reaped the natural abundance of their realm, and in the process they have transformed the environment, created a highly affluent society, and extended their global economic, cultural, and political reach. In a remarkably short time period, a unique mix of varied cultural groups contributed to the settlement of a huge and resource-rich continent that is now the world's most urbanized region. Along the way, North Americans produced two societies that are closely intertwined, yet face distinctive national political and cultural issues. For both, the twenty-first century brings uncertainties. In Canada, the nation's very existence remains problematic as it works through the persisting challenges of its bicultural character and the costs and benefits of its proximity to its continental neighbor. For the United States, social problems linked to ethnic diversity and continuing poverty remain central concerns, along with the environmental price tag associated with high rates of resource consumption. In addition, the pivotal role of the United States in the international economy and in the global balance of power makes the country a continuing target for groups focused on destabilizing the country. How the United States responds to those challenges will directly shape its global role and its impact on globalization processes in the first decade of the twenty-first century.

KEY TERMS

acid rain (page 51)
boreal forest (page 48)
concentric zone model (page 55)
connectivity (page 69)
counterurbanization (page 55)
cultural assimilation (page 57)

digital divide (page 72)
ethnicity (page 57)
federal state (page 66)
gentrification (page 56)
Group of Seven (G-7) (page 71)
location factor (page 70)

Megalopolis (page 52)
North America Free Trade
   Agreement (NAFTA) (page 64)
prairie (page 48)
sectoral transformation (page 69)
tundra (page 48)

unitary state (page 66)
urban decentralization (page 55)
urban realms model (page 55)

Seventy-four percent of the region's 490 million people live in cities, making it the most urbanized region of the developing world. It is noted for its production of primary exports; however, rates of economic development vary greatly between states.

LATIN AMERICA
Political Map

⊗ ● Over 1,000,000

○ ● 500,000–1,000,000
(selected cities)

★ ● Selected smaller cities

Elevation in meters

4000+
2000–4000
500–2000
200–500
0–200
Below sea level

Sea Level

SETTING THE BOUNDARIES

The concept of Latin America as a distinct region has been commonly accepted for nearly a century. The boundaries of this region are straightforward, beginning at the Rio Grande (called the Rio Bravo in Mexico) and ending at Tierra del Fuego (see Figure 4.7). French geographers are credited with coining the term "Latin America" in the nineteenth century to distinguish the Spanish- and Portuguese-speaking republics of the Americas plus Haiti from the English-speaking territories. There is nothing particularly "Latin" about the area, other than the prevalence of romance languages. The term stuck because it was vague enough to include areas with different colonial histories while also offering a clear cultural boundary from Anglo-America, the region referred to as North America in this text.

Latin America is one of the world regions that North Americans are most likely to visit. The trend is to visit the northern edge of this region. Tourism is strong, especially along Mexico's northern border and coastal resorts such as Cancun. Unfortunately, there is a tendency to visit one area in the region and generalize for all of it. Although it is historically sound to think of Latin America as a major world region, there are extreme variations in the physical environment, levels of social and economic development, and the influences of native peoples.

Because the region is so large, geographers often divide Latin America. The continent of South America is typically distinguished from Middle America (which includes Central America, Mexico, and the Caribbean). The term "Middle America" was created to identify an area culturally distinct from North America but physically part of it, because most of Mexico and all of Cuba rest on the North American Plate. Such a division has its advantages, but it also separates countries with very similar histories (such as Mexico and Peru) while joining countries that have very little in common (such as El Salvador and Jamaica). In this text the Americas are divided slightly differently. Latin America consists of the Spanish- and Portuguese-speaking countries of Central and South America, including Mexico. Chapter 5 examines the Caribbean, consisting of the islands of the Antilles, the Guianas, and Belize. So divided, the important Indian and Iberian influences of mainland Latin America are emphasized. Similarly, the Caribbean's unique colonial and demographic history will be discussed in Chapter 5.

Beginning with Mexico and extending to the tip of South America, Latin America's regional unity stems largely from its shared colonial history, rather than from its present-day level of development. More than 500 years ago, the Iberian countries of Spain and Portugal began their conquest of the Americas. Iberia's mark is still visible throughout the area: Officially, two-thirds of the population speaks Spanish, and the rest speak Portuguese. Iberian architecture and town design add homogeneity to the colonial landscape. The vast majority of the population is Catholic. These European traits blended with those of different Amerindian peoples. The Indian presence remains especially strong in Bolivia, Peru, Ecuador, Guatemala, and southern Mexico, where large and diverse Amerindian populations maintain their native languages, dress, and traditions. After the initial conquest, other cultural groups were added to this mix of native and Iberian peoples.

The legacy of slavery brought several million Africans to the region, primarily on the coasts of Colombia and Venezuela and throughout Brazil. In the nineteenth and twentieth centuries new waves of settlers came from Spain, Italy, Germany, Japan, and Lebanon. The end result is one of the world's most racially mixed regions. The modern states of Latin America are multiethnic, with very different rates of social and economic development (Figure 4.1; see also "Setting the Boundaries").

Through colonialism, immigration, and trade, the forces of globalization have been at work in Latin America for decades. The early Spanish Empire concentrated on mining precious metals, sending ships laden with silver and gold across the Atlantic. The Portuguese became important producers of natural dyes, sugar products, gold, and later coffee. In the late nineteenth and early twentieth centuries, exports to North America and Europe fueled the region's economy. Most countries specialized in one or two products: bananas and coffee, meats and wool, wheat and corn, petroleum and copper. Such a primary export tradition, led to an unhealthy economic dependence on a handful of exports. They argued in the 1960s that Latin American economies were too specialized on primary exports to foster development.

Since the 1960s the countries of the region have industrialized and diversified their production, but they continue to be major producers of primary goods for North America, Europe, and Japan. Changes in the economic landscape, however, are reflected in increased levels of manufacturing for internal consumption and export. Intraregional trade within Latin America, as well as the prospect of a **Free Trade Area of the Americas (FTAA)** that would include all of Latin America, the Caribbean (minus Cuba), and North America in a hemispheric free-trade zone by 2005, are indicators of increased economic integration.

FIGURE 4.2 • TROPICAL WILDERNESS The biological diversity of Latin America—home to the world's largest rain forest—is increasingly seen as a genetic and economic asset. These forested mesas (called *tepuis*) in southern Venezuela are representative of the wild lands that many conservationists seek to protect. *(Rob Crandall)*

Roughly equal in area to North America, Latin America has a much larger and faster-growing population of 490 million people. Its most populated state, Brazil, has more than 170 million people, making it the fifth largest country in the world. The next largest state, Mexico, has a population of 100 million. The typical Latin American woman now has three children, compared with six in the 1960s. Unlike most areas of the developing world today, Latin America is decidedly urban. Prior to World War II, most people lived in rural settings and worked as farmers. Today three-quarters of Latin Americans are city dwellers. Even more startling is the number of **megacities**. São Paulo, Mexico City, Buenos Aires, and Rio de Janeiro all have more than 10 million residents. In addition, there are more than 40 cities of at least 1 million residents.

Despite the region's growing industrial and urban characteristics, industries focusing on natural resource extraction will continue to be important, in part because of the area's impressive resource base. Latin America is home to Earth's largest rain forest, the greatest river by volume, and massive reserves of natural gas, oil, and copper. With its extensive territory, its tropical location, and its relatively low population density (Latin America has half the population of India in nearly seven times the area), the region is also recognized as one of the world's great reserves of biological diversity (Figure 4.2). How this diversity will be managed in the face of global demand for natural resources is an increasingly important question for the countries of this region.

Environmental Geography: Neotropical Diversity and Urban Degradation

Much of the region is characterized by its tropical nature. Travel posters of Latin America showcase lush forests and brightly colored parrots. The diversity and uniqueness of the **neotropics** (tropical ecosystems of the Western Hemisphere) have long been attractive to naturalists eager to understand their unique flora and fauna. It is no accident that Charles Darwin's theory of evolution was inspired by his two-year journey in tropical America. Even today, scientists throughout the region work to understand complex ecosystems, discover new species, protect them, and interpret the impact of human settlement, especially in neotropical forests. Not all of the region is tropical. Important population centers extend below the Tropic of Capricorn, most notably Buenos Aires, Argentina, and Santiago, Chile. Much of northern Mexico, including the city of Monterrey, is north of the Tropic of Cancer. Yet it is Latin America's tropical climate and vegetation that most affect popular images of the region. Given the territory's large size and relatively low population density, Latin America has not experienced the same levels of environmental degradation witnessed in East Asia and Europe. The region's biggest environmental concerns are related to deforestation and loss of biodiversity and the livability of urban areas (Figure 4.3). Mexico City, in particular, suffers from a host of environmental problems and is a good example of the kinds of environmental challenges facing modern Latin American cities.

Yet huge areas of Latin America remain relatively untouched, supporting an incredible diversity of plant and animal life. Throughout the region, national parks offer some protection to unique communities of plants and animals. A growing environmental movement in countries such as Costa Rica and Brazil has yielded both popular and political support for conservation efforts. In short, Latin Americans enter the twenty-first century with a real opportunity to avoid many of the environmental mistakes seen in other regions of the world. At the same time, global market forces are driving governments to exploit minerals, fossil fuels, forests, and soils. The region's biggest natural resource management challenge is to balance the economic benefits of extraction with the ecological soundness of conservation. Another major challenge is to improve the environmental quality of Latin American cities.

The Destruction of Tropical Rainforests

Perhaps the environmental issue most commonly associated with Latin America is deforestation. The Amazon Basin and portions of the eastern lowlands of Central America and Mexico still maintain unique and impressive stands of tropical forest.

Pine-Oak Forests of the Sierra Madre Occidental, Mexico. This is one of the world's most extensive subtropical coniferous forests. Commercial logging, conversion of land for agriculture, and overgrazing threaten the viability of the ecosystem.

Brazilian Amazon. Over the last 30 years 14 percent of this region has been deforested, mostly along the Amazonian highways. It is hoped that extractive reserves, natural parks, and sustainable forestry practices can preserve the world's largest rain forest.

Cloud (or Montaine) Forest of the Eastern Andean Piedmont (especially Peru and Bolivia). Wildlands increasingly under pressure from the production of coca leaf. Home of the Andean spectacled bear.

Curitiba. One of the urban planning success stories of Latin America. This city of 2 million is considerably less polluted than other cities. City officials have emphasized public transportation, open space, and recycling.

Pampas of Argentina. One of the great natural grasslands of Latin America that is steadily being converted into cropland and pasture. Burning and draining now threaten remaining natural ecosystems.

The Brazilian Coastal Atlantic Forest. One of the most degraded ecosystems in all of Latin America. Virtually destroyed in the nineteenth and twentieth centuries with the expansion of agriculture, urbanization, industrialization, and household fuel wood consumption. The Atlantic forests were characterized by extraordinary biodiversity, with high levels of regional and local endemism.

Legend:
- Tropical forest
- Forest destroyed
- Desert
- Desertification
- Coastal pollution
- Polluted rivers
- Poor urban air quality

FIGURE 4.3 • ENVIRONMENTAL ISSUES IN LATIN AMERICA
Tropical forest destruction, desertification, water pollution, and poor urban air quality are some of the pressing environmental problems facing Latin America. Yet vast areas of tropical forest are still present, supporting a wealth of biological diversity. *(Adapted from DK World Atlas, 1997, pp. 7, 55. London: DK Publishing)*

Other woodland areas, such as the Atlantic coastal forests of Brazil and the Pacific forests of Central America, have nearly disappeared as a result of agriculture, settlement, and ranching. In the midlatitudes, the ecologically unique evergreen rain forest of southern Chile (the Valdivian forest) is being cleared to export wood chips to Asian markets (Figure 4.4). The coniferous forests of northern Mexico are also

FIGURE 4.4 • CHILEAN WOOD CHIPS A mountain of wood chips awaits shipment to Japanese paper mills from the southern Chilean port of Punta Arenas. The development of wood products from both native and plantation forests supports Chile's booming export economy. *(Rob Crandall/Rob Crandall Photography)*

being cut, in part because of new commercial logging opportunities stimulated by the North American Free Trade Association (NAFTA) agreement.

The loss of tropical rain forests is most critical in terms of biological diversity. Tropical rain forests cover only 6 percent of Earth's landmass, but at least 50 percent of the world's species are found in this biome. Moreover, the Amazon contains the largest undisturbed stretches of rain forest in the world. Unlike Southeast Asian forests, where hardwood extraction drives deforestation, Latin American forests are usually seen as an agricultural frontier that state governments divide in an attempt to give land to the landless and reward political elites. Thus, forests are cut and burned, with settlers and politicians carving them up to create permanent settlements, slash-and-burn plots, or large cattle ranches. In addition, some tropical forest cutting has been motivated by the search for gold (Brazil, Venezuela, and Costa Rica) and the production of coca leaf for cocaine (Peru, Bolivia, and Colombia).

Brazil has been criticized more than other countries for its Amazon forest policies. Through these policies, some 15 percent of the Brazilian Amazon has been deforested. In states such as Rondônia, where settlers streamed in along a popular road known as BR364, close to 60 percent of the state has been deforested (Figure 4.5). What most alarmed environmentalists and forest dwellers (Indians and rubber tree tappers) was the dramatic increase in the rate of clearing, especially in the 1980s. Rates of deforestation have slowed in response to both pressure from environmental, native, and development agencies and Brazil's economic crisis, which limited resources for expansion into the region. During the 1990s forest loss was averaging 0.5 percent annually. However, there is growing interest in harvesting tropical hardwoods from the Brazilian Amazon; Southeast Asian logging firms have sought permits from the Brazilian government for logging. At the same time, tropical forestry experts argue that only through sustainable management can commercial logging support the region's economy while maintaining its resource base.

The conversion of tropical forest into pasture, called **grassification**, is another practice that has contributed to deforestation. Particularly in southern Mexico, Central America, and the Brazilian Amazon, an assortment of development policies from the 1960s through the 1980s encouraged deforestation to make room for cattle. Although there are many natural grasslands such as the Llanos (Venezuela and Colombia), the Chaco (Bolivia, Paraguay and Argentina), and the Pampas (Argentina) suitable for grazing, it was the rush to convert forest into pasture that made ranching environmentally destructive. All told, nearly 300 million acres of tropical forest were destroyed in the 1980s, and much of this went to pasture.

Urban Environmental Challenges: The Valley of Mexico

At the southern end of the great Central Plateau of Mexico lies the Valley of Mexico, the cradle of Aztec civilization and the site of Mexico City, a metropolitan area of approximately 18 million people. The severe problems facing this urban ecosystem underscore how environment, population, technology, and politics are interwoven in the effort to make Mexico City livable. This high-altitude basin with mild temperatures, fertile soils, and ample water surrounded by snow-capped volcanoes was an early center of Amerindian settlement and plant domestication, and the site of one of Spain's most important colonial cities. Given these features, it is no wonder that Mexico City has kept its primacy into the twenty-first century, even though

FIGURE 4.5 • FRONTIER SETTLEMENT IN RONDÔNIA, BRAZIL Colonization along Amazonian highways, such as BR364, initiated a wave of forest clearing in the 1980s. Paired satellite images from Rondônia, Brazil, in 1975 (left) and 1992 (right) reveal the extent of forest clearing that occurred along this major highway and its side roads. *(EROS Data Center, U.S. Geological Survey)*

the environmental setting that made it attractive centuries ago is now severely degraded. Some of the most pressing problems are air quality, adequate water, and subsidence (soil sinkage) caused by overdrawing the valley's aquifer (groundwater).

Air Pollution   The smog of Mexico City is so bad that most modern-day visitors have no idea that mountains surround them because they are rarely visible. Air quality has been a major issue for Mexico City since the 1960s, driven in part by the city's unusually high rate of growth. (Between 1950 and 1980 the city's annual rate of growth was 4.8 percent.) It is hard to imagine a better setting for creating air pollution. The city sits in a bowl 7,400 feet (2,250 meters) above sea level, where a layer of warm air (called a *thermal inversion*) regularly traps exhaust, industrial smoke, garbage, and fecal matter (Figure 4.6). During the worst pollution emergencies in the winter when a layer of warm air traps pollutants, schoolchildren are required to stay indoors, and high-polluting vehicles are banned from the streets. Steps were finally taken in the late 1980s to reduce emissions from factories and cars. Unleaded gas is now widely available for the 3 million cars in the metropolitan area, and some of the worst polluting factories in the Valley of Mexico have closed. In fact, a study in 2001 suggested that the high levels of lead, carbon dioxide, and sulfur dioxide in the atmosphere were starting to decline. Still, the health costs of breathing such contaminated air are real, as elevated death rates due to heart disease, influenza, and pneumonia suggest.

Water Resources   One of Mexico City's most significant environmental problems is water. Mexican President Vicente Fox declared water (both scarcity and quality) a national security issue, not just for the capital but for the entire country. Ironically, it was the abundance of water that made this site attractive for settlement. Large shallow lakes once filled the valley, but over the centuries most were drained to expand agricultural land. As surface water became scarce, wells were dug to tap the basin's massive freshwater aquifer. Today approximately 70 percent of the water used in the metropolitan area is drawn from the valley's aquifer. There is troubling evidence that the aquifer is being overdrawn and at risk of contamination, especially in areas where unlined drainage canals can leak pollutants into the surrounding soil, which then leach into the aquifer. To reduce reliance on the aquifer, the city now pumps water nearly a mile uphill from more than 100 miles (160 kilometers) away.

The Sinking Land   A lesser-known but equally worrisome problem is that Mexico City is sinking. As the metropolis grows and pumps more water from its aquifer, subsidence worsens. Mexico City sank 30 feet (9 meters) during the twentieth century. By comparison, Venice, an Italian city knows for its subsidence problems, sank only 9 inches (23 centimeters) during the same period. Subsidence is a huge problem that will not go away because the city is still so dependent on groundwater. Although the amount of subsidence varies across the metropolitan area, its impact is similar throughout the city. Building foundations are destroyed, and water and sewer lines rupture. Repair crews race through city streets, patching some 40,000 ruptures in the water lines every year.

These serious urban environmental problems are made worse by poverty and governmental inaction. During the 1970s and 1980s, politicians and industrialists denied that there were problems. For most of the century, one-party rule in Mexico reduced the likelihood of meaningful environmental reforms. When steps were finally taken to introduce unleaded gasoline and catalytic converters, open dumps, aging cars and minibuses, and unregulated factories continued to spew pollutants into the air. Mexico City's poorest citizens suffer the most from urban contamination. For the vast majority of Latin American urban poor, the lack of reliable water and clean air is the most pressing environmental problem.

Modern Urban Challenges   For most Latin Americans, air pollution, water availability and quality, and garbage removal are the pressing environmental problems of everyday life. Consequently, many environmental activists from the

FIGURE 4.6 • AIR POLLUTION IN MEXICO CITY   Mexico City is known for its poor air quality, and its high elevation and immense size make management of air quality difficult. While lead levels have declined with the introduction of unleaded gasoline, respiratory illnesses are on the rise. *(Tom Owen Edmunds/Getty Images, Inc.)*

region focus their efforts on making urban environments cleaner by introducing "green" legislation and calling people to action. In this most urbanized region of the developing world, city dwellers do have better access to water, sewers, and electricity than their counterparts in Asia and Africa. Moreover, the density of urban settlement seems to encourage the widespread use of mass transportation—both public and private bus and van routes make getting around cities fairly easy. Yet the usual environmental problems that come from dense urban settings ultimately require expensive remedies such as new power plants and better sewer and water lines. The money for such projects is never enough, thanks to currency devaluation, inflation, and foreign debt. Since many urban dwellers tend to reside in unplanned squatter settlements, servicing these communities with utilities after they are built is difficult and costly. These settlements are especially vulnerable to natural hazards, as witnessed by the catastrophic landslides in Caracas, Venezuela, in December 1999. Unusually heavy rains on the coastal mountains north of the city triggered massive mudflows and landslides that buried entire communities. Death toll estimates range from 10,000 to 40,000 people, making it one of Latin America's worst natural disasters in the twentieth century.

There are, however, examples of sustainable cities in Latin America. Curitiba is the celebrated "green city" of Brazil because of some relatively simple yet progressive planning decisions. More than 2 million people live in this industrial and commercial center, yet it is significantly less polluted than similar-sized cities. Because the city's location was vulnerable to flooding, city planners built drainage canals and set aside the remaining natural drainage areas as parks in the 1960s, well before explosive growth would have made such a policy difficult. This action added green space and reduced the negative impacts of flooding. Next, public transportation became a top priority. An extensive bus system that featured rapid loading and unloading of passengers made Curitiba a model for transportation in the developing world. And lastly, a low-tech but effective recycling program has greatly reduced solid waste. Cities such as Curitiba demonstrate that designing with nature makes sense both ecologically and economically. To appreciate the diversity of environments Latin Americans must cope with when making development decisions, an understanding of the region's physical geography is important.

Western Mountains and Eastern Shields

Latin America is a region of diverse landforms, including high mountains and extensive upland plateaus. The movement of tectonic plates explains much of the region's basic topography, including the formation of its geologically young western mountain ranges, such as the Andes and the Volcanic Axis of Central America (Figure 4.7). This area is also geologically active, especially with regard to earthquakes that threaten people and damage property. In January 2001, for example, a major earthquake struck El Salvador, killing nearly 900 people and destroying more than 100,000 homes. In contrast, the Atlantic side of South America is characterized by humid lowlands interspersed with large upland plateaus called **shields**. The Brazilian shield is the largest, followed by the Patagonian and Guiana shields. Across these lowlands meander some of the great rivers of the world, including the Amazon.

Historically, the most important areas of settlement in tropical Latin America were not along the region's major rivers, but across its shields, plateaus, and fertile mountain valleys. In these places the combination of arable land, mild climate, and sufficient rainfall produced the region's most productive agricultural areas and its densest settlement. The Mexican Plateau, for example, is a massive upland area ringed by the Sierra Madre mountains. The southern end of the plateau is where the Valley of Mexico is located. Similarly, the elevated and well-watered basins of Brazil's southern mountains provide an ideal setting for agriculture. These especially fertile areas are able to support high population densities, so it is not surprising that the region's two largest cities, Mexico City and São Paulo, emerged in these settings. The Latin American highlands also lend a special character to the region. Lush tropical valleys nestled below snow-covered mountains hint at the

Mexico City earthquake, 1985. The 8.1 quake caused many high-rise buildings to collapse. Nearly 10,000 people died.

Caracas 1999. Unusually heavy rains in the coastal zone north of Caracas caused devastating landslides that knocked high-rises off their foundations and buried entire villages. The official death toll was 10,000, yet unofficial figures were four times that.

El Salvador earthquake. A 7.6 earthquake in January 2001 killed nearly 900 people and destroyed more than 100,000 homes.

Eruption of Nevado de Ruiz. In 1985 this volcano suddenly erupted and triggered a massive mudslide that buried the village of Armero, killing more than 20,000 people.

Altiplano. A large, elevated plateau between eastern and western branches of the Andes is found on the Bolivian–Peruvian border. Averaging more than 12,000 feet (3,600 meters) in elevation, this unique landform supports two large lakes, Titicaca and Poopó.

Paraná flooding. Unusually heavy rains during the 1997–98 El Niño resulted in record flooding of the Paraná River, forcing thousands of Argentines to flee their homes.

FIGURE 4.7 • PHYSICAL GEOGRAPHY OF LATIN AMERICA
Centered in the tropics but extending into the midlatitudes, Latin American landforms include mountains, shields, highland plateaus, vast river basins, and grassy plains. The region is geologically active, especially on the Pacific coast, where the region's highest mountains are found. The lower landforms on the Atlantic side include the Amazon and Plata river basins, as well as the highly eroded surfaces of the Brazilian, Guiana, and Patagonian shields.

diversity of ecosystems found near one another. The most dramatic of these highland areas, the Andes, runs like a spine down the length of the South American continent (see Figure 4.7).

The Andes   Beginning in northwestern Venezuela and ending at Tierra del Fuego, the Andes are relatively young mountains that extend nearly 5,000 miles (8,000 kilometers). They are an ecologically and geologically diverse mountain chain with some 30 peaks higher than 20,000 feet (6,000 meters). Many rich veins of precious

FIGURE 4.8 • BOLIVIAN ALTIPLANO The Altiplano is an elevated plateau straddling the Bolivian and Peruvian Andes. This stark, windswept land is inhabited mostly by Amerindians. Considered one of the poorer areas of the Andes, mining and pastoral activities dominate. *(Loren McIntyre/Woodfin Camp & Associates)*

FIGURE 4.9 • GUATEMALAN FARMER A peasant farmer works his small plot of subsistence and cash crops. Behind him rises one of the region's many volcanoes. The fertile volcanic soils, ample rainfall, and temperate climate of the Guatemala highlands have supported dense populations for centuries. *(Sean Sprague/Panos Pictures)*

metals and minerals are found in these mountains. In fact, the initial economic wealth of many Andean countries came from mining silver, gold, tin, copper, and iron.

Given the length of the Andes, the mountain chain is typically divided into northern, central, and southern components. In Colombia the northern Andes actually split into three distinct mountain ranges before merging near the border with Ecuador. High-altitude plateaus and snow-covered peaks distinguish the central Andes of Ecuador, Peru, and Bolivia. The Andes reach their greatest width here. Of special interest is the treeless high plain of Peru and Bolivia, called the **Altiplano**. The floor of this elevated plateau ranges from 11,800 feet (3,600 meters) to 13,000 feet (4,000 meters) in altitude, and it has limited usefulness for grazing (Figure 4.8). The highest peaks are found in the southern Andes, shared by Chile and Argentina, including the highest peak in the Western Hemisphere, Aconcagua, at almost 23,000 feet (6,958 meters).

The Uplands of Mexico and Central America The Mexican Plateau and the Volcanic Axis of Central America are the most important uplands in terms of settlement. Most major cities of Mexico and Central America are found here. The southern end of the Mexican plateau, the Mesa Central, contains a number of flat-bottomed basins interspersed with volcanic peaks. Mexico's megalopolis—a concentration of the largest population centers such as Mexico City, Guadalajara, and Puebla—is in the Mesa Central. (The Valley of Mexico, discussed earlier, is one of the basins of the Mesa Central.)

Along the Pacific coast of Central America lies a chain of volcanoes that stretches from Guatemala to Costa Rica. The Volcanic Axis of Central America is a handsome landscape of rolling green hills, elevated basins with sparkling lakes, and volcanic peaks. More than 40 volcanoes are found here, many of them still active, which have produced a rich volcanic soil that yields a wide variety of domestic and export crops. Most of Central America's population is also concentrated in this zone, in the capital cities or the surrounding rural villages. The bulk of the agricultural land is tied up in large holdings that produce beef, cotton, and coffee for export. Yet most of the farms are small subsistence properties that produce corn, beans, squash, and assorted fruits (Figure 4.9).

The Shields As mentioned earlier, South America has three major shields—large upland areas of exposed crystalline rock that are similar to upland plateaus found in Africa and Australia. (The Guiana shield will be discussed in Chapter 5.) The Brazilian and Patagonian shields vary in elevation between 600 and 5,000 feet (200 and 1,500 meters). The Brazilian shield is the larger and more important in terms of natural resources and settlement. Far from a uniform land surface, the Brazilian shield covers much of Brazil from the Amazon Basin in the north to the Plata Basin in the south. In the southeast corner of the plateau is the city of São Paulo, the largest urban conglomeration in South America. The other major population centers are on the coastal edge of the plateau, where large protected bays made the sites of Rio de Janeiro and Salvador attractive to Portuguese colonists. Finally, the Paraná basalt plateau, located on the southern end of the Brazilian shield, is famous for its fertile red soils (*terra roxa*), which yield coffee, oranges, and soybeans. So fertile is this area that the economic rise of São Paulo is attributed to the expansion of commercial agriculture, especially coffee, into this area (Figure 4.10).

River Basins and Lowlands

Three great river basins drain the Atlantic lowlands of South America: the Amazon, Plata, and Orinoco (see Figure 4.7). Within these basins are vast interior lowlands, less than 600 feet (200 meters) in elevation. Most of these lowlands are sparsely settled and offer limited agricultural potential except for grazing livestock. Yet the pressure to open new areas for settlement and to develop natural resources has created pockets of intense economic activity in the lowlands. Areas such as the Amazon and the Chaco have witnessed significant increases in resource extraction and settlement since the 1970s.

Amazon Basin   The Amazon drains an area of roughly 2.4 million square miles (6.1 million square kilometers), making it the largest river system in the world by volume and area and the second longest by length. Everywhere in the basin, rainfall is more than 60 inches (150 centimeters) a year and in many places more than 80 inches (200 centimeters) (Figure 4.11). The basin's largest city, Belém, averages close to 100 inches (250 centimeters) a year. Although there is no real dry season, there are definitely drier and wetter times of year, with August and September being the driest months. In the basin, rainfall is likely most days, but showers often pass quickly, leaving bright blue skies. The extent of this watershed and its hydrologic cycle is highlighted by the fact that 20 percent of all freshwater discharged into the oceans comes from the Amazon.

Since the Amazon Basin draws from nine countries, it would seem that this watershed would be an ideal network to integrate the northern half of South America. Ironically, compared with the other great rivers of the world, settlement in the basin continues to be sparse, in large part due to the poor quality of the forest soils. Active colonization of the Brazilian portion of the Amazon since the 1960s has made the population soar. According to the 2000 census, 12 million people live in the Brazilian Amazon, which is only 7 percent of the country's total population. Still, the population in the Amazonian states is increasing at nearly 4 percent a year.

Plata Basin   The region's second largest watershed begins in the tropics and discharges into the Atlantic in the midlatitudes. Three major rivers make up this system: the Paraná, the Paraguay, and the Uruguay. The Paraguay River and its tributaries drain the eastern Andes of Bolivia, the Brazilian shield, and the Chaco. The Paraná primarily drains the Brazilian uplands before the Paraguay River joins it in northern Argentina.

Unlike the Amazon, much of the Plata Basin is now economically productive through large-scale mechanized agriculture, especially soybean production. Arid areas such as the Chaco and seasonally flooded lowlands such as the Pantanal support livestock. The Plata Basin contains several major dams, including the region's largest hydroelectric plant, the Itaipú on the Paraná, which generates all of Paraguay's electricity and much of southern Brazil's. As agricultural output in this watershed grows, sections of the Paraná River are being dredged to improve the river's capacity for barge and boat traffic.

Climate

In tropical Latin America average monthly temperatures in settings such as Managua (Nicaragua), Quito (Equador), or Manaus (Brazil) show little variation (see Figure 4.11). Precipitation patterns, however, are variable and create distinct wet and dry seasons. In Managua, for example, January is typically a dry month and October is a wet one. The tropical lowlands of Latin America, especially east of the Andes, are usually classified as tropical humid climates that support forest or savanna, depending on the amount of rainfall. The region's desert climates are found along the Pacific coasts of Peru and Chile, Patagonia, northern Mexico, and the Bahia of Brazil. Thus, a city such as Lima, Peru, which is clearly in the tropics, averages only 1.5 inches (4 centimeters) of rainfall a year due to the extreme aridity of the Peruvian coast.

Midlatitude climates, with hot summers and cold winters, prevail in Argentina, Uruguay, and parts of Paraguay and Chile. Of course, the midlatitude temperature shifts in the Southern Hemisphere are the opposite of those in the Northern Hemisphere (cold Julys and warm Januarys). In the mountain ranges, complex climate patterns result from changes in elevation.

Perhaps the most talked-about weather phenomenon in the world, **El Niño** (named after the Christ child) occurs when a warm Pacific current arrives along the normally cold coastal waters of Ecuador and Peru in December, around Christmastime. This change in ocean temperature happens every few years and produces torrential rains, signaling the arrival of an El Niño year. The 1997–98 El Niño was especially bad. Devastating floods occurred in Peru and Ecuador. Heavy May rains in Paraguay and Argentina caused the Paraná River to rise 26 feet (8 meters) above normal.

FIGURE 4.10 • BRAZILIAN ORANGES   Most estate-grown oranges in Brazil are processed into frozen concentrate and exported. São Paulo and Paraná have some of the finest soils in Brazil. In addition to oranges, coffee and soybeans are widely cultivated. *(Stephanie Maze/NGS Image Collection)*

FIGURE 4.11 • CLIMATE MAP OF LATIN AMERICA Latin America includes the world's largest rain forest (Af) and driest desert (BWh), as well as nearly every other climate classification. Latitude, elevation, and rainfall play important roles in determining the region's climates. Note the contrast in rainfall patterns between humid Quito and arid Lima. *(Temperature and precipitation data from Pearce and Smith, 1984,* The World Weather Guide, *London: Hutchinson)*

Other than flooding, the less-talked-about result of El Niño is drought. While the Pacific coast of South and North America experienced record rainfall in the 1997–98 El Niño, Colombia, Venezuela, northern Brazil, Central America, and Mexico battled drought. In addition to crop and livestock losses, estimated to be in the billions of dollars, hundreds of brush and forest fires also left their mark on the region's landscape.

Population and Settlement: The Dominance of Cities

Great river basin civilizations like those in Asia never existed in Latin America. In fact, the great rivers of the region are surprisingly underused as areas of settlement or corridors for transportation. While the major population clusters of Central America and Mexico are in the interior plateaus and valleys, the interior lowlands of South America are relatively empty. Historically, the highlands supported most of the region's population during the pre-Hispanic and colonial eras. In the twentieth century population growth and immigration to the Atlantic lowlands of Argentina and Brazil, along with continued growth of Andean coastal cities such as Guayaquil, Barranquilla, and Maracaibo, have reduced the demographic importance of the highlands. Major highland cities such as Mexico City, Guatemala City, Bogotá, and La Paz still dominate their national economies, but the majority of large cities are on or near the coasts (Figure 4.12).

Like the rest of the developing world, Latin America experienced dramatic population growth in the 1960s and 1970s. In 1950 its population totaled 150 million people, which equaled the population of the United States at that time. By 1995 the population had tripled to 450 million; in comparison, the United States will probably not reach 300 million until 2010. Latin America outpaced the United States because its infant mortality rate declined and life expectancy soared, while its birthrate remained higher than that of the United States. In 1950 Brazilian life expectancy was only 43 years; by the 1980s it was 63. In fact, most countries in the region experienced a 15- to 20-year improvement in life expectancy between 1950 and 1980, which pushed up growth rates. Four countries account for 75 percent of the region's population: Brazil with more than 170 million, Mexico with 100 million, Colombia with 43 million, and Argentina with 38 million.

The Latin American City

A quick glance at the population map of Latin America shows a concentration of people in cities. One of the most significant demographic shifts has been the movement out of rural areas to cities, which began in earnest in the 1950s. In 1950 just one-quarter of the region's population was urban; the rest lived in small villages and the countryside. Today the pattern is reversed, with three-quarters of the population living in cities. In the most urbanized countries, such as Argentina, Chile, Uruguay, and Venezuela, more than 85 percent of the population lives in cities. This preference for urban life is attributed to cultural as well as economic factors. Under Iberian rule, people residing in cities had higher social status and greater economic opportunity. Initially, only Europeans were allowed to live in the colonial cities, but this exclusivity was not strictly enforced. Over the centuries colonial cities became the hubs for transportation and communication, making them the primary centers for economic and social activities.

Latin American cities are noted for high levels of **urban primacy**, a condition in which a country has a primate city three to four times larger than any other city in the country. Examples of primate cities are Lima,

FIGURE 4.12 • POPULATION MAP OF LATIN AMERICA
The concentration of population in urban and coastal settlements is evident in this map. Population density in central and southern Mexico, as well as Central America, is quite high. In South America, the majority of people live on or near the coasts, leaving the interior of the continent lightly populated.

Caracas, Guatemala City, Santiago, Buenos Aires, and Mexico City. Primacy is often viewed as a problem, since too many national resources are concentrated into one urban center. In three cases the growth of urbanized regions has led to the emergence of a megalopolis: the Mexico City–Puebla–Toluca–Cuernavaca area on the Mesa Central, the Niterói–Rio de Janeiro–Santos–São Paulo–Campinas axis in southern Brazil, and the Rosario–Buenos Aires–Montevideo–San Nicolás corridor in Argentina and Uruguay's lower Rio Plata Basin (see Figure 4.12).

Urban Form Latin American cities have a distinct urban form that reflects both their colonial origins and their present-day growth (Figure 4.13). Usually a clear central business district (CBD) exists in the old colonial core. Radiating out from the central business district is older middle- and lower-class housing found in the zones of maturity and *in situ* accretion (an area of mixed levels of housing and services). In this model, residential quality declines as one moves from the center to the periphery. The exception is the elite spine, a newer commercial and business strip

FIGURE 4.13 • LATIN AMERICAN CITY MODEL This urban model highlights the growth of Latin American cities and the class divisions within them. While the central business district (CDB), elite spine, and residential sectors may have excellent access to services and utilities, life in the zone of peripheral squatter settlements is much more difficult. In many Latin American cities, one-third of the population resides in squatter settlements. [*Model reprinted from Ford, 1996, "New and Improved Model of Latin American City Structure," Geographical Review 86(3), 437–40; photos by Rob Crandall/Rob Crandall, Photographer*]

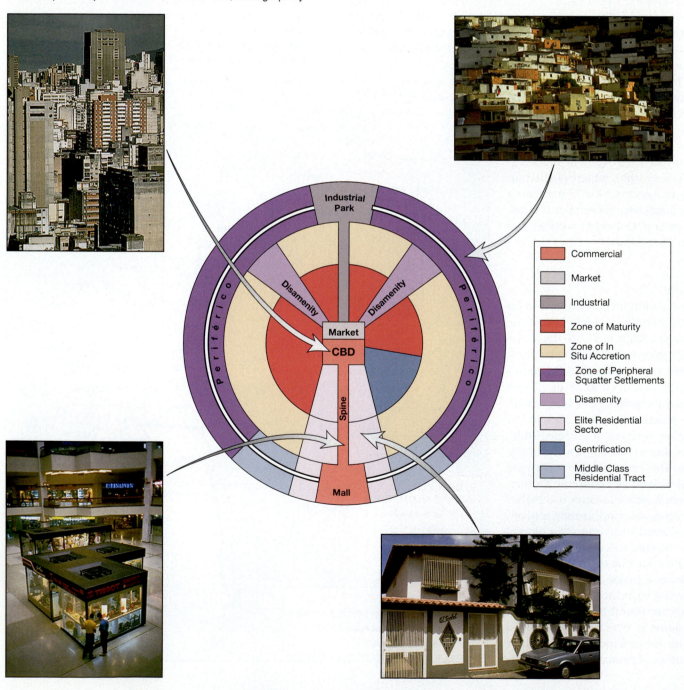

that extends from the colonial core to newer parts of the city. Along the spine one finds superior services, roads, and transportation. The city's best residential zones, as well as shopping malls, are usually on either side of the spine. Close to the elite residential sector, a limited area of middle-class tract housing is typically found. Most major urban centers also have a *periférico* (a ring road or beltway highway) that encircle the city. Industry is located in isolated areas of the inner city and in larger industrial parks outside the ring road.

Straddling the periférico is a zone of peripheral **squatter settlements** where many of the urban poor live in the self-built housing on land that does not belong to them. Services and infrastructure are extremely limited: Roads are unpaved, water is often trucked in, and sewer systems are nonexistent. The dense ring of squatter settlements that encircle Latin American cities reflect the speed and intensity with which these zones were created. In some cities more than one-third the population lives in these self-built homes of marginal or poor quality. These kinds of dwellings are found throughout the developing world, yet the practice of building one's home on the "urban frontier" has a longer history in Latin America than in most Asian and African cities. The combination of a rapid inflow of migrants (at times reaching 1,000 people per day), the inability of governments to meet pressing housing needs, and the eventual official recognition of many of these neighborhoods with land titles and utilities meant that this housing strategy was rarely discouraged. Each successful settlement on the urban edge encouraged more.

Rural-to-Urban Migration   Beginning in the 1950s, peasants began to pour into the cities of Latin America in a process referred to as **rural-to-urban migration**. As rural jobs disappeared, many households would send family members to the cities for employment as domestics, construction workers, artisans, and vendors. Once in the cities, rural migrants generally found conditions better, especially access to education, health care, electricity, and clean water. It was not poverty alone that drove people out of rural areas, but individual choice and an urban preference. Migrants believed in, and often realized, greater opportunities in cities, especially the capital cities. Those who came were usually young (in their twenties) and better educated than those who stayed behind. Women slightly outnumbered men in this migrant stream. The move itself was made easier by extended family networks formed by earlier migrants who settled in distinct areas of the city and aided new arrivals. The migrants maintained their links to their rural communities by periodically sending money and making return visits.

Patterns of Rural Settlement

Throughout Latin America a distinct rural lifestyle exists, especially among peasant subsistence farmers. While the majority of people live in cities, approximately 130 million people do not. In Brazil alone at least 30 million people live in rural areas. Interestingly, the absolute number of people living in rural areas today is roughly equal to the number in the 1960s. Yet rural life has definitely changed. The links between rural and urban areas are much improved, making rural people less isolated. In addition to village-based subsistence production, in most rural areas highly mechanized capital-intensive farming occurs. Much like the region's cities, the rural landscape is divided by extremes of poverty and wealth. A source of social and economic tension in the countryside is the uneven distribution of arable land.

Rural Landholdings   The control of land in Latin America was the basis for political and economic power. Historically, colonial authorities granted large tracts of land to the colonists, who were also promised the service of Indian laborers. These large estates typically took up the best lands along the valley bottoms and coastal plains. The owners were often absentee landlords, spending most of their time in the city and relying on a mixture of hired and slave labor to run their rural operations. Passed down from one generation to the next, many estates can trace their ownership back several centuries. The establishment of large blocks of estate land meant that peasants were denied access to territory of their own, so they were forced to work for the estates. This long-observed practice of maintaining large estates is called **latifundia**.

Although the pattern of estate ownership is well documented, peasants have always farmed small plots for their subsistence. This practice of **minifundia** can lead to permanent or shifting cultivation. Small farmers typically plant a mixture of crops for subsistence as well as for trade. Peasant farmers in Colombia or Costa Rica, for example, grow corn, fruits, and various vegetables alongside coffee bushes that produce beans for export. Strains on the minifundia system occur when rural populations grow and land becomes scarce, forcing farmers to divide their properties into smaller and less-productive parcels.

Much of the turmoil in twentieth-century Latin America surrounded the question of land, with peasants demanding its redistribution through the process of **agrarian reform**. Governments have addressed these concerns in different ways. The Mexican Revolution in 1910 led to a system of communally held lands called *ejidos*. In the 1950s Bolivia crafted agrarian reform policies that led to the government appropriation of estate lands and redistributing them to small farmers. As part of the Sandinista revolution in Nicaragua in 1979, lands were taken from the political elite and converted into collective farms. Each of these programs met with resistance and proved to be politically and economically difficult. Eventually the path chosen by most governments was to make frontier lands available to land-hungry peasants.

Agricultural Frontiers The creation of agricultural frontiers served several purposes: providing peasants with land, tapping unused resources, and filling in blank spots on the map with settlers. Although the dominant demographic trend has been a rural-to-urban movement, an important rural-to-rural flow has changed undeveloped areas into agricultural communities (Figure 4.14).

The opening of the Brazilian Amazon for settlement was the most ambitious frontier colonization scheme in the region. In the 1960s Brazil began its frontier expansion by constructing several new Amazonian highways, a new capital (Brasília), and state-sponsored mining operations. It was the Brazilian military who directed the opening of the Amazon to provide an outlet for landless peasants and to extract the region's many resources. Yet the generals' plans did not deliver as intended. Throughout the basin, nutrient-poor tropical soils could not support permanent agricultural colonies. Government-promised land titles, agricultural subsidies, and credit were slow to reach small farmers. Instead, too much money went to subsidizing large cattle ranches through tax breaks and improvement deals where "improved" land meant cleared land. Despite a concerted effort by Brazilian officials to settle the entire region, most commercial activities and residents are concentrated in a few sites. Four times more people lived in the Amazon in the 1990s than in the 1960s. The 12 million people living in the Brazilian Amazon in 2000 is projected to reach 15 million by 2010; thus, increased human modification of the Brazilian Amazon is inevitable (Figure 4.15).

Population Growth and Movements

The high growth rates in Latin America throughout the twentieth century are attributed to natural increase as well as immigration. The 1960s and 1970s were decades

FIGURE 4.14 • PRINCIPAL LATIN AMERICAN MIGRATION FLOWS Internal and international migrations have opened frontier zones and created transnational communities. Over the past two decades, the flow of Latin Americans to the United States has grown. In the 2000 U.S. Census, 35 million people of Hispanic ancestry were counted. Most of these people were either born in or have ancestral ties to Latin America. (Adapted from Clawson, 2000, Latin America and the Caribbean: Lands and People, 2nd ed., Boston: McGraw-Hill)

of tremendous growth resulting from high fertility rates and increasing life expectancy. In the 1960s, for example, a typical Latin American woman had six or seven children. By the 1980s family sizes were half this level. A number of factors explain this: more urban families, which tend to be smaller than rural ones; increased participation of women in the workforce; higher education levels of women; state support of family planning; and better access to birth control. The exceptions to this trend are the poor and more rural countries, such as Guatemala and Bolivia, where the average woman has four or five children. Cultural factors may also be at work, as Amerindian peoples in the region tend to have more children.

Even with family sizes shrinking and nearing replacement rates in Uruguay and Chile, there is built-in potential for continued growth because of the relative demographic youth of these countries. The average percentage of the population below the age of 15 is 32 percent. In North America that same group is 21 percent of the population, and in western Europe it is just 17 percent. This means that a proportionally larger segment of the population has yet to enter into its childbearing years.

Waves of immigrants into Latin America and migrant streams within Latin America have influenced population size and patterns of settlement. Beginning in the late nineteenth century, new immigrants from Europe and Asia added to the region's size and ethnic complexity. Important population shifts within countries have also occurred in recent decades, as witnessed by the growth of Mexican border towns and the new settlements in eastern Bolivian lowlands. In an increasingly globalized economy, even more Latin Americans live and work outside the region, especially in the United States.

European Migration   After gaining their independence from Iberia, Latin America's new leaders sought to develop their territories through immigration. Firmly believing that "to govern is to populate," many countries set up immigration offices in Europe to attract hardworking peasants to till the soils and "whiten" the **mestizo** (people of mixed European and Indian ancestry) population. The Southern Cone countries of Argentina, Chile, Uruguay, Paraguay, and southern Brazil were the most successful in attracting European immigrants from the 1870s until the depression of the 1930s. During this period, some 8 million Europeans arrived (more than came during the entire colonial period), with Italians, Portuguese, Spaniards, and Germans being the most numerous.

Asian Migration   Less well-known are the Asian immigrants who also arrived during the late nineteenth and twentieth centuries. Although considerably fewer, over time they established an important presence in the large cities of Brazil, Peru, Argentina, and Paraguay. Beginning in the mid-nineteenth century, the Chinese and Japanese who settled in Latin America were contracted to work on the coffee estates in southern Brazil and the sugar estates and coastal mines of Peru. In the 1990s a Japanese-Peruvian, Alberto Fujimori, was president of Peru. Between 1908 and 1978 a quarter-million Japanese immigrated to Brazil; today the country is home to 1.3 million people of Japanese descent. Initially, most Japanese were landless laborers, yet by the 1940s they had built up enough wealth so that three-quarters of the migrants had their own land in the rural areas of São Paulo and Paraná states. As a group, the Japanese have been closely associated with the expansion of soybean and orange production. Increasingly, second- and third-generation Japanese have taken professional and commercial jobs in Brazilian cities; many of them have married outside their ethnic group and are losing their fluency in Japanese. Even with intermarriage, they continue to maintain a strong sense of their ethnic identity, with more than two-thirds of Japanese-Brazilians speaking both Portuguese and some Japanese (Figure 4.16).

Latino Migration and Hemispheric Change   Movement within Latin America and between Latin America and North America has had a significant impact on sending and receiving communities alike. Within Latin America, international migration is shaped by shifting economic and political realities. For example, Venezuela's oil

FIGURE 4.15 • SETTLERS IN THE AMAZON   Newly arrived settlers in the Brazilian Amazon have cleared a patch of land from the forest and constructed their new home out of wood and palm. Thousands of such homesteads are found throughout the Amazon. *(Brian Godfrey)*

FIGURE 4.16 • JAPANESE-BRAZILIANS   Retired Japanese-Brazilians play the board game "Go" in a city plaza. Most Brazilians of Japanese ancestry live in the southern states of São Paulo, Paraná, and Santa Catarina. The majority are descended from Japanese who immigrated to Brazil in the first half of the twentieth century. *(Gary Payne/Getty Images, Inc.)*

FIGURE 4.17 • MEXICAN–U.S. BORDER CROSSING
Mexican day workers cross the border into El Paso, Texas, from Ciudad Juárez. Mexicans have long used these busy border crossings to enter the United States. Other Latino immigrants, especially from Central America, now join them. *(Rob Crandall/ Rob Crandall, Photographer)*

wealth during the 1960s and 1970s attracted between 1 and 2 million Colombian immigrants who tended to work as domestics or agricultural laborers. Argentina has long been a destination for Bolivian and Paraguayan laborers. And farmers in the United States have depended on Mexican laborers for most of the twentieth century (see Figure 4.14).

Political turmoil also sparked waves of international migrants. The bloody civil wars in El Salvador and Guatemala, for example, sent waves of refugees into neighboring countries, such as Mexico and the United States. Since most countries in the region have democratically elected governments, most of today's immigrants are classified as economic migrants, not political refugees seeking a safe haven.

Presently, Mexico is the largest country of origin of legal immigrants to the United States, followed by the Philippines, China, Korea, and Vietnam (Figure 4.17). Twenty-two million people claimed Mexican ancestry in the 2000 U.S. Census, of whom approximately 8 million were immigrants. Mexican labor migration to the United States dates back to the late 1800s when relatively unskilled labor was recruited to work in agriculture, mining, and railroads. Today roughly 60 percent of the Hispanic population (both foreign-born and native-born) in the United States claims Mexican ancestry. Mexican immigrants are most concentrated in California and Texas, but increasingly they are found throughout the country. Although Mexicans continue to have the greatest presence among Latinos in the United States, the number of immigrants from El Salvador, Guatemala, Nicaragua, Colombia, Ecuador, and Brazil has steadily grown. The 2000 Census counted 35 million Hispanics in the United States (both foreign- and native-born). Most of this population has ancestral ties with peoples from Latin America and the Caribbean (see Chapter 5 on Caribbean migration).

Patterns of Cultural Coherence and Diversity: Repopulating a Continent

The Iberian colonial experience brought political and cultural unity to Latin America that makes it recognizable today as a world region. Yet this was not a simple transplanting of Iberia across the Atlantic. Often a process unfolded in which European and Indian traditions blended as native groups were added into either the Spanish or the Portuguese empires. In some areas such as southern Mexico, Guatemala, Bolivia, Ecuador, and Peru, Indian cultures have showed remarkable resilience, as evidenced by the survival of Amerindian languages. Yet the prevailing pattern is one of forced assimilation in which European religion, languages, and political organization were imposed upon surviving Amerindian societies. Later other cultures, especially more than 10 million African slaves, added to the cultural mix of Latin America, the Caribbean, and North America. The legacy of the African slave trade will be examined in greater detail in Chapters 5 and 6. For Latin America, perhaps the single most important factor in the dominance of European culture was the demographic collapse of native populations.

The Decline of Native Populations

It is hard to grasp the enormity of cultural change and human loss due to this encounter between two worlds (the Americas and Europe). Throughout the region archaeological sites are reminders of the complexity of Amerindian civilizations prior to contact with Europe. Dozens of stone temples found throughout Mexico and Central America, where the Mayan and Aztec civilizations flourished, attest to the ability of these societies to thrive in the area's tropical forests and upland plateaus. In the Andes, stone terraces built by the Incas are still being used by Andean farmers; earthen platforms for village sites and raised fields for agriculture are still being discovered and mapped. Ceremonial centers such as Cuzco—the core of the great Incan empire that was nearly leveled by the Spanish—and the Incan site of Machu Picchu—unknown to most of the world until U.S. archaeologist Hiram Bingham investigated the site in the early 1900s—are evidence of the complexity of precontact civilizations (Figure 4.18). The Spanish, too, were impressed by the sophistication and wealth they saw around them, especially in Tenochtitlán, where Mexico City sits today. Tenochtitlán was the political and ceremonial center of the Aztecs, supporting a complex metropolitan area with some 300,000 residents. The largest city in Spain at the time was considerably smaller.

The Demographic Toll   The most telling figures of the impact of European expansion in Latin America are demographic. It is widely believed that the precontact Americas had 54 million inhabitants; by comparison, western Europe in 1500 had approximately 42 million. Of the 54 million, about 47 million were in what is now Latin America, and the rest were in North America and the Caribbean. By 1650, after a century and a half of colonization, the indigenous population was one-tenth its precontact size. The human tragedy of this population loss is difficult to comprehend. The relentless elimination of 90 percent of the native population was largely caused by epidemics of influenza and smallpox, but warfare, forced labor, and starvation due to a collapse of food production systems also contributed to the death rate.

Indian Survival   Presently, Mexico, Guatemala, Ecuador, Peru, and Bolivia have the largest indigenous populations. Not surprisingly, these areas had the densest native populations at contact. Indigenous survival also occurs in isolated settings where the workings of national and global economies are slow to break through.

In many cases Indian survival comes down to one key resource—land. Indigenous peoples who are able to maintain a territorial home, formally through land title or informally through long-term occupancy, are more likely to preserve a distinct ethnic identity. Because of this close association between identity and territory, native peoples are increasingly insisting on a recognized space within their countries. These efforts to define indigenous territory are seldom welcomed by the state.

Today the state of Panama recognizes four *comarcas* that encompass six native groups; the most successful is Comarca San Blas on the Caribbean coast, where 40,000 Kuna live. A *comarca* is a loosely defined territory similar to a province or a homeland. What distinguishes the *comarca* from an Indian reservation is that the native people have defined the territory and assert political and resource control within its boundaries. From Amazonia to the highlands of Chiapas in Mexico, many native groups are demanding formal political and territorial recognition as a means to address centuries of injustice. Whether these efforts will actually reshape political and cultural space for Latin America's Amerindians is still uncertain.

Patterns of Ethnicity and Culture

The Indian demographic collapse enabled Spain and Portugal to reshape Latin America into a European likeness. Yet instead of a neo-Europe rising in the tropics, a complex ethnic blend evolved. Beginning with the first years of contact, unions between European sailors and Indian women began the process of racial mixing that over time became a defining feature of the region. The courts of Spain and Portugal officially discouraged racial mixing, but such positions could not be realistically enforced in the colonial territories.

After generations of intermarriage, four broad racial categories resulted: *blanco* (European ancestry), *mestizo* (mixed ancestry), *indio* (Indian ancestry), and *negro* (African ancestry). The *blancos* (or Europeans) continue to be well represented among the elites, yet the vast majority of people are of mixed racial ancestry. *Dia de la Raza*, the region's observance of Columbus Day, recognizes the emergence of a new *mestizo* race as the legacy of European conquest. Throughout Latin America, more than other regions of the world, miscegenation (or racial mixing) is the norm, which makes the process of mapping racial or ethnic groups especially difficult.

Languages   Roughly two-thirds of Latin Americans are Spanish speakers, and one-third speak Portuguese. These colonial languages were so widespread by the nineteenth century that they were the unquestioned languages of government and

FIGURE 4.18 • MACHU PICCHU   This complex, ancient city near Cuzco was not known to the outside world until the early 1900s. Located above the humid Urubamba River valley, Machu Picchu is one of Peru's major tourist destinations. *(Rob Crandall/ Rob Crandall, Photographer)*

FIGURE 4.19 • LANGUAGE MAP OF LATIN AMERICA The dominant languages of Latin America are Spanish and Portuguese. Nevertheless, there are significant areas in which native languages still exist and, in some cases, are recognized as official languages. Smaller language groups exist in Central America, the Amazon Basin, and southern Chile. *(Adapted from the Atlas of the World's Languages, 1994, New York: Routledge)*

education for the newly independent Latin American republics. In fact, until recently many countries actively discouraged, and even repressed, Indian tongues. It took a constitutional amendment in Bolivia in the 1990s to legalize native-language instruction in primary schools and to recognize the country's multiethnic heritage (more than half the population is Indian, and Quechua, Aymara, and Guaraní are widely spoken) (Figure 4.19).

Dominant/Official* Languages
- Spanish
- Portuguese

Indigenous Languages
1 Aymara
2 Embera
3 Garifuna
4 Guaraní
5 Quechua
6 Kuna
7 Mapuche
8 Mayan
9 Miskito
10 Mixtec
11 Nawan/Spanish
12 Pemong
13 Zapotec
14 Wahiro
15 Yanomama
Scattered indigenous language communities

*Multiple Official Languages:
Bolivia: Spanish, Quechua, Aymara, Guaraní
Peru: Spanish, Quechua

Because Spanish and Portuguese dominate, there is a tendency to overlook the influence of indigenous languages in the region. Mapping the use of native languages, however, reveals important areas of Indian resistance and survival. In the Central Andes of Peru, Bolivia, and southern Ecuador, more than 10 million people still speak Quechua and Aymara, along with Spanish. In Paraguay and lowland Bolivia there are 4 million Guaraní speakers, and in southern Mexico and Guatemala at least 6 to 8 million speak Mayan languages. Small groups of native-language speakers are found scattered throughout the sparsely settled interior of South America and the more isolated forests of Central America, but many of these languages have fewer than 10,000 speakers.

Blended Religions Like language, the Roman Catholic faith appears to have been imposed upon the region without challenge. Most countries report 90 percent or more of their population as Catholic. Every major city has dozens of churches, and even the smallest village maintains a graceful church on its central square (Figure 4.20). In some countries, such as El Salvador and Uruguay, a sizable portion of the population attends Protestant churches, but the Catholic core of this region is still intact.

Exactly what native peoples absorbed of the Christian faith is unclear. Throughout Latin America **syncretic religions**, blends of different belief systems, enabled animist practices to be included in Christian worship. These blends took hold and endured, in part because Christian saints were easy replacements for pre-Christian gods and because the Catholic Church tolerated local variations in worship as long as the process of conversion was under way. The Mayan practice of paying tribute to spirits of the underworld seems to be replicated today in Mexico and Guatemala through the practice of building small cave shrines to favorite Catholic saints and leaving offerings of fresh flowers and fruits. One of the most celebrated religious symbols in Mexico is the Virgin of Guadeloupe, a dark-skinned virgin seen by an Indian shepherd boy who became the patron saint of Mexico.

Syncretic religious practices also evolved and endured among African slaves. By far the greatest concentration of slaves was in the Caribbean, where slaves were used to replace the indigenous population, which was wiped out by disease (see Chapter 5). Within Latin America the Portuguese colony of Brazil received the most Africans—at least 4 million. In Brazil, where the volume and the duration of the slave trade were the greatest, the transfer of African-based religious and medical systems is most evident.

The blend of Catholicism with African traditions is most obvious in the celebration of carnival, Brazil's most popular festival and one of the major components of Brazilian national identity. The three days of carnival known as the Reign of Momo combines pre-Lenten celebrations with African musical traditions represented by the rhythmic samba bands. By the 1960s, carnival became an important symbol for Brazil's multiracial national identity. Today the festival—which is most associated with Rio de Janeiro—draws thousands of participants from all over world.

FIGURE 4.20 • THE CATHOLIC CHURCH Churches, such as the Dolores Church in Tegucigalpa, are important religious and social centers. The majority of people in Latin America define themselves as Catholic. Many churches built in the colonial era are valued as architectural treasures and are beautifully preserved. *(Rob Crandall/Rob Crandall, Photographer)*

Geopolitical Framework: Redrawing the Map

Latin America's colonial history, more than its present condition, unifies this region geopolitically. For the first 300 years after Columbus's arrival, Latin America was a territorial prize sought by various European countries but effectively settled by Spain and Portugal. By the nineteenth century the independent states of Latin America had formed, but they continued to experience foreign influence and, at times, overt political pressure, especially from the United States. At other times a more neutral hemispheric vision of American relations and cooperation has held sway, represented by the formation of the **Organization of American States (OAS)**. The present organization was officially formed in 1948, but its origins date back to 1889. Yet there is no doubt that U.S. policies toward trade, economic assistance, political development, and at times military intervention are often seen as undermining the independence of these states.

Within Latin America there have been cycles of intraregional cooperation and antagonism. Neighboring countries have fought over territory, closed borders, imposed high tariffs, and cut off diplomatic relations. The 1990s witnessed a revival in the trade block concept with the formation of **Mercosur** (the Southern Cone Common Market, which includes Brazil, Uruguay, Argentina, and Paraguay as full members and Bolivia and Chile as associate members) and NAFTA (Mexico, the United States, and Canada). It is possible that, as these economic ties strengthen, these trade blocks could form the basis for a new alignment of political and economic interests in the region.

Iberian Conquest and Territorial Division

Because it was Christopher Columbus who claimed the Americas for Spain, the Spanish were the first active colonial agents in the Western Hemisphere. In contrast, the Portuguese presence in the Americas was the result of the **Treaty of Tordesillas**, brokered by the pope in 1493-94. By that time Portuguese navigators had charted much of the coast of Africa in an attempt to find a water route to the Spice Islands (Moluccas) in Southeast Asia. With the help of Christopher Columbus, Spain sought a western route to the Far East. When Columbus discovered the Americas, Spain and Portugal asked the pope to settle how these new territories should be divided. Without consulting other European powers, the pope divided the Atlantic world in half—the eastern half containing the African continent was awarded to Portugal, the western half with most of the Americas was given to Spain. The line of division established by the treaty actually cut through the eastern part of South America, placing it under Portuguese rule. This treaty was never recognized by the French, English, or Dutch, who also claimed territory in the Americas, but it did provide the legal justification for the creation of Portuguese Brazil. This state would later become the largest and most populous in Latin America (Figure 4.21).

Six years after the treaty was signed, Portuguese navigator Alvares Cabral accidentally reached the coast of Brazil on a voyage to southern Africa. The Portuguese soon realized that this territory was on their side of the Tordesillas line. Initially they were unimpressed by what Brazil had to offer; there were no spices or major native settlements. Quickly, however, they came to appreciate the utility of the coast as a provisioning site as well as a source for brazilwood, used to produce a valuable dye. Portuguese interest in the territory intensified in the late sixteenth century with the development of sugar estates and the expansion of the slave trade and in the seventeenth century with the discovery of gold in the Brazilian interior.

Spain, in contrast, aggressively pursued the conquest and settlement of its new American territories from the very start. After discovering little gold in the Caribbean, by the mid-sixteenth century Spain's energy was directed toward developing the silver resources of Central Mexico and the Central Andes (most notably Potosí in Bolivia). Gradually the economy diversified to include some agricultural exports, such as cacao (for chocolate) and sugar, as well as a variety of livestock. In terms of foodstuffs, the colonies were virtually self-sufficient. Manufacturing, however, was forbidden in the Spanish American colonies in order to keep them dependent on Spain.

Revolution and Independence It was not until the 1800s, with a rise of revolutionary movements between 1810 and 1826, that Spanish authority on the mainland was challenged. Ultimately European elites born in the Americas gained control, displacing those leaders loyal to the crown. In Brazil the evolution from Portuguese colony to independent republic was a slower and less violent process that spanned eight decades (1808-89). In the nineteenth century Brazil was declared a separate kingdom from Portugal with its own king, and later became a republic.

FIGURE 4.21 • SHIFTING POLITICAL BOUNDARIES The evolution of political boundaries in Latin America began with the 1494 Treaty of Tordesillas, which gave much of the Americas to Spain and a slice of South America to Portugal. The larger Spanish territory was gradually divided into viceroyalties and audiencias, which formed the basis for many modern national boundaries. The 1830 borders of these newly independent states were far from fixed. Bolivia would lose its access to the coast; Peru would gain much of Ecuador's Amazon; and Mexico would be stripped of its northern territory by the United States. *(From Lombardi, Cathryn L., and John V. Lombardi. Latin American History: A Teaching Atlas. © 1993. Reprinted by permission of the University of Wisconsin Press.)*

Later, the territorial division of Spanish and Portugese America into administrative units provided the legal basis for the modern states of Latin America (see Figure 4.21). The Spanish colonies were first divided into two viceroyalties (the administrative units of New Spain and Peru) and within these were various subdivisions that later became the basis for the modern states. (In the eighteenth century the Viceroyalty of Peru, which included all of Spanish South America, was divided to form three viceroyalties: La Plata, Peru, and New Granada.) Unlike Brazil, which evolved from a colony into a single republic, the former Spanish colonies experienced fragmentation in the nineteenth century.

Today the former Spanish mainland colonies include 16 states (plus three Caribbean islands), with a total population of 340 million. If the Spanish colonial territory had remained a unified political unit, it would now have the third largest population in the world, following China and India.

Persistent Border Conflicts As the colonial administrative units turned into states, it became clear that the territories were not clearly demarcated, especially the borders that stretched into the sparsely populated interior of South America. This would later become a source of conflict as new states struggled to define their territorial boundaries. Numerous border wars erupted in the nineteenth and twentieth centuries, and the map of Latin America had to be redrawn many times. Some of the more noted conflicts were the War of the Pacific (1879–82), in which Chile expanded to the north and Bolivia lost its access to the Pacific; warfare between Mexico and the United States in the 1840s, which resulted in the present border under the Treaty of Hidalgo (1848) and Mexico's loss of what became the southwestern United States; and the War of the Triple Alliance (1864–70), the bloodiest war of the postcolonial period, which occurred when Argentina, Brazil, and Uruguay allied themselves to defeat Paraguay in its claim to control the upper Paraná River Basin. In the 1980s Argentina lost a war with Great Britain over control of the Falkland, or Malvinas, Islands in the South Atlantic. And as recently as 1998, Peru and Ecuador fought over a disputed boundary in the Amazon Basin (Figure 4.22).

The Trend Toward Democracy Early in the twenty-first century, most of the 17 countries in this region will celebrate their bicentennials. Compared with most of the developing world, Latin Americans have been independent for a long time. Yet political stability is not a characteristic of the region. Among the countries in the region, some 250 constitutions have been written since independence, and military takeovers have been alarmingly frequent. Since the 1980s, however, the trend has been toward democratically elected governments, the opening of markets, and broader public participation in the political process. Where dictators once outnumbered elected leaders, by the 1990s each country in the region had a democratically elected president. (Cuba, the one exception, will be discussed in Chapter 5.)

Democracy may not be enough for the millions frustrated by the slow pace of political and economic reform. In survey after survey, Latin Americans reveal their dissatisfaction with politicians and governments. Most of the newly elected democratic leaders are also free-market reformers who are quick to eliminate state-backed social safety nets, such as food subsidies, government jobs, and pensions. Many of the poor and middle class have grown doubtful about whether this brand of democracy could make their lives better. The political left, however, has yet to produce an alternative to privatization and market-driven policies. For now, the status quo continues, although popular frustration with falling incomes, rising violence, and continual underemployment are a recipe for political instability (Figure 4.23).

FIGURE 4.22 • BORDER DISPUTES AND TRADE BLOCKS IN LATIN AMERICA Of the five economic trade blocks shown Mercosur and NAFTA are the most dynamic. In fact, several members of the Andean Group have expressed interest in joining Mercosur, while members of the Central American Common Market would like to be included in NAFTA. Most of Latin America's disputed boundaries are not being actively challenged. Yet these long-standing border disputes often contribute to tense relations between neighbors. *(Data from Child, 1985, Geopolitics and Conflict in South America, New York: Praeger Press; and Allcock, 1992, Border and Territorial Disputes, 3rd ed., Harlow, Essex, UK: Longman Group)*

Regional Organizations

At the same time democratically elected leaders struggle to address the pressing needs of their countries, political developments at the supranational and subnational levels pose new challenges to their authority. The most discussed **supranational organizations** (governing bodies that include several states) are the trade blocks. **Subnational organizations** (groups that represent areas or people within the state) often form along ethnic or ideological lines (such as socialist or communist groups) and can provoke serious internal divisions. Native groups seeking territorial recognition (such as the Kuna in Panama) and insurgent groups (such as the FARC in Colombia) have challenged the authority of the states. Finally, the financial and political force of drug cartels (mafia-like organizations in charge of the drug trade) goes beyond state boundaries and undermines judicial systems.

Trade Blocks Beginning in the 1960s, regional trade alliances were formed in an effort to promote internal markets and reduce trade barriers. The Latin American Free Trade Association (LAFTA), the Central American Common Market (CACM), and the Andean Group have existed for decades, but their ability to influence economic trade and growth is limited at best. In the 1990s Mercosur and NAFTA emerged as supranational structures that could influence development (see Figure 4.22). For Latin America, the lessons of Mercosur and NAFTA are causing politicians to rethink the value of regional trade.

Mercosur was formed in 1991 with Brazil and Argentina, the two largest economies in South America, and the smaller states of Uruguay and Paraguay as members. Since its formation, trade among these countries has grown tremendously, so much so that Chile and Bolivia have now joined the group as associate members. This is significant in two ways: It reflects the growth of these economies and the willingness to put aside old rivalries (especially long-standing antagonisms between Argentina and Brazil) for the economic benefits of cooperation. The size and productivity of this market have not gone unnoticed. Mercosur's leaders negotiate directly with the European Union to develop separate trade agreements. With Chile on board, expansion into Asian-Pacific markets is also likely.

NAFTA took effect in 1994 as a free trade area that would gradually eliminate tariffs and ease the movement of goods among the member countries (Mexico, the United States, and Canada). NAFTA has increased intraregional trade (estimated at more than $700 billion in 1999), but there is considerable controversy about costs to the environment and to employment (see Chapter 3). NAFTA did prove, however, that a free trade area combining industrialized and developing states was possible. Moreover, it stimulated interest in the hemispheric vision of the Free Trade Area of the Americas (FTAA), which was first proposed in 1994 at the Miami Summit of the Americas. Thirty-four states from Canada, the United States, and all of Latin America (except Cuba, which was not invited) have agreed to conclude negotiations by 2005. If an agreement is reached, FTAA will be the largest free trade area in the world, including more than 800 million people. Yet criticism of such an agreement is strong and a successful treaty by 2005 is uncertain.

Insurgencies and Drug Traffickers Guerrilla groups such as the FARC (Revolutionary Armed Forces of Colombia) in Colombia have controlled large territories of their countries through the support of those loyal to the cause, along with theft, kidnapping, and violence. The FARC, along with the ELN (National Liberation Army) gained wealth through the drug trade in the 1990s, and there is now a recognized FARC-controlled zone in the southern Llanos and a proposed ELN-zone in the central Magdalena river basin. The level of violence in Colombia has escalated further with the rise of paramilitary groups—armed private groups that terrorize those sympathetic to insurgency. The paramilitary groups are blamed for some 6,000 politically motivated murders in 2000. Colombia also has the distinction of the highest murder rate in the world, ten times that of the United States. Perhaps as many as 2 million Colombians have been internally displaced by violence since the late 1980s, most

FIGURE 4.23 • PROTESTING POLITICAL VIOLENCE Women protestors demand justice for the alleged killing of "disappeared" labor organizers in Honduras. Political violence at the hands of state, paramilitary, and guerrilla organizations affects the lives of many Latin Americans. *(Rob Crandall/Rob Crandall, Photographer)*

FIGURE 4.24 • COCA ERADICATION Removing coca plant by hand has been practiced throughout the Bolivian Chapare and the Upper Huallaga Valley of Peru. In this photo an eradication line is destroying a coca field near Tinga Maria, Peru. As part of the U.S.-backed war on drugs, aid monies have gone to try and reduce coca production in the Andes.

fleeing rural areas for towns and cities. More alarming still is the number of Colombians who are leaving the country. Between 1996 and 2001, nearly 1 million Colombians fled to Venezuela, the United States, Ecuador, and Central America.

The drug trade is often seen as the root of many of the region's problems, but this illegal trade also generates billions of dollars for Latin Americans (from the small coca farmer in Bolivia to the cartel leader in Colombia). The drug trade began in earnest in the 1970s with Colombia at its center. By the 1980s the Medellín Cartel was a powerful and wealthy crime syndicate that used drug-dollars to bribe or murder anyone who got in its way. The organization was weakened by the death of its leader, Pablo Escobar, in 1993, but by then the Cali Cartel was poised to take control. Its power was reduced by arrests of key leaders in 1995. No one cartel dominates the drug trade today; instead, it has decentralized into dozens of smaller productive syndicates.

Initially, most Latin governments cared little about controlling the drug trade as it brought in much-needed hard currency. Within the region, drug consumption was scarcely a problem. Some drug lords even became popular folk heroes, spending large amounts of money on housing, parks, and schools for their communities. The social costs of the drug trade to Latin America became evident by the 1980s when the region was crippled by a badly damaged judicial system. By paying off police, the military, judges, and politicians, the drug syndicates exert incredible political power that threatens the social order of the states in which they work. Years of counternarcotics work have done little to reduce the overall flow of drugs to North America and Europe, but the programs have changed the areas of production (Figure 4.24).

Economic and Social Development: Dependent Economic Growth

Most Latin American economies fit into the broad middle-income category set by the World Bank. Clearly part of the developing world, their people are much better off than those in Sub-Saharan Africa, South Asia, and China. Still, the economic contrasts are sharp both between states and within them. Generally, the Southern Cone states (including southern Brazil and excluding Paraguay) are the richest. Argentina leads the way with a per capita GNI of $7,550, followed by Uruguay at $6,220. Brazil and Mexico, the largest countries in Latin America, have per capita GNIs around $4,400. The poorest countries in the region lie in Central America and the Andes. In 2001 Nicaragua, Honduras, and Bolivia had per capita income figures of less than $1,000. Even with its middle-income status, extreme poverty is evident throughout the region; nearly one-third of all Latin Americans live on less than $2 a day.

Development Strategies

Development policies of the 1960s and 1970s increased industrialization and infrastructural development. Latin America experimented with various policies, from closed economies dependent on **import substitution** (policies that encourage domestic industry by imposing high tariffs on all imports) to state-run nationalized industries and various attempts at agrarian reform. In the 1960s Brazil, Mexico, and Argentina all seemed poised to enter the ranks of the industrialized world. Multilateral agencies such as the World Bank and the Inter-American Development Bank loaned money for big development projects: continental highways, dams, mechanized agriculture, and power plants.

Yet the dream of the 1960s became a nightmare by the 1980s. Argentina's economy, which was larger than Japan's in 1960, ranked seventeenth in 1999, while

Japan's was ranked second. Oil wealth helped develop a large middle class in Mexico that was badly shaken by the debt crisis in the early 1980s. And Brazil, the eighth largest economy in the world, also has the largest debt of any developing country. Brazil also maintains the region's worst income disparity: 10 percent of the country's richest people control nearly half the country's wealth, while the bottom 40 percent control only 10 percent.

Industrialization   Since the 1960s, most government development policies have emphasized manufacturing. The results have been mixed. Today at least 25 to 30 percent of the male labor force in Mexico, Argentina, Brazil, Chile, Colombia, Peru, Uruguay, and Venezuela is employed in industry (including mining, construction, and energy). Yet this is far short of the hoped-for levels of industrial manufacturing, especially when the size of urban populations is taken into account. In addition, the most industrialized areas tend to be around the capitals or in the **growth poles** (planned industrial cities) such as Ciudad Guayana in Venezuela and the Mexican border cities of Ciudad Juárez and Tijuana.

There are cases in which industry has thrived in noncapital cities and without direct state support. Monterrey, Mexico; Medellín, Colombia; and São Paulo, Brazil, all developed important industrial sectors initially from local investment. Long before Medellín (2 million people) was associated with cocaine, it was a major center of textile production, more industrialized than the larger capital of Bogotá. Similarly Monterrey (3 million people), a city that is not well known outside Mexico, is recognized as economically strong with a solidly middle-class citizenry.

The industrial giant of Latin America is metropolitan São Paulo in Brazil. Rio de Janeiro has greater name recognition and was the capital before Brasília was built, but it does not have the economic muscle of São Paulo. This city of 18 million, which competes with Mexico City for the title of Latin America's largest, began to industrialize in the early 1900s when the city's coffee merchants started to diversify their investments. Since then, a combination of private and state-owned industries have concentrated around São Paulo. Within a 60-mile radius of the city center, automobiles, aircraft, chemicals, processed foods, and construction materials are produced. There are also heavy industry and industrial parks. With the port of Santos nearby and the city of Rio de Janeiro a few hours away, São Paulo is the uncontested financial center of Brazil.

Maquiladoras and Foreign Investment   The Mexican assembly plants that line the border with the United States, called **maquiladoras**, are characteristic of manufacturing systems in an increasingly globalized economy. More than 4,000 maquiladoras exist, employing 1.3 million people who assemble automobiles, consumer electronics, and apparel. Between 1994 and 2000, 3 out of every 10 new jobs in Mexico were in the maquiladoras, which account for nearly half of Mexico's exports. As part of a border development program that began in the 1960s, the Mexican government allowed the duty-free import of machinery, components, and supplies from the United States to be used for manufacturing goods for export back to the United States. Initially, all products had to be exported, but changes in the law in 1994 now allow up to half of the goods to be sold in Mexico.

Considerable controversy surrounds this form of industrialization on both sides of the border. Organized labor in the United States complains that well-paying manufacturing jobs are being lost to low-cost competitors, while environmentalists point out serious industrial pollution resulting from lax government regulation. Mexicans worry that these plants are poorly integrated with the rest of the economy and that many of the workers are young unmarried women who are easily taken advantage of. With NAFTA, foreign-owned manufacturing plants are no longer restricted to the border zone and are increasingly being constructed near the population centers of Monterrey, Puebla, and Veracruz. Mexican workers and foreign corporations, however, continue to locate in the border zone because there are unique advantages to being positioned next to the U.S. border.

FIGURE 4.25 • HIGH-TECH IN COSTA RICA   A Costa Rican worker straps on protective clothing before entering the manufacturing area of Intel's plant in Belen, not far from the capital city of San José. In 2000 Costa Rica earned more foreign exchange from exporting computer chips than it did from coffee, its traditional export. *(AP/Wide World Photos)*

Other Latin American states are attracting foreign companies through tax incentives and low labor costs. Assembly plants in Honduras, Guatemala, and El Salvador are drawing foreign investors, especially in the apparel industry. A recent report from El Salvador claims that not one of its 229 apparel factories has a union. Making goods for American labels such as the Gap, Liz Claiborne, and Nike, many Salvadoran garment workers complain they do not make a living wage, work 80-hour weeks, and will lose their jobs if they become pregnant. The situation in Costa Rica, which is now a major computer chip manufacturer for Intel and exported more than $2 billion in chips in 1999, is quite different (Figure 4.25). With a well-educated population, low crime rate, and stable political scene, Costa Rica is now attracting other high-tech firms. Hopeful officials claim that Costa Rica is transitioning from a banana republic (bananas and coffee were the country's long-standing exports) to a high-tech manufacturing center. As a result, the Costa Rican economy averaged 3 percent annual growth from 1990 to 1999.

The Informal Sector   Even in prosperous San José, Costa Rica, a short drive to the urban periphery shows large neighborhoods of self-built housing filled with street traders and family-run workshops. Such activities make up the informal sector, which is the provision of goods and services without the benefit of government regulation, registration, or taxation. Most people in the informal economy are self-employed and receive no wages or benefits except the profits they clear. The most common informal activities are housing construction (in many cities half of all residents live in self-built housing), manufacturing in small workshops, street vending, transportation services (messenger services, bicycle delivery, and collective taxis), garbage picking, street performing, and even paid line-waiters (Figure 4.26).

No one is sure how big this economy is, in part because it is difficult to separate formal activities from informal ones. One estimate in 1998 showed that nearly 60 percent of the total non-agricultural employment was in the informal sector. Visit Lima, Belém, Guatemala City, or Guayaquil and it is easy to get the impression that the informal economy *is* the economy. From self-help housing that dominates the landscape to hundreds of street vendors that crowd the sidewalks, it is impossible to avoid. There are advantages in the informal sector—hours are flexible, children can work with their parents, and there are no bosses. As important as this sector may be, however, widespread dependence on it signals Latin America's poverty, not its wealth. It reflects the inability of the formal economies of the region, especially in industry, to provide enough jobs for the many people seeking employment.

FIGURE 4.26 • PERUVIAN STREET VENDORS   A street vendor selling produce in Huancayo, Peru. Street vending plays a critical role in the distribution of goods and the generation of income. Contrary to popular opinion, it is often regulated by city governments and by the vendors themselves. *(Rob Crandall/Rob Crandall, Photographer)*

Primary Exports

Historically, Latin America's abundant natural resources were its wealth. In the colonial period silver, gold, and sugar generated great wealth for the colonists. With independence in the nineteenth century, a series of export booms introduced commodities such as bananas, coffee, cacao, grains, tin, rubber, copper, wool, and petroleum to an expanding world market. One of the legacies of this export-led development was a tendency to specialize in one or two major commodities, a pattern that continued into the 1950s. During that decade, 90 percent of Costa Rica's export earnings came from bananas and coffee; 70 percent of Nicaragua's came from coffee and cotton; 85 percent of Chilean export income came from copper; and half of Uruguay's export income came from wood. Even Brazil generated 60 percent of its export earnings from coffee in 1955; by 1998 coffee accounted for less than 5 percent of the country's exports, yet Brazil remained the world leader in coffee production.

Agricultural Production   Since the 1960s, the trend in Latin America has been to diversify and to mechanize agriculture. Nowhere is this more evident than in the Plata Basin, which includes southern Brazil, Uruguay, northern Argentina, Paraguay, and eastern Bolivia. Soybeans, used for oil and animal feed, transformed these lowlands in the 1980s and early 1990s. Added to this are acres of rice, cotton, and orange groves, as well as the more traditional plantings of wheat and sugar. The speed with which the plains were converted into fields alarmed environmentalists. Throughout the 1990s, efforts to conserve some of the ecosystem increased.

Similar large-scale agricultural frontiers exist along the piedmont zone of the Venezuelan Llanos (mostly grains) and the Pacific slope of Central America (cotton and some tropical fruits). In northern Mexico water supplied from dams along the Sierra Madre Occidental has turned the valleys in Sinaloa into intensive agricultural centers of fruits and vegetables for consumers in the United States. The relatively mild winters in northern Mexico allow growers to produce strawberries and tomatoes during the winter months.

In each of these cases, the agricultural sector is capital intensive. By using machinery, hybrid crops, chemical fertilizers, and pesticides, many corporate farms are extremely productive and profitable. What these operations fail to do is employ many rural people, which is especially problematic in countries where a third or more of the population depends on agriculture for its livelihood. As industrialized agriculture becomes the norm in Latin America, subsistence peasant producers are further marginalized.

Mining and Forestry   The exploitation of silver, zinc, copper, iron ore, bauxite, gold, oil, and gas is the economic mainstay for many countries in the region. The oil-rich nations of Venezuela, Mexico, and Ecuador are able to meet their own fuel needs and to earn vital state revenues from oil exports. Venezuela is most dependent on revenues from oil, earning up to 90 percent of its foreign exchange from crude petroleum and petroleum products. In 1998 Venezuela also exported more oil to the United States than any other country, including oil-rich Saudi Arabia (Figure 4.27). Vast oil reserves also exist in the Llanos of Colombia, yet a costly and vulnerable pipeline that connects the oil fields to the coast is a regular target of guerrilla groups. By the late 1990s Colombian oil production had improved, but it is far from reliable.

Like agriculture, mining has become more mechanized and less labor intensive. Even Bolivia, a country dependent on tin production, cut 70 percent of its miners from state payrolls in the 1990s. The measure was part of a broad-based cutback in public spending, yet it suggests that the majority of the miners were not needed. Similarly, the vast copper mines of northern Chile are producing record amounts of copper with fewer miners. Gold mining, in contrast, continues to be labor intensive, offering employment for thousands of prospectors. Gold rushes are occurring in remote tropical regions of Venezuela, Brazil, Colombia, and Costa Rica. The famous Serra Pelada mine in Brazil, an anthill of workers with ladders, shovels, and burlap sacks filled with gold-bearing soil, is the most dramatic example of labor-intensive mining (Figure 4.28). Generally, gold miners tend to work in small groups. Moreover, many gold discoveries occur illegally on Indian lands or within the borders of national parks (as in the case of Costa Rica). Invariably, if a large enough gold strike occurs, the state steps in to regulate mining operations and to get its share of the profits.

Logging is another important, and controversial, resource-based activity. Several countries rely on plantation forests of introduced species of pines, teak, and eucalyptus to supply domestic fuelwood, pulp, and board lumber.

FIGURE 4.27 • OIL PRODUCTION   A portable drilling platform is shown on Lake Maracaibo, Venezuela. Oil production increased the pace of development for countries like Venezuela, Mexico, and Ecuador, but these economies struggled in the 1990s as oil prices declined. *(Rob Crandall/Rob Crandall, Photographer)*

FIGURE 4.28 • AMAZONIAN GOLD   Serra Pelada, in the Brazilian state of Pará, is one of the most productive gold mines in Latin America. *(Peter Frey/Getty Images, Inc.)*

These plantation forests grow single species and fall far short of the complex ecosystems occurring in natural forests. Still, growing trees for paper or fuel reduces the pressure on other forested areas. Leaders in plantation forestry are Brazil, Venezuela, Chile, and Argentina. Considered Latin America's economic star in the 1990s, Chile relied on timber and wood chips to boost its export earnings. Thousands of hectares of exotics (eucalyptus and pine) have been planted, systematically harvested, cut into boards, or chipped for wood pulp (see Figure 4.4). Japanese capital is heavily involved in this sector of the Chilean economy. The recent expansion of the wood chip business, however, led to a dramatic increase in the logging of native forests. By 1992, more than half of all wood chips were from native species.

There is growing interest in certification programs that designate when a wood product has been produced sustainably. This is due to consumer demand for certified wood, mostly in Europe. Unfortunately, such programs are small and the lure of profit usually is stronger than the impulse to conserve for future generations.

Latin America in the Global Economy

In order to conceptualize Latin America's place in the world economy, **dependency theory** was advanced in the 1960s by scholars from the region. The basis of the theory is that expansion of European capitalism created the region's underdevelopment. For the developed "cores" of the world to prosper, the "peripheries" became dependent and impoverished. Dependent economies, such as those in Latin America, were export-oriented and vulnerable to fluctuations in the global market. Even when they experienced economic growth, it was secondary to the economic demands of the core (North America and Europe).

Economists who accepted this interpretation of Latin America's history were convinced that economic development could occur only through self-sufficiency, growth of internal markets, agrarian reform, and greater income equality. In short, they argued for strong state intervention and less trade with the economic cores of Europe and North America. Policies such as import substitution industrialization (developing a country's industrial sector by making imported products extremely expensive and domestic manufactured goods cheaper), and nationalization of key industries were partially influenced by this view. Dependency theory has its critics. In its simplest form it becomes a means to blame forces external to Latin America for the region's problems. Unspoken in dependency theory is also the notion that the path to development taken by Europe and North America cannot be easily copied. This was a radical idea for its time because most economists saw development as occurring in stages that all countries would eventually follow.

A look at trade in the Western Hemisphere reveals both increased integration within Latin America as well as the dominance of the North American market. In terms of regional integration, the formation of Mercosur in 1991 stimulated new levels of trade, especially between Brazil and Argentina. Even with this increase, the level of intraregional trade among the Mercosur countries ($18.3 billion) is a small fraction of the intraregional NAFTA trade ($700 billion). Generally, the increase in intraregional trade is widely recognized as a positive sign of greater economic independence for Latin America. Nevertheless, the historic pattern of dependence on North America still exists as the United States continues to be the major trading partner for most states in the region.

Neoliberalism as Globalization By the 1990s governments and the World Bank had become champions of neoliberalism as a sure path to economic development. **Neoliberal policies** stress privatization, export production, direct foreign investment, and few restrictions on imports. They summarize the forces of globalization by turning away from policies that emphasize state intervention and self-sufficiency. Most Latin American political leaders embrace neoliberalism and the benefits that come with it, such as increased trade and more favorable terms for debt repayment. Yet there are signs of discontent with neoliberalism throughout the region. The usually tranquil city of Porto Alegre, Brazil, hosted the World Social

Forum in 2001 as an antiglobalization counterpoint to the World Economic Forum's meeting in Switzerland. This leftist international forum staged rallies criticizing neoliberalism and the FTAA and supporting reduction of poverty and inequality as Latin America's top priorities.

Chile is an outspoken defender of neoliberalism. Its growth rate between 1990 and 1999 averaged 5.6 percent, the region's healthiest. In 1995 alone the Chilean economy grew 10.4 percent, placing it in the same league as the Asian "tiger economies." Consequently, it is the most studied and watched country in Latin America. By the numbers, Chile's 15.4 million people are doing well, but the country's accomplishments are not readily transferable to other countries. To begin with, the radical move to privatize state-owned business and open the economy occurred under an oppressive military dictatorship that did not tolerate opposition. Much of Chile's export-led growth has been based on primary products: fruits, seafood, copper, and wood. Although many of these are renewable, Chile will need to develop more manufactured goods if it hopes to be labeled "developed." Furthermore, its relatively small and homogeneous population in a resource-rich land does not have the same ethnic divisions that hinder so many states in Latin America. Experiments in neoliberalism are still new, so all the social and environmental costs associated with this laissez-faire economic policy are not known.

Dollarization   As financial crises spread through Latin America in the late 1990s, governments began to consider the economic benefits of **dollarization**, a process by which a country adopts—in whole or in part—the U.S. dollar as its official currency. In a totally dollarized economy, the U.S. dollar becomes the only medium of exchange and the country's national currency ceases to exist. This was the radical step taken by Ecuador in 2000 to address the dual problems of currency devaluation and hyperinflation rates of more than 1,000 percent annually. El Salvador adopted dollarization in 2001 as a means to reduce the cost of borrowing money. Dollarization is not a new idea; back in 1904 Panama dollarized its economy the year after it had gained independence from Colombia. Until 2000, however, Panama was the only fully dollarized state in Latin America.

A more common strategy in Latin America is limited dollarization, in which U.S. dollars circulate and are used alongside the country's national currency. Limited dollarization exists in many countries around the world but is widespread in Latin America. Since the economies of Latin America are prone to currency devaluation and hyperinflation, limited dollarization is a type of insurance. Many banks in Latin America, for example, allow customers to maintain accounts in dollars to avoid the problem of capital flight should a local currency be devalued.

Social Development

Significant improvements in life expectancy, child survival, and educational attainment have occurred in Latin America the past three decades. One telling indicator is the mortality rate for children younger than 5 years old, which was cut by 50 percent or more in most Latin America countries between 1980 and 1999. This indicator is important because when children younger than 5 years of age are surviving, it suggests that basic nutritional and health-care needs are being met. One can also conclude that resources are being used to sustain women and their children. Despite economic downturns, the region's social networks have been able to lessen the negative effects on children.

Grassroots and nongovernmental organizations (NGOs) play a fundamental role in contributing to social well-being. Initiated by international humanitarian organizations, church organizations, and community activists, these popular groups provide many services that state and local governments cannot. Catholic Relief Services and Caritas, for example, work with the rural poor throughout the region to improve their water supplies, health care, and education. Other groups lobby local governments to build schools or recognize squatters' claims. Grassroots organizations also develop cooperatives that market anything from sweaters to cheeses. Other important indicators for social development are life expectancy, literacy, and access to improved water sources. In total, 80 percent of the people in the region have access to an adequate amount of water from an improved source; literacy rates are higher than 85 percent; and life expectancy (men and women) is 71 years. Masked by this combined data are extreme variations between rural and urban areas, between regions, and along racial and gender lines.

Within Mexico and Brazil tremendous internal differences exist in socioeconomic indicators (Figure 4.29). The northeast part of Brazil lags behind the rest of the country in every social indicator. The country has a literacy rate of 85 percent, but in the northeast it is only 60 percent. Moreover, 70 percent of city residents in the northeast are literate, but only 40 percent of rural ones are. In Mexico the levels of poverty are highest in the Indian states of Chiapas and Oaxaca. In contrast, Mexico City, the state of Nuevo Leon (Monterrey is the capital), and the state of Quintana Roo (home to Mexico's largest resort, Cancun) have a per capita GDP of more than

FIGURE 4.29 • MAPPING POVERTY AND PROSPERITY
Mexico and Brazil are the largest states in the region, and yet the levels of poverty vary greatly within each country. In Mexico, the southern states of Chiapas and Oaxaca have the lowest GDP per capita. In Brazil, the northeastern states of Paraíba, Alagoas, Tocantins, Piauí, Ceará, and Maranhão have the country's lowest per capita GDP. (© 1999 and 2000 The Economist Newspaper Group, Inc. Reprinted with permission. Further reproduction prohibited.)

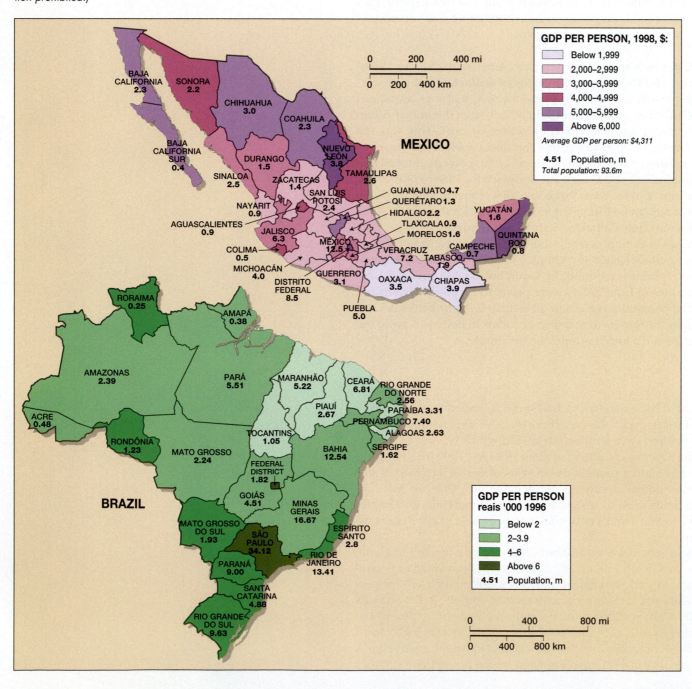

$6,000. All countries have spatial inequalities regarding income and availability of services, but the contrasts tend to be sharper in the developing world. In the cases of Mexico and Brazil, it is hard to ignore ethnicity and race when trying to explain these patterns.

Race and Inequality   There is much to admire about race relations in Latin America. The complex racial and ethnic mix that was created in Latin America encouraged tolerance for diversity. That said, Indians and blacks are more likely to be counted among the poor of the region. More than ever, racial discrimination is a major political issue in Brazil. Reports of organized killings of street children, most of them Afro-Brazilian, make headlines. For decades, Brazil championed its vision of a color-blind racial democracy. True, residential segregation by race is rare in Brazil and interracial marriage is common, but certain patterns of social and economic inequality seem best explained by race.

There are major problems with trying to assess racial inequalities in Brazil. The Brazilian census asks few racial questions, and all are based on self-classification. In 1987 only 11 percent of the population called itself black and 53 percent white. Some Brazilian sociologists, however, claim that more than half the population is of African ancestry, making Brazil the second largest "African state" after Nigeria. Racial classification is always highly subjective and relative, but there are patterns that support the existence of racism. Evidence from northeastern Brazil, where Afro-Brazilians are the majority, shows death rates approaching those of some of the world's poorest countries. Throughout Brazil, blacks suffer higher rates of homelessness, landlessness, illiteracy, and unemployment. There is even a movement in the northeastern state of Bahia to extend the same legal protection given Indians to hundreds of Afro-Brazilian villages that were once settled by runaway slaves.

In areas of Latin America where Indian cultures are strong, one also finds indicators of low socioeconomic position. In most countries areas where native languages are widely spoken regularly correspond with areas of persistent poverty. In Mexico the Indian south lags behind the booming north and Mexico City. Prejudice is embedded in the language. To call someone an *indio* (Indian) is an insult in Mexico. It is difficult to separate status divisions based on class from those based on race. From the days of conquest, being European meant an immediate elevation in status over the Indian, African, and *mestizo* populations. Class awareness is very strong. Race does not necessarily determine one's economic standing, but it certainly influences it. Most people recognize the power of the elite and envy their lifestyle.

The Status of Women   Many contradictions exist with regard to the status of women in Latin America. Many Latina women work outside the home. In most countries the formal figures are between 30 and 40 percent of the workforce, not far off from many European countries but lower than in the United States. Legally speaking, women can vote, own property, and sign for loans, although they are less likely to do so than men, reflecting the patriarchal (male-dominated) tendencies in the society. Even though Latin America is predominantly Catholic, divorce is legal and family planning is promoted. In most countries, however, abortion remains illegal.

Overall, access to education in Latin America is good compared to other developing regions, and thus illiteracy rates tend to be low. The rates of adult illiteracy are slightly higher for women than for men, but usually by only a few percentage points. Throughout higher education in Latin America, male and female students are equally represented today. Consequently, in the fields of education, medicine, and law, women are regularly employed.

The biggest changes for women are the trends toward smaller families, urban living, and educational equality with men. These factors have greatly improved the participation of women in the labor force. Since 1970 in 11 of the 17 Latin American countries, the average family size has decreased by at least two children. In the countryside, however, serious inequalities remain. Rural women are less likely to be educated and tend to have larger families. In addition, they are often left to care for their families alone as husbands leave in search of seasonal employment. In most cases, the conditions facing rural women have been slow to improve.

Women are increasingly playing an active role in politics. In 1990 Nicaragua elected the first woman president in Latin America, Violeta Chamorro, the owner of an opposition newspaper. Nine years later, Panamanians voted Mireya Moscoso into power. A Mayan woman from Guatemala, Rigoberta Menchú, won the Nobel Peace Prize in 1992 for denouncing human rights abuses in her native country. Women are also active organizers and participants in cooperatives, small businesses, and unions. In a relatively short period, urban women have won a formal place in the economy and a political voice. Moreover, evidence suggests that this trend will continue, reflecting an improved status for women in the region.

CONCLUSION

Latin America and the Caribbean were the first world regions to be fully colonized by Europe. In the process, perhaps 90 percent of the native population died from disease, cruelty, and forced resettlement. The slow demographic recovery of native peoples and the continual arrival of Europeans and Africans resulted in an unusual level of racial and cultural mixing. It took nearly 400 years for the populations of Latin America and the Caribbean to reach 54 million again, their precontact level. During this long period, European culture, technology, and political systems were transplanted and modified. In short, a syncretic process unfolded in which many native customs were integrated into Iberian ones. Over time, a blending of indigenous and Iberian societies, with some African influences, gave distinction to this part of the world.

Compared with Asia, this is still a region rich in natural resources and relatively lightly populated. Yet as population continues to grow along with economic expectations, there is considerable concern that much of the region's resource base will be destroyed for short-term economic gains. In the midst of a boom in natural resource extraction, popular concern for the state of the environment is growing. Not only was Brazil the site of the 1992 United Nations Earth Summit, but hundreds of locally based environmental groups have formed to try to protect forests, grasslands, Amerindian peoples, and freshwater supplies. This brand of environmentalism is pragmatic; it recognizes the need for economic development but hopes to improve urban environmental quality and sustainable resource use.

Unlike other developing areas, most Latin Americans live in cities. This shift started early and reflects a cultural bias toward urban living with roots in the colonial past. Not everyone who came to the city found employment, however. Thus, the informal sector was established. Even though population growth rates have declined, the overall makeup of the population is young. Serious challenges lie ahead in educating and finding employment for the population younger than age 15. In this matter some countries are faring better than others. Industrial employment in Mexico and the Southern Cone is up, yet in many parts of the Central Andes subsistence farming is still widely practiced.

The 1990s were better years for much of Latin America if measured by real annual growth. What is deceptive about economic growth is that the percentage of people living in poverty (as well as the total number of people) was higher in 2000 than in 1980. So too was the widening gap between rich and poor in the region. Latin America has the world's highest inequality when it comes to the distribution of wealth. Brazil's measures of income distribution are nearly the same as South Africa's when it was still under apartheid (see Chapter 6). Moreover, a number of important economies (namely, Brazil, Argentina, Colombia, and Venezuela) are faltering in the new millennium as they wrestle with problems such as currency devaluation, debt burdens, inflation, and growing unemployment. How new waves of economic turmoil will influence attitudes toward neoliberalism, globalization, and hemispheric trade remains to be seen.

KEY TERMS

agrarian reform (page 88)
Altiplano (page 82)
dependency theory (page 102)
dollarization (page 103)
El Niño (page 83)
Free Trade Area of the Americas (FTAA) (page 75)
grassification (page 78)

growth poles (page 99)
import substitution (page 98)
informal sector (page 100)
latifundia (page 88)
maquiladora (page 99)
megacities (page 76)
Mercosur (page 94)
mestizo (page 89)

minifundia (page 88)
neoliberal policies (page 102)
neotropics (page 76)
Organization of American States (OAS) (page 94)
rural-to-urban migration (page 87)
shields (page 80)
squatter settlements (page 87)

subnational organizations (page 97)
supranational organizations (page 97)
syncretic religion (page 93)
Treaty of Tordesillas (page 94)
urban primacy (page 85)

The Caribbean

FIGURE 5.1 The Caribbean
Named for one of its former native peoples—the Carib Indians—the modern Caribbean is a culturally complex and economically marginal world region. Settled by various colonial powers, most of the region's residents are a mix of African, European, and South Asian peoples. Twenty-five countries and dependent territories make up the Caribbean, yet most of the population resides on the four islands of the Greater Antilles: Cuba, Hispaniola, Jamaica, and Puerto Rico. Home to 38 million people, the Caribbean is demographically larger than Australia and Oceania.

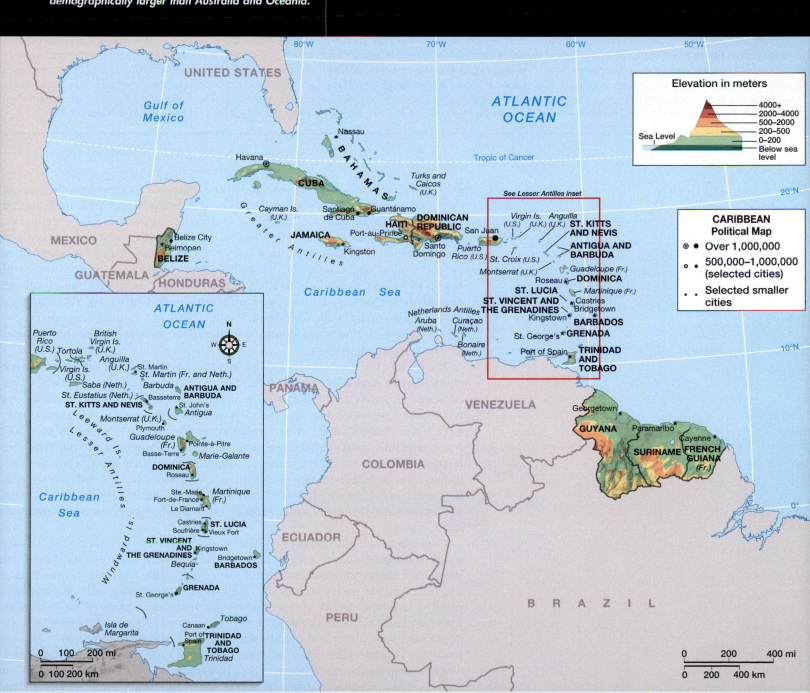

SETTING THE BOUNDARIES

This culturally diverse area of the world does not fit neatly into current world regional divisions. In fact, in most textbooks it is scarcely mentioned, being folded into Latin America or appearing as part of Middle America (along with Mexico and Central America). By focusing on the Caribbean as a distinct world region, however, one can more clearly see the long-term processes of colonization and globalization that have created it.

For much of the colonial period, the islands themselves were referred to as "the Indies" or "the Spanish Main," and the sea surrounding them was called *Mar del Norte* (North Sea). It was not until the late eighteenth century that an English cartographer first applied the name "Caribbean" to the sea (a reference to the Carib Indians who once inhabited the islands). It took another 200 years for the term *Caribbean* to be generally applied to this region of islands and rimland.

The basis for treating the Caribbean as a distinct area lies within its particular cultural and economic history. Culturally, the islands and portions of the coast can be distinguished from the largely Iberian-influenced mainland of Latin America because of their more diverse European colonial history and much stronger African influence. Demographically, the native population of the Caribbean was virtually eliminated within the first 50 years following Columbus's arrival. Thus, unlike Latin America and North America, the Amerindians of the Caribbean islands have virtually no legacy, save a few place names. The dominance of plantation agriculture, such as sugar production, explains many of the social, economic, and environmental patterns in the region. African slaves were introduced as replacements for lost native labor. As Caribbean forests and savannas became sugarcane fields, a pattern was set in which foreign species of plants replaced native ones. The long-term effect of plantation life is visible today in the hugely uneven distribution of land and resources among the people of the region.

Today, organizations such as the Caribbean Common Market exist to promote the common interests of island and mainland states. Yet internal alliances often split along colonial lines, so that former Spanish, British, French, or Dutch colonies at times have more in common with each other than with the region as a whole. All of the Caribbean states with the exception of Cuba are eager to join the proposed Free-Trade Area of the Americas. Given the marginal economic status of the Caribbean and its location as the crossroads of the Americas, it is likely that the region will experience greater economic integration with the rest of the hemisphere.

The Caribbean was the first region of the Americas to be extensively explored and colonized by Europeans. Yet its modern regional identity is unclear, often merged with Latin America but also viewed as apart from it (see "Setting the Boundaries"). Today the region is home to 38 million people scattered across 25 countries and dependent territories. They range from the small British dependency of the Turks and Caicos, with 12,000 people, to the island of Hispaniola, with nearly 16 million. In addition to the Caribbean Islands, Belize of Central America and the three Guianas—Guyana, Suriname, and French Guiana—of South America are included as part of the Caribbean. For historical and cultural reasons, the peoples of these mainland states identify with the island nations and are thus included in this chapter (Figure 5.1).

Historically, the Caribbean was a battleground between rival European powers competing for territorial control of these tropical lands. In the early 1900s, the United States took over as the dominant geopolitical force in the region, maintaining what some have called a neocolonial presence. As in many developing areas, external control of the Caribbean by foreign governments and companies produced highly dependent and inequitable economies. The plantation was the dominant production system and sugar the leading commodity. Over time other products began to have economic importance, as did the international tourist industry. Increasingly, governments have sought to diversify their economies by expanding into banking services and manufacturing to reduce the region's dependence on agriculture and tourism.

Generally, when one thinks of the Caribbean, images of white sandy beaches and turquoise tropical waters come to mind. Tucked between the Tropic of Cancer and the equator, with year-round temperatures averaging in the high 70s, the hundreds of islands and picturesque waters of the Caribbean have often inspired comparisons to paradise. Columbus began the tradition by describing the islands of the New World as the most marvelous, beautiful, and fertile lands he had ever known, filled with flocks of parrots, exotic plants, and friendly natives. Writers today are still lured by the sea, sands, and swaying palms of the Caribbean. Since the 1960s, the Caribbean has earned much of its international reputation as a playground for northern vacationers. But there is another Caribbean that is far poorer and economically more dependent than the one portrayed on travel posters. Haiti, which has the third largest population in the region with 7 million people, is the poorest country in the Western Hemisphere.

The two largest countries—Cuba (11 million) and the Dominican Republic (8 million)— also suffer from serious economic problems.

The majority of Caribbean people are poor, living in the shadow of North America's vast wealth. The concept of **isolated proximity** has been used to explain the region's unique position in the world. The *isolation* of the Caribbean sustains the area's cultural diversity (Figure 5.2) but also explains its limited economic opportunities. Caribbean writers note that this isolation fosters a strong sense of place and a tendency to focus inward. Yet the relative *proximity* of the Caribbean to North America (and, to a lesser extent, Europe) ensures its international connections and economic dependence. For example, each year Dominican workers in the United States send more than $1 billion to family and friends in the Dominican Republic who rely on this money for their sustenance. Through the years, the Caribbean has evolved as a distinct but economically marginal world region. This position expresses itself today as workers flee the region in search of employment, while foreign companies are attracted to the Caribbean for its cheap labor. The economic well-being of most Caribbean countries is uncertain. Despite such uncertainty, an enduring cultural richness and attachment to place is witnessed here that may explain a growing counter-current of immigrants back to the Caribbean.

Environmental Geography: Paradise Undone

The levels of environmental degradation and poverty found in Haiti are unmatched in the Western Hemisphere. Haiti's story, although extreme, illustrates how social and economic inequalities contribute to misguided natural resource choices, which in turn lead to more poverty. What was once considered France's richest colony—its tropical jewel—now has a per capita income of $460. Haiti's life expectancy (49 years) is the lowest in the Americas, while its levels of child malnutrition and infant mortality are the highest in the hemisphere. The explanations for Haiti's misfortune are complex, but political economy, demography, and natural resources are critical factors.

In the late eighteenth century, Haiti's plantation economy yielded vast wealth for several thousand Europeans (mostly French) who exploited the labor of half a million African slaves. During the colonial period, the lowlands were cleared to make way for sugarcane fields, but most of the mountains remained in tropical forest.

By the mid-twentieth century, a destructive cycle of environmental and economic impoverishment was established in Haiti. While under the rule of corrupt dictators from 1957 to 1986, Haiti's elite benefited as the conditions for the country's poor majority worsened. More than two-thirds of Haiti's people are peasants who work small hillside plots and seasonally labor on large estates. As the population grew, people sought more land. They cleared the remaining hillsides, subdivided their plots into smaller units, and abandoned the practice of fallowing land (leaving land untilled for several seasons) in an effort to survive. When the heavy tropical rains came, the exposed and easily eroded mountain soils were washed away, leaving rocky and semi-arable surfaces for the next season. Downstream, soil collected in irrigation ditches and behind dams, which negatively affected agriculture, electricity production, and water supply throughout the country. The state of the environment was further aggravated by dependence upon wood fuel. Because of their poverty and limited electricity supplies, most Haitians use charcoal (made from trees) to cook meals and heat water. By the late 1990s it was estimated that only 3 percent of Haiti remained forested. The decline of the resource base is evident in aerial photos, which reveal a sharp boundary between a treeless Haiti and a forested Dominican Republic (Figure 5.3).

Environmental Issues

Ecologically speaking, it is difficult to picture a landscape that has been so completely altered as the Caribbean. For nearly five centuries the destruction of forests and the unrelenting cultivation of soils resulted in the extinction of many endemic (native) Caribbean plants and animals, including various shrubs and trees, songbirds, large mammals, and monkeys. This severe depletion of biological resources helps explain

FIGURE 5.2 • CARNIVAL DANCER A woman in elaborate costume celebrates Carnival in Port of Spain, Trinidad. Linking a Christian pre-Lenten celebration with African musical rhythms, Carnival has become an important symbol of Caribbean identity. *(Rob Crandall/Rob Crandall, Photographer)*

FIGURE 5.3 • DEFORESTATION ON HISPANIOLA The border between Haiti and the Dominican Republic is easily seen in this aerial photo. Haitians, the poorest people in the Western Hemisphere, rely almost exclusively on firewood and charcoal for their energy needs. As trees are removed and soils are exposed to erosion, agricultural harvests often significantly decline. Successful reforestation efforts have not yet been put into practice. *(James Blair/NGS Image Collection)*

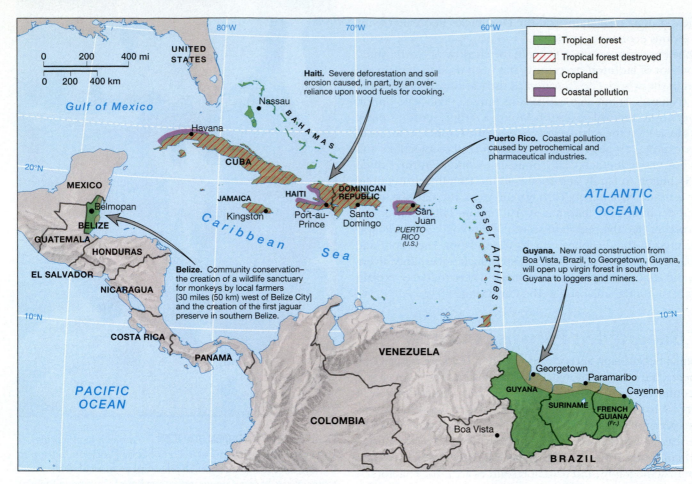

FIGURE 5.4 • ENVIRONMENTAL ISSUES IN THE CARIBBEAN
It is hard to imagine a region in which the environment has been so completely altered. Most of the island forests were removed long ago for agriculture or fuel wood, and soil erosion is a chronic problem. Coastal pollution is serious around the largest cities and industrial zones. The forest cover of the rimland states, however, is largely intact and is attracting the interest of environmentalists. As tourism becomes increasingly important for the Caribbean, efforts to protect the beaches and reefs along with the fauna and flora are growing. (*Adapted from* DK World Atlas, *1997, pp. 7, 55, London: DK Publishing*)

some of the present economic and social instability of the region (Figure 5.4). Most of the environmental problems in the region are associated with agricultural practices, soil erosion, excessive reliance upon wood fuels, and water and air pollution associated with sprawling and impoverished cities.

Agriculture's Legacy of Deforestation Much of this region was covered in tropical rain forests and deciduous forests prior to the arrival of Europeans. The great clearing of the Caribbean's forests began on European-owned plantations on the smaller islands of the eastern Caribbean in the seventeenth century and spread westward. The island forests were removed not only to make room for sugarcane, but also to provide the fuel necessary to turn the cane juice into sugar, as well as to provide lumber for housing, fences, and ships. Primarily, however, tropical forests were removed because they were seen as unproductive; the European colonists valued cleared land. The newly exposed tropical soils easily eroded and ceased to be productive after several harvests, a situation that led to two distinct land-use strategies. On the larger islands of Cuba and Hispaniola and on the mainland, new lands were constantly cleared and older ones abandoned or fallowed in an effort to keep up sugar production. On the smaller islands where land was limited, such as Barbados and Antigua, labor-intensive efforts to conserve soil and maintain fertility were employed. In either case, the island forests were replaced by a landscape devoted to crops for world markets.

While Haiti has lost most of its forest cover, on Jamaica and the Dominican Republic nearly 30 percent of the land is still forested. On Puerto Rico roughly one-quarter of the land has forest cover, whereas on Cuba the figure has dropped to 20 percent. Cuba has experienced a surge in charcoal production brought on by the economic and energy crises that began in 1990. Since much of the country's electricity is generated by imported fuel, locals have turned to Cuba's forests to make charcoal for domestic energy needs. The government recently reported that up to 60 percent of the national territory is being desertified by mechanized agriculture and a surge in charcoal production.

Managing the Rimland Forests   The Caribbean **rimland** is the coastal zone of the mainland, beginning with Belize and extending along the coast of Central America to northern South America. In general, the biological diversity and stability of the rimland states are less threatened than in the rest of the Caribbean. Thus, current conservation efforts could produce important results. Even though much of Belize was selectively logged for mahogany in the nineteenth and twentieth centuries, healthy forest cover still supports a diversity of mammals, birds, reptiles, and plants. Public awareness of the negative consequences of deforestation is also greater now. Many protected areas, most notably the first jaguar reserve in the Americas, have been established in Belize in the last decade.

In Guyana, the relatively pristine interior forests are becoming a battleground between conservationists and developers. During the 1990s, the Guyanese government made the wood processing industry a priority by encouraging private investment and negotiating with companies from Malaysia and China about forest concessions and sawmill operations. A new dry-season highway traverses the length of the country, connecting Boa Vista, Brazil, with Georgetown, Guyana. While governments in both states are encouraged by the economic possibilities of this road, an alliance of conservationists (local and foreign) and indigenous peoples is attempting to establish a vast national park in part of this region where no commercial logging or mining could occur.

Failures in Urban Infrastructure   The growth of Caribbean cities has produced environmental strains as well, especially water contamination and waste disposal. The urban poor are the most at risk for health problems associated with overly strained or nonexistent water and sewage services. According to 2001 World Bank estimates, only half of Haiti's urban population has access to clean water sources, typically through shared neighborhood faucets. The figures are better for Jamaica and the Dominican Republic, where more than 80 percent of city dwellers have access to treated water supplies. In rural areas, however, many people drink surface water or collected rainwater, taking their chances on its quality. Existing freshwater supplies fall far short of domestic needs. As tourism, manufacturing, and a growing urban population demand more water and produce more waste, Caribbean countries will be forced to make hard decisions about meeting these increased demands on their infrastructure.

In addition to being a public health concern, water contamination poses serious economic problems for states dependent on tourism. Governments are caught between a desire to fix the problem before tourists notice and a tendency not to discuss it at all. Doing nothing, however, may not be an option for a tourism-based economy. In industrialized Puerto Rico, it is estimated that half the country's coastline is unfit for swimming, mostly due to contamination from sewage.

Throughout the Caribbean it is increasingly recognized that protecting the environment is not a luxury but a question of economic livelihood. Various environmental groups lobby their governments to create new laws or comply with existing ones. To appreciate the environmental complexity of the region, an understanding of the physical geography of the islands and rimland is essential.

The Sea, Islands, and Rimland   It is the Caribbean Sea itself, that body of water enclosed between the *Antillean* islands (the arc of islands that begins with Cuba and ends with Trinidad) and the mainland of Central and South America, that links the states of the region (Figure 5.5). Historically the sea connected people through its trade routes and sustained them with its marine resources of fish, green turtle, manatee, lobster, and crab. While the sea is noted for its clarity and biological diversity, it has never supported large commercial fishing because the quantities of any one species are not great. The surface temperature of the sea ranges from 73°

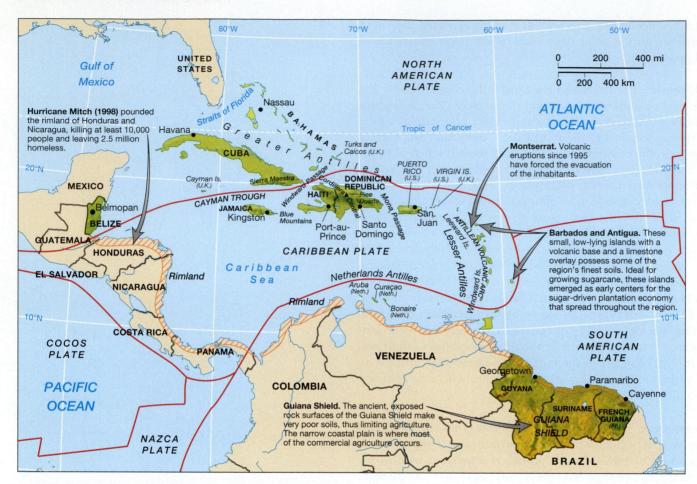

FIGURE 5.5 • PHYSICAL GEOGRAPHY OF THE CARIBBEAN
Centered on the Caribbean Sea, the region falls neatly within the neotropical belt north of the equator. In addition to a warm year-round climate and beautiful waters, the Greater Antilles have several important mountain ranges, whereas some of the Lesser Antilles have active volcanoes.

FIGURE 5.6 • CARIBBEAN SEA Noted for its calm turquoise waters, steady breezes, and treacherous shallows, the Caribbean Sea has both sheltered and challenged sailors for centuries. This aerial photograph shows the southern Caribbean islands of Los Roques, off the Venezuelan coast. *(Rob Crandall/ Rob Crandall, Photographer)*

to 84°F (23° to 29°C), over which forms a warm tropical marine air mass that influences daily weather patterns. This warm water and tropical setting continue to be a key resource for the region as millions of tourists visit the Caribbean each year (Figure 5.6).

The arc of islands that stretches across the Caribbean Sea is its most distinguishing physical feature. The Antillean islands are divided into two groups: the Greater and Lesser Antilles. The majority of the region's population live on the islands. The rimland (the Caribbean coastal zone of the mainland) includes Belize and the Guianas, as well as the Caribbean coast of Central and South America. In contrast to the islands, the rimland has low population densities.

Greater Antilles The four large islands of Cuba, Jamaica, Hispaniola (shared by Haiti and the Dominican Republic), and Puerto Rico make up the **Greater Antilles**. On these islands are found the bulk of the region's population, arable lands, and large mountain ranges. Given the popular interest in the Caribbean coasts, it surprises many people that Pico Duarte in the Cordillera Central of the Dominican Republic is more than 10,000 feet tall (3,000 meters), Jamaica's Blue Mountains top 7,000 feet (2,100 meters), and Cuba's Sierra Maestra is more than 6,000 feet tall (1,800 meters). Historically, the mountains of the Greater Antilles were of little economic interest because plantation owners preferred the coastal plains and valleys. Yet the mountains were an important refuge for runaway slaves and subsistence farmers, and thus are important in the cultural history of the region.

Lesser Antilles The **Lesser Antilles** form a double arc of small islands stretching from the Virgin Islands to Trinidad. Smaller in size and population than the Greater Antilles, they were important early footholds for rival European colonial powers. The islands from St. Kitts to Grenada form the inner arc of the Lesser Antilles. These mountainous islands, with peaks ranging from 4,000 to 5,000 feet (1,200 to 1,500 meters), have volcanic origins. Erosion of the island peaks and the accumulation of ash from volcanic eruptions have created small pockets of arable soils, although the steepness of the terrain places limits on agricultural development. The latest round of volcanic activity began in July 1995 on Montserrat. A series of volcanic eruptions of ash and rock have taken several lives and forced most of the island's 10,000 inhabitants to nearby islands, and even to London. Residents of Plymouth, the capital, were forced to evacuate in 1996 (Figure 5.7).

Just east of this volcanic arc are the low-lying islands of Barbados, Antigua, Barbuda, and the eastern half of Guadeloupe. Covered in limestone that overlays volcanic rock, these lands were much more inviting for agriculture. In particular, such soils were ideal for growing sugarcane, causing these islands to become important early settings for the plantation economy that diffused throughout the region.

Rimland States   Unlike the rest of the Caribbean, the rimland states of Belize and the Guianas still contain significant amounts of forest cover. In Suriname alone, 96 percent of the landscape is forest or woodland. Timber continues to be an important export for this region. As on the islands, agriculture in these states is closely tied to local geology and soils. Much of low-lying Belize is limestone. Sugarcane is the dominant crop in the drier north, while citrus is produced in the wetter central portion of the state. The Guianas, however, are characterized by the rolling hills of the Guiana Shield. The shield's crystalline rock is responsible for the area's overall poor soil quality. Thus, most agriculture in the Guianas occurs on the narrow coastal plain, with sugar and rice as the primary crops.

Climate and Vegetation

Coconut palms framed by a blue sky evoke the postcard image of the Caribbean. In this tropical region, it is warm all year and rainfall is abundant. Much of the Antillean islands and rimland receive more than 80 inches (200 centimeters) of rainfall annually and can support tropical forests. Amid the forests are pockets of naturally occurring grasslands in parts of Cuba, Hispaniola, and southern Guyana. Distinctly dry areas exist, such as the rain shadow basin in western Hispaniola. As explained earlier, much of the natural vegetation on the islands has been removed to accommodate agriculture and fuel needs, and only small forest fragments remain today. Tropical ecosystems in the rimland are largely intact.

As in many tropical lowlands, seasonality in the Caribbean is defined by changes in rainfall more than temperature. Although some rain falls throughout the year, the rainy season is from July to October (Figure 5.8). This is when unstable atmospheric conditions sometimes cause the formation of hurricanes. During the slightly cooler months of December through March, rainfall declines. This time of year corresponds with the peak tourist season.

Hurricanes   Each year several **hurricanes** pound the Caribbean as well as Central and North America with heavy rains and fierce winds. Beginning in July, westward-moving low-pressure disturbances form off the coast of West Africa, picking up moisture and speed as they move across the Atlantic. Usually no more than 100 miles across, these disturbances achieve hurricane status when wind speeds reach 75 miles per hour. Hurricanes may take several paths through the region, but they typically enter through the Lesser Antilles. They then curve north or northwest and collide with the Greater Antilles, Central America, Mexico, or southern North America before moving to the northeast and breaking up in the Atlantic Ocean. The hurricane zone lies just north of the equator on both the Pacific and Atlantic sides of the Americas. Typically a half dozen to a dozen hurricanes form each season and move through the region, causing limited damage.

There are, of course, exceptions, and most longtime residents of the Caribbean have felt the full force of at least one major hurricane in their lifetime. The destruction caused by these storms is not just from the high winds, but also from the heavy downpours that can cause severe flooding and deadly coastal tidal surges. In 1998 the torrential rains of Hurricane Mitch, the most deadly tropical storm in a

FIGURE 5.7 • VOLCANIC DESTRUCTION   The eruptions of the Soufrière Hills volcano on the island of Montserrat sent mud-flows and volcanic ash into the capital city of Plymouth, partially submerging a clock tower and a telephone booth in the town center. Damage was widespread on the island, and many of the 10,000 residents of this British colony had to be evacuated. Eruptions began on July 18, 1995, and are ongoing. *(Rob Huibers/Panos Pictures)*

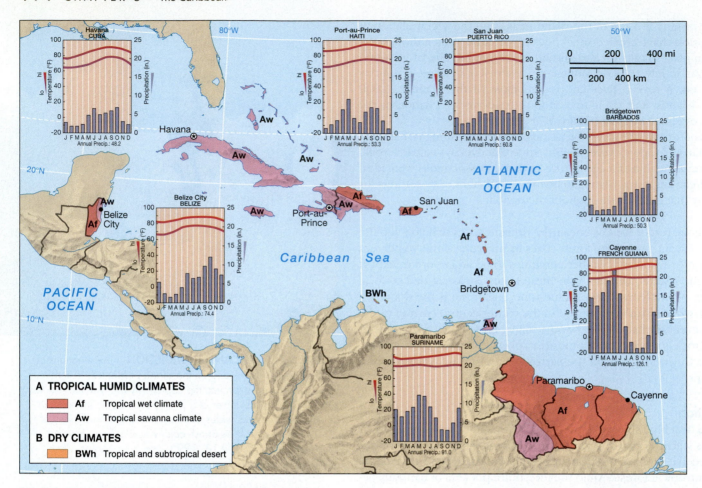

FIGURE 5.8 • CLIMATE MAP OF THE CARIBBEAN Most of the region is classified as having either a tropical wet (Af) or tropical savanna (Aw) climate. Temperature varies little across the region, with highs slightly above 80 degrees and lows around 70 degrees. Important differences in total rainfall and the timing of the dry season distinguish different places. In the Guianas, for example, the dry season is September through October, whereas the drier months for the islands are December through March. *(Temperature and precipitation data from Pearce and Smith, 1984,* The World Weather Guide, *London: Hutchinson.)*

century, resulted in the death of at least 10,000 people in Honduras, Nicaragua, and El Salvador. Mud slides and flooding destroyed structures and roads, leaving an estimated 2.5 million people homeless. Modern tracking equipment has improved hurricane forecasting and reduced the number of fatalities, primarily by allowing early evacuation of areas in the hurricane's path. However, improved forecasting cannot reduce the damage to crops, forests, or infrastructure. A badly timed storm can destroy a banana harvest or shut down resorts for a season or more. The sheer force of a hurricane can transform the landscape as well, turning a palm-strewn sandy beach into a treeless rocky shore covered with debris.

Population and Settlement: Densely Settled Islands and Rimland Frontiers

In the Caribbean, the population density is generally quite high and, as in neighboring Latin America, increasingly urban. Eighty-six percent of the region's population is concentrated on the four islands of the Greater Antilles (Figure 5.9). Add to this Trinidad's 1.3 million and Guyana's 700,000, and most of the population of the Caribbean is accounted for by six countries and one U.S. territory (Puerto Rico).

In terms of total population, few people inhabit the Lesser Antilles; nevertheless, some of these island states are densely settled. The small island of Barbados is an extreme example. With only 166 square miles of territory, it has 1,620 people per square mile. Population densities on St. Vincent, Martinique, and the Netherlands Antilles, while not as high, are still more than 700 people per square mile. Since productive land is scarce on some of these islands, access to agricultural land has become a basic resource problem for many Caribbean peoples. The growth in the region's population coupled with its scarcity of land has forced many people into the cities or abroad. It also has forced many Caribbean states to import more food than they export.

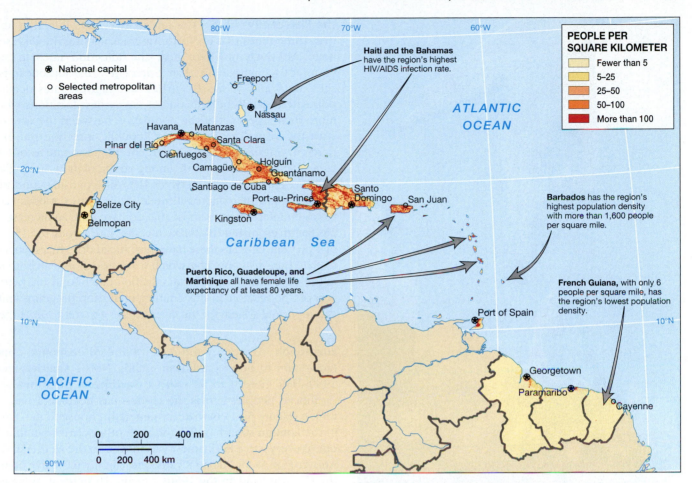

FIGURE 5.9 • POPULATION OF THE CARIBBEAN The major population centers are on the islands of the Greater Antilles. The pattern here, like the rest of Latin America, is a tendency toward greater urbanism. The largest city of the region is Santo Domingo, followed by Havana. In comparison, the rimland states are very lightly settled. All of Suriname has just 400,000 people, and Belize has fewer than 300,000.

In contrast to the islands, the mainland territories of Belize and the Guianas are lightly populated; Guyana averages 8 people per square mile, and Suriname averages only 7. These areas are sparsely settled, in part because the relatively poor quality and accessibility of arable land made them less attractive to colonial enterprises.

Demographic Trends

During the years of sugar production based on slave labor, mortality rates were extremely high due to disease, inhumane treatment, and malnutrition. Consequently, the only way population levels could be maintained was by continuing to import African slaves. With the end of slavery in the mid- to late nineteenth century and the gradual improvement of health and sanitary conditions on the islands, natural population increases began to occur. In the 1950s and 1960s many states achieved peak growth rates of 3.0 or higher, causing population totals and densities to soar. Over the past 20 years, however, growth rates have come down or stabilized. As noted earlier, the current population of the Caribbean is 38 million. The population is now growing at an annual rate of 1.3 percent, and projected population in 2025 is 48 million.

Fertility Decline The most significant demographic trend in the Caribbean is that of fertility decline (Figure 5.10). Cuba and Barbados have the region's lowest rates of natural increase. In socialist Cuba the education of women combined with the availability of birth control and abortion means that the average woman has 1.6 children (compared to 2.1 in the United States). In capitalist Barbados similar results have been achieved. Population pressures in Barbados have also been eased by a steady out-migration of young Barbadians overseas, especially to Great Britain. This movement of the population, in combination with a preference for smaller families, has contributed to slower growth rates throughout the region. Even states with relatively high TFR rates, such as Haiti, have seen a decline in family size. Haiti's total fertility rate fell from 6.0 in 1980 to 4.7 in 2003.

FIGURE 5.10 • FEWER CHILDREN A mother from the Bahamas with her two children. The average Caribbean woman has far fewer children now than 30 years ago. Higher levels of education, improved availability of contraception, an increase in urban living, and the large percentage of women in the labor force contribute to slower population growth rates in many countries. *(Tony Arruza/Tony Arruza Photography)*

The Rise of HIV/AIDS The rate of HIV/AIDS infection in the Caribbean is more than three times that in North America, making the disease an important regional issue. Although nowhere near the infection rates in Sub-Saharan Africa (see Chapter 6), slightly more than 2 percent of the Caribbean population between the ages of 15 and 49 has HIV/AIDS. The highest rates occur on some of the most populated islands. In Haiti, one of the earliest locations where AIDS was detected, more than 5 percent of the population between the ages of 15 and 49 is infected with the virus. The Dominican Republic infection rate has climbed to nearly 3 percent; Guyana also had a 3 percent infection rate as of 1999. AIDS is already the largest single cause of death among young men in the English-speaking Caribbean.

There is a relationship between HIV/AIDS transmission and prostitution that serves both local and international clientele. While many Caribbean countries do not keep statistics on infection rates, those that do have seen an increase. The Bahamas, for example, has an estimated HIV/AIDS infection rate among the 15- to 49-year-old population of 4 percent. Officials there have introduced drug treatments to help prevent mother-to-child transmission. Cuba, which witnessed a surge in both tourism and prostitution in the 1990s, claims to have an infection rate of only 0.2 percent among its 15- to 49-year-old population. In this country, education programs and an effective screening and reporting system for the disease have apparently kept the infection rate down.

Emigration Driven by the region's limited economic opportunities, a pattern of emigration to other Caribbean islands, North America, and Europe began in the 1950s. For more than 50 years a **Caribbean diaspora**—the economic flight of Caribbean peoples across the globe—has become a way of life for much of the region (Figure 5.11). Barbadians generally choose to move to England, most settling in the London suburb of Brixton with other Caribbean immigrants. In contrast, one out of every three Surinamese has moved to the Netherlands, with most residing in Amsterdam. As for Puerto Ricans, only slightly more live on the island than reside on the U.S. mainland. In the 1980s roughly 10 percent of Jamaica's population legally immigrated to North America (some 200,000 to the United States and 35,000 to Canada). Cubans have made the city of Miami their destination of choice since the 1960s. Today they are a large percentage of that city's population.

Movement within the region is also important. Haitians, one-fifth of whom do not live in their country of birth, most commonly relocate to the neighboring Dominican Republic, followed by the United States, Canada, and French Guiana. Dominicans also leave; the vast majority come to the United States, settling in New York. Others, however, simply cross the Mona Passage and settle in Puerto Rico. The economic implications of this labor-related migration are significant and will be discussed later.

The Rural–Urban Continuum

Initially, plantation agriculture and subsistence farming shaped Caribbean settlement patterns. Low-lying arable lands were dedicated to export agriculture and controlled by the colonial elite. Only small amounts of land were set aside for subsistence production. Over time, villages of freed or runaway slaves were established, especially in remote island interiors. Still the vast majority of people lived on estates as owners, managers, or slaves. The few cities existed to serve the administrative and social needs of the colonizers, but most were small, containing a small fraction of a colony's population. The colonists who linked the Caribbean to the world economy saw no need to develop major urban centers.

Even today the structure of Caribbean communities reflects the plantation legacy. Many of the region's subsistence farmers are ancestors of former slaves who continue to farm their small plots and work part-time as wage-labor on estates. The social and economic patterns generated by slavery still mark the landscape. Rural communities tend to be loosely organized; labor is temporary; and small farms are scattered wherever pockets of available land are found. Since men tend to leave home for seasonal labor, female-headed households are common.

FIGURE 5.11 • CARIBBEAN DIASPORA Emigration has long been a way of life for Caribbean peoples. With relatively high education levels but limited professional opportunities, the region commonly loses residents to North America, Great Britain, France, and the Netherlands. Migration within the region also occurs, mainly between Haiti and the Dominican Republic or the Dominican Republic and Puerto Rico. *(Data from Levin, 1987,* Caribbean Exodus, *New York: Praeger Publishers)*

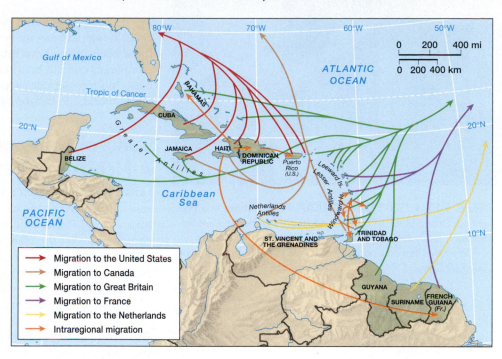

Caribbean Cities Since the 1960s, the mechanization of agriculture, offshore industrialization, and rapid population growth have caused a surge in rural-to-urban migration. Cities have grown accordingly, and today 60 percent of the region is classified as urban. Of the large countries, Cuba is the most urban (75 percent) and Haiti the least (35 percent). Caribbean cities are not large by world standards, as only four have 1 million or more residents: Santo Domingo (2.6 million), Havana (2.2 million), Port-au-Prince (1.5 million), and San Juan (1.0 million). All but Port-au-Prince were former Spanish colonies.

Like their counterparts in Latin America, the Spanish Caribbean cities were laid out on a grid with a central plaza. Vulnerable to raids by rival European powers and pirates, these cities were usually walled and extensively fortified. The oldest continually occupied European city in the Americas is Santo Domingo in the Dominican Republic, settled in 1496, and today the largest city in the region. Havana emerged as the most important colonial city in the region, serving as a port for all incoming and outgoing Spanish ships. Strategically situated on Cuba's north coast at a narrow opening to a natural deep-water harbor, Havana became an essential city for the Spanish empire. Consequently, Havana possesses a handsome collection of colonial architecture, especially from the eighteenth and nineteenth centuries (Figure 5.12). Only recently did Santo Domingo edge out Havana as the Caribbean's largest city.

Other colonial powers left their mark on the region's cities. For example, Paramaribo, the capital of Suriname, has been described as a tropical, tulipless extension of Holland. In the British colonies a preference for wooden whitewashed cottages with shutters was evident. Yet the British and French colonial cities tended to be unplanned afterthoughts; these port cities were built to serve the rural estates, not the other way around. Most of them have grown dramatically over the last 40 years. No longer small ports for agricultural exports, increasingly the focus of these cities is on welcoming cruise ships and sun-seeking tourists.

Caribbean cities and towns do have their charms and reflect a blend of cultural influences. Throughout the region, houses are often simple structures (made of wood, brick, or stucco), raised off the ground a few feet to avoid flooding and painted in soft pastels. Most people still get around by foot, bicycle, or public transportation; neighborhoods are filled with small shops and services that are within easy walking distance. Streets are narrow, and the pace of life is markedly slower than in North America and Europe. Even when space is tight in town, most settlements are close to the sea and its cooling breezes. An afternoon or evening stroll along the waterfront is a common activity.

Housing The sudden surge in urbanization in the Caribbean is best explained by decrease in rural jobs, rather than a rise in urban opportunities. Thousands poured into the cities as economic migrants, erecting shantytowns and filling the ranks of the informal sector. Squatter settlements in Port-au-Prince and Santo Domingo are especially bad, with residents living under miserable conditions without the benefit of sewers and running water. Electricity is usually taken illegally from nearby power lines.

FIGURE 5.12 • OLD HAVANA A Cuban family strolls down a narrow cobble-stoned street near the Cathedral in Old Havana, Cuba. The best examples of eighteenth- and nineteenth-century colonial architecture in the Caribbean are found in Havana. In 1982 UNESCO declared Old Havana a World Heritage Site, and new funds became available to aid its restoration. *(Rob Crandall/Rob Crandall, Photographer)*

FIGURE 5.13 • GOVERNMENT HOUSING Cubans in a state-built apartment block on the outskirts of greater Havana. Under socialism, thousands of standardized apartment blocks were built to ensure that all Cubans had access to basic modern housing. *(Rob Crandall/Rob Crandall, Photographer)*

The one place that dramatically breaks from this pattern is Cuba. Forged in a socialist mode, Cubans are housed in uniform government-built apartment blocks like those seen throughout Russia and eastern Europe (Figure 5.13). These uninspired complexes contain hundreds of identical one- and two-bedroom apartments where basic modern services are provided. Compared to other large cities of the developing world, the lack of squatter settlements makes Havana unusual.

Cultural Coherence and Diversity: A Neo-Africa in the Americas

Linguistic, religious, and ethnic differences thrive in the Caribbean. The presence of several former European colonies, millions of descendents of ethnically distinct Africans and indentured workers from India and China, and isolated Amerindian communities on the mainland challenges any notion of cultural coherence.

What holds this region together are the common historical and cultural processes. In particular, this section will focus on three cultural influences shared throughout the Caribbean. They are the European colonial presence, the African influence, and the mix of European and African cultures referred to as **creolization**.

The Cultural Impact of Colonialism

The arrival of Columbus in 1492 triggered a devastating chain of events that depopulated the Caribbean region within 50 years. A combination of Spanish brutality, enslavement, warfare, and disease reduced the densely settled islands, which supported up to 3 million Caribs and Arawaks, into an uninhabited territory ready for the colonizer's hand. By the mid-sixteenth century, as rival European states competed for Caribbean territory, the lands they fought for were virtually uninhabited. In many ways this simplified their task, as they did not have to recognize native land claims or work amid Amerindian societies. Instead, the Caribbean territories were reorganized to serve a plantation-based production system. The critical missing element was labor. Once slave labor from Africa, and later contract labor from Asia, was secured, the small Caribbean colonies became surprisingly profitable. Much of Caribbean culture and society today can be traced to the same processes that created "plantation America."

FIGURE 5.14 • SUGAR PLANTATION A historical illustration (1823) of slaves harvesting sugarcane on a plantation in Antigua. Sugar production was profitable but physically demanding work. Several million Africans were enslaved and forcibly relocated into the region. *(Michael Holford/Michael Holford Photographs)*

Plantation America The term **plantation America** was coined by anthropologist Charles Wagley to designate a cultural region that extends from midway up the coast of Brazil through the Guianas and the Caribbean into the southeastern United States. Ruled by a European elite dependent on an African labor force, this society primarily existed in coastal regions and produced agricultural exports. It relied on **mono-crop production** (a single commodity, such as sugar) under a plantation system that concentrated land in the hands of elite families. Such a system created rigid class lines, as well as the formation of a multiracial society in which people with lighter skin were privileged. The term *plantation America* is not meant to describe a race-based division of the Americas, but rather a production system that brought about specific ecological, social, and economic relations (Figure 5.14).

Asian Immigration Before detailing the pervasive influence of Africans on the Caribbean, the lesser-known Asian presence deserves mention. By the mid-nineteenth century, most colonial governments in the Caribbean had begun to free their slaves. Fearful of labor shortages, they sought **indentured labor** (workers contracted to labor on estates for a set period of time, often several years) from South and Southeast Asia.

The legacy of these indentured arrangements is clearest in Suriname, Guyana, and Trinidad and Tobago. In Suriname, a former Dutch colony, more than one-third of the population is of South Asian descent, and 16 percent are Javanese (from Indonesia). Guyana and Trinidad were British colonies, and most of their contract labor came from India. Today half of Guyana's population and 40 percent of Trinidad and Tobago's claim South Asian ancestry. Hindu temples are found in the cities and villages, and many families speak Hindi in the home. Such racial diversity has led to tension. During Guyana's general election in 2001, heightened tensions between Afro- and Indo-Guyanese caused international election observers to be called in to monitor the vote. In the end an Indo-Guyanese candidate was peacefully elected, but many fear that Guyana is teetering on the brink of serious ethnic conflict (Figure 5.15).

Creating a Neo-Africa

The introduction of African slaves to the Americas began in the sixteenth century and continued into the nineteenth century. This forced migration of Africans to the Americas was only part of a much more complex **African diaspora**—the forced removal of Africans from their native area. The slave trade also crossed the Sahara to include North Africa, and linked East Africa with a slave trade in the Middle East (see Chapter 6). The best documented slave route was the transatlantic one; at least 10 million Africans landed in the Americas, and it is estimated that another 2 million died en route. More than half of these slaves were sent to the Caribbean (Figure 5.16).

This influx of slaves, combined with the elimination of nearly all the native inhabitants, remade the Caribbean as the area with the greatest concentration of relocated African people in the Americas. The African source areas extended from Senegal to Angola, and slave purchasers intentionally mixed tribal groups in order to weaken ethnic identities. Consequently, intact transfer of religion and languages into the Caribbean did not occur; instead languages, customs, and beliefs were blended.

Maroon Societies   Communities of runaway slaves—termed **maroons**—offer the most interesting examples of African cultural diffusion across the Atlantic. Hidden settlements of escaped slaves existed wherever slavery was practiced. Called maroons in English, *palenques* in Spanish, and *quilombos* in Portuguese, many of

FIGURE 5.15 • AFRO- AND INDO-GUYANESE POLITICIANS During a 2001 campaign rally, the Indo-Guyanese President, Bharrat Jagdeo and the Afro-Guayanese Prime Minister, Samuel Hinds, stand side by side. In Guyana, peoples of Indian (South Asian) and African ancestry make up the majority of the population. *(AFB/CORBIS)*

FIGURE 5.16 • TRANSATLANTIC SLAVE TRADE   At least 10 million Africans landed in the Americas during the four centuries in which the Atlantic slave trade operated. Most of the slaves came from West Africa, especially the Gold Coast (now Ghana) and the Bight of Biafra (now Nigeria). Angola, in southern Africa, was also an important source area. *(Data based on Curtin, 1969, The Atlantic Slave Trade, A Census. Madison: University of Wisconsin Press, p. 268.)*

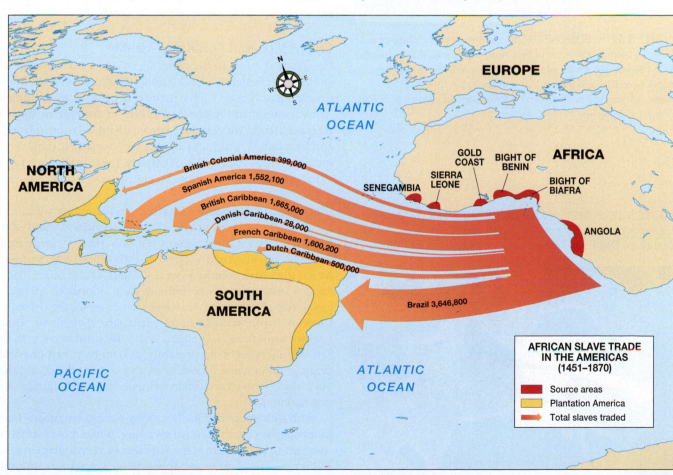

these settlements were short-lived. However, others have endured and allowed for the survival of African traditions, especially farming practices, house designs, community organization, and language.

The maroons of Jamaica formed dozens of isolated and independent communities in the forested mountains of the island's interior. A constant source of aggravation for the white plantation owners because slaves sought to escape to them, these small communities endured and are well documented. The most striking characteristic of these settlements (with names like Me No Seen and You No Come) was their ability to maintain some African language and religious practices, especially the belief in Obi (an African god) and Obeah (black magic). Today villages that began as maroon settlements might look like any other rural community in Jamaica, except that there is a local appreciation for the origin of these communities.

The so-called Bush Negros of Suriname still show clear links to West Africa. Whereas other maroon societies gradually blended into their local populations, to this day the Bush Negros maintain a distinct identity. Six distinct Bush Negro tribes formed, ranging in size from a few hundred to 20,000 (Figure 5.17). Living relatively undisturbed for 200 years, these rainforest inhabitants fashioned a rich ritual life for themselves involving prophets, spirit possession, and witch doctors.

African Religions   Linked to maroon societies, but more widely diffused, is the transfer of African religious and magical systems to the Caribbean. These patterns are another reflection of neo-Africa in the Americas and are most closely associated with northeastern Brazil and the Caribbean. Millions of Brazilians practice the African-based religions of Umbanda, Macuba, and Candomblé, along with Catholicism. Likewise, Afro-religious traditions in the Caribbean have evolved into unique forms that have clear ties to West Africa. The most widely practiced are Voodoo (also Vodoun) in Haiti, Santería in Cuba, and Obeah in Jamaica. These religions have their own priesthood and unique patterns of worship. Their impact is considerable; the father and son dictators of Haiti, the Duvaliers, were known to hire Voodoo priests to scare off government opposition (Figure 5.18).

Creolization and Caribbean Identity

Creolization refers to the blending of African, European, and even some Amerindian cultural elements into the unique cultural systems found in the Caribbean. The Creole identities that have formed over time are complex; they illustrate the dynamic cultural and national identities of the region. Today Caribbean writers (V. S. Naipaul, Derek Walcott, and Jamaica Kinkaid), musicians (Bob Marley, Celia Cruz, and Juan Luís Guerra), and artists (Trinidadian costume designer Peter Minshall) are internationally regarded. Collectively, these artists are representative of their individual islands and of Caribbean culture as a whole.

Language   The dominant languages in the region are European: Spanish (24 million speakers), French (8 million), English (6 million), and about half a million Dutch (Figure 5.19). Yet these figures tell only part of the story. In Cuba, the Dominican Republic, and Puerto Rico, Spanish is the official language, and it is universally spoken. As for the other countries, local variants of the official language exist, especially in spoken form, that can be difficult for a nonnative speaker to understand. In some cases completely new languages emerge; in the islands of Aruba, Bonaire and Curaçao, Papiamento (a trading language that blends Dutch, Spanish, Portuguese, English, and African languages) is the *lingua franca*, with usage of Dutch declining. Similarly, French Creole, or *patois*, in Haiti has been given official status as a distinct language. In practice, French is used in higher education, government, and the courts, but *patois* (with clear African influences) is the language of the street, the home, and oral tradition.

With independence in the 1960s, Creole languages became politically and culturally charged with national meaning. While most formal education is taught using standard language forms, the richness of vernacular expression and its ability to instill a sense of identity are appreciated. Locals rely on their ability to switch from

FIGURE 5.17 • BUSH NEGROS   A man weaves a basket in the Bush Negro settlement of Bolopasi, Suriname. The so-called Bush Negros fled Dutch plantations in the eighteenth and nineteenth centuries and settled in the interior rain forests of Suriname, where they continue many West African cultural practices. *(Martha Cooper/The Viesti Collection, Inc.)*

FIGURE 5.18 • AFRICAN RELIGIOUS INFLUENCES   A voodoo practitioner in a trance during a ceremony honoring the God of the Dead in a Port-au-Prince cemetery. *(AFB/CORBIS)*

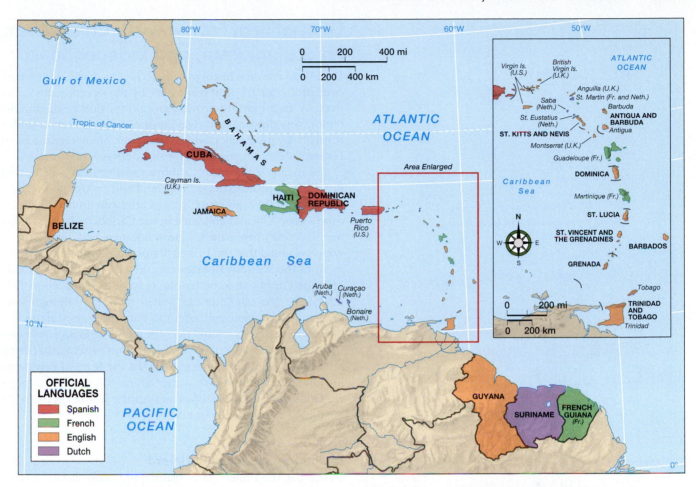

FIGURE 5.19 • CARIBBEAN LANGUAGE MAP   Since this region has no significant Amerindian population (except on the mainland), the dominant languages are European: Spanish (24 million), French (8 million), English (6 million), and Dutch (0.5 million). However, many of these languages have been creolized, making it difficult for outsiders to understand them.

standard to vernacular forms of speech. Thus, a Jamaican can converse with a tourist in standard English and then switch to a Creole variant when a friend walks by, effectively excluding the outsider from the conversation. While this ability to switch is evident in many cultures, it is widely used in the Caribbean.

Music   The rhythmic beats of the Caribbean might be the region's best-known product. This small area is the home of reggae, calypso, merengue, rumba, zouk, and scores of other musical forms. The roots of modern Caribbean music reflect a combination of African rhythms with European forms of melody and verse. These diverse influences coupled with a long period of relative isolation sparked distinct local sounds. As movement among the Caribbean population increased, especially during the twentieth century, musical traditions were blended, but characteristic sounds remained.

The famed steel pan drums of Trinidad were created from oil drums discarded from a U.S. military base there in the 1940s. The bottoms of the cans are pounded with a sledgehammer to create a concave surface that produces different tones. During Carnival (a pre-Lenten celebration), racks of steel pans are pushed through the streets by dancers while drummers play. So skilled are these musicians that they even perform classical music, and government agencies encourage troubled teens to learn steel pan (Figure 5.20).

The distinct sound and the ingenious rhythms make Caribbean music so popular. Yet it is more than good dancing music; the music is closely tied to Afro-Caribbean religions and is a popular form of political protest. In Haiti, *ra-ra* music mixes percussion instruments, saxophones, and bamboo trumpets, while weaving in funk and reggae basslines. The songs are always performed in French Creole and typically celebrate Haiti's African ancestry and the use of Voodoo. The lyrics deal with difficult issues, such as political oppression or poverty.

FIGURE 5.20 • CARNIVAL DRUMMER   A steel pan drummer performs while his drum cart is pushed through the streets during Carnival. Steel drums originated from discarded oil cans that local peoples fashioned into drums by hammering the tops into a concave surface. Many steel drum bands perform internationally, playing everything from calypso to classical music. *(Rob Crandall/Rob Crandall, Photographer)*

Geopolitical Framework: Colonialism, Neocolonialism, and Independence

Caribbean colonial history is a patchwork of competing European powers fighting over profitable tropical territories. By the seventeenth century, the Caribbean had become an important proving ground for European ambitions. Spain's grip on the region was slipping, and rival European nations felt confident that they could gain territory by gradually pushing Spain out. Many territories, especially islands in the Lesser Antilles, changed European rulers several times.

Europeans viewed the Caribbean as a strategically located and profitable region in which to produce sugar, rum, and spices. Geopolitically, rival European powers also felt that their presence in the Caribbean limited Spanish authority there. Yet Europe's geopolitical dominance in the Caribbean began to diminish by the mid-nineteenth century just as the U.S. presence increased. Inspired by the **Monroe Doctrine**, which claimed that the United States would not tolerate European military involvement in the Western Hemisphere, the U.S. government made it clear that it considered the Caribbean to be within its sphere of influence. This view was highlighted during the Spanish-American War in 1898. Even though several English, Dutch, and French colonies remained after this date, the United States indirectly (and sometimes directly) asserted its control over the region, bringing in a period of **neocolonialism**. In an increasingly global age, however, even neocolonial ties can be short-lived or sporadic. The Caribbean has not attracted the level of private foreign investment seen by other regions. Moreover, as the Caribbean's strategic importance in a post-Cold War era fades and old colonial ties unravel, the leaders of the region openly worry about becoming even more marginal.

Life in the "American Backyard"

To this day, the United States maintains a controlling attitude toward the Caribbean, referring to it as "the American backyard." Initial foreign policy objectives were to free the region from European authority and encourage democratic governance. Yet time and again, American political and economic ambitions undermined those goals. President Theodore Roosevelt made his priorities clear with imperialistic policies that extended the influence of the United States beyond its borders. Policies such as the construction of the Panama Canal and the maintenance of open sea-lanes benefited the United States but did not necessarily support social, economic, or political gains for the Caribbean people. The United States later offered benign-sounding development packages such as the Good Neighbor Policy (1930s), the Alliance for Progress (1960s), the Caribbean Basin Initiative (1980s), and, most recently, the proposed Free Trade Area of the Americas (2005). The Caribbean view of these initiatives has been wary at best. Rather than feeling liberated, many residents believe that one kind of political dependence was being traded for another—colonialism for neocolonialism.

In the early 1900s the role of the United States in the Caribbean was visibly military and political. The Spanish-American War (1898) secured Cuba's freedom from Spain and also resulted in Spain's giving up the Philippines, Puerto Rico, and Guam to the United States; the latter two are still U.S. territories. The U.S. government also purchased the Danish Virgin Islands in 1917, renaming them the U.S. Virgin Islands and developing the harbor of St. Thomas. French, English, and Dutch colonies were tolerated as long as these allies recognized the supremacy of the United States in the region. Outwardly against colonialism, the United States had become much like an imperial force.

One of the requirements of an empire is the ability to impose one's will, by force if necessary. When a Caribbean state refused to abide by U.S. trade rules, U.S. Navy vessels would block its ports. Marines landed and U.S.-backed governments were installed throughout the Caribbean basin. These were not short-term engagements; U.S. troops occupied the Dominican Republic from 1916 to 1924, Haiti from 1913 to 1934, and Cuba from 1906 to 1909 and 1917 to 1922 (Figure 5.21). Even today, several important military bases are in the region, including Guantánamo in eastern Cuba.

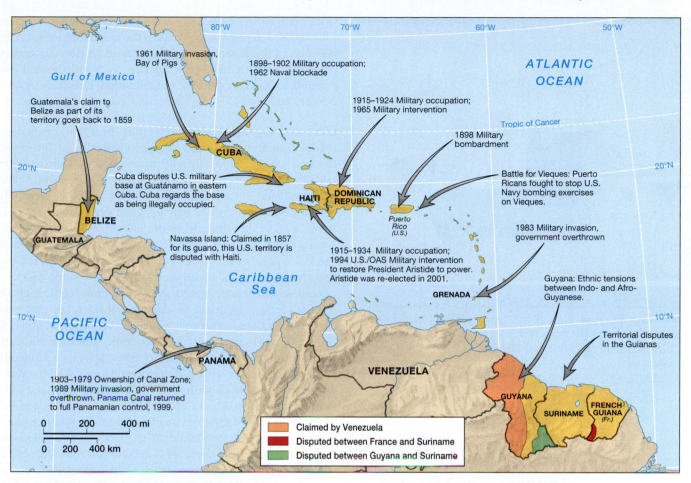

FIGURE 5.21 • U.S. MILITARY INVOLVEMENT AND REGION- AL DISPUTES The Caribbean was regarded as the geopolitical backyard of the United States, and U.S. military occupation was a common occurrence in the first half of the twentieth century. Border and ethnic conflicts also exist, most notably in the Guianas. (Data sources: Barbara Tenenbaum, ed., 1996 Encyclopedia of Latin American History and Culture, vol. 5, p. 296, with permission of Charles Scribner's Sons; and John Allcock, 1992, Border and Territorial Disputes, 3rd ed., Harlow, Essex, UK: Longman Group)

Many critics of U.S. policy in the Caribbean complain that business interests overwhelm democratic principles when foreign policy is determined. U.S. banana companies settled the coastal plain of the Caribbean rimland and operated as if they were independent states. Sugar and rum manufacturers from the United States bought the best lands in Cuba, Haiti, and Puerto Rico. Meanwhile, truly democratic institutions remained weak, and there was little improvement in social development. True, exports increased, railroads were built, and port facilities improved; but levels of income, education, and health remained dreadfully low throughout the first half of the twentieth century.

The Commonwealth of Puerto Rico Puerto Rico is both within the Caribbean and apart from it because of its status as a commonwealth of the United States. Throughout the twentieth century, various Puerto Rican independence movements sought to separate the island from the United States. Even today, residents of the island are divided about their island's political future. At the same time, Puerto Rico depends on U.S. investment and welfare programs; U.S. food stamps are a major source of income for many Puerto Rican families. Commonwealth status also means that Puerto Ricans can freely move between the island and the U.S. mainland, a right they actively assert. In other ways Puerto Ricans symbolically display their independence; for example, they support their own "national" sports teams and send a Miss Puerto Rico to international beauty pageants.

Puerto Rico led the Caribbean in the transition from an agrarian economy to an industrial one, beginning in the 1950s. For some U.S. officials, Puerto Rico became the model for the rest of the region. Puerto Rican President Muñoz Marín advocated an industrialization program called "Operation Bootstrap." Through tax incentives and cheap labor, hundreds of U.S. textile and clothing firms relocated to Puerto Rico. Over the next two decades, 140,000 industrial jobs were added, resulting in a marked increase in per capita GNP. In the 1970s, when Puerto Rico faced stiff competition

FIGURE 5.22 • PUERTO RICAN FACTORY A pharmaceutical plant near San Juan, Puerto Rico. "Operation Bootstrap" helped launch the rapid industrialization of Puerto Rico. The island is the Caribbean's most industrialized location. *(Robert Frerck/Odyssey Productions)*

from Asian clothing manufacturers, the government encouraged petrochemical and pharmaceutical plants to relocate to the island (Figure 5.22). By the 1990s, Puerto Rico was one of the most industrialized places in the region, with a significantly higher per capita income than its neighbors. Yet it still showed many signs of underdevelopment, including extensive out-migration, low rates of education, and widespread poverty and crime.

Cuba and Regional Politics   The most profound challenge to U.S. authority in the region came from Cuba and its superpower ally, the former Soviet Union. Historically, Havana was the dominant city of the region. Cubans formally challenged Spain's colonial rule in 1895. Over a three-year period, 400,000 Cubans died in their fight for independence. Even though the United States is credited with winning what is commonly known as the Spanish-American War in 1898, U.S. forces were relative latecomers to this struggle. In the end, many Cubans resented the presence of the United States on their island because it weakened their hard-won independence. In the 1950s another revolutionary effort began in Cuba, led by Fidel Castro against the pro-American Batista government. Cuba's economic productivity had soared but its people were still poor, uneducated, and increasingly angry. The contrast between the lives of average sugarcane workers and the foreign elite was sharp. Castro tapped a deep vein of Cuban resentment against U.S. policy, and in 1959 Castro took power.

After Castro's government nationalized American industries and took ownership of all foreign-owned properties, the United States responded by refusing to buy Cuban sugar and ultimately ending diplomatic relations with the state. Various U.S. embargoes (laws forbidding trade with a particular country) against Cuba have existed for more than four decades. When Cuba established strong diplomatic relations with the Soviet Union in 1960 during the height of the Cold War, the island state became a geopolitical enemy of the United States. With the Soviet Union financially and militarily backing Castro, a direct U.S. invasion of Cuba was too risky. The fall of 1962 produced one of the most dangerous episodes of the Cold War when Soviet missiles were discovered on Cuban soil. Ultimately, the Soviet Union removed its weapons; in return, the United States promised not to invade Cuba.

The Cuban missile crisis signaled that American dominance in the Caribbean was not universally accepted. Cuba by itself was not a threat to the United States, but the geopolitical significance of its turn toward the Soviet Union could not be ignored. Moreover, Castro wanted to extend his socialist vision abroad, supporting at different times revolutionaries in Colombia, Venezuela, Guyana, Suriname, Bolivia, and across the Atlantic in Angola. Other Caribbean states watched the Cuba socialist experiment with interest, especially its remarkable improvements in literacy and public health. By the mid-1970s Cuba had the lowest infant mortality rate in all of the Caribbean and Latin America.

Cuban-style socialism did not readily transfer to other countries, however, because the United States was determined to prevent it, and Soviet support of Cuba was expensive. Even with the end of the Cold War, Cuba maintains an identity apart from its regional neighbors and in defiance of the United States. By the 1990s, every other independent country in the region had a democratically elected leader. And as the geopolitical landscape of the region is being reworked with the Free Trade Area of the Americas treaty, Cuba has not been part of negotiations.

Independence and Integration

Given the repressive colonial history of the Caribbean, it is no wonder that the struggle for political independence began more than 200 years ago. Haiti was the second colony in the Americas to gain independence in 1804; the United States was the first in 1776. However, the political independence of many states in the region has not guaranteed economic independence. Many Caribbean states struggle to meet the basic needs of their people. Surprisingly, today some Caribbean territories maintain their colonial status as an economic asset. For example, the French territories of Martinique, Guadeloupe, and French Guiana are overseas departments of France; residents have full French citizenship and social welfare benefits.

Independence Movements   Haiti's revolutionary war began in 1791 and ended in 1804. During this conflict the island's population was cut in half by casualties and emigration; ultimately, the former slaves became the rulers. Independence, however, did not allow this crown of the French Caribbean to prosper. Slowed down by economic and political problems, it was ignored by the European powers and never accepted by the states of the Spanish mainland when they became independent in the 1820s.

Several revolutionary periods followed in the nineteenth century. In the Greater Antilles, the Dominican Republic finally gained independence in 1844 after taking control of the territory from Spain and Haiti. Cuba and Puerto Rico were freed from Spanish colonialism in 1898, but their independence was weakened by greater U.S. involvement. The British colonies also faced revolts, especially in the 1930s; yet it was not until the 1960s that independent states emerged from the English Caribbean. First the larger colonies of Jamaica, Trinidad and Tobago, Guyana, and Barbados gained their independence. Other British colonies followed throughout the 1970s and early 1980s. Suriname, the only Dutch colony on the rimland, became a self-governing territory in 1954 but remained part of the Kingdom of the Netherlands until 1975, when it declared itself an independent republic.

Regional Integration   Perhaps the hardest task facing the Caribbean is to increase economic integration. Scattered islands, a divided rimland, different languages, and limited economic resources hinder the formation of a meaningful regional trade block. It is more common to see economic cooperation between groups of islands with a shared colonial background than between, for example, former French and English colonies.

During the 1960s, the Caribbean began to experiment with regional trade associations as a means to improve its economic competitiveness. The goal of regional cooperation was to improve employment rates, increase intraregional trade, and ultimately reduce economic dependence. The countries of the English Caribbean took the lead in this development strategy. In 1963 Guyana proposed an economic integration plan with Barbados and Antigua. In 1972 the integration process intensified with the formation of the **Caribbean Community and Common Market (CARICOM)**. Representing the former English colonies, CARICOM proposed an ambitious regional industrialization plan and the creation of the Caribbean Development Bank to assist the poorer states. As important as this trade group is as an economic symbol of regional identity, it has produced limited improvements in intraregional trade. There are 13 full member states—all of the English Caribbean and French-speaking Haiti. Other dependencies, such as Anguilla, Turks and Caicos, and the British Virgin Islands, are associate members. Still, the predominance of English-speaking territories in CARICOM illustrates the deep linguistic divides in the Caribbean.

The dream of regional integration as a way to produce a more stable and self-sufficient Caribbean has never been realized. One scholar of the region argues that a limiting factor is a "small-islandist ideology." For example, islanders tend to keep their backs to the sea, oblivious to the needs of neighbors. At times such isolationism results in suspicion, distrust, and even hostility toward nearby states. Yet economic necessity dictates engagement with partners outside the region. And so this peculiar status of isolated proximity unfolds in the Caribbean, expressing itself in uneven social and economic development trends.

Economic and Social Development: From Cane Fields to Cruise Ships

Collectively, the population of the Caribbean, although poor by U.S. standards, is economically better off than most of Sub-Saharan Africa, South Asia, and China. Despite periods of economic stagnation in the Caribbean, social gains in education, health, and life expectancy are significant. Historically the Caribbean's links to the world economy were through tropical agricultural exports, yet several specialized industries, such as tourism, offshore banking, and assembly plants, have challenged the dominance of agriculture. These industries grew because of the region's proximity to North America and Europe, the availability of cheap labor, and the presence of policies that created a nearly tax-free environment for foreign-owned companies. Unfortunately, growth in these industries does not employ all the region's displaced rural laborers, so the lure of jobs in North America and Europe is still strong.

From Fields to Factories and Resorts

Agriculture used to dominate the economic life of the Caribbean. Decades of unstable crop prices and decline in special trade agreements with former colonial masters have produced more hardship than prosperity. Ecologically, the soils are overworked, and there are no frontier areas to expand production, except for areas of the rimland. Moreover, agricultural prices have not kept pace with rising production costs, so wages and profits remain low. With the exception of a few mineral-rich territories, such as Trinidad, Guyana, Suriname, and Jamaica, most countries have tried to diversify their economies, relying less on their soils and more on manufacturing and services.

Comparing export figures over time demonstrates the shift away from monocrop dependence. In 1955 Haiti earned more than 70 percent of its foreign exchange through the export of coffee; by 1990 coffee accounted for only 11 percent of its export earnings. Similarly, in 1955 the Dominican Republic earned close to 60 percent of its foreign exchange through sugar, but 35 years later sugar earned less than 20 percent of the country's foreign exchange, and pig iron exports nearly equaled that of sugar. The one exception to this trend is Cuba, which earned approximately 80 percent of its foreign exchange through sugar production from the 1950s to 1990. Cuba, however, was forced to diversify in the 1990s when Russia would no longer guarantee the above-market price it paid for sugarcane.

Sugar The economic history of the Caribbean cannot be separated from the production of sugarcane. Even relatively small territories such as Antigua and Barbados yielded fabulous profits because there was no limit to the demand for sugar in the eighteenth century. Once considered a luxury crop, it became a popular necessity for European and North American laborers by the 1750s. It sweetened tea and coffee and made jams a popular spread for stale bread. In short, it made the meager and bland diets of ordinary people tolerable, and it also boosted caloric intake. Distilled into rum, sugar produced a popular intoxicant. Though it is hard to imagine today, consumption of a pint of rum a day was not uncommon in the 1800s.

Sugarcane is still grown throughout the region for domestic consumption and export. Its economic importance has declined, however, mostly due to increased competition from sugar beets grown in the midlatitudes. The Caribbean and Brazil are the world's major sugar exporters. Up until 1990, Cuba alone accounted for more than 60 percent of the value of world sugar exports. Cuba's dominance in sugar exports had more to do with its subsidized and guaranteed markets in eastern Europe and Russia than with unmatched productivity. Since 1990 the Cuban sugar harvest has been reduced by half.

The Banana Wars The major banana exporters are in Latin America, not the Caribbean. In fact, the success of banana plantations is mixed in this region, as banana plants are especially vulnerable to hurricanes. Still, several small states in the Lesser Antilles, most notably Dominica, St. Vincent, and St. Lucia, have become economically dependent on bananas, gaining as much as 60 percent of their export earnings from the yellow fruit. Bananas have not made people rich, yet their production for export has led to greater economic and social development. In the eastern Caribbean, where most bananas are grown on small farms of five acres, the landowners are the laborers and thus earn two to four times more than banana plantation workers in Ecuador and Central America. Moreover, for these small states, banana exports are their link with the global economy. Pressures on the

European Union to drop the higher prices given to banana growers from the former colonies has reduced the economic viability of this crop in the Caribbean (Figure 5.23). Where once a market and a minimum price were guaranteed, under the new global order neither is certain.

The case of the eastern Caribbean shows what happens to the losers in globalization. In 1996 the United States, Ecuador (the world's leading banana exporter), Mexico, Guatemala, and Honduras challenged the EU's banana trade agreement in the World Trade Organization court. The agreement was denounced as unfair, and the EU was told to eliminate it by 1998. To make matters worse, consumer tastes had changed, with buyers preferring the uniformly large and unblemished yellow banana typical of the Latin American plantations rather than these grown in places like St. Lucia. To survive in this newly competitive environment, the growers of the eastern Caribbean will have to produce a more standardized fruit and increase their yield per acre. While island governments are trying to aid in this transition, local farmers have experimented with new crops, such as okra, tomatoes, avocados, and even marijuana. Reportedly, a few well-tended marijuana plants will earn 30 times more per pound than bananas. Just what will happen to family-run banana farms is hard to tell, but many Caribbean growers fear that the days of the banana economy are numbered. The banana story also reflects another major shift for the region: the decline in the economic importance of agriculture as well as the number of people employed by it.

Assembly-plant Industrialization One regional strategy to deal with the unemployed agricultural workers was to invite foreign investors to set up assembly plants and thus create jobs. This was first tried successfully in Puerto Rico in the 1950s and was copied throughout the region. During Puerto Rico's "Operation Bootstrap," island leaders encouraged U.S. investment by offering cheap labor, local tax breaks, and, most important, federal tax exemptions (something only Puerto Rico can do because of its special status as a commonwealth of the United States). Initially the program was a tremendous success, so that by 1970 nearly 40 percent of the island's gross domestic product (GDP) came from manufacturing. Today about 13 percent of the Puerto Rican labor force is employed in manufacturing, and this sector accounts for nearly half of the island's GDP. Yet competition from other states with even lower wages and the U.S. Congress's decision in 1996 to phase out many of the tax exemptions may threaten Puerto Rico's ability to maintain its specialized industrial base.

Through the creation of **free trade zones (FTZs)**— duty free and tax-exempt industrial parks for foreign corporations—the Caribbean is an increasingly attractive location to assemble goods for North American consumers. Manufacturing in the Dominican Republic demonstrates this production trend. The Dominican Republic took advantage of tax incentives and guaranteed access to the U.S. market offered through the Caribbean Basin Initiative. The number of operational free trade zones in the Dominican Republic in 2001 was 16, including an FTZ on the island's north shore, near the Haitian border, optimistically named "Hong Kong of the Caribbean." Plans exist to double the number of FTZs. Firms from the United States and Canada are the most frequent investors in these zones, followed by Dominican, South Korean, and Taiwanese firms. Traditional manufacturing on the island was tied to sugar refining, whereas up to 60 percent of production in the FTZs is in garments and textiles (Figure 5.24). Manufacturing is the largest economic sector in the Dominican Republic, accounting for nearly one-fifth of the country's GDP.

FIGURE 5.23 • BANANA FARM Small banana farms struggle to be economically feasible in a globalizing marketplace. At one time, small island producers such as St. Lucia, shown here, were guaranteed access to the European market. As this special status is removed, many banana-dependent Caribbean nations fear for their livelihood. *(Bob Krist/Corbis)*

FIGURE 5.24 • FREE TRADE ZONES IN THE DOMINICAN REPUBLIC Another sign of globalization is the increase in duty-free and tax exempt industrial parks in the Caribbean. Currently 16 FTZs are operational, with foreign investors from the United States, Canada, South Korea, and Taiwan. *(Source: Modified from Warf, 1995, "Information Services in the Dominican Republic," Yearbook, Conference of Latin Americanist Geographers, p. 15)*

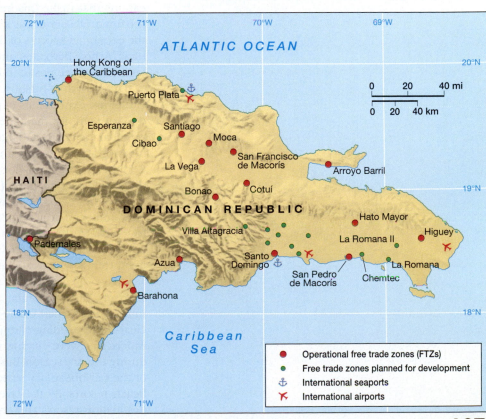

The growth in manufacturing depends on national and international policies that support export-led development through foreign investment. Certainly, new jobs are being created and national economies are diversifying in the process, yet critics believe that foreign investors gain more than the host countries. Since most goods are assembled from imported materials, there is little development of national suppliers. Often higher than local averages, the wages are still low compared to the developed world—sometimes just two to three dollars a day. Moreover, as other developing countries compete with the Caribbean for the establishment of FTZs, this development strategy may become less significant over time.

Offshore Banking The rise of offshore banking in the Caribbean is most closely associated with the Bahamas, which began this industry back in the 1920s. **Offshore banking** centers appeal to foreign banks and corporations by offering specialized services that are confidential and tax-exempt. Places that provide offshore banking make money through registration fees, not taxes. The Bahamas were so successful in developing this sector that by 1976 the country was the third largest banking center in the world. Its dominance began to decline due to competitors from the Caribbean, Hong Kong, and Singapore and because there was no longer a tax advantage in arranging large international loans offshore. Concerns about corruption and laundering of drug money also hurt the islands' financial status in the 1980s, and major reforms were introduced to reduce the presence of funds gained from illegal activities. By 1998 the Bahamas' ranking in terms of global financial centers had dropped to fifteenth. Still, offshore banking remains an important part of the Bahamian economy. In the 1990s, the Cayman Islands emerged as the region's leader in financial services. With a population of 30,000, this crown colony of Britain has some 50,000 registered companies and the highest per capita income of the region.

Each of the offshore banking centers in the Caribbean tries to develop special financial services to attract clients, such as banking, functional operations, insurance, or trusts. The Caribbean is an attractive locality for such services because of its closeness to the United States (home of many of the registered firms), client demand for these services in different countries, and the steady improvement in telecommunications that make this industry possible. The resource-poor islands of the region see financial services as a way to bring foreign capital to state treasuries. Envious of the economic success of the Bahamas and the Cayman Islands, countries such as Antigua, Aruba, Barbados, and Belize hope that greater prosperity will come from establishing close ties to international finance.

The growth of offshore financial centers and the manufacturing zones discussed earlier are expressions of globalization in the developing world. Ironically, banking does not employ many people and can quickly disappear with the slightest sign of political instability. Offshore banking also attracts money tied to the drug trade. A major trafficking area for cocaine bound for North America and Europe, the Caribbean is also a major money laundering center where "dirty" dollars from the drug trade become part of legitimate financial institutions. This brings its own problems to the region, such as increases in drug consumption, corruption of local officials, and even drug-related murders. Thus the financial services industry continues to evolve in the Caribbean, altering many island economies and highlighting this region's ties to the global economy. What is less certain is whether such development trends will greatly improve local earnings and standards of living.

Tourism Environmental, locational, and economic factors converge to support tourism in this part of the world. The earliest visitors to this tropical sea admired its clear and sparkling turquoise waters. By the nineteenth century, wealthy North Americans were fleeing winter to enjoy the healing warmth of the Caribbean during the dry season. Businessmen later realized that the simultaneous occurrence of the Caribbean dry season with the Northern Hemisphere's winter was ideal for beach resorts. By the twentieth century, tourism was well established with both destination resorts and cruise lines. By the 1950s the leader in tourism was Cuba,

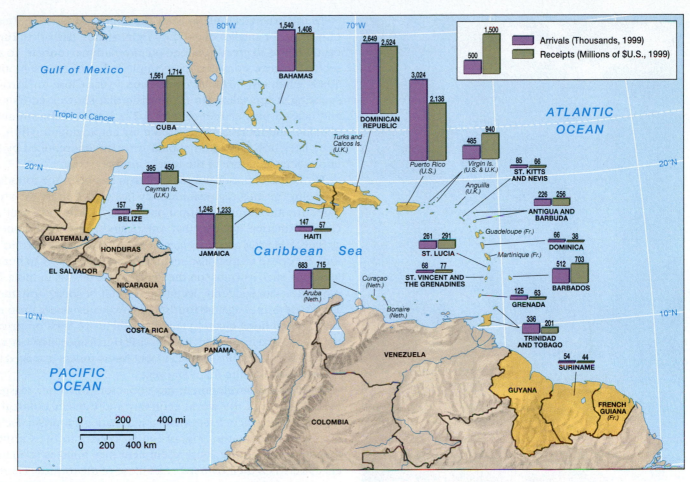

FIGURE 5.25 • GLOBAL LINKAGES: INTERNATIONAL TOURISM IN THE CARIBBEAN   Tourism directly links the Caribbean to the global economy. Each year 14 million tourists come to the islands, mostly from North America, Latin America, and Europe. The most popular destinations are the Dominican Republic, Puerto Rico, Cuba, the Bahamas, and Jamaica. *(Data from* World Bank Atlas, *2001)*

and the Bahamas was a distant second. Castro's rise to power, however, eliminated this sector of the island's economy for nearly three decades and opened the door for other islands to develop their own tourism economies.

Five islands hosted 70 percent of the 14 million international tourists who came to the Caribbean in 1999: Puerto Rico, the Bahamas, the Dominican Republic, Jamaica, and Cuba (Figure 5.25). Puerto Rico saw its tourist sector begin to grow with commonwealth status in 1952. San Juan is now the largest home port for cruise lines and the second largest cruise-ship port in the world in terms of total visitors. The cruise business, in combination with hotel visitors, totaled more than 3 million arrivals in 1999. The Bahamas attributes most of its economic development and high per capita income to tourism. With more than 1.5 million hotel guests in 1999 and 1.7 million cruise ship passengers, the Bahamas is another major hub for tourism in the region. Some 30 percent of the Bahamian population is employed in tourism, and tourism represents nearly half the country's GDP (Figure 5.26). Because the Bahamas is closer to the United States than most Caribbean destinations, the majority of its international visitors are American (about 80 percent), followed by Europeans and Canadians.

After years of neglect, Cuba is reviving tourism in an attempt to earn badly needed foreign currency. Tourism represented less than 1 percent of the national economy in the early 1980s. By 1999 more than 1.5 million tourists (mostly Canadians and Europeans) yielded more than $2 billion in profits. Conspicuous in their absence are travelers from the United States, forbidden to travel to Cuba because of the U.S.-imposed sanctions. The Helms-Burton Law of 1996 even tried to reduce the flow of foreign investment by threatening to sue major foreign investors in Cuba. While some investors delayed their plans because of this law, others, such as a leading Spanish hotel group, are busy building the island's tourist capacity in the hope that the U.S. ban will someday be lifted.

FIGURE 5.26 • CARIBBEAN CRUISE SHIPS   Cruise ships in port at Nassau, the Bahamas. Tourism is vital to many Caribbean states, but most cruise ships are owned by companies outside the region and offer relatively little employment for Caribbean workers. *(Galen Rowell/Corbis)*

FIGURE 5.27 • ECOTOURISM A scuba diver explores the Belizean reef. Many Caribbean and Latin American countries are betting on ecotourism as a means of generating income and better managing their environments. In Belize, for example, part of the reef was declared a marine park in hope of protecting species diversity and attracting more tourists. *(Doug Perrine/ SeaPics.com)*

As important as tourism is for the larger islands, it is often the principal source of income for smaller ones. The Virgin Islands, Barbados, Turks and Caicos, and, recently, Belize all greatly depend on international tourists. To show how quickly this sector can grow, consider this example: Belize began promoting tourism in the early 1980s when just 30,000 arrivals came a year. Close to North America and English-speaking, it specialized in ecotourism that showcased its interior tropical forests and coastal barrier reef (Figure 5.27). By 1994 the number of tourists topped 300,000, and tourism was credited for employing one-fifth of the workforce. Five years later, however, international arrivals were reduced by half, thus showing the unpredictable nature of the tourist industry.

For more than four decades, tourism has been the foundation of the Caribbean economy. Yet this regional industry has grown more slowly in recent years compared to other tourist destinations in the Middle East, southern Europe, and even Central America. It seems that Americans are favoring domestic destinations, such as Hawaii, Florida, and Las Vegas, or are going to more "exotic" settings, such as Costa Rica. European tourists also seem to be sticking closer to home or venturing to new locations, such as Dubai on the Persian Gulf or Goa in India. Further, a destination's reputation can suddenly decline as a result of social or natural forces. In the 1980s a wave of crimes against tourists in Jamaica caused a drop in visitors. It took a locally owned hotel chain to market all-inclusive trips on private beaches to restore the island's image as a safe place to visit. Hurricane Hugo virtually shut down St. Croix to visitors for an entire season in 1989, but islanders managed to rebuild most of their infrastructure by the following year.

Tourism-led development has detractors for other reasons. It is subject to the overall health of the world economy and current political affairs. Thus if North America experiences a recession or international tourism declines due to heightened fears of terrorism, the flow of tourist dollars to the Caribbean dries up. Where tourism is on the rise, local resentment may build as residents confront the differences between their own lives and those of the tourists. There is also a serious problem of **capital leakage**, which is the huge gap between gross income and the total tourist dollars that remain in the Caribbean. Since many guests stay in hotel chains with corporate headquarters outside the region, leakage of profits is expected. On the plus side, tourism tends to promote stronger environmental laws and regulation. Countries quickly learn that their physical environment is the foundation for success. And while tourism does have its costs (higher energy and water consumption and demand for more imports), it is environmentally less destructive than traditional export agriculture and at present more profitable.

Social Development

While the record for economic growth in the region is inconsistent, measures of social development are stronger, in part due to Cuba's accomplishments in health care and education. With the exception of Haiti's extremely low life expectancy of 49 years, most Caribbean peoples have an average life expectancy of 70 years or more. Levels of adult illiteracy are very low, especially in Cuba, the Bahamas, Trinidad, Grenada, Guyana, the Cayman Islands, Barbados, and the Netherlands Antilles, where only 2 to 3 percent of the population cannot read. Better education levels and out-migration have contributed to a marked decline in natural increase rates over the last 30 years, so that the average Caribbean woman has two or three children. Despite real social gains, many people are underemployed, poorly housed, and dependent on foreign remittances or subsistence agriculture. For rich and poor alike, the temptation to leave the region in search of better opportunities is always present.

Education Many Caribbean states have excelled in educating their citizens. Literacy is the norm, and the expectation is for most people to receive at least a high school degree. In many respects, Cuba's educational accomplishments are the most impressive, given the size of the country and its high illiteracy rates in the 1960s. Today 80 percent of the population receives secondary education, and illiteracy among adults is just 3 to 4 percent. Cuba also excels in training medical personnel and advancing biomedical technology. Hispaniola is the obvious contrast to Cuba's success. Less than 15 percent of Haitians and about half of the Dominican Republic's citizens acquire secondary education. While the Dominican Republic has taken steps to improve adult literacy (83 percent of adults are literate), more than half of all Haitian adults are illiterate. Political stability and economic growth have helped the Dominican Republic better its social conditions over the past decade. In fact, many Haitians have fled to the Dominican Republic because conditions, although far from ideal, are much better than in their homeland.

Education is expensive for these nations, but it is considered essential for development. Ironically, many states express frustration about training professionals for the benefit of developed countries in a phenomenon called **brain drain**. In the early 1980s the prime minister of Jamaica complained that 60 percent of his country's newly trained workers left for the United States, Canada, and Britain, representing a subsidy to these economies far greater than the foreign aid Jamaica received from them. During the economically depressed 1980s, Barbados, Guyana, and even the Dominican Republic and Haiti lost up to 20 percent of their most educated citizens. Brain drain occurs throughout the developing world, especially between former colonies and the mother countries. Yet given the small population of many Caribbean territories, each professional person lost to emigration can negatively impact local health care, education, and enterprise. For the time being, stronger economic performance in the Caribbean has slowed the outflow of professionals.

Status of Women   The matriarchal (female-dominated families) basis of Caribbean households is often singled out as a distinguishing characteristic of the region. The rural custom of men leaving home for seasonal employment tends to nurture strong and self-sufficient female networks. Women typically run the local street markets. With men absent for long periods of time, women tend to make household and community decisions. While giving women local power, this position does not always imply higher status. In rural areas female status is often undermined by the relative exclusion of women from the cash economy; men earn wages while women provide subsistence.

As Caribbean society urbanizes, more women are being employed in assembly plants (the garment industry, in particular, prefers to hire women), in data-entry firms, and in tourism. With new employment opportunities, female labor force participation has surged; in countries such as the Bahamas, Barbados, Jamaica, and Martinique, more than 40 percent of the workforce is female. Increasingly, women are the principal earners of cash, which challenges established gender roles. One impact of greater female participation in the formal economy seems to be smaller families.

Long before offshore assembly plants and resorts began hiring women, Cuba's educational and labor policies produced the most educated and professional women in the Caribbean. Here female doctors outnumber their male counterparts, and women have achieved near equality in the sciences, which is not true for many developed countries. Yet economic hardship may undermine these achievements as Cuban professionals see their wages and purchasing power erode and many decide to leave the country.

Labor-Related Migration   After World War II, better transportation and political developments in the Caribbean produced a surge of migrants to North America. This trend began with Puerto Ricans going to New York in the early 1950s and intensified in the 1960s with the arrival of nearly half a million Cubans. Since then, large numbers of Dominicans, Haitians, Jamaicans, Trinidadians, and Guyanese have also migrated to North America, typically settling in Miami, New York, Los Angeles, and Toronto.

Crucial in this exchange of labor from south to north is the flow of cash **remittances** (monies sent back home). Immigrants are expected to send something back, especially when immediate family members are left behind. Collectively, remittances add up; it is estimated that nearly $1 billion is annually sent to the Dominican Republic by immigrants in the United States, making remittance income the country's second leading source of income. Governments and individuals alike depend on these transnational family networks. Families carefully select the member most likely to succeed abroad in the hopes that money will flow back and a base for future immigrants will be established. A Caribbean nation in which no one left would soon face financial crisis because migration has become a necessity for livelihood in the region.

CONCLUSION

As a world region, the Caribbean is more integrated into the global economy than most areas of the developing world, although as a dependent economic periphery. Even though this is the smallest world region discussed, it offers some of the best examples of the long-term effects of globalization. This region was shaped five centuries ago by Europeans who eliminated the native population and imported millions of Africans to support a plantation-based economy. Consequently, most of the human and economic resources of the region (such as sugarcane) enriched Europe and later North America more than it did the people of the Caribbean. Today the region includes 20 independent countries, which each have strong international connections. Rather than harvesting sugarcane, Caribbean laborers are more likely to sew garments, assemble telephones, manage offshore banks, and cater to the needs of the 14 million tourists who visit each year. The economies of the region have diversified, although they still depend on foreign investment, exports, and tourism. Perhaps more

significant are the region's advances in social development, especially in education, health, and the status of women, which distinguish it from other developing areas.

There is a definite transnational quality to life in this region as social and economic influences from Latin America, Africa, North America, and Europe are woven into the lifestyle. Rap music blares from boom boxes in the streets of Havana and Kingston, while African-based religions are mixed in with with Catholicism in the countryside. Cricket is a popular sport in the eastern Caribbean, while baseball dominates in Cuba and the Dominican Republic. Similarly, as Caribbean residents emigrate in search of opportunities, they introduce their own traditions to host countries. In Toronto, for example, Caribbean immigrants hold Canada's largest Carnival. The input of so many cultures has produced a dynamic and rich Caribbean culture.

The Caribbean also maintains important ties with Europe and Africa. European colonies still exist in the region and maintain special trade

relations with their former colonizers. Even independent states look to their former rulers for help; for example, since the fall of the Soviet Union, Spain is one of Cuba's largest foreign investors. In contrast, Caribbean links with Africa are more cultural and political than economic. The cultural transfers of music, religion, and dance from West Africa to the Caribbean are well documented.

With the end of the Cold War, states in the region fear that their lack of strategic importance may result in neglect by the United States and Europe and may limit their ability to participate in world trade. It seems that the recent intervention of the United States in the Caribbean region has been driven more by a fear of thousands of refugees seeking sanctuary in Florida than by the cold war ideologies of the past. Perhaps it is best to view the Caribbean as the crossroads of the Americas. Whether through offshore banking, the drug trade, or commercial airline traffic, the Caribbean is an exciting intersection for the Western Hemisphere.

KEY TERMS

African diaspora (page 119)
brain drain (page 131)
capital leakage (page 130)
Caribbean Community and
    Common Market (CARICOM)
    (page 125)

Caribbean diaspora (page 116)
creolization (page 118)
free trade zones (page 127)
Greater Antilles (page 112)
hurricanes (page 113)
indentured labor (page 118)

isolated proximity (page 109)
Lesser Antilles (page 112)
maroons (page 119)
mono-crop production (page 118)
Monroe Doctrine (page 122)

neocolonialism (page 122)
offshore banking (page 128)
plantation America (page 118)
remittances (page 131)
rimland (page 111)

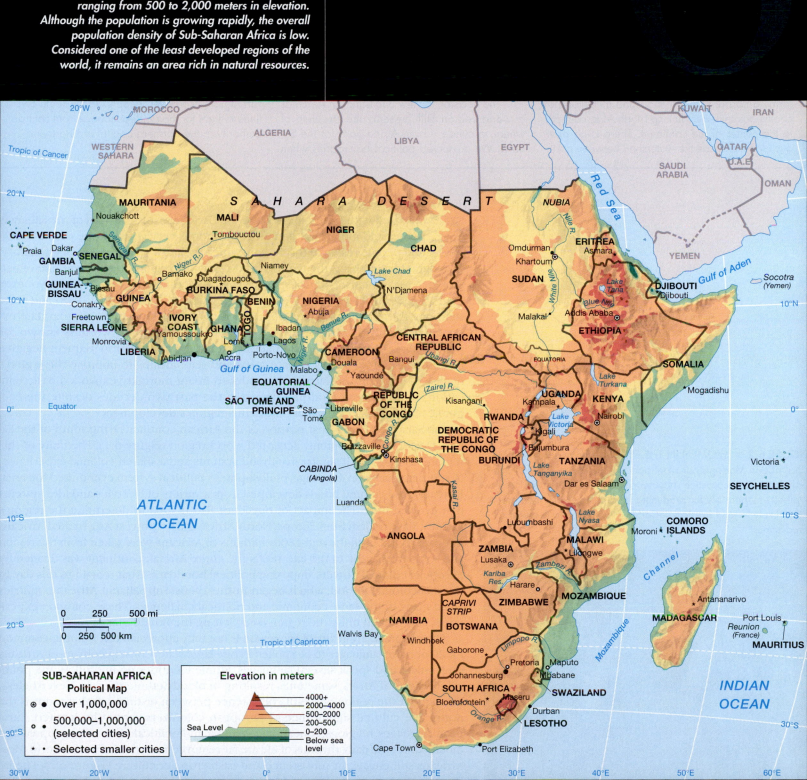

SETTING THE BOUNDARIES

Sub-Saharan Africa includes 43 mainland states plus the island nations of Madagascar, Cape Verde, São Tomé and Principe, Seychelles, Mauritius, and the French territory of Reunion. When setting this particular regional boundary, the major question one faces is how to treat North Africa.

Many scholars argue for keeping the African continent intact as a culture region. The Sahara has never formed a complete barrier between the Mediterranean north and the remainder of the African landmass. Moreover, the Nile River forms a corridor several thousand miles long of continuous settlement linking North Africa directly to the center of the continent. There is no obvious place to divide the watershed between north-

ern and Sub-Saharan Africa. Political units, such as the Organization of African Unity, are modern examples of the continent's unity.

The lack of a clear divide across Africa does not mean it cannot be divided into two world regions. North Africa is generally considered more closely linked, both culturally and physically, to Southwest Asia. Arabic is the dominant language and Islam the dominant religion of North Africa. Consequently, North Africans feel more closely connected to the Arab hearth in Southwest Asia than to the Sub-Saharan world.

The decision to view Sub-Saharan Africa as a world region still presents the problem of where to divide it from Africa north of the Sahara. We preserved political boundaries when

drawing the line so that the Mediterranean states of North Africa are discussed with Southwest Asia. Sudan is discussed in both Chapters 6 and 7 because it shares characteristics common to both regions. Sudan is Africa's largest state in terms of area; it is one-fourth the size of the United States. In the more populous and powerful north, Muslim leaders have crafted an Islamic state that is culturally and politically oriented toward North Africa and Southwest Asia. Southern Sudan, however, has more in common with the animist and Christian groups of the Sub-Saharan region. Moreover, peoples in the south continue to fight for their independence from central authority in the northern capital of Khartoum.

FIGURE 6.2 • WOMEN IN THE FIELDS Women from the Merina tribe in Madagascar plant paddy rice. The majority of Sub-Saharan people practice subsistence agriculture. Women are typically responsible for producing food crops, while men are more likely to tend livestock. *(Carl D. Walsh/Aurora & Quanta Productions)*

Compared with Latin America and the Caribbean, Africa south of the Sahara is poorer and more rural, and its population is very young. More than 670 million people reside in this region, which includes 48 states and one territory. Demographically, this is the world's fastest-growing region (2.5 percent rate of natural increase); in most countries, nearly half the population (45 percent) is younger than 15 years old. Income levels are low (a per capita GNI of $520 in 1998), and more than one-third of the region's population lives on less than $1 per day. Life expectancy is only 51 years. For many people, this part of the world is known for poverty, disease, violence, and refugees. Overlooked in the all-too-frequent negative headlines are programs by private, local, and state groups to improve the region's quality of life. Many countries have reduced infant mortality, expanded basic education, and increased food production in the past two decades despite the economic and political crises that hinder this region's development.

Sub-Saharan Africa—that portion of the African continent lying south of the Sahara Desert—is a commonly accepted world region (Figure 6.1). The unity of this region has to do with similar livelihood systems and a shared colonial experience. No common religion, language, philosophy, or political system ever united the area. Instead, loose cultural bonds developed from a variety of lifestyles and idea systems that evolved here. The impact of outsiders also helped to determine the region's identity. Slave traders from Europe, North Africa, and Southwest Asia treated Africans as property; up until the mid-1800s millions of Africans were taken from the region and sold into slavery. In the late 1800s the entire African continent was divided by European colonial powers, imposing political boundaries that remain to this day. In the postcolonial period, which began in the 1960s, Sub-Saharan African countries faced many of the same economic and political challenges. These common experiences also helped unify the region (see "Setting the Boundaries").

The cultural complexity of the region is not fully appreciated. It is not uncommon for 20 or more languages to be spoken in one large state. Consequently, most Africans understand and speak several languages. Ethnic identities do not follow the political divisions of Africa, sometimes resulting in bloody ethnic warfare. Nevertheless, throughout the region peaceful coexistence between distinct ethnic groups is the norm. The cultural significance of European colonizers cannot be ignored: European languages, religions, educational systems, and political ideas were adopted and modified. Yet the daily rhythms of life are far removed from the industrial world. Most Africans still engage in subsistence and cash-crop agriculture (Figure 6.2). The influence of African peoples outside the region is great, especially when one considers that human origins are traceable to this part of the world. In historic times, the legacy of the slave trade resulted in the transfer of African peoples, religious systems, and musical traditions throughout the Western Hemisphere. Even today, African-based religious systems are widely practiced in the Caribbean and Latin America, especially in Brazil.

The African economy, however, is marginal when compared with the rest of the world. According to the World Bank, Sub-Saharan Africa's economic output in 1999 amounted to just 1 percent of global output, even though the region contains 11 percent of the world's population. Moreover, the gross national product of just one country, South Africa, accounts for nearly half (44 percent) of the region's total economic output. Many scholars feel that Sub-Saharan Africa has benefited little from its integration (both forced and voluntary) into the global economy. Slavery, colonialism, and export-oriented mining and agriculture served the needs of consumers outside the region but failed to improve domestic food supplies.

Foreign assistance in the postindependence years initially improved agricultural and industrial output but also led to mounting foreign debt and corruption, which over time undercut the region's economic gains. Ironically, many of these same scholars and politicians worry that negative global attitudes about the region have produced a pattern of neglect. Private capital investment in Sub-Saharan Africa lags far behind investment rates in Latin America, the Caribbean, and East and Southeast Asia. It is hard to imagine how the region can economically recover without new private investment. To recognize the scale of what is possible for this region, familiarity with its environment and resource base is necessary.

Environmental Geography: The Plateau Continent

The largest landmass straddling the equator, Sub-Saharan Africa is vast in scale and remarkably beautiful. Called the plateau continent, the African interior is dominated by extensive areas of geological uplift that formed huge elevated plateaus. The highest areas are found on the eastern edge of the continent, where the Great Rift Valley forms a complex upland area of lakes, volcanoes, and deep valleys. In contrast, lowlands prevail in West Africa. Despite this region's immense biodiversity, vast water resources, and wealth of precious minerals, one finds relatively poor soils, widespread disease, and vulnerability to drought.

Africa's Environmental Issues

The prevailing perception of Africa south of the Sahara is one of environmental scarcity and degradation, no doubt encouraged by televised images of drought-ravaged regions and starving children. Single explanations such as rapid population growth or colonial exploitation cannot fully capture the complexity of African's environmental issues or the ways that people have adapted to living in marginal ecosystems. Because much of Sub-Saharan Africa's population is rural and poor, earning its livelihood directly from the land, sudden environmental changes have dramatic effects and can cause mass migrations, famine, and even death. As Figure 6.3 illustrates, deforestation and **desertification**, the expansion of desertlike conditions as a result of human activies such as poor agricultural practices and overgrazing, are commonplace. Sub-Saharan African is also vulnerable to drought, most notably in the Horn of Africa, parts of southern Africa, and the Sahel.

As Sub-Saharan cities grow in size and importance, urban environments increasingly face problems of air and water pollution as well as sewage and waste disposal. At the same time, wildlife tourism is an increasingly important source of income for many African states. Throughout the region, national parks have been created in an effort to strike a balance between humans' and animals' competing demands for land.

The Sahel and Desertification   In the 1970s the **Sahel** became a popular symbol for the dangers of unchecked population growth and human-induced environmental degradation when a relatively wet period came to an abrupt end and six years of

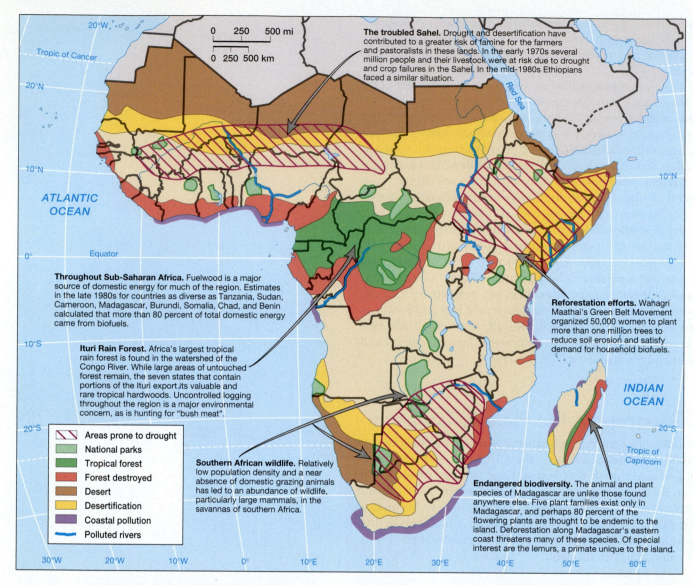

The troubled Sahel. Drought and desertification have contributed to a greater risk of famine for the farmers and pastoralists in these lands. In the early 1970s several million people and their livestock were at risk due to drought and crop failures in the Sahel. In the mid-1980s Ethiopians faced a similar situation.

Throughout Sub-Saharan Africa. Fuelwood is a major source of domestic energy for much of the region. Estimates in the late 1980s for countries as diverse as Tanzania, Sudan, Cameroon, Madagascar, Burundi, Somalia, Chad, and Benin calculated that more than 80 percent of total domestic energy came from biofuels.

Ituri Rain Forest. Africa's largest tropical rain forest is found in the watershed of the Congo River. While large areas of untouched forest remain, the seven states that contain portions of the Ituri export its valuable and rare tropical hardwoods. Uncontrolled logging throughout the region is a major environmental concern, as is hunting for "bush meat".

Reforestation efforts. Wanagri Maathai's Green Belt Movement organized 50,000 women to plant more than one million trees to reduce soil erosion and satisfy demand for household biofuels.

Southern African wildlife. Relatively low population density and a near absence of domestic grazing animals has led to an abundance of wildlife, particularly large mammals, in the savannas of southern Africa.

Endangered biodiversity. The animal and plant species of Madagascar are unlike those found anywhere else. Five plant families exist only in Madagascar, and perhaps 80 percent of the flowering plants are thought to be endemic to the island. Deforestation along Madagascar's eastern coast threatens many of these species. Of special interest are the lemurs, a primate unique to the island.

Legend:
- Areas prone to drought
- National parks
- Tropical forest
- Forest destroyed
- Desert
- Desertification
- Coastal pollution
- Polluted rivers

FIGURE 6.3 • ENVIRONMENTAL ISSUES IN SUB-SAHARAN AFRICA Given the immense size of Sub-Saharan Africa, it is difficult to generalize about environmental problems. Dependence on trees for fuel places strains on forests and wooded savannas throughout the region. In semiarid regions, such as the Sahel, population pressures and land-use practices seem to have led to desertification. Yet Sub-Saharan Africa also supports the most impressive array of wildlife, especially large mammals, on Earth. The rapid growth of cities in the past three decades has resulted in new concerns about water quality and pollution. *(Adapted from DK World Atlas, 1997, p. 75. London: DK Publishing)*

drought (1968–74) ravaged the land. The Sahel is a zone of ecological transition between the Sahara to the north and wetter savannas and forest in the south (see Figure 6.3). Life depends on a delicate balance of limited rain, drought-resistant plants, and a pattern of animal **transhumance** (the movement of animals between wet-season and dry-season pasture). During the drought, rivers in the area diminished and desertlike conditions began to move south. Unfortunately, tens of millions of people lived in this area, and farmers and pastoralists whose livelihoods had come to depend on the more abundant precipitation of the relatively wet period were temporarily forced out.

Considerable disagreement continues over the basic causes of the Sahelian drought. Was it a matter of too many humans degrading their environment, unsound settlement schemes encouraged by European colonizers, or a failure to understand global atmospheric cycles? Certainly human activity in the region greatly increased in the mid-twentieth century, making the case for human-induced desertification more likely. But it is worth noting that parts of the Sahel were important areas of settlement long before European colonization. What appears to be desert wasteland in April or May is transformed into productive fields of millet, sorghum, and peanuts after the drenching rains of June. Relatively free of the tropical diseases found in the wetter zones to the south, Sahelian soils are also quite fertile, which helps to explain why people continue to live there despite the unreliable rainfall patterns of recent decades.

The main practices in the case for desertification are the expansion of agriculture and overgrazing, leading to the loss of natural vegetation and declines in soil fertility. Through most of the Sahel, French colonial authorities forced villagers to grow peanuts as an export crop, a policy continued by the newly independent states of the region. However, peanuts tend to use up several key soil nutrients, which means that peanut farms are often abandoned after a few years as farmers move on to fresh sites. This process leads to further vegetation removal and soil erosion.

Overgrazing may also be a cause of Sahelian desertification. Livestock is a traditional product of the region, and animal production dramatically expanded after World War II as the population increased and the need for export earnings grew. In many areas, natural pasture was degraded by intensified grazing pressure, leading to increased wind erosion. The problem became more widespread when national and international developmental agencies, hoping to increase production further, began to dig deep wells in areas that had previously been unused by herders through most of the year. The new supplies of water, in turn, allowed year-round grazing in places that, over time, could not support livestock (Figure 6.4).

Deforestation   Although Sub-Saharan Africa still contains extensive forests, much of the region is either grasslands or agricultural lands that were once forest. Lush forests that existed in places such as highland Ethiopia were long ago reduced to a few remnant patches. Local populations have relied on such woodlands throughout history for their daily needs. Tropical savannas, which cover large portions of the region to the north and south of the tropical rain forest zone, are grassland areas with scattered wooded areas of trees and shrubs. The woodlands that dot this extensive African vegetation biome are few, and unlike tropical America, savanna deforestation is of greater local concern than the limited commercial logging of the rain forest. North of the equator, for example, only a few wooded areas remain in a landscape dominated by grasslands and cropland. Loss of woody vegetation has resulted in extensive hardship, especially for women and children who must spend many hours a day looking for wood (Figure 6.5). Deforestation, especially in the scattered woodlands of the savannas, aggravates problems of soil erosion and shortages of **biofuels** (wood and charcoal used for household energy needs, especially cooking).

In some countries, village women have organized into community-based nongovernmental organizations (NGOs) to plant trees and create greenbelts to meet future fuel needs. One of the most successful efforts is in Kenya under the leadership of Wangari Maathai. Maathai's Green Belt Movement has more than 50,000 members, mostly women, organized into 2,000 local community groups. Since the group's beginning in 1977, millions of trees have been successfully planted. In those areas, village women now spend less time collecting fuel, and local environments have improved. Kenya's success has drawn interest from other African countries, spurring a Pan-African Green Belt Movement largely organized through nongovernmental organizations interested in biofuel generation, protection of the environment, and the empowerment of women.

Destruction of tropical rain forest for logging is most evident in the fringes of Central Africa's Ituri forest (see Figure 6.3). Given the great size of this forest and the relatively small number of people living there, however, it is less threatened than other forest areas. Two smaller rain forests, one along the Atlantic coast from Sierra Leone to western Ghana and the other along the eastern coast of the island of Madagascar, have nearly disappeared. These rain forests have been severely degraded by commercial logging and agricultural clearance. Madagascar's eastern rain forests, as well as its western dry forests, have suffered serious degradation in the past three decades (Figure 6.6). Deforestation in Madagascar is especially worrisome because the island forms a unique environment with a large number of endemic (plants and animals native to a particular area) species.

FIGURE 6.4 • DESERTIFICATION IN THE SAHEL   In southern Niger, pastoralists rely upon wells to water their herds of goats and cattle. The relationship between livestock, farming, and the Sahelian ecosystem is complex. Many researchers believe that increased grazing pressure brought on by more livestock and wells may have made the Sahel more vulnerable to desertification. *(Victor Englebert/Englebert Photography, Inc.)*

FIGURE 6.5 • GATHERING FUEL   A woman loads a wagon with firewood in rural Mali. Villagers and city dwellers alike rely upon wood for household energy needs, placing an enormous strain on Africa's woodlands. In areas with the greatest scarcity, women and children may spend many hours each day gathering fuel. *(Betty Press/Woodfin Camp & Associates)*

FIGURE 6.6 • DEFORESTATION IN MADAGASCAR Forest clearing for agriculture and fuel has stripped much of the countryside in Madagascar of vegetation. Serious problems with soil erosion exist throughout the country, limiting agricultural productivity. During the rainy season, the discharge of eroded topsoil into the Indian Ocean is visible from space. *(Bios/M. Gunther/ Peter Arnold, Inc.)*

FIGURE 6.7 • SOUTH AFRICAN WILDLIFE An African elephant crosses the savannas of Kruger National Park in South Africa, one of the region's oldest wildlife parks. South Africa, Botswana, Zimbabwe, and Zambia have the largest elephant populations. As a result of poaching and habitat destruction, other countries have seen a troubling decline in elephant numbers. *(Rob Crandall/Rob Crandall, Photographer)*

Wildlife Conservation   Sub-Saharan Africa is famous for its wildlife. In no other region of the world can one find such abundance and diversity of large mammals. The survival of wildlife here reflects, to some extent, the historically low human population density and the fact that sleeping sickness (transferred by the tsetse fly) and other diseases have kept people and their livestock out of many areas. In addition, many African peoples have developed various ways of successfully coexisting with wildlife.

But as is true elsewhere in the world, wildlife is quickly declining in much of Sub-Saharan Africa. The most noted wildlife reserves are in East Africa; in Kenya and Tanzania these reserves are major tourist attractions and are economically important. Even there, however, population pressure, political instability, and poverty make the maintenance of large wildlife reserves difficult. Poaching is a major problem, particularly for rhinoceroses and elephants; the price of a single horn or tusk in distant markets represents several years' wages for most Africans. Ivory is sought in East Asia, especially in Japan. In China powdered rhino horn is used as a traditional medicine, whereas in Yemen rhino horn is prized for dagger handles. Wildlife reserves in southern Africa now seem to be the most secure (Figure 6.7). In fact, elephant populations are considered to be too large for the land to sustain in countries such as Zimbabwe. Some wildlife experts contend that herds could be reduced to prevent overgrazing and that the ivory and rhino horn should be legally sold in the international market in order to generate revenue for further conservation. Many environmentalists, not surprisingly, disagree strongly.

In 1989 a worldwide ban on the legal ivory trade was imposed as part of the Convention on International Trade in Endangered Species (CITES). While several African states, such as Kenya, lobbied hard for the ban, others, such as Zimbabwe, Namibia, and Botswana, complained that their herds were growing and the sale of ivory had helped to pay for conservation efforts. Conservationists feared that lifting the ban would bring on a new wave of poaching and illegal trade. In the late 1990s the ban was lifted so some southern African states could sell down their inventories of ivory confiscated from poachers and continue with limited sales. The long-term effects of this policy change for elephant survival are not yet known, although political leaders from Kenya and India have voiced strong opposition to all ivory sales. The ivory controversy shows how differences within the region make it difficult to come up with a consistent conservation strategy.

Plateaus and Basins

A series of plateaus and elevated basins dominate the African interior and explain much of the region's unique physical geography (Figure 6.8). Generally, elevations increase toward the south and east of the continent. Most of southern and eastern Africa lies well above 2,000 feet (600 meters), and sizable areas sit above 5,000 feet (1,500 meters). This is typically referred to as High Africa; Low Africa includes West Africa and much of Central Africa. Steep escarpments form where plateaus abruptly end, as illustrated by the majestic Victoria Falls on the Zambezi River (Figure 6.9). Much of southern Africa is rimmed by a landform called the **Great Escarpment** (a high cliff separating two comparatively level areas), which begins in southwestern Angola and ends in northeastern South Africa, creating a barrier to coastal settlement.

Though Sub-Saharan Africa is an elevated landmass, it has few significant mountain ranges. The one extensive area of mountainous topography is in Ethiopia, which lies in the northern portion of the Rift Valley zone. Yet even there the dominant features are high plateaus intercut with deep valleys rather than actual mountain ranges. Receiving heavy rains in the wet season, the Ethiopian Plateau forms the headwaters of several important rivers, most notably the Blue Nile, which joins the White Nile at Khartoum, Sudan.

Watersheds   Africa south of the Sahara does not have the broad, alluvial lowlands that influence patterns of settlement throughout other regions. The four major river systems are the Congo, Niger, Nile and Zambezi (the Nile will be discussed in Chapter 7). Smaller rivers, such as the Orange in South Africa; the Senegal, which

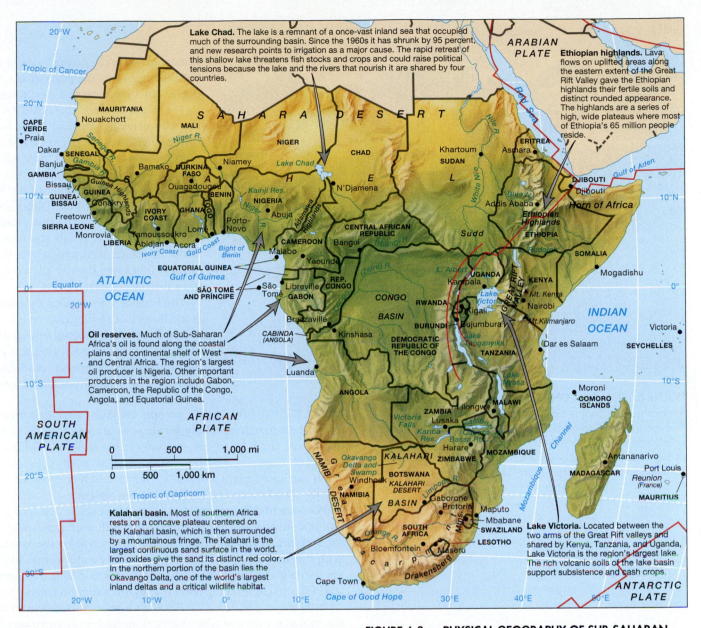

Lake Chad. The lake is a remnant of a once-vast inland sea that occupied much of the surrounding basin. Since the 1960s it has shrunk by 95 percent, and new research points to irrigation as a major cause. The rapid retreat of this shallow lake threatens fish stocks and crops and could raise political tensions because the lake and the rivers that nourish it are shared by four countries.

Ethiopian highlands. Lava flows on uplifted areas along the eastern extent of the Great Rift Valley gave the Ethiopian highlands their fertile soils and distinct rounded appearance. The highlands are a series of high, wide plateaus where most of Ethiopia's 65 million people reside.

Oil reserves. Much of Sub-Saharan Africa's oil is found along the coastal plains and continental shelf of West and Central Africa. The region's largest oil producer is Nigeria. Other important producers in the region include Gabon, Cameroon, the Republic of the Congo, Angola, and Equatorial Guinea.

Kalahari basin. Most of southern Africa rests on a concave plateau centered on the Kalahari basin, which is then surrounded by a mountainous fringe. The Kalahari is the largest continuous sand surface in the world. Iron oxides give the sand its distinct red color. In the northern portion of the basin lies the Okavango Delta, one of the world's largest inland deltas and a critical wildlife habitat.

Lake Victoria. Located between the two arms of the Great Rift valleys and shared by Kenya, Tanzania, and Uganda, Lake Victoria is the region's largest lake. The rich volcanic soils of the lake basin support subsistence and cash crops.

FIGURE 6.8 • PHYSICAL GEOGRAPHY OF SUB-SAHARAN AFRICA Although this region, the so-called plateau continent, consists of many large elevated plateaus, there are relatively few mountains. Four major river basins are associated with the region: the Congo, Nile, Niger, and Zambezi. The Congo flows through the largest rain forest in the world after the Amazon. The Nile, the lifeline for Sudan and Egypt, is the world's longest river.

divides Mauritania and Senegal; and the Limpopo in Mozambique, are locally important but drain much smaller areas. Ironically, most people think of Africa south of the Sahara as suffering from water scarcity and tend to discount the size and importance of the watersheds (or catchment areas) drained by these river systems.

The Congo River (or Zaire) is the largest watershed in the region, both in terms of drainage area and the volume of river flow produced. It is second only to South America's Amazon River in terms of annual flow. The Congo flows across a relatively flat basin that lies more than 1,000 feet (300 meters) above sea level, meandering through Africa's largest tropical forest, the Ituri. Entry from the Atlantic into the Congo Basin is prevented by a series of rapids and falls, making the Congo River only partially navigable. Despite these limitations, the Congo River has been the major corridor for travel within the Republic of the Congo and the Democratic Republic of the Congo (formerly Zaire); the capitals of both countries, Brazzaville and

FIGURE 6.9 • VICTORIA FALLS The Zambezi River descends over Victoria Falls in the southern part of the region. The Zambezi has never been important for navigation, but it is a vital supply of hydroelectricity for Zimbabwe, Zambia, and Mozambique. *(Rob Crandall/Rob Crandall, Photographer)*

FIGURE 6.10 • THE CONGO RIVER Heavily laden barges ferry cargo and people between the cities of Kinshasa and Kisangani in the Democratic Republic of the Congo. The Congo is Sub-Saharan Africa's largest river by volume. Meandering through a relatively flat basin, the river is an important transportation corridor. *(Georg Gerster/NGS Image Collection)*

Kinshasa, rest on opposite sides of the river (Figure 6.10). Political turmoil of the past decade, however, has greatly reduced commerce on this vital waterway.

The Niger River is the critical source of water for the arid countries of Mali and Niger. Beginning in the humid Guinea highlands, the Niger flows first to the northeast and then spreads out to form a huge inland delta in Mali before making a great bend southward at the margins of the Sahara near Gao. On the banks of the Niger River are the capitals of Mali (Bamako) and Niger (Niamey), as well as the historic city of Tombouctou (Timbuktu). After flowing through the desert north, the Niger River returns to the humid lowlands of Nigeria, where the Kainji Reservoir temporarily blocks its flow to produce electricity for Africa's most heavily populated state.

The considerably smaller Zambezi River begins in Angola and flows east, spilling over an escarpment at Victoria Falls, and finally reaching Mozambique and the Indian Ocean. More than other rivers in the region, the Zambezi is a major supplier of commercial energy. Sub-Saharan Africa's two largest hydroelectric installations are located on this river.

Soils With a few major exceptions, Sub-Saharan Africa's soils are relatively infertile. Generally speaking, fertile soils are young soils, those deposited in recent geological time by rivers, volcanoes, glaciers, or windstorms. In older soils—especially those located in moist tropical environments—natural processes tend to wash out most plant nutrients over time. Over most of Sub-Saharan Africa, the agents of soil renewal have largely been absent.

Portions of Sub-Saharan Africa are, however, noted for their natural soil fertility, and not surprisingly these areas support denser settlement. Some of the most fertile soils are in the Rift Valley, made productive by the volcanic activity associated with the area. The population densities of rural Rwanda and Burundi, for example, are partially explained by the highly productive soils. The same can be said for highland Ethiopia, which supports the region's second largest population of more than 65 million people. The Lake Victoria lowlands and central highlands of Kenya are also noted for their sizable populations and productive agricultural bases.

Climate and Vegetation

Most of Sub-Saharan Africa lies in the tropical latitudes and it is the largest tropical landmass on the planet. Only the far south of the continent extends into the subtropical and temperate belts. Much of the region averages high temperatures from 70°F to 80°F (22° to 28°C) year-round (Figure 6.11). Rainfall, more than temperature, determines the different vegetation belts that characterize the region. Addis Ababa, Ethiopia, and Walvis Bay, Namibia, have similar average temperatures, but the former is in the moist highlands and receives nearly 50 inches (127 centimeters) of rainfall annually, while Walvis Bay rests on the Namibian Desert and receives less than 1 inch (2.5 centimeters). The three main biomes of the region are tropical forests, savannas, and deserts.

Tropical Forests The center of Sub-Saharan Africa falls in the tropical wet climate zone. The world's second-largest expanse of humid equatorial rain forest, the Ituri, lies in the Congo Basin, extending from the Atlantic Coast of Gabon two-thirds of the way across the continent, including the northern portions of the Republic of the Congo and the Democratic Republic of the Congo (Zaire). The conditions here are constantly warm to hot and precipitation falls year-round (see graph for Kisangani, Figure 6.11).

Commercial logging and agricultural clearing have degraded the western and southern fringes of this vast forest, but much of it is still intact. Considering the high

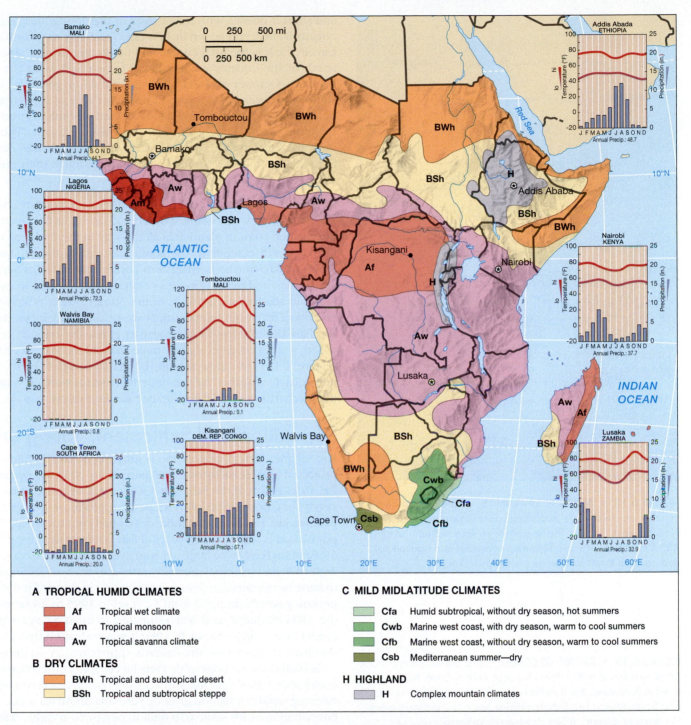

A TROPICAL HUMID CLIMATES

	Af	Tropical wet climate
	Am	Tropical monsoon
	Aw	Tropical savanna climate

B DRY CLIMATES

	BWh	Tropical and subtropical desert
	BSh	Tropical and subtropical steppe

C MILD MIDLATITUDE CLIMATES

	Cfa	Humid subtropical, without dry season, hot summers
	Cwb	Marine west coast, with dry season, warm to cool summers
	Cfb	Marine west coast, without dry season, warm to cool summers
	Csb	Mediterranean summer—dry

H HIGHLAND

	H	Complex mountain climates

rates of tropical deforestation in Southeast Asia and Latin America, the Central African case is an exception. Certainly the area's low population of subsistence farmers does not place much strain on the resource base. Moreover, Gabon and the Republic of the Congo export oil, making commercial logging less important to local economies. The political chaos in the Democratic Republic of the Congo has made large-scale logging a difficult proposition. In the future, however, it seems likely that Central Africa's rain forests could suffer the same kind of degradation experienced in other equatorial areas.

Savannas   Wrapped around the Central African rain forest belt in a great arc lie Africa's vast tropical wet and dry savannas. Savannas are dominated by a mixture of trees and tall grasses in the wetter zones immediately adjacent to the forest belt

FIGURE 6.11 • CLIMATE MAP OF SUB-SAHARAN AFRICA
Much of the region lies within the tropical humid and tropical dry climatic zones; thus, the seasonal temperature changes are not great. Precipitation, however, varies significantly from month to month. Compare the distinct rainy seasons in Lusaka and Lagos: Lagos is wettest in June and Lusaka receives most of its rain in January. Although there are important tropical forests in West and Central Africa (coinciding with the tropical wet and monsoon climate zones), vegetation in much of the region is tropical savanna. [*Temperature and precipitation data from Pearce and Smith, 1984, The World Weather Guide, London: Hutchinson*]

FIGURE 6.12 • AFRICAN SAVANNAS A line of wildebeests marches across the tree-studded savannas found in Masai Mara National Park, Kenya. The savannas, an essential habitat for Africa's wildlife, are steadily being converted into agricultural and pastoral lands for human needs. *(Nik Wheeler)*

FIGURE 6.13 • KALAHARI DESERT Sand dunes are one of the attractions found in Kalahari Gemsbok Park in South Africa. Centered in Botswana, the Kalahari extends into parts of Namibia and South Africa. Sub-Saharan Africa has many semiarid areas but few true deserts. *(Clem Haagner/Photo Researchers, Inc.)*

and shorter grasses with fewer trees in the drier zones (Figure 6.12). North of the equatorial belt, rain generally falls only from May to October. The farther north one travels, the less the total rainfall and the longer the dry season. Climatic conditions south of the equator are similar, only reversed, with the wet season occurring between October and May and precipitation generally decreasing toward the south (see Lusaka in Figure 6.11). A larger area of wet savanna exists south of the equator, with extensive woodlands located in southern portions of the Democratic Republic of the Congo, Zambia, and eastern Angola. These savannas are also a critical habitat for the region's large fauna. Elephants, zebras, rhinoceroses, and lions are found in the wooded grasslands, although their numbers have declined in most areas since the 1980s.

Deserts Major deserts exist in the southern and northern boundaries of the region. The Sahara, the world's largest desert and one of its driest, crosses the landmass from the Atlantic coast of Mauritania all the way to the Red Sea coast of Sudan. A narrow belt of desert extends to the south and east of the Sahara, wrapping around the **Horn of Africa** (the northeastern corner that includes Somalia, Ethiopia, Djibouti, and Eritrea) and pushing as far as eastern and northern Kenya. An even drier zone is found in southwestern Africa. In the Namib Desert of coastal Namibia, rainfall is a rare event, although temperatures are usually mild. Inland from the Namib lies the Kalahari Desert. Most of the Kalahari is not dry enough to be classified as a true desert, since it receives slightly more than 10 inches (25 centimeters) of rain a year (Figure 6.13). Its rainy season, however, is brief. Most of the precipitation is immediately absorbed by the underlying sands. Surface water is thus scarce, giving the Kalahari a desertlike appearance for most of the year.

Population and Settlement: Young and Restless

Sub-Saharan Africa's population is growing quickly. While the global population is projected to increase by nearly 50 percent by 2050, the projected population growth for Africa south of the Sahara is 130 percent over the same time period. It is also a very young population, with 45 percent of the people younger than age 15, compared to just 18 percent for more-developed countries. Only 3 percent of the region's population is older than 65 years of age, whereas in Europe 15 percent is older than 65. Families tend to be large, with an average woman having five or six children. Yet high child and maternal mortality rates also exist, reflecting disturbingly low access to basic health services. The most troubling indicator for the region is its low life expectancy, which dropped from 52 years in 1995 to 48 years in 1998, in part due to the AIDS epidemic, and was estimated to be at 51 years in 2001. The growth of cities is also a major trend. In 1980 it was estimated that 23 percent of the population lived in cities; now the figure is approximately 30 percent.

Behind these demographic facts lie complex differences in settlement patterns, livelihoods, belief systems, and access to health care. Although the region is experiencing rapid population growth, Sub-Saharan Africa is not densely populated. The entire region holds some 670 million persons—roughly half the population that is crowded into the much smaller land area of South Asia. In fact, the overall population density of the region (72 people per square mile) is similar to that of the United States (77 people per square mile). Just six states account for half of the region's population: Nigeria, Ethiopia, the Democratic Republic of Congo, South Africa, Tanzania, and Sudan. True, there are states with very high population densities while others are sparsely inhabited. Chad has just 18 people per square mile (7 people per square kilometer) and Namibia has only 6 people per square mile (2 people per square kilometer). Many of the governments of the more densely settled territories, however, began to seriously encourage family planning policies in the 1980s.

Population Trends and Demographic Debates

It is the combination of population growth in particular areas of Sub-Saharan Africa and decline in some economic and social indicators that make demographers concerned about the region's overall well-being. The Sahel, for example, is not crowded by European or Asian standards, but it may already contain too many people for the land to support given the unpredictability of its limited rainfall.

Some believe that the region could support many more people than it presently does. Pessimists, however, argue that the region is a demographic time bomb and that unless fertility is quickly reduced, Sub-Saharan Africa will face massive famines in the near future. The majority of African states officially support lowering rates of natural increase and are slowly promoting modern contraception practices both to reduce family size and to protect people from sexually transmitted diseases.

Family Size   A preference for large families is the basis for the region's demographic growth. In the 1960s many areas in the developing world had comparable total fertility rates (TFR) of 6.0 or higher, but by the mid-1990s only people in Sub-Saharan Africa and Southwest Asia continued to have such large families. For Southwest Asia the dominance of Islam is used to explain high fertility rates. In Sub-Saharan Africa a combination of cultural practices, rural lifestyles, and economic realities encourage large families (Figure 6.14).

Throughout the region large families guarantee a family's lineage and status. Even now most women marry young, typically when they are teenagers, which increases their opportunity to have children. Demographers often point to the limited formal education available to women as another factor contributing to high fertility. Religious affiliation has little bearing on the region's fertility rates; Muslim, Christian, and animist communities all have similarly high birthrates.

The everyday realities of rural life make large families an asset. Children are an important source of labor; from tending crops and livestock to gathering fuelwood, they add more to the household economy than they take. Also, for the poorest places in the developing world, such as Sub-Saharan Africa, children are seen as social security. When the parent's health falters, they expect their grown children to care for them.

By the 1980s a shift in national policies occurred. For the first time, government officials argued that smaller families and slower population growth were needed for social and economic development. Other factors are bringing down the growth rate. As African states slowly become more urban, there is a corresponding decline in family size—a pattern seen throughout the world. Tragically, declines in natural increase are also occurring as a result of AIDS.

The Impact of AIDS on Africa

Southern Africa is ground zero for the AIDS epidemic that is devastating the region (Figure 6.15). In South Africa, the most populous state in southern Africa, nearly 5 million people (one in five people aged 15–49) are infected with HIV/AIDS. The rate of infection in neighboring Botswana is a staggering 36 percent for the same age group. Although the rate of infection is highest in southern Africa, East Africa and West Africa are seeing increases. In Nigeria it is estimated that 5 percent of the 15–49 age group has HIV or AIDS. For Kenya the figure is 14 percent. By comparison, only 0.6 percent of the same age group in North America is infected. As of 2000, two-thirds of the HIV/AIDS cases in the world were found in Sub-Saharan Africa, and it was estimated that the epidemic had already claimed 17 million lives in the region. The virus is thought to have originated in the forests of the Congo, possibly crossing over from chimpanzees to humans sometime in the 1950s. Yet it was not until the late 1980s that the impact of the disease was widely felt in some of the more populated parts of the region.

Infection rates are so high that demographers anticipate a slower population growth rate for the entire region. Sub-Saharan Africa will still grow, but population estimates for 2025 have been trimmed by as much as 200 million. Sadly, the social and economic implications of this epidemic are hard to measure. AIDS typically hits the portion of the population that is most active economically. Time lost to care for sick family members and the cost of workers' compensation benefits could reduce economic productivity and overwhelm public services in hard-hit areas. Infection rates among newborns are high, and many areas struggle to care for children orphaned by AIDS.

FIGURE 6.14 • LARGE FAMILIES   A Gambian father poses with his 10 children. Large families are still common in Sub-Saharan Africa. Growth rates soared in the 1970s and 1980s, but they began to slow slightly in the 1990s. In southern Africa total fertility rates have dropped to 3 and 4 children per woman. The overall average for the region is 5.6 children per woman. *(Mark Boulton/Photo Researchers, Inc.)*

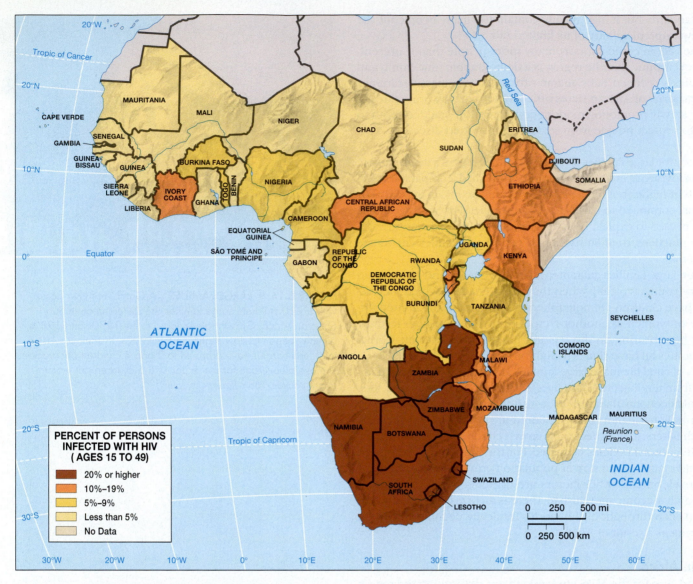

FIGURE 6.15 • HIV PREVALENCE Two-thirds of the world's people who are infected with HIV/AIDS are in Sub-Saharan Africa. As of 2000, more than 17 million lives have been lost in the region due to AIDS. Infection rates at the end of 1999 were highest in southern Africa, especially Botswana (35.8 percent) and Zimbabwe (25.1 percent). [Data from Population Reference Bureau, World Population Data Sheet, 2001]

Unlike the developed world, where expensive and powerful drug therapies have prolonged the lives of people with HIV/AIDS, Sub-Saharan Africa does not have the money for these treatments. The government of South Africa insisted on its right to use cheap generic drugs to prolong the lives of those with HIV/AIDS, even though some of these drugs are in violation of drug company patents. The world's major drug companies promptly sued South Africa but backed down in 2001 in the face of growing international pressure (and the realization that pharmaceutical companies in developing countries such as India would happily provide generic alternatives). Through a UN-led agreement, generic drugs are now available, but a year's supply costs close to $1,000, which is still beyond the economic reach of most Africans.

For now, the surest way to stem the epidemic is through prevention, mostly through educating people about how the virus is spread and convincing them to change their sexual behavior. In Uganda, state agencies, along with NGOs, began a national no-nonsense campaign for AIDS awareness in the 1980s that focused on the schools. As a result of explicit materials, role-playing games, and frank discussion, the frequency of HIV among women in prenatal clinics had declined by the late 1990s.

Patterns of Settlement and Land Use

Because of the dominance of rural settlements in Sub-Saharan Africa, people are widely scattered throughout the region (Figure 6.16). Population concentrations are the highest in West Africa, highland East Africa, and the eastern half of South

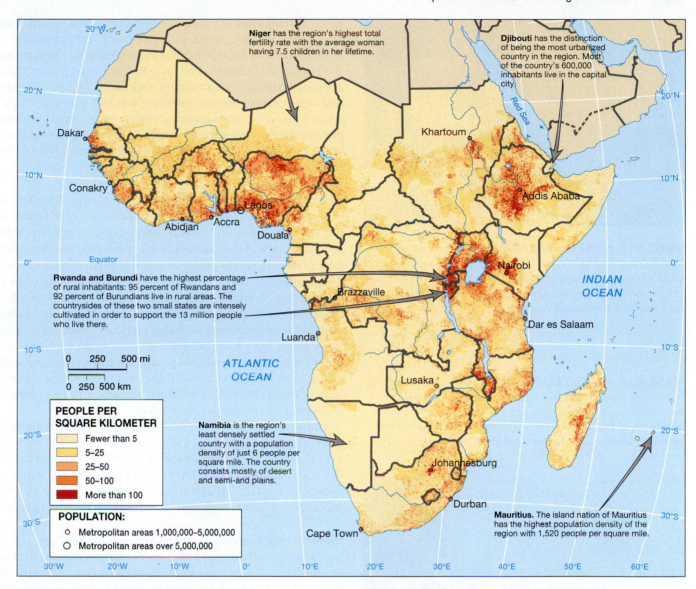

Niger has the region's highest total fertility rate with the average woman having 7.5 children in her lifetime.

Djibouti has the distinction of being the most urbanized country in the region. Most of the country's 600,000 inhabitants live in the capital city.

Rwanda and Burundi have the highest percentage of rural inhabitants: 95 percent of Rwandans and 92 percent of Burundians live in rural areas. The countrysides of these two small states are intensely cultivated in order to support the 13 million people who live there.

Namibia is the region's least densely settled country with a population density of just 6 people per square mile. The country consists mostly of desert and semi-arid plains.

Mauritius. The island nation of Mauritius has the highest population density of the region with 1,520 people per square mile.

PEOPLE PER SQUARE KILOMETER

	Fewer than 5
	5–25
	25–50
	50–100
	More than 100

POPULATION:

○ Metropolitan areas 1,000,000–5,000,000

◯ Metropolitan areas over 5,000,000

FIGURE 6.16 • POPULATION DISTRIBUTION The majority of people in Sub-Saharan Africa live in rural areas. Some of these rural zones, however, are densely settled, such as West Africa and the East African highlands. Major urban centers, especially in South Africa and Nigeria, support millions. Overall, the region has few large cities with more than 1 million people.

Africa. The first two areas have some of the region's best soils, and native systems of permanent agriculture developed there. In the last case, an urbanized economy based on mining, as well as the forced concentration of black South Africans into eastern homelands, contributed to the region's overall density.

As more Africans move to cities, patterns of settlement are becoming more concentrated. Towns that were once small administrative centers for colonial elites grew into major cities. The region even has its own mega-city; Lagos topped 10 million in the 1990s. Throughout the continent, African cities are growing faster than rural areas. But before examining the Sub-Saharan urban scene, a more detailed discussion of rural subsistence is needed.

Agricultural Subsistence The staple crops over most of Sub-Saharan Africa are millet, sorghum, and corn (maize), as well as a variety of tubers and root crops such as yams. Irrigated rice is widely grown in West Africa and Madagascar. Wheat and barley are grown in parts of South Africa and Ethiopia. Intermixed with subsistence foods are a variety of export crops—coffee, tea, rubber, bananas, cacao, cotton, and peanuts—that are grown in distinct ecological zones and often in some of the best soils.

In areas that support annual cropping, population densities are greater. In parts of humid West Africa, for example, the yam became the king of subsistence crops. The Ibos' mastery of yam production allowed them to produce more food and live in denser permanent settlements in the southeastern corner of present-day Nigeria.

FIGURE 6.17 • HARVESTING PALM OIL   A man harvests palm oil near Ibadan, Nigeria. Palm oil is a major export for western Nigeria. Other important Sub-Saharan plantation crops include cocoa, rubber, tea, coffee, cotton, and peanuts. *(M. & E. Bernheim/Woodfin Camp & Associates)*

FIGURE 6.18 • MASAI PASTORALISTS   A Masai man holds a cow steady while a woman collects blood being drained from the animal's neck. Pastoral groups such as the Masai live in the drier areas of Sub-Sahara Africa. The Masai live in Kenya and Tanzania. Other pastoral groups are found in the Sahel and the Horn of Africa.   *(© Kennan Ward/Corbis)*

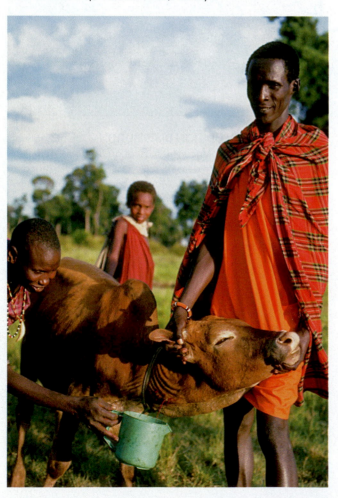

Much of traditional Ibo culture is tied to the demanding tasks of clearing the fields, tending the delicate plants, and celebrating the harvest.

Over much of the continent, African agriculture remains relatively unproductive, and population densities tend to be low. Amid the poorer tropical soils, cropping usually entails shifting cultivation (or **swidden**). This process involves burning the natural vegetation to release fertilizing ash and planting crops such as maize, beans, sweet potatoes, banana, papaya, manioc, yams, melon, and squash. Each plot is temporarily abandoned once its source of nutrients has been exhausted. Swidden cultivation is often a very finely tuned adaptation to local environmental conditions, but it is unable to support high population densities.

Plantation Agriculture   Plantation agriculture, designed to produce crops for export, is critical to the economies of many states. If African countries are to import the modern goods and energy resources they require, they must sell their own products on the world market. Since the region has few competitive industries, the bulk of its exports are primary products derived from farming, mining, and forestry.

A number of African countries rely heavily on one or two export crops. Coffee, for example, is vital for Ethiopia, Kenya, Rwanda, Burundi, and Tanzania. Peanuts have historically been the primary source of income in the Sahel, while cotton is tremendously important for Sudan and the Central African Republic. Ghana and the Ivory Coast have long been the world's main suppliers of cocoa (the source of chocolate); Liberia produces plantation rubber; and many farmers in Nigeria specialize in palm oil (Figure 6.17). The export of such products can bring good money when commodity prices are high, but when prices collapse, as they periodically do, economic devastation may follow.

Unfortunately, the region's share of world primary exports (such as copper, cacao, and coffee) has fallen since 1980. Most troubling of all, Sub-Saharan Africa's per capita food production, which made gains in the 1960s and early 1970s, has been below the region's 1961 output level for the last 25 years. Rapid population growth has greatly increased food demands. Even though not all subsistence food production is counted, it is hard to ignore this troubling decline in productivity.

Herding and Livestock   Animal husbandry (the care of livestock) is extremely important in Sub-Saharan Africa, particularly in the semiarid zones. Camels and goats are the principal animals in the Sahara and its southern fringe, but farther south cattle are primary. Many African peoples have traditionally specialized in cattle raising and are often tied into mutually beneficial relationships with neighboring farmers. Such **pastoralists** typically graze their stock on the stubble of harvested fields during the dry season and then move them to drier uncultivated areas during the wet season when the pastures turn green. Farmers thus have their fields fertilized by the manure of the pastoralists' stock, while the pastoralists find good dry-season grazing. At the same time, the nomads can trade their animal products for grain and other goods of the sedentary world. Several pastoral peoples of East Africa, however, are noted for their extreme reliance on cattle and general (but never complete) independence from agriculture. The Masai of the Tanzanian-Kenyan borderlands traditionally obtain a large percentage of their nutrition from drinking a mixture of milk and blood (Figure 6.18). The blood is obtained periodically from the animal's jugular veins, a procedure that evidently causes little harm to the cattle.

Large expanses of Sub-Saharan Africa have been off-limits to cattle because of infestations of **tsetse flies**, which spread sleeping sickness to cattle, humans, and some wildlife. Where wild animals, which harbor the disease but are immune to it, were present in large numbers, especially in environments containing brush or woodland (which are necessary for tsetse fly survival), cattle simply could not be raised. At present, tsetse fly eradication programs are reducing the threat, and cattle raising is spreading into areas that were previously forbidden. This process is beneficial for African peoples, but it may endanger the continued survival of many wild animals. When people and their stock move into new areas in large numbers, wildlife almost inevitably declines.

Urban Life

Although Sub-Saharan Africa is considered the least urbanized region in the developing world, most Sub-Saharan cities are growing at twice the national growth rates. If present trends continue, half of the region's population may well be living in cities by 2025. One of the consequences of this surge in city living is urban sprawl. Rural-to-urban migration, industrialization, and refugee flows are forcing the cities of the region to absorb more people and use more resources. As in Latin America, the tendency is toward urban primacy, the condition in which one major city is dominant and at least three times larger than the next largest city. Kinshasa, the capital of the Democratic Republic of the Congo, is an example. In the 1960s less than half a million people resided there; by the late 1990s it was a city of more than 4 million, making it by far the country's primate city.

Sub-Saharan Africa's largest city, Lagos, has some 12 million inhabitants (Figure 6.19). In 1960 it was a city of only 1 million. Unable to keep up with the surge of rural migrants, Lagos's streets are clogged with traffic; for those living on the city's periphery, three- and four-hour commutes (one way) are common. City officials struggle to build enough roads and provide electricity, water, sewage service, and employment for all of these people. In many cases the informal sector (unregulated services and trade) provides urban employment and services. Crime is another major problem for Lagos. The chances of being attacked on the streets or robbed in one's home are quite high.

European colonialism greatly influenced urban form and development in the region. Africans, however, had an urban tradition prior to the colonial era, although a very small percentage of the population lived in cities. Ancient cities, such as Axum in Ethiopia, thrived 2,000 years ago. Similarly, in the Sahel, major trans-Saharan trade centers, such as Timbuktu (Tombouctou) and Gao, have existed for more than a millennium. In East Africa, an urban trading culture emerged that was rooted in Islam and the Swahili language. West Africa, however, had the most developed precolonial urban network, reflecting both native and Islamic traditions. It also supports some of the region's largest cities today.

FIGURE 6.19 • DOWNTOWN LAGOS, NIGERIA   Lagos is home to 12 million people, making it the region's largest city. A classic primate city, its streets are loaded with buses, collective taxis, thousands of pedestrians, and street vendors. Infrastructure has not kept up with the city's rapid growth. The demand for roads and utilities is far greater than the city's government is able to provide. *(Daniel Lainé/Corbis)*

West African Urban Traditions   The West African coastline is dotted with cities, from Dakar, Senegal, in the far north to Lagos, Nigeria, in the east. More than one-third of Nigerians live in cities, and the country has half a dozen metropolitan areas with populations of more than 1 million. Historically, the Yoruba cities in southwestern Nigeria are the best documented. Developed in the twelfth century, cities such as Ibadan were walled and gated, with a palace encircled by large rectangular courtyards at the city center. An important center of trade for an extensive surrounding area, Ibadan was also a religious and political center. Lagos was also a Yoruba settlement. Founded on a coastal island on the Bight of Benin, most of the modern city has spread onto the nearby mainland. Its coastal setting and natural harbor made this relatively small native city attractive to colonial powers. When the British took control in the mid-nineteenth century, the city's size and importance grew.

Urban Industrial South Africa   The major cities of southern Africa, unlike those of West Africa, are colonial in origin. Most of these cities grew as administrative or mining centers such as Lusaka, Zambia, or Harare, Zimbabwe. The nation of South Africa is one of the most urbanized states in the entire region, and it is certainly the most industrialized. The foundations of South Africa's urban economy rest largely on its incredibly rich mineral resources (diamonds, gold, chromium, platinum, tin, uranium, coal, iron ore, and manganese). Eight metropolitan areas have more than 1 million people; the largest of them are Johannesburg, Durban, and Cape Town.

The form of South African cities continues to be imprinted by the legacy of **apartheid** (an official policy of racial segregation that shaped social relations in South Africa for nearly 50 years). Even though apartheid was abolished in 1994, it is still evident in the landscape. Under apartheid rules, cities such as Cape Town

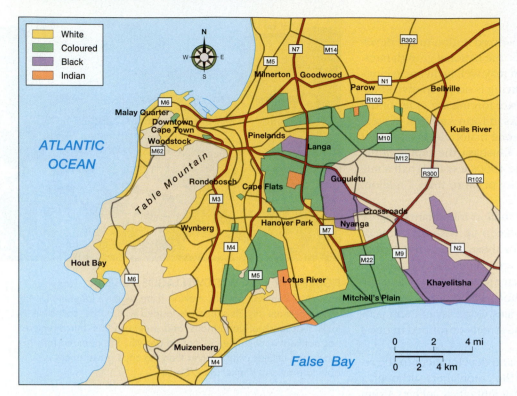

FIGURE 6.20 • RACIAL SEGREGATION IN CAPE TOWN
Under apartheid, most land in Cape Town was designated for white use. Coloureds, blacks, and Indians (South Asians) were crammed into far smaller areas on the less desirable flat and sandy soils east of Table Mountain. (*From Christopher, 1994, The Atlas of Apartheid, p. 156, London: Routledge*)

FIGURE 6.21 • CAPE TOWN TOWNSHIP Gugulethu is a densely settled black township in the sandy flats east of downtown Cape Town. One of the older black townships, it was considered an undesirable area by whites because of fierce winter winds that whip through the area. Residents stack heavy rocks onto their metal roofs to keep them from blowing away. (*Rob Crandall/Rob Crandall, Photographer*)

were divided into residential areas according to racial categories: white, **coloured** (a South African term describing people of mixed African and European ancestry), Indian (South Asian), and African (black). Whites occupied the largest and most desirable portions of the city, especially the scenic areas below Table Mountain (Figure 6.20). Blacks were crowded into the least desired areas, forming squatter settlements called *townships* in places such as Gugulethu (Figure 6.21). Today blacks, coloureds, and Indians are legally allowed to live anywhere they want. Yet the economic differences between racial groups as well as long-standing prejudices make residential integration difficult.

Cultural Coherence and Diversity: Unity Through Adversity

No world region is culturally homogeneous, but most have been partially unified in the past by widespread systems of belief and communication. The lack of traditional cultural and political coherence across Sub-Saharan Africa is not surprising if one considers the region's huge size. Sub-Saharan Africa is more than four times larger than Europe or South Asia. Had foreign imperialism not carved up the region, it is quite possible that West Africa and southern Africa would have developed into their own distinct world regions.

An African identity south of the Sahara was created through a common history of slavery and colonialism as well as struggles for independence and development. More telling, the people of the region often define themselves as African, especially to the outside world. That Sub-Saharan Africa is poor, no one will argue. And yet the cultural expressions of its people—its music, dance, and art—are joyous. Africans share a resilience and optimism that visitors to the region often comment on. The cultural diversity of the region is obvious, yet there is unity among the people drawn from surviving many adversities.

Language Patterns

In most Sub-Saharan countries, as in other former colonies, multiple languages are used that reflect tribal, colonial and national affiliations. Indigenous languages, many from the Bantu subfamily, are often localized to relatively small rural areas. More widely spoken African trade languages, such as Swahili or Hausa, serve as a lingua franca over broader areas. Overlaying native languages are Indo-European and Afro-Asiatic ones (French and English; Arabic and Somali). Figure 6.22 illustrates the complex pattern of language families and major languages found in Africa today. Contrast the larger map with the inset that shows current "official" languages. A comparison of the two shows that most African countries are multilingual, which can be a source of tension within states. In Nigeria, for example, the official language is English, yet there are 25 million Hausa speakers, 24 million Yoruba, 20 million Igbo (or Ibo), 11 million Ful (or Fulani), and 6 million Efik, as well as speakers of dozens of other languages.

African Language Groups

Three of the six language groups mapped in Figure 6.22 are unique to the region (Niger-Congo, Nilo-Saharan, and Khoisan), while the other three (Afro-Asiatic, Austronesian, and Indo-European) are more closely associated with other parts of the world. Afro-Asiatic languages, especially Arabic, dominate North Africa and are understood in Islamic areas of Sub-Saharan Africa as well. Amharic in Ethiopia and Somali of Somalia are also Afro-Asiatic languages. The Malayo-Polynesian language family is limited to the island of Madagascar, which many believe was first settled by seafarers from Indonesia some 1,500 years ago.

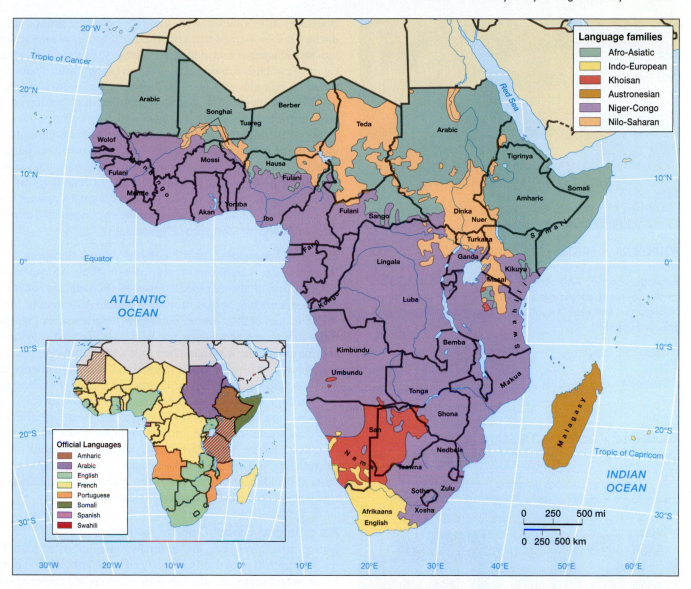

FIGURE 6.22 • AFRICAN LANGUAGE GROUPS AND OFFICIAL LANGUAGES Mapping language is a complex task for Sub-Saharan Africa. There are languages with millions of speakers, such as Swahili, and there are languages spoken by a few hundred people living in isolated areas. Six language families are represented in the region. Among these families are scores of individual languages (see the labels on the map). Since most modern states have many native languages, the colonial language often became the "official" language. English and French are the most common official languages in the region (see inset).

Indo-European languages, especially French, English, Portuguese, and Afrikaans, are a legacy of colonialism and widely used today.

The Niger-Congo language group is by far the most important one in the region. This linguistic group originated in West Africa and includes Mandingo, Yoruba, Ful(ani), and Igbo, among others. Around 3,000 years ago a people of the Niger-Congo language group began to expand out of western Africa into the equatorial zone. This group, called the Bantu, began one of the most far-ranging migrations in human history, which introduced agriculture into large areas of central and southern Africa. Most individual Sub-Saharan languages are limited to relatively small areas and are significant only at the local level. One language in the Bantu subfamily, Swahili, eventually became the most widely spoken Sub-Saharan language. Swahili originated as a trade language on the East African coast, where a number of merchant colonies from Arabia were established around A.D. 1100. A hybrid society grew up in a narrow coastal band of modern Kenya and Tanzania, one speaking a language of Bantu structure enriched with many Arabic words. While Swahili became the primary language only in the narrow coastal belt, it spread far into the interior as the language of trade. After independence was achieved, both Kenya and Tanzania adopted Swahili as an official language.

Language and Identity Ethnic identity as well as linguistic affiliation have historically been highly unstable over much of the region. The tendency was for new language groups to form when people threatened by war fled to less-settled areas, where they often mixed with peoples from other places. In such situations, new

FIGURE 6.23 • MULTILINGUAL SOUTH AFRICA A South African sign warns pedestrians not to cross a highway in three languages: English, Afrikaans, and Zulu. Throughout Sub-Saharan Africa, many people are multilingual because of the diversity of languages that exist in each state. *(Rob Crandall/Rob Crandall, Photographer)*

FIGURE 6.24 • ERITREAN CHRISTIANS AT PRAYER Coptic Christians gather in front of St. Mary's Church in Asmara, Eritrea, on Good Friday, a holy day for Christians. Half of the populations in Eritrea and Ethiopia belong to the Coptic Church, which has ties to the Christian minority in Egypt. *(Ed Kashi/Corbis)*

languages arise quickly and divisions between groups are blurred. Nevertheless, distinct **tribes** formed that consisted of a group of families or clans with a common kinship, language, and definable territory. European colonial administrators were eager to establish a fixed social order to better control native peoples. In this process a flawed cultural map of Sub-Saharan Africa evolved. Some tribes were artificially divided; meaningless names were applied; and territorial boundaries were often misinterpreted.

Social boundaries between different ethnic and linguistic groups have become more stable in recent years, and a number of individual languages have become particularly important for communication at the national scale. With the exception of Swahili, none has the status of being the official language of any country. Indeed, there are only a handful of Sub-Saharan countries in which any single language has a clear majority status, which is partially explained by the arbitrary territorial boundaries imposed by Europeans (Figure 6.23).

European Languages In the colonial period, European countries used their own languages for administrative purposes in their African empires. Education in the colonial period also stressed literacy in the language of the imperial power. In the postindependence period, most Sub-Saharan African countries have continued to use the languages of their former colonizers for government and higher education. Few of these new states had a clear majority language that they could employ, and picking any minority tongue would have met with opposition from other peoples. The one exception is Ethiopia, which maintained its independence during the colonial era. The official language is Amharic, although other native languages are also spoken.

Two large blocks of European language exist in Africa today: Francophone Africa, including the former colonies of France and Belgium, where French serves the main language of administration; and Anglophone Africa, where the use of English prevails (see inset, Figure 6.22). Early Dutch settlement in South Africa resulted in the use of Afrikaans (a Dutch-based language) by several million South Africans. In Mauritania and Sudan, Arabic serves as the main language, although in the case of Sudan, this has resulted in severe internal tensions.

Religion and Global Cultural Exchange

Native African religions are generally classified as animist. This is a somewhat misleading catchall term used to classify all local faiths that do not fit into one of the handful of "world religions." Most animist religions are centered on the worship of nature and ancestral spirits, but the internal diversity within the animist tradition is great. Classifying a religion as animist says more about what it is not than what it actually is.

Both Christianity and Islam actually entered the region early in their histories, but they advanced slowly for many centuries. Since the beginning of the twentieth century, both religions have spread rapidly—more rapidly, in fact, than in any other part of the world. But tens of millions of Africans still follow animist beliefs, and many others combine animist practices and ideas with their observances of Christianity and Islam.

The Introduction and Spread of Christianity Christianity came first to northeast Africa. Kingdoms in both Ethiopia and central Sudan were converted by A.D. 300—the earliest conversions outside of the Roman Empire. The peoples of northern and central Ethiopia adopted the Coptic form of Christianity and have thus historically looked to Egypt's Christian minority for their religious leadership (Figure 6.24). At present, roughly half of the population of both Ethiopia and Eritrea is Coptic Christian; most of the rest are Muslim, but there are still some animist communities, especially in Ethiopia's western lowlands.

European settlers and missionaries introduced Christianity to other parts of Sub-Saharan Africa beginning in the 1600s. The Dutch, who began to colonize South Africa at this time, brought their Calvinist Protestant faith. Later European immigrants to South Africa brought Anglicanism and other Protestant beliefs, as well as Catholicism. Most black South Africans eventually converted to one or another form of Christianity as well. In fact, churches in South Africa were instrumental in the long fight against white racial supremacy. Religious leaders such as Bishop Desmond Tutu were outspoken critics of the injustices of apartheid and worked to bring down the system.

Elsewhere in Africa, Christianity came with European missionaries, most of whom arrived after the mid-1800s. As was true in the rest of the world, missionaries had little success where Islam had preceded them, but they eventually made numerous conversions in animist areas. As a general rule, Protestant Christianity prevails in areas of former British colonization, while Catholicism is more important in former French, Belgian, and Portuguese territories. In the postcolonial era, African Christianity has spread out, at times taking on a life of its own independent from foreign missionary efforts. Still active in the region are various Pentecostal, Evangelical, and Mormon missionary groups, mostly from the United States. It is difficult to map the distribution of Christianity in Africa, however, since it has spread irregularly across the entire non-Islamic portion of the region.

The Introduction and Spread of Islam

Islam began to advance into Sub-Saharan Africa 1,000 years ago (Figure 6.25). Berber traders from North Africa and the Sahara introduced the religion to the Sahel, and by 1050 the Kingdom of Tokolor in modern Senegal emerged as the first Sub-Saharan Muslim state. Somewhat later, the ruling class of the powerful Mande-speaking trading empires of Ghana and Mali converted as well. In the fourteenth century the emperor of Mali astounded the Muslim world when he and his royal court made the pilgrimage to Mecca, bringing with them so much gold that they set off a brief period of high inflation throughout Southwest Asia.

Mande-speaking traders, whose networks spanned the area from the Sahel to the Gulf of Guinea, gradually introduced the religion to other areas of West Africa. There, however, many peoples remained committed to animism, and Islam made slow and unsteady progress. Today orthodox Islam prevails through most of the Sahel. Farther south Muslims are mixed with Christians and animists, but their numbers continue to grow and their practices tend to be orthodox as well (Figure 6.26).

Interaction Between Religious Traditions

The southward spread of Islam from the Sahel, coupled with the northward spread of Christianity from the port cities, has generated a complex religious frontier across much of West Africa. In Nigeria the Hausa are firmly Muslim, while the southeastern Igbo are largely Christians. The Yoruba of the southwest are divided between Christians and Muslims. In the more remote parts of Nigeria, moreover, animist traditions remain strong. But despite this religious diversity, there has been little religious conflict in Nigeria—or elsewhere in West Africa, for that matter. Most of West Africa's regional conflicts continue to be framed in ethnic and linguistic, rather than religious, terms.

Religious conflict has historically been far more serious in northeastern Africa, where Muslims and Christians have struggled against each other for centuries. Sudan is currently the scene of an intense conflict that is both religious and ethnic in origin. Here Islam was introduced in the 1300s by an invasion of Arabic-speaking pastoralists who destroyed the indigenous Coptic Christian kingdoms of the area. Within a few hundred years, central and northern Sudan had become completely Islamic. The southern equatorial province of Sudan, however, where tropical diseases and extensive wetlands prevented Arab advances, remained animist. During the British colonial era, many of the Dinka, Nuer, and other peoples of this area converted to Christianity. In the postcolonial period, the Arabic-speaking Muslims of the north and center quickly became the country's dominant group, and in the 1970s they began

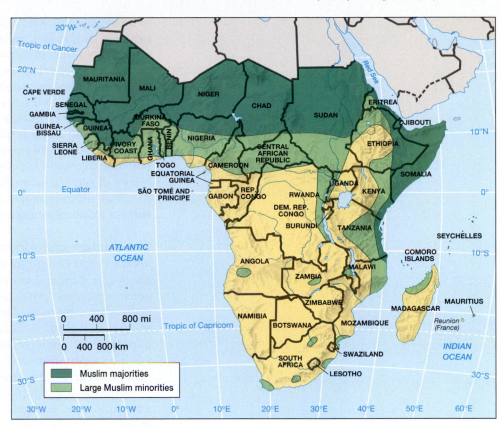

FIGURE 6.25 • EXTENT OF ISLAM   Muslim majorities prevail in the Sahelian states that border North Africa, as well as Somalia and Djibouti. Throughout West and East Africa there are also large Muslim minorities. Except in Sudan, religion has not been a source of political tension in the region. *(From Phillips, 1984, The African Political Dictionary, p. 196, Santa Barbara, CA: Clio Press Ltd.)*

FIGURE 6.26 • WEST AFRICAN MOSQUE   A mosque rises above the houses in the town of Mankono, Ivory Coast. Islam entered this part of West Africa more than 600 years ago, but conversions were limited, with many Ivorians retaining their animist religious practices. Today Christian, Muslim, and animist faiths are practiced in the Ivory Coast, which reflects a pattern of religious tolerance seen throughout much of Sub-Saharan Africa. *(Victor Englebert/Englebert Photography, Inc.)*

to build an Islamic state. Experiencing both religious discrimination and economic exploitation, the peoples of the south launched a massive rebellion. Human rights groups report that some southerners are being enslaved by people in northern and central Sudan. Fighting since the 1980s has been intense, with the government generally controlling the main towns and roads and the rebels maintaining power in the countryside. The war has periodically prevented the distribution of food, resulting in terrible famines and the massive flight of refugees.

Sub-Saharan Africa is a land of religious vitality. Both Christianity and Islam are spreading rapidly, but animism continues to hold widespread appeal. Many new and syncretic (blended) forms of religious expression are also emerging. With such a diversity of faiths, it is fortunate that religion is not typically the cause of overt conflict. With the exception of Sudan, a pattern of peaceful coexistence among different faiths is a distinguishing feature of Sub-Saharan Africa that deserves further study.

Globalization and African Culture The slave trade that linked Africa to the Americas and Europe set in process patterns of cultural diffusion that transferred African peoples and cultures across the Atlantic. Tragically, slavery damaged the demographic and political strength of African societies, especially in West Africa, from where the most slaves were taken. An estimated 12 million Africans were shipped to the Americas as slaves from the 1500s until 1870 (Figure 6.27). Slavery impacted the entire region, sending Africans not just to the Americas, but to Europe, North Africa, and Southwest Asia. The vast majority, however, worked on plantations across the Americas.

Out of this tragic displacement of people came a blending of African cultures with Amerindian and European ones. African rhythms are at the core of various musical styles from rumba to jazz, the blues, and rock and roll. Brazil, the largest country in Latin America, is claimed to be the second largest "African state" (after Nigeria) because of its huge Afro-Brazilian population. Thus the forced migration of Africans as slaves had a huge cultural influence on many areas of the world.

However, cultural exchanges almost never occur in one direction only. Foreign language, religion, and dress were adopted by Africans from the beginning of European

FIGURE 6.27 • AFRICAN SLAVE TRADE The slave trade had a devastating impact on Sub-Saharan societies. From ship logs, it is estimated that 12 million Africans were shipped to the Americas to work as slaves on sugar, cotton, and rice plantations; the majority went to Brazil and the Caribbean. Yet other slave routes existed, although the data are less reliable. Africans from south of the Sahara were used as slaves in North Africa. Others were traded across the Indian Ocean into Southwest Asia and South Asia.

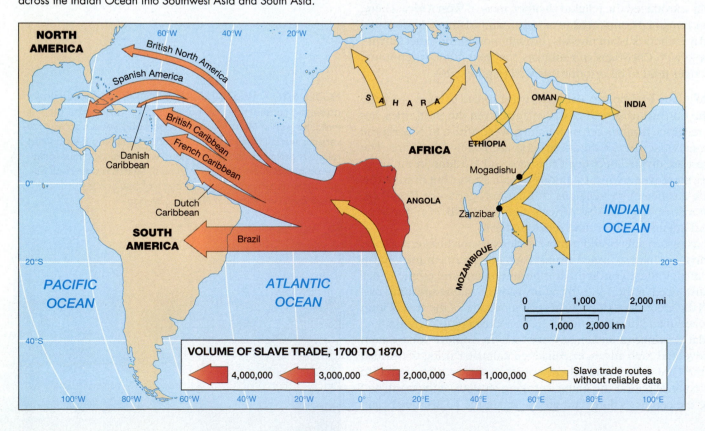

exploration. With independence in the 1960s and 1970s, several states tried to rediscover their ancestral roots by openly rejecting European cultural elements. Ironically, the search for traditional religions found African scholars traveling to the Caribbean and Brazil to consult with Afro-religious practitioners about ceremonial elements lost to West Africa. Musical styles from the United States, the Caribbean, and Latin America were imitated and transformed into new African sounds. Kwaito, the latest musical craze in South Africa, sounds a lot like gangsta rap from the United States. A closer listen, however, reveals an inclusion of local rhythms, lyrics in Zulu and Xhosa, and themes about life in the post-apartheid townships. Thus, just as the slave trade spread African influences to other parts of the world, the effects of colonization and globalization are apparent in the cultures of Sub-Saharan Africa.

Geopolitical Framework: Legacies of Colonialism and Conflict

The duration of human settlement in Sub-Saharan Africa is unmatched by any other region. Evidence shows that humankind originated there, and many diverse ethnic groups formed in the region over the past few thousand years. Although conflicts among these groups existed, cooperation and coexistence among different peoples also occurred. Some 3,000 years ago the state of Nubia emerged in central and northern Sudan. One thousand years later the Kingdom at Axum arose in northern Ethiopia and Eritrea. Both of these states were strongly influenced by political models derived from Egypt and Arabia. The first wholly indigenous African states were founded in the Sahel around A.D. 700. Over the next several centuries, a variety of other states emerged in West Africa. By the 1600s, the states located near the Gulf of Guinea took advantage of the opportunities presented by the slave trade, namely the selling of slaves to Europeans.

Thus, prior to European colonization, Sub-Saharan Africa presented a complex mosaic of kingdoms, states, and tribal societies. With the arrival of Europeans, patterns of social organization and ethnic relations were changed forever. As Europeans rushed to carve up the continent to serve their imperial ambitions, they set up various policies that heightened ethnic tensions and promoted hostility. Many of the region's modern conflicts can trace their roots back to the colonial era, especially the drawing of political boundaries.

European Colonization

Unlike the relatively rapid colonization of the Americas, Europeans needed centuries to gain effective control of Sub-Saharan Africa. Portuguese traders arrived along the coast of West Africa in the 1400s, and by the 1500s they were well established in East Africa as well. Initially the Portuguese made large profits, converted a few local rulers to Christianity, established several defensive trading posts, and gained control over the Swahili trading cities of the east. They stretched themselves too thin, however, and failed in many of their activities. Only along the coasts of modern Angola and Mozambique, where a sizeable population of mixed African and Portuguese peoples emerged, did Portugal maintain power. Along the Swahili, or eastern, coast they were eventually expelled by Arabs from Oman, who then established their own trade-based empire in the area.

The Disease Factor   One of the main reasons for the Portuguese failure was the disease environment of Africa. With no resistance to malaria and other tropical diseases, roughly half of all Europeans who remained on the African mainland died within a year. Protected both by their armies and by the diseases of their native lands, African states were able to maintain an upper hand over European traders and adventurers well into the 1800s. Unlike the Americas, where European conquest was made easy by the introduction of Old World diseases that devastated native populations (see Chapters 4 and 5), in Sub-Saharan Africa endemic disease limited European settlement until the mid-nineteenth century.

Colonists risked the hazards of malaria and other tropical diseases such as sleeping sickness because they were drawn to the region's wealth, and soon other European traders followed the Portuguese. By the 1600s Dutch, British, and French firms dominated the profitable export of slaves, gold, and ivory from the Gulf of Guinea. The Dutch also established a settler colony in South Africa, safely outside of the tropical disease zone, to supply their ships bound for Indonesia. For the next 200 years European traders came and went, occasionally building fortified coastal posts, but they almost never ventured inland and seldom had any real influence on African rulers. By exporting millions of slaves, however, they had a strongly negative impact on African society.

In the 1850s European doctors discovered that a daily dose of quinine would offer protection against malaria, completely changing the balance of power in Africa. Explorers immediately began to penetrate the interior of the continent, while merchants and expeditionary forces began to move inland from the coast. The first imperial claims soon followed. The French quickly grabbed power along the easily navigated Senegal River, while the British established protectorates over the indigenous states of the Gold Coast (modern-day Ghana).

The Scramble for Africa In the 1880s European colonization of the region quickly accelerated, leading to the so-called scramble for Africa. By this time, after the invention of the machine gun, no African state could long resist European force.

As the colonization of Africa intensified, tensions among the colonizing forces of Britain, France, Belgium, Germany, Italy, Portugal, and Spain mounted. Rather than risk war, 13 countries convened in Berlin at the invitation of the German chancellor Bismarck in 1884 in a gathering known as the **Berlin Conference**. During the conference, which no African leaders attended, rules were established as to what determined "effective control" of a territory, and Sub-Saharan Africa was carved up and traded like properties in a game of Monopoly® (Figure 6.28). While European arms were by the 1880s far superior to anything found in Africa, several indigenous states did organize effective resistance campaigns. For example, in central Sudan an Islamic-inspired force held out against the British until 1900.

Eventually European forces prevailed everywhere, with one major exception: Ethiopia. The Italians had conquered the Red Sea coast and the far northern highlands (modern Eritrea) by 1890, and they quickly set their sights on the large Ethiopian kingdom, called Abyssinia, which had itself been vigorously expanding for several decades. In 1896, however, Abyssinia defeated the invading Italian army, earning the respect of the European powers. In the 1930s fascist Italy launched a major invasion of the country, now renamed Ethiopia, to redeem its earlier defeat, and with the help of poison gas and aerial bombardment, it quickly prevailed.

Although Germany was a principal instigator of the scramble for Africa, it lost its own colonies after suffering defeat in World War I. Britain and France then partitioned most of Germany's African empire between themselves. Figure 6.28 shows the colonial status of the region in 1913, prior to Germany's territorial loss.

Establishment of South Africa While the Europeans were strengthening their rule over Africa after World War I, South Africa was inching toward political freedom,

FIGURE 6.28 • EUROPEAN COLONIZATION IN 1913

Before 1880 few areas of Africa were under direct European control. By 1884, when the Berlin Conference began, Africa was carved up and traded between European powers. France and Britain controlled the most territory, but Germany, Portugal, Belgium, Spain, and Italy all had their claims. By 1913 the entire continent, except Ethiopia, Liberia, and South Africa, were under European colonial control. (From The Times Atlas of World History, 1989, Maplewood, NJ: Hammond Inc.)

at least for its white population. South Africa was one of the oldest colonies in Sub-Saharan Africa, and it became the first to obtain its political independence from Europe in 1910. Its economy is the most productive and influential of the region, yet the legacy of apartheid made it an international exile, especially within Africa.

The original body of Dutch settlers in South Africa had grown slowly, expanding to the north and east of its original center around Cape Town. As a farming and pastoral people largely isolated from the European world, these Afrikaners, or Boers, developed an extremely conservative cultural outlook marked by an increasing belief in their racial superiority over the native population. In 1806 the British, then the world's unchallenged maritime power, seized a large area of land from the Dutch. Relations between the British rulers and their new Afrikaner subjects quickly soured. In the 1830s the bulk of the Afrikaner community opted to leave British territory and strike out for new lands. By the late 1800s it had become clear that the new Boer territory was sitting above one of the world's greatest reserves of mineral wealth, enticing the British to claim the territory for their own. English-speaking people were increasingly drawn to the area, and they grew resentful of Afrikaner power. The result was the Boer War, which turned into a long and brutal guerrilla struggle between the British Army and mobile Afrikaner bands. By 1905 the Boers relented, and their territory was joined with that of the British to form the Union of South Africa. Five years later South Africa was given its independence from Britain. Britons and other Europeans continued to settle there, but the Afrikaners remained the majority white group. Black and coloured South Africans, however, greatly outnumbered whites.

In 1948, when the Afrikaner's National Party gained control of the government, they introduced the policy of "separateness" known as *apartheid*. British South Africans had enacted a series of laws that were prejudicial to nonwhite groups, but it was under Afrikaner leadership that racial separation become more formalized and systematic. Operating on three scales—petite, meso-, and grand—apartheid managed social interaction by controlling space. Petite apartheid, like Jim Crow laws in the United States, created separate service entrances for government buildings, bus stops, and restrooms based on skin color. Meso-apartheid divided the city into residential sectors by race, giving whites the best and largest urban zones. Finally, grand apartheid was the construction of black **homelands** by ethnic group. The intention was to make all blacks citizens of these semi-independent homelands. Likened to the reservations created for Native Americans in the United States, homelands were rural, overcrowded, and on marginal land. Some 3 million blacks were forcibly relocated into homelands during apartheid, and residence outside of the homelands was strictly regulated.

Granted its independence in 1910, South Africa was the first state in the region freed from colonial rule. Yet because of its formalized system of discrimination and racism, it was hardly a symbol of liberty. Ironically, at the same time that the Afrikaners tightened their political and social control over the nonwhite population, the rest of the continent was preparing for political independence from Europe.

Decolonization and Independence

Decolonization of the region happened rather quickly and peacefully beginning in 1957. Independence movements, however, had sprung up throughout the continent, some dating back to the early 1900s. Workers' unions and independent newspapers became voices for African discontent and the hope for freedom.

By the late 1950s, Britain, France, and Belgium decided that they could no longer maintain their African empires and began to withdraw. (Italy had already lost its colonies during World War II, and Britain gained Somalia and Eritrea.) Once started, the decolonization process moved rapidly. By the mid-1960s, virtually the entire region had achieved independence. In most cases the transition was relatively peaceful and smooth.

FIGURE 6.29 • A MONUMENT TO KWAME NKRUMAH
Charismatic independence leader Kwame Nkrumah is remembered with this monument in Accra, Ghana. Nkrumah led Ghana to an early independence in 1957; he was also a founder of the Organization of African Unity (OAU). *(Victor Englebert/Englebert Photography, Inc.)*

FIGURE 6.30 • DISPLACED ANGOLANS   Since 1999 an estimated 3 million Angolans have been internally displaced from their villages due to fighting between government troops and the rebel group UNITA. Technically speaking, they are not refugees, as they have not left their country, but they live in refugee-like conditions. *(Malcolm Linton/Getty Images, Inc.)*

Dynamic African leaders put their mark on the region during the early decades after independence. Men such as Kenya's Jomo Kenyatta, Ivory Coast's Felix Houphuët-Boigney, and Ghana's Kwame Nkrumah became powerful father figures who molded their new nations (Figure 6.29). President Nkrumah's vision for Africa was the most expansive. After helping to secure independence for Ghana in 1957, his ultimate aspiration was the political unity of Africa. While his dream was never realized, it set the stage for the founding of the **Organization of African Unity (OAU)** in 1963. The OAU is a continent-wide organization. Its main role has been to mediate disputes between neighbors. Certainly in the 1970s and 1980s it was a constant voice of opposition to South Africa's minority rule, and the OAU intervened in some of the more violent independence movements in southern Africa.

Southern Africa's Independence Battles   Independence did not come easily to southern Africa. In Southern Rhodesia (modern-day Zimbabwe) the problem was the presence of some 250,000 white residents, most of whom owned large farms. Unwilling to see power pass to the country's black majority, then some 6 million strong, these settlers declared themselves the rulers of an independent, white-supremacist state in 1965. The black population continued to resist, however, and in 1978 the Rhodesian government was forced to give up power. The renamed country of Zimbabwe was henceforth ruled by the black majority, although the remaining whites still form an economically privileged community. Since the mid-1990s, disputes over government land reform (splitting up the large commercial farms mostly owned by whites and giving the land to black farmers) have resulted in serious racial and political tensions.

In the former Portuguese colonies, independence came violently. Unlike the other imperial powers, Portugal refused to hand over its colonies in the 1960s. As a result, the people of Angola and Mozambique turned to armed resistance. The most powerful rebel movements adopted a socialist orientation and received support from the Soviet Union and Cuba. A new Portuguese government came to power in 1974, however, and it withdrew suddenly from its African colonies. At this point Marxist regimes quickly came to power in both Angola and Mozambique. The United States, and especially South Africa, responded to this perceived threat by supplying arms to rebel groups that opposed the new governments. Fighting dragged on for several decades in Angola and Mozambique. The countryside in both states is now so heavily laden with land mines that it can hardly be used. With the end of the Cold War, however, outsiders lost their interest in continuing these conflicts, and sustained efforts have been under way to negotiate a peace settlement. At present Mozambique is at peace, but Angola's attempts at establishing peace have been less successful. As of 2001 the military strength of UNITA, a rebel group, has weakened considerably, but it still wages war against the government of Angola. It is estimated that from 1999 to 2001, more than 3 million Angolans were displaced from their homes by fighting (Figure 6.30).

Apartheid's Demise in South Africa   While fighting continued in the former Portuguese zone, South Africa underwent a remarkable transformation. Through the 1980s, its government had remained firmly committed to white supremacy. Under apartheid only whites enjoyed real political freedom, while blacks were denied even citizenship in their own country—technically, they were citizens of homelands.

The first major change came in 1990, when South Africa withdrew from Namibia, which it had controlled as a protectorate (when one state assumes authority over a dependent political unit) since the end of World War I. South Africa now stood alone as the single white-dominated state in Africa. A few years later the leaders of the Afrikaner-dominated political party decided they could no longer resist the pressure for change. In 1994 free elections were held in which Nelson Mandela, a black leader who had been imprisoned for 27 years by the old regime, emerged as the new president. Black and white leaders pledged to put the past behind them and work together to build a new, multiracial South Africa.

Continuing Political Conflict

Although most Sub-Saharan countries made a relatively peaceful transition to independence, virtually all of them immediately faced a difficult set of institutional and political problems. In several cases the old authorities had done almost nothing to prepare their colonies for independence. Lacking an institutional framework for independent government, countries such as the Democratic Republic of the Congo faced a chaotic situation from the beginning. Only a handful of Congolese had received higher education, let alone been trained for administrative posts. The indigenous African political framework had been essentially destroyed by colonization, and in most cases very little had been built in its place.

Even more problematic, in the long run, was the political geography of the newly independent states. Civil servants could always be trained and administrative systems built, but little could be done to rework the region's basic political map. The problem was the fact that the European colonial powers had essentially ignored indigenous cultural and political boundaries, both in dividing Africa among themselves and in creating administrative subdivisions within their own imperial territories.

The Tyranny of the Map   All over Africa, different ethnic groups found themselves forced into the same state with peoples of distinct linguistic and religious backgrounds, many of whom had recently been their enemies. At the same time, a number of the larger ethnic groups of the region found their territories split between two or more countries. The Hausa people of West Africa, for example, were divided between Niger (formerly French) and Nigeria (formerly British), each of which they had to share with several former enemy groups.

Given the imposed political boundaries, it is no wonder that many African countries struggle to generate a common sense of national identity or establish stable political institutions. **Tribalism**, or loyalty to the ethnic group rather than to the state, has been the most difficult African political problem. Especially in rural areas, tribal identities are usually more important than national ones. Because nearly all of Africa's countries inherited an inappropriate set of colonial borders, one might assume that they would have been better drawing a new political map based on indigenous identities. Such a strategy was impossible, as all the leaders of the newly independent states realized. Any new territorial divisions would have created winners and losers, and thus would have resulted in more conflict. Moreover, since ethnicity in Sub-Saharan Africa was traditionally fluid, and since many ethnic groups were intermixed, it would have been difficult to generate a clear-cut system of division. Finally, most African ethnic groups were considered too small to form viable countries. With such complications in mind, the new African leaders, meeting in 1963 to form the Organization of African Unity, agreed that colonial boundaries should remain. The violation of this principle, they argued, would lead to pointless wars between states and endless civil struggles within them.

Ethnic and secessionist conflicts, as well as military governments, have undermined African hopes for peaceful democracy in the postcolonial period (Figure 6.31). The human cost of this turmoil is several million refugees. **Refugees** are people who flee their state because of a well-founded fear of persecution based on race, ethnicity, religion, or political orientation. More than 3 million Africans were considered refugees in 2000. Added to this figure are at least another 13 million **internally displaced persons**, who have fled from conflict but still reside in their country of origin. At the end of 2000, Sudan had the largest number of internally displaced people (4 million), followed by Angola (estimates range from 2 to 4 million) and the Democratic Republic of the Congo (2 million). These populations are not technically considered refugees, making it difficult for humanitarian nongovernmental organizations and the United Nations to assist them

Ethnic Conflict in Rwanda   In 1994 the world was shocked by the genocide in Rwanda among the Hutus and Tutsis. Rwanda and its neighbor Burundi inherited a potentially explosive ethnic mix from the precolonial period. Both countries had been indigenous states in which a Tutsi aristocracy ruled a Hutu peasantry. During

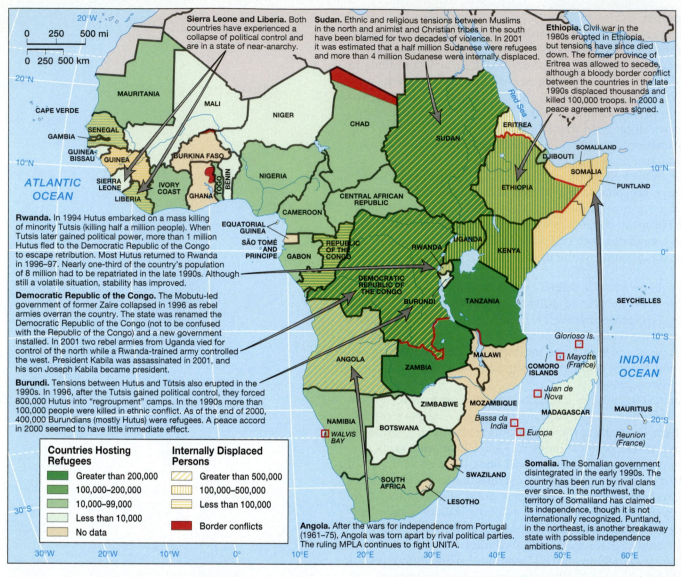

Sierra Leone and Liberia. Both countries have experienced a collapse of political control and are in a state of near-anarchy.

Sudan. Ethnic and religious tensions between Muslims in the north and animist and Christian tribes in the south have been blamed for two decades of violence. In 2001 it was estimated that a half million Sudanese were refugees and more than 4 million Sudanese were internally displaced.

Ethiopia. Civil war in the 1980s erupted in Ethiopia, but tensions have since died down. The former province of Eritrea was allowed to secede, although a bloody border conflict between the countries in the late 1990s displaced thousands and killed 100,000 troops. In 2000 a peace agreement was signed.

Rwanda. In 1994 Hutus embarked on a mass killing of minority Tutsis (killing half a million people). When Tutsis later gained political power, more than 1 million Hutus fled to the Democratic Republic of the Congo to escape retribution. Most Hutus returned to Rwanda in 1996–97. Nearly one-third of the country's population of 8 million had to be repatriated in the late 1990s. Although still a volatile situation, stability has improved.

Democratic Republic of the Congo. The Mobutu-led government of former Zaire collapsed in 1996 as rebel armies overran the country. The state was renamed the Democratic Republic of the Congo (not to be confused with the Republic of the Congo) and a new government installed. In 2001 two rebel armies from Uganda vied for control of the north while a Rwanda-trained army controlled the west. President Kabila was assassinated in 2001, and his son Joseph Kabila became president.

Burundi. Tensions between Hutus and Tutsis also erupted in the 1990s. In 1996, after the Tutsis gained political control, they forced 800,000 Hutus into "regroupment" camps. In the 1990s more than 100,000 people were killed in ethnic conflict. As of the end of 2000, 400,000 Burundians (mostly Hutus) were refugees. A peace accord in 2000 seemed to have little immediate effect.

Angola. After the wars for independence from Portugal (1961–75), Angola was torn apart by rival political parties. The ruling MPLA continues to fight UNITA.

Somalia. The Somalian government disintegrated in the early 1990s. The country has been run by rival clans ever since. In the northwest, the territory of Somaliland has claimed its independence, though it is not internationally recognized. Puntland, in the northeast, is another breakaway state with possible independence ambitions.

Countries Hosting Refugees
- Greater than 200,000
- 100,000–200,000
- 10,000–99,000
- Less than 10,000
- No data

Internally Displaced Persons
- Greater than 500,000
- 100,000–500,000
- Less than 100,000
- Border conflicts

FIGURE 6.31 • POSTCOLONIAL CONFLICTS   The independence period has witnessed many ethnic conflicts as well as civil wars between competing political groups. In contrast to other world regions, Sub-Saharan Africa has had only one successful secessionist movement, which led to the creation of Eritrea. One consequence of political turmoil has been several million refugees and internally displaced persons. *(Data from U.S. Committee for Refugees, 2001, World Refugee Survey; and Allcock et al., 1992, Border and Territorial Disputes, 3rd ed., Harlow, Essex, UK: Longman Group)*

the colonial period, the practice of giving one group privileges over the other intensified hostility between the two groups. The Hutus gained political power in Rwanda after independence in 1962. In 1994 the Hutu president died in a plane crash under somewhat mysterious conditions, which Hutu leaders interpreted as the work of Tutsi rebels. In a planned wave of genocide, half a million Tutsis were slaughtered over the next several months. Rebel Tutsi soldiers soon thereafter managed to overthrow the government and gain power, after which more than a million Hutus, fearing revenge, fled the country (see Figure 6.31).

Images of a million refugees living in camps on the Rwandan border filled the international media. Efforts to send home the refugees were unsuccessful for two years. Ultimately, political turmoil in the Democratic Republic of the Congo (a major host country) caused a massive and sudden repatriation in 1996. In four days, a half million Rwandans left the Congolese camps and walked back to their homes.

A fortified Tutsi-led Rwandan army attacked the mostly Hutu refugee camps in the Democratic Republic of the Congo in 1996, forcing the repatriation of Hutus back to Rwanda. The new army proved more capable than anyone expected. It headed west, picking up allies from the Democratic Republic of the Congo as well as Uganda and eventually took over the capital, Kinshasa, installing Laurent Kabila as president in 1997. Under Kabila's rocky leadership, which ended with his assassination in 2001, two countries invaded the Democratic Republic of the Congo, Rwanda and Uganda, and new rebel groups formed. By 2001 the Kinshasa-based government controlled the western and southern portions of the country, while two rebel groups from Uganda and one from Rwanda controlled the northern and eastern portions. As civil unrest continues, the number of displaced people has soared while economic output has plummeted. Meanwhile, the Democratic Republic of the Congo's nine neighboring states fear spreading destabilization and a potential outflow of refugees.

Secessionist Movements   Knowing how problematic African political boundaries are, one would expect territories to secede and form new states. The Shaba (or Katanga) province in what was then the state of Zaire tried to leave its national territory soon after independence. The rebellion was crushed a couple of years after it started with the help of France and Belgium. Zaire was unwilling to give up its copper-rich territory, and the former colonialists had economic interests in the territory worth defending. Similarly, the Igbo in oil-rich southeastern Nigeria declared an independent state of Biafra in 1967. After a short but brutal war during which Biafra was essentially starved into submission, Nigeria was reunited.

Only one territory in the region has successfully seceded. In 1993 Eritrea gained independence from Ethiopia after two decades of civil conflict. What is striking about this territorial secession is that Ethiopia gave up its access to the Red Sea, making it landlocked. Yet the creation of Eritrea still did not bring about peace. After years of fighting, the transition to Eritrean independence began remarkably well. Unfortunately, border disputes between the two countries erupted in 1998, resulting in some 100,000 troop deaths. In 2000 a peace accord was reached and the fighting stopped. If Ethiopia can accept Eritrea's independence, it might be possible for other areas torn by ethnic warfare to do so. The bloody civil war in southern Sudan, which has produced a half million refugees and 4 million internally displaced persons, might be resolved by granting independence to the province of Equatoria. Still, any major transformation of Africa's political map should not be expected.

Economic and Social Development: The Struggle to Rebuild

By almost any measure, Sub-Saharan Africa is the poorest and least developed world region. Only a few African countries, such as Gabon, South Africa, Seychelles, Botswana, and Mauritius, are counted among the middle-income economies of the world (with per capita GNIs more than $3,000). The rest are ranked at the bottom of most lists comparing per capita GNI or purchasing power parity. Using social indicators such as life expectancy or adult illiteracy, the region also fares poorly. Interestingly, female labor force participation far exceeds rates in other parts of the world. Yet the overall status of African women is mixed, and labor force participation is not necessarily an indicator of equality.

The most troubling economic indicators show that the region's economic productivity actually declined during the 1990s. Most countries had low to negative growth rates between 1990 and 1999. The impact of falling economic output was made worse by rapid population growth. (The region's population growth rate of 2.5 percent in 2001 was considerably higher than Latin America's 1.7 percent rate.) This means that the national economies of Africa needed to grow rapidly just to maintain their living standards; instead, they declined. The economic and debt crisis that followed prompted the introduction of **structural adjustment programs**. Promoted by the International Monetary Fund (IMF) and the World Bank, structural adjustment programs typically reduce government spending, cut food subsidies, and encourage private-sector initiatives. Yet these same policies caused immediate hardships for the poor, especially women and children, and led to social protest, most notably in cities. The idea of debt forgiveness for Africa's poorest states, argued for by the unlikely duo of economist Jeffrey Sachs and rock star Bono from U2, is gradually gaining acceptance as a strategy to reduce human suffering in the region.

Over the past few years there have been some signs of an approaching economic turnaround. The economy of the region as a whole seems to have experienced some growth beginning in 1995, but this growth has been at a slow pace. The late 1990s witnessed African economies growing at a slightly higher rate than their population, something that has not happened in many years. A few African countries, moreover, have done much better over the past decade, most notably Uganda and Equatorial Guinea.

Roots of African Poverty

In the past outside observers often attributed Africa's poverty to its environment. Favored explanations included the infertility of its soils, the erratic patterns of its rainfall, the scarcity of its navigable rivers, and the persistence of its tropical diseases. Most contemporary scholars, however, argue that such handicaps are not prevalent throughout the region, and that even where they do exist they can be—and have often been—overcome by human labor and creativity. The explanations for African poverty now look much more to historical and institutional factors than to environmental circumstances.

Numerous scholars have singled out the slave trade for its debilitating effect on Sub-Saharan African economic life. Large areas of the region were depopulated, and many people were forced to flee into poor, inaccessible refuges. Colonization was another blow to Africa's economy. European powers invested little in infrastructure, education, or public health and were instead interested mainly in developing mineral and agricultural resources for their own benefit. Several plantation and mining zones did achieve some prosperity under colonial regimes, but strong national economies failed to develop. In almost all cases, the basic transport and communications systems were designed to link administration centers and zones of extraction directly to the colonial powers, rather than to their own surrounding areas. As a result, after achieving independence, Sub-Saharan African countries faced economic challenges that were as serious as their political problems.

Failed Development Policies The first decade or so of independence was a time of relative prosperity and optimism for many African countries. Most of them relied heavily on the export of mineral and agricultural products, and through the 1970s commodity prices generally remained high. Some foreign capital was attracted to the region, and in many cases the European economic presence actually increased after decolonization.

A period of economic growth in the 1960s and early 1970s ceased to exist in the 1980s as most commodity prices began to decline. A large amount of foreign debt began to weigh down many Sub-Saharan countries. By the end of the 1980s, most of the region was in serious economic decline. By the early 1990s, the region's foreign debt was around $200 billion. Although low compared to other developing regions (such as Latin America), as a percentage of its economic output, Sub-Saharan Africa's debt was the highest in the world.

Many economists argue that Sub-Saharan African governments introduced economic policies that were counterproductive and thus brought some of their misery on themselves. Eager to build their own economies and reduce their dependency on the former colonial powers, most African countries followed a course of economic nationalism. More specifically they set about building steel mills and other forms of heavy industry that were simply not competitive (Figure 6.32).

Corruption Although prevalent through most of the world, corruption also seems to have been particularly widespread in several African countries. Because civil servants are not paid a living wage they are after forced to solicit bribes. According to a recent poll of international businesspeople, Nigeria ranks as the world's most corrupt country. (Skeptical observers, however, point out that several Asian nations with highly successful economies, such as China, are also noted for high levels of corruption, so that corruption alone may not be the problem.)

With millions of dollars in loans and aid pouring into the region, officials at various levels were tempted to take something for themselves. Some African states, such as the former Zaire, were given the name *kleptocracies*. A **kleptocracy** is a state

FIGURE 6.32 • INDUSTRIALIZATION Heavy industry, such as this chemical plant in Kafue, Zambia, has failed to deliver Sub-Saharan Africa from poverty. In the worst cases, these industrial enterprises were unable to produce competitive products for world and domestic markets. *(Marc & Evelyne Bernheim/Woodfin Camp & Associates)*

in which corruption is so institutionalized that politicians and government bureaucrats siphon off a huge percentage of the country's wealth. President Mobutu of the Democratic Republic of the Congo was a legendary kleptocrat. While his country was saddled with an enormous foreign debt, he reportedly maintained several billion dollars in Belgian banks.

Links to the World Economy

In terms of trade, Sub-Saharan Africa's connection with the world is limited. The level of overall trade is low both within the region and outside it. Most exports are bound for the European Union, especially the former colonial powers of England and France. The United States is the second most common destination. Exports to Asia are surprisingly few—mostly oil and tropical hardwoods. The import pattern mirrors exports. Despite four decades of independence, the majority of African countries turn to Europe for imports. Increasingly, however, cheaper goods from Asia are making their way across the Indian Ocean to African consumers.

By most measures of connectivity, Sub-Saharan Africa lags behind other developing regions. Telephone lines are scarce; in Nigeria there are only 4 lines per 1,000 persons. For every 1,000 people in the region there are 40 television sets, whereas the number of televisions in high-income countries averages 700 per 1,000. One hopeful change is the expansion of mobile telephones in Africa. No longer reliant on expensive fixed telephone lines for communication, multinational providers are now competing for mobile-phone customers. South Africa alone has more than 8 million mobile phone subscribers (Figure 6.33).

Africa also lacks the infrastructure and goods to facilitate more intraregional trade. Only southern Africa has a modern telecommunications network. Of the few roads that exist, especially in West and East Africa, the majority are not paved. Nothing comparable to Mercosur (the trade bloc in South America) is found in Sub-Saharan Africa. This, too, is a legacy of colonialism, and one that keeps Africans from nurturing economic linkages with each other.

Aid Versus Investment As a poor region, Sub-Saharan Africa is linked to the global economy more through the flow of financial aid and loans than through the flow of goods. As Figure 6.34 reveals, for several states aid accounts for 10 percent of the GNI. In the extreme case of Malawi, economic assistance in 1999 accounted for one-quarter of the GNI. Most of this aid comes from a handful of developed countries (see the insert in Figure 6.34). France has provided the most aid, closely followed by the United States and then Japan, Germany, and the United Kingdom. By contrast, net flows of private capital are extremely low. Countries such as South Africa, Equatorial Guinea, Gabon, and Angola are the exceptions because of their oil or mineral wealth. The majority of states in the region attract less than $10 per capita in private foreign investment, whereas the figure for Latin America is 10 times greater.

The reasons foreign investors avoid Africa are fairly obvious. The region is generally perceived to be too poor and unstable to merit much attention, and most foreign investors eager to take advantage of low wages put their money in Asia or Latin America. Sub-Saharan Africa, some economists suggest, is therefore starved for capital. Other scholars, however, see foreign investment as more of a trap—and one that the region would be wise to avoid. Recently in a number of African countries a small boom has occurred in mineral exploration and production, attracting considerable overseas interest. Optimists interpret this as a sign of Africa's economic renewal, but others counter that in the past mineral extraction has failed to lead to broad-based and lasting economic gains.

FIGURE 6.33 • MOBILE PHONES FOR AFRICA A shopkeeper has steady customers for mobile phones in Abidjan, Ivory Coast. Mobile telephones are quickly becoming the preferred instrument of communication in Sub-Saharan Africa. They do not require expensive fixed lines and multinational firms are competing for the region's business. *(Andre Ramasore/Galbe.com)*

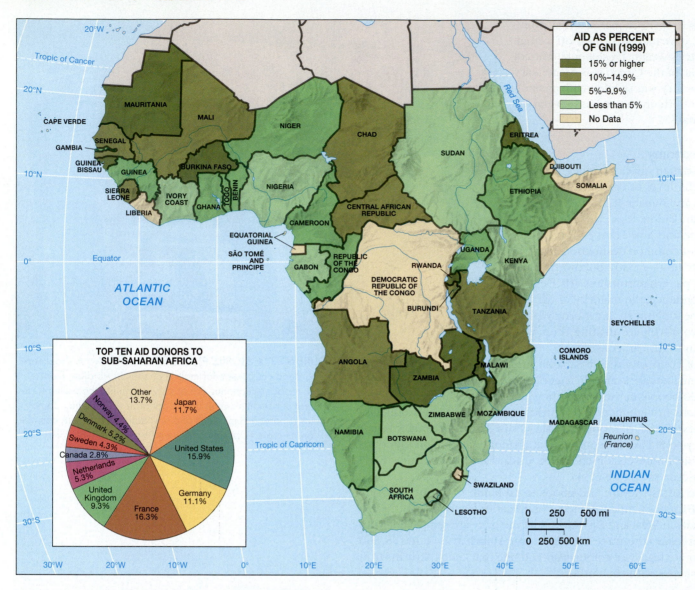

FIGURE 6.34 • GLOBAL LINKAGES: AID DEPENDENCY Many states in Sub-Saharan Africa are dependent on foreign aid as their primary link to the global economy. This figure maps aid as a percentage of GNI, which ranges from less than 1 percent to more than 25 percent in Guinea-Bissau and Malawi. *(From World Development Indicators, 2001. Washington, D.C.: World Bank.)*

Debt Relief The World Bank and the IMF proposed in 1996 to reduce debt levels for heavily indebted poor countries, many of which are in Sub-Saharan Africa. Most Sub-Saharan states are indebted to official creditors such as the World Bank, not to commercial banks (as is the case in Latin America and Southeast Asia). Under this new World Bank/IMF program, substantial debt reduction will be given to Sub-Saharan countries that are determined to have "unsustainable" debt burdens. Mauritania, for example, spends six times more money on repaying its debts than it does on health care. States qualify for different levels of debt relief provided they present a poverty reduction strategy. Uganda was the first state to qualify for the program, using the money it saved on debt repayment to expand primary schooling. Other countries will also benefit from debt reduction.

Economic Differentiation Within Africa

As in most regions, the large difference in levels of economic and social development is a lasting problem. For example, the small island nations of Mauritius and the Seychelles have a high per capita GNI, life expectancies averaging in the low 70s, and economies built on tourism. In contrast, Chad has a per capita GNI of $210, and more than half the adult population is illiterate. In general, the most prosperous states are those that exploit mineral and oil resources, such as South Africa and Gabon.

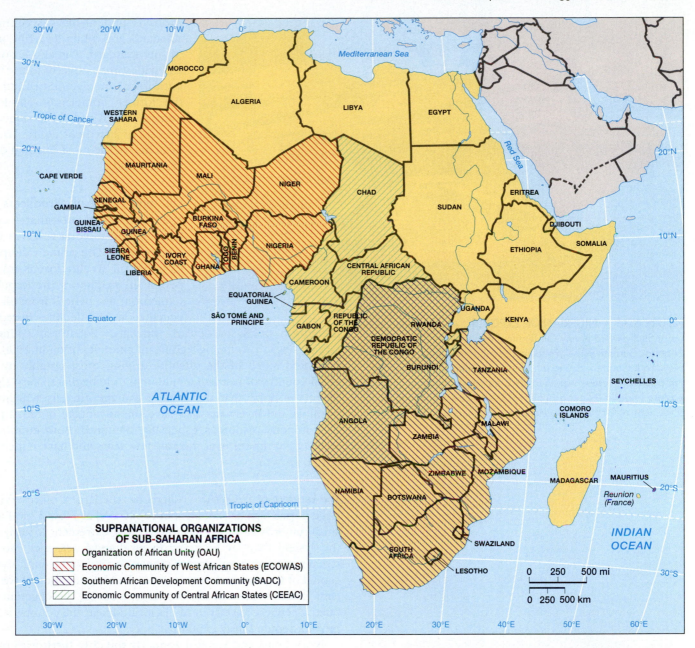

FIGURE 6.35 • SUPRANATIONAL ORGANIZATIONS OF SUB-SAHARAN AFRICA Political affiliations in Sub-Saharan Africa are both continental and regional. The OAU includes all African countries. Smaller organizations, such as SADC, CEEAC, and ECOWAS, represent regional affiliations. Of these, SADC shows the most economic promise.

Given the scale of the African continent, it is not surprising that groups of states formed trade blocs to facilitate intraregional exchange and development. The two most active regional organizations are the Southern African Development Community (SADC) and the Economic Community of West African States (ECOWAS). Both were founded in the 1970s but became more important in the 1990s (Figure 6.35). SADC and ECOWAS are anchored by the region's two largest economies: South Africa and Nigeria. Another regional trade block, the Economic Community of Central African States (CEEAC), has been unable to foster economic integration, in part because of the political instability of the area.

South Africa South Africa is the unchallenged economic powerhouse of the region. The GNI of Nigeria, the next largest economy, is just one-fourth that of South Africa. Only South Africa has a well-developed and well-balanced industrial economy. It also boasts a healthy agricultural sector and, more important, it stands as one of the world's mining superpowers. South Africa remains unchallenged in gold production and is a leader in many other minerals as well. But while South

FIGURE 6.36 • NIGERIAN OIL EXPORTS An oil tanker loads up at Shell-owned Bonny Oil Terminal in the Niger Delta area of Nigeria. Currently, Shell Petroluem of Nigeria is being sued in New York Court for taking land for oil development without paying adequate compensation to the Ogoni tribe who live in southeastern Nigeria. Nigeria is the region's largest oil exporter and a member of OPEC. *(Chris Hondros/Getty Images, Inc.)*

Africa is undeniably a wealthy country by African standards and while its white minority is prosperous by any standard, it is also a country beset by severe and widespread poverty. In the townships lying on the outskirts of the major cities and in the rural districts of the former homelands, employment opportunities remain limited and living standards marginal. Despite the end of apartheid, South Africa continues to suffer from one of the most unequal distributions of income in the world.

Oil and Mineral Producers Another group of relatively well-off Sub-Saharan countries benefits from large oil and mineral reserves and small populations (Figure 6.36). The prime example is Gabon, a country of noted oil wealth that is inhabited by 1.2 million people. Its neighbors, the Republic of the Congo and Cameroon, also benefit from oil, but both have experienced economic declines in recent years, with that of Cameroon being particularly dramatic. Equatorial Guinea, a former Spanish colony, began producing significant amounts of oil in 1998. American investment is behind much of the offshore drilling, most notably by Exxon Mobil. As this small country of just half a million people begins to earn significant oil revenues, observers hope that the lives of average citizens will improve. Farther south, Namibia and Botswana also have the advantage of small populations and abundant mineral resources, especially diamonds. Both countries have also enjoyed stable government over the past few years and have experienced significant economic growth.

The Leaders of ECOWAS Nigeria has the largest oil reserves in the region, but its huge population has kept its per capita GNI at a low $260. Oil money has allowed a small minority of its population to grow extremely wealthy more by manipulating the system than by engaging in productive activities. Most Nigerians, however, remain trapped in poverty. Oil money also led to the explosive growth of the former capital of Lagos, which by the 1980s had become one of the most expensive—and least livable—cities in the world. As a result, the Nigerian government chose to build a new capital city in Abuja, located near the country's center, a move that has proved tremendously expensive.

Ivory Coast and Senegal, formerly the core territories of the French Sub-Saharan empire, still function as commercial centers, but they also suffered economic downturns in the 1980s. In the mid-1990s, the Ivorian economy again began to grow. Supporters within the country call it an emerging "African elephant" (comparing it to the successful "economic tigers" of eastern Asia). Yet its real annual growth throughout the 1990s was less than 1 percent. Ghana, a former British colony and ECOWAS member, also began an economic recovery in the 1990s. In 2001 it negotiated with the IMF and World Bank for debt relief to reduce its nearly $6 billion foreign debt.

The Poorest States The poorest parts of Sub-Saharan Africa include the Sahel, the Horn, and several war-torn states. Sahelian countries such as Mali, Burkina Faso, Niger, and Chad have suffered from prolonged droughts and serious environmental degradation, and they have poorly developed infrastructures and commercial networks. More so than the rest of Africa, their economies remain dominated by agriculture, with very little urban or industrial development. In the Horn of Africa, Ethiopia, Eritrea, and Somalia have all been the victims of drought as well as war. As of 1999, Ethiopia had the distinction of being the region's poorest country on a per capita basis ($100 per year). Currently there are no estimates for Somalia's GNI.

Conflict has destroyed the economies of several African states, such as Burundi, Rwanda, Sierra Leone, Liberia, Angola, and the Democratic Republic of the Congo. Yet conflict alone cannot explain why more than half the states in the region average a GNI per capita of less than $1 a day. Many of these impoverished states, such as Malawi, Guinea Bissau, Gambia, and Zambia, will be eligible for debt relief. Still, with limited infrastructure and economic output, it is hard to imagine the conditions in these countries improving any time soon.

Measuring Social Development

By world standards, measures of social development in Africa are extremely low. Yet unlike Africa's economic indicators, at least there are some positive trends in social development. Rates of child survival, which improved dramatically in the 1960s, have continued to get better since 1980. Still, compared to the rest of the world, Sub-Saharan Africa's child mortality rates are high. Levels of adult illiteracy vary greatly in the region, but several states, such as Kenya, Zimbabwe, and the Republic of the Congo, have invested in education (Figure 6.37). Troubling gaps do exist between male and female literacy rates. In most states, illiteracy is higher among women than men, and in some cases, such as Ghana and Eritrea, adult illiteracy among women is twice as high as that among men.

Life Expectancy   Sub-Saharan Africa's figures on life expectancy are, overall, the world's lowest. Only the poorest Asian countries, such as Afghanistan, Nepal, and Laos, stand at the average African level. Angola's troubling figures of 37 years for men and 39 for women reflect, in part, years of conflict and political disorder. Despite these figures, some progress has been made in enhancing life expectancy in Sub-Saharan Africa. While the region's average life expectancy of 51 years may seem incredibly short, it must be remembered that infant and childhood mortality figures depress these numbers; average life expectancy for adults alone is quite a bit higher. States such as Kenya, Botswana, and Zimbabwe established relatively high life expectancy figures in the 1980s, but the growing AIDS epidemic has taken away from these accomplishments.

The causes of low life expectancy are generally related to extreme poverty, environmental hazards (such as drought), and various environmental and infectious diseases (malaria, schistosomiasis, cholera, AIDS, and measles). Often these factors work in combination. Malaria, for example, kills a half million African children each year. The death rate is also affected by poverty, undernourished children being the most vulnerable to the effects of high fevers. Cholera outbreaks occur in crowded slums and villages where food or water is contaminated by the feces of infected persons. Such unsanitary practices result from lack of basic infrastructure. Tragically, diseases that are preventable, such as measles, occur when people have no access to or cannot afford vaccines.

Health Issues   A lack of doctors and health facilities, especially in rural areas, also helps to explain Sub-Saharan Africa's high mortality levels. The more successful countries in the region have discovered that inexpensive rural clinics dispensing basic treatments, such as oral rehydration therapy for infants with severe diarrhea, can substantially improve survival rates. Another problem is the severity of the African disease environment itself. Malaria is making a comeback in many areas as the disease-causing organisms develop resistance to common drugs and the mosquito carriers develop resistance to insecticides. While most Africans have partial immunity to malaria, it remains a major killer, and many who survive infections may never regain their full strength.

Many other tropical diseases also continue to plague the region. Schistosomiasis is carried by freshwater snails and can cause chronic diarrhea and cramping. The construction of dams, reservoirs, and irrigation ditches has greatly increased the environment for these snails, yet basic eradication programs are often not implemented. Consequently, 200 million Africans are affected by schistosomiasis, which can lead to dehydration and susceptibility to other diseases.

FIGURE 6.37 • EDUCATING CHILDREN   An all-girls school in Kenya reflects an important investment in the country's social development. Education of girls, in particular, is seen as a way to raise the status of women, lower birthrates, and improve economic opportunities. *(Betty Press/Woodfin Camp & Associates)*

Women and Development

Development gains cannot be made in Africa unless the economic contributions of African women are recognized. Officially, women are the invisible contributors to local and national economies. In agriculture, women account for 75 percent of the labor that produces more than half the food consumed in the region. Tending subsistence plots, taking in extra laundry, and selling surplus produce in local markets all contribute to household income. Yet because many of these activities are considered informal economic activities, they are not counted. For many of Africa's poorest people, however, the informal sector *is* the economy. Within this sector, women dominate.

Status of Women The social position of women is difficult to measure for Sub-Saharan Africa. Women traders in West Africa, for example, have considerable political and economic power. By such measures as female labor force participation, many Sub-Saharan African countries show relative gender equality. And women in most Sub-Saharan societies do not suffer the kinds of traditional social restrictions encountered in much of South Asia, Southwest Asia, and North Africa so that women work outside the home, conduct business, and own property.

By other measures, however, such as the prevalence of polygamy, the practice of the "bride-price," and the tendency for males to inherit property over females, African women do suffer discrimination. Perhaps the most controversial issue regarding women's status is the practice of female circumcision, or genital mutilation. In Sudan, Ethiopia, Somalia, and Eritrea, as well as parts of West Africa, almost 80 percent of girls are subjected to this practice, which is extremely painful and can have serious health consequences. Yet because the practice is considered traditional, most African states are unwilling to ban it.

Regardless of their social position, most African women still live in remote villages where educational and wage-earning opportunities remain limited and bearing numerous children remains a major economic contribution to the family. As educational levels increase and urban society expands—and as reduced infant mortality provides greater security—one can expect fertility in the region to gradually decrease. Governments can greatly quicken the process by providing birth control information and cheap contraceptives—and by investing more money in health and educational efforts aimed at women. As the economic importance of women receives greater attention from national and international organizations, more programs are being directed exclusively toward them.

Building from Within Major shifts in the way development agencies view women and women view themselves have the potential to transform the region. All across the continent, support groups and networks have formed, raising women's awareness, offering women micro-credit loans for small businesses, and organizing their economic power. From farm-labor groups to women's market associations, investment in the organization of women has paid off. In Kenya, for example, hundreds of women's groups organize tree plantings to prevent soil erosion and ensure future fuel supplies.

Whether inspired by feminism, African socialism, or the free market, community organizations have made a difference in meeting basic needs in sustainable ways. No doubt the majority of the groups fall short of all of their objectives. Yet for many people, especially women, the message of creating local networks to solve community problems is an empowering one.

CONCLUSION

By simply looking at the numbers, it is easy to be pessimistic about Africa's future. Population growth has been exceeding economic expansion for some years, and the living standards of most of the region's inhabitants have steadily declined. Many Sub-Saharan cities are now growing rapidly, but few jobs await the rural migrants or the youngsters born and raised in the urban environment. In many respects, however, opportunities in the villages are even more limited. Frustration is often particularly strong for young people who have struggled to obtain an education, only to find that jobs are scarce.

Some pessimists think that Sub-Saharan Africa could experience an almost complete political and economic breakdown within the next few decades. Large numbers of unemployed and disillusioned young people, they remind us, often lead to political instability. In the worst-case situation—such as currently exists in Liberia—young men may see joining private armed bands as their only option. As various armed forces vie for power, an entire country can be quickly devastated—as occurred in the 1990s in the Democratic Republic of the Congo. Ethnic conflicts and the spread of AIDS worsen the region's many problems. Meanwhile, desertification and other forms of environmental degradation, which are themselves linked to runaway population growth, threaten the very survival of tens of millions of Africans, as well as the economic and political foundation of the entire region. For the confirmed pessimists, Rwanda—densely populated and torn by an ethnic conflict of genocidal proportions—is the bleak image of Africa's future.

The optimists' vision of Africa's future, however, can be equally convincing. Most of the region remains lightly populated, and large expanses of land could be productively farmed. Countries such as the Central African Republic, Gabon, and Angola could keep growing for many generations before crowding becomes a serious problem. In other words, most of the region still enjoys a crucial breathing space in which to address its problems.

Many recent developments give cause for hope. These include early signs of a fertility decline, improvements in health and education, and the tentative movement away from authoritarianism and toward democracy. Equally important has been the seemingly successful transition to majority rule and a multiracial society in South Africa. Now that peace seems to have come at long last to Mozambique, southern Africa as a whole has emerged as a promising area. As the Southern African Development Community, a group of 14 countries pledged to economic cooperation, matures, it could quickly improve infrastructure and attract more foreign investment. Positive examples include the new Trans-Kalahari highway, which links Johannesburg, South Africa, with Windhoek, Namibia. Also planned is an improved highway between Johannesburg, South Africa, and Maputo, Mozambique, as well as new investment in Maputo's war-ravaged port.

In focusing on Africa's traumas, the U.S. media miss many of the region's success stories. We have heard much about Somalia, Liberia, and Rwanda in recent years, but very little about Botswana. Yet the story of Botswana is in many respects just as instructive. It has a stable, democratic government, and it enjoys a strong and growing economic base that averaged 1.8 percent real annual growth in the 1990s. Botswana's levels of social development (under age five mortality and adult literacy) are among the best in the region. The country still faces a number of significant problems—most notably the AIDS epidemic—but thus far it has demonstrated an impressive ability to address them in a constructive manner. Let us hope Botswana, rather than Rwanda, becomes the best symbol for Sub-Saharan Africa at the beginning of the new millennium.

KEY TERMS

apartheid (page 147)
Berlin Conference (page 154)
biofuels (page 137)
coloured (page 148)
desertification (page 135)
Great Escarpment (page 138)
homelands (page 155)

Horn of Africa (page 142)
internally displaced persons
   (page 157)
kleptocracy (page 160)
Organization of African Unity
   (OAU) (page 156)

pastoralists (page 146)
refugee (page 157)
Sahel (page 135)
structural adjustment programs
   (page 159)

swidden (page 146)
transhumance (page 136)
tribalism (page 157)
tribe (page 150)
tsetse fly (page 146)

Southwest Asia and North Africa

FIGURE 7.1 Southwest Asia and North Africa
This vast region extends from the shores of the Atlantic Ocean to the Caspian Sea. Within its ▮▮▮aries, major cultural differences and globally important petroleum reserves have contributed to recent political tensions. The region's economic geography reveals great differences in wealth, as it includes some of the world's richest and poorest nations. In addition, rapid population growth and limited water supplies pose future challenges.

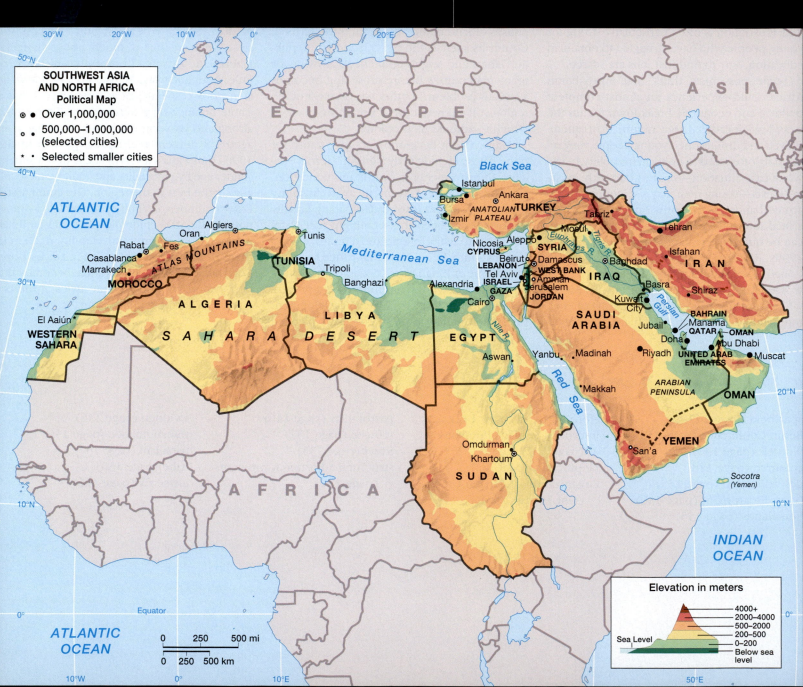

SETTING THE BOUNDARIES

"Southwest Asia and North Africa" is both an awkward term and a complex region. Often the same area is simply called the "Middle East," but some experts would exclude the western parts of North Africa as well as Turkey and Iran from such a region. In addition, the "Middle East" suggests a European point of view—Lebanon is in the "middle of the east" only from the perspective of the western Europeans who colonized the region and still shape the names we give the world today. Instead, "Southwest Asia and North Africa" offers a straightforward way to describe the general limits of the region. Largely arid climates, an Islamic religious tradition, abundant fossil fuel reserves, and persistent political instability are present in many areas. Still, local variations in physical and human patterns make it difficult to generalize about the region.

There are also problems with simply defining the geographical limits of the region. In the northwest corner of Turkey, a small piece of the country actually sits west of the Bosporus Strait, generally considered to be the dividing line between Europe and Asia (see Figure 7.1). Nearby Cyprus, located in the eastern Mediterranean, also resists easy regional definition because the population is divided between ethnic Greeks and Turks. We include it in Southwest Asia and North Africa because of its proximity to the Turkish and Syrian coasts. To the northeast, the largely Islamic peoples of Central Asia share many religious and linguistic ties with Turkey and Iran, but we have chosen to treat these groups in a separate Central Asia chapter (see Chapter 10).

African borders are also problematic. The conventional regional division of "North Africa" from "Sub-Saharan Africa" often cuts directly through the middle of modern Mauritania, Mali, Niger, Chad, and Sudan. We place all of these transitional countries in Chapter 6, Sub-Saharan Africa but also include some discussion of Sudan within our Southwest Asia and North Africa material because of its status as an "Islamic republic," a political and cultural title that ties it strongly to the Muslim world.

Climate, culture, and oil all help define the complex Southwest Asia and North Africa world region (see "Setting the Boundaries"). Located at the historic meeting ground between Europe, Asia, and Africa, the region includes thousands of miles of parched deserts, rugged plateaus, and oasis-like river valleys. It extends 4,000 miles (6,400 kilometers) between Morocco's Atlantic coastline and Iran's eastern boundary with Pakistan. More than two dozen nations are included within its borders, with the largest populations found in Egypt, Turkey, and Iran (Figure 7.1). Climates are generally arid, although the region's diverse physical geography causes precipitation to vary greatly. Diverse languages, religions, and ethnic identities have influenced the region for centuries and these cultural patterns have had profound social and political implications. One traditional zone of conflict is the Middle East, where Jewish, Christian, and Islamic peoples have yet to resolve longstanding cultural tensions and political differences. In particular, conflict continues within Israel, focused on the key Palestinian issue. Iraq has also been a setting for instability as a recent American-led invasion toppled the regime of Saddam Hussein. In addition, Southwest Asia and North Africa's extensive petroleum resources place the area in the global economic spotlight. The value of oil in the global economy has combined with ongoing ethnic and religious conflicts to produce one of the world's least-stable political settings, one prone to geopolitical conflicts both within and between the countries of the region.

No world region better exemplifies the theme of globalization than Southwest Asia and North Africa. A key global **culture hearth**, the region witnessed many new cultural ideas that subsequently spread out and diffused widely to other portions of the world (Figure 7.2). As an early center for agriculture, several great civilizations, and three major world religions, the region has been a key human crossroads for thousands of years. Important long-distance trade routes have connected North Africa with the Mediterranean and Sub-Saharan Africa. Southwest Asia has also had historic ties to Europe, the Indian subcontinent, and Central Asia. As a result, new ideas within the region have often spread far beyond its bounds. For example, the domestication of wheat and cattle in Southwest Asia had far-reaching global impacts. Urban-based civilizations that began in this region formed models for city-building that spread to Europe and beyond. In addition, religions born within the region (Judaism, Christianity, and Islam) have shaped many other parts of the world.

Particularly within the past century, globalization has also operated in the opposite direction: The region's strategic importance has made it increasingly vulnerable to outside influences. The twentieth-century development of the petroleum industry, largely initiated by U.S. and European investment, has had enormous consequences

FIGURE 7.2 • ALEPPO, SYRIA For centuries, the northern Syrian city of Aleppo has served as an important regional trading center. Today, the city has more than 1.5 million residents and reveals traditional mosques and minarets next to newer apartments and office buildings. *(Jean-Leo Dugast/Panos Pictures)*

FIGURE 7.3 • MODERN ISTANBUL Turkey's largest city has been an urban settlement for more than 2,000 years. The city was a center for Christianity before Islam arrived in the fifteenth century. Modern Western influences now add cultural diversity to this scene on Independence Avenue. *(Rob Crandall/Rob Crandall, Photographer)*

for economic development within the oil-rich countries of the region. Global demand for oil and natural gas has driven rapid industrial change within the region, defining its central role in world trade. Many key members of **OPEC (Organization of Petroleum Exporting Countries)** are found within the region, and these countries greatly influence global prices and production levels for petroleum. Politically, petroleum resources have also increased the region's strategic importance, adding to many traditional cultural tensions existing within the realm. As a result, the region was a key area of conflict in the Cold War between the United States and the Soviet Union, and international tensions remain focused there.

More recently, **Islamic fundamentalism** has argued for a return to more traditional practices within the religion and has challenged the intrusion of global popular culture. Fundamentalists have also advocated merging civil and religious authority and rejecting many characteristics of modern, Western-style consumer culture (Figure 7.3). While often harshly criticized in the Western press, fundamentalism can be seen

FIGURE 7.4 • PHYSICAL GEOGRAPHY OF SOUTHWEST ASIA AND NORTH AFRICA Vast deserts stretch across much of the region and dominate the physical setting. Major mountains and highland zones complicate the scene by altering local climate and vegetation patterns and offering distinctive environments for human settlement. In addition, the region is geologically active and the earthquake hazard is high across much of its extent.

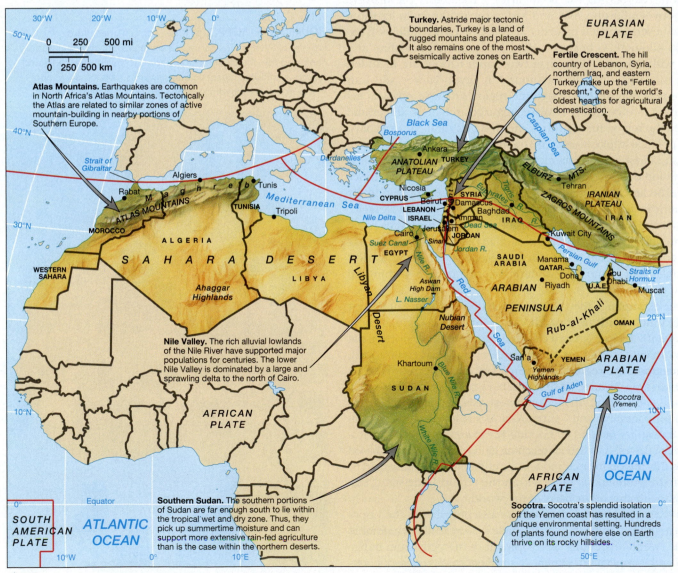

as a cultural reaction to the disruptive role that forces of globalization have played across the region. Recently, political expressions of fundamentalism have also impacted the entire world and have placed it increasingly in the global spotlight.

The realities of the region's environment provide additional challenges for the populations of Southwest Asia and North Africa. The availability of water in this largely dry portion of the world has shaped both the physical and human geographies of the region. Water remains a daily and ever-present need. Biologically, the region's plants and animals must adapt to the aridity of long dry seasons and short, often unpredictable rainy periods. Similarly, the geography of human settlement is linked to water. Whether it comes from precipitation, underground aquifers, or rivers, water has shaped patterns of human settlement and has placed severe limits on agricultural development across huge portions of the region. In the future, the region's growing population of 400 million people will stress these available resources even further and water issues will no doubt increase economic and political instability.

Environmental Geography: Life in a Fragile World

In the popular imagination, much of Southwest Asia and North Africa is a land of shifting sand dunes, searing heat, and scattered oases. Although examples of those stereotypes certainly can be found across the region, the actual physical setting, in terms of both landforms and climate, is considerably more complex (Figure 7.4). One theme is dominant, however: A lengthy legacy of human settlement has left its mark on a fragile environment, and the entire region will be faced with increasingly difficult ecological problems in the decades ahead.

Regional Landforms

A quick tour of the region reveals a surprising diversity of environmental settings and landforms. In North Africa, the **Maghreb** region (meaning "western island") includes the nations of Morocco, Algeria, and Tunisia and is dominated near the Mediterranean coastline by the Atlas Mountains. The rugged flanks of the Atlas rise like a series of islands above the narrow coastal plains to the north and the vast stretches of the lower Saharan deserts to the south (Figure 7.5). South and east of the Atlas Mountains, interior North Africa varies between rocky plateaus and extensive lowlands. In northeast Africa, the Nile River dominates the scene as it flows north through Sudan and Egypt.

Southwest Asia is more mountainous than North Africa. In the **Levant**, or eastern Mediterranean region, mountains rise within 20 miles (32 kilometers) of the sea, and the highlands of Lebanon reach heights of more than 10,000 feet (3,048 meters). Farther south, the Arabian Peninsula forms a massive tilted plateau, with western highlands higher than 5,000 feet (1,524 meters) gradually sloping eastward to extensive lowlands in the Persian Gulf area. North and east of the Arabian Peninsula lie the two great upland areas of Southwest Asia: the Iranian and Anatolian plateaus (*Anatolia* refers to the large peninsula of Turkey, sometimes called Asia Minor) (Figure 7.4 and 7.6). Both of these plateaus, averaging between 3,000 and 5,000 feet (915 to 1,524 meters) in elevation, are geologically active and prone to earthquakes. One dramatic quake in western Turkey in August 1999 measured 7.8 on the Richter scale, killed more than 17,000 people, and left 350,000 residents homeless. Another quake near the Iranian city of Bam in late 2003 measured 6.8 on the Richter scale and claimed more than 40,000 lives.

Smaller lowlands characterize other portions of Southwest Asia. Narrow coastal strips are common in the Levant, along both the southern (Mediterranean) and northern (Black Sea) Turkish coastlines, and north of the Iranian Elburz Mountains near the Caspian Sea. Iraq contains the most extensive alluvial lowlands in Southwest Asia, dominated by the Tigris and Euphrates rivers, which flow southeast to empty into the Persian Gulf. Although much smaller, the distinctive Jordan River Valley is also a notable lowland that straddles the strategic borderlands of Israel, Jordan, and Syria and drains southward to the Dead Sea.

FIGURE 7.5 • ATLAS MOUNTAINS Residents of Morocco's Atlas Mountains adapt to the region's steep slopes, and their Mediterranean-style agriculture takes advantage of higher rates of annual precipitation than are found in the arid deserts below. *(Sean Sprague/Stock Boston)*

FIGURE 7.6 • ANATOLIAN PLATEAU Turkey's vast interior is a mix of high mountains and elevated plateaus. The Anatolian Plateau provides a complex environment in which traditional agriculture includes midlatitude crops and livestock adapted to the region's semiarid setting. *(Adam Woolfitt/Woodfin Camp & Associates)*

Patterns of Climate

Although often termed the "dry world," a closer look at Southwest Asia and North Africa reveals a more complex climatic pattern (Figure 7.7). Both latitude and altitude come into play. Aridity dominates large portions of the region. A nearly continuous belt of desert lands stretches eastward from the Atlantic coast of southern Morocco across the continent of Africa, through the Arabian Peninsula, and into central and eastern Iran (Figure 7.8). Indeed, much of the central Sahara Desert receives less than 1 inch (2.5 centimeters) of rain a year. Throughout this vast dry zone, plant and animal life have adapted to extreme conditions. Deep or extensive root systems allow desert plants to benefit from the limited moisture they receive. Similarly, animals adjust by efficiently storing water, hunting at night, or migrating seasonally to avoid the worst of the dry cycle.

Elsewhere, altitude and latitude dramatically alter the desert environment and produce a surprising amount of climatic variety. The Atlas Mountains and nearby lowlands of northern Morocco, Algeria, and Tunisia experience a Mediterranean climate

FIGURE 7.7 • CLIMATE MAP OF SOUTHWEST ASIA AND NORTH AFRICA Dry climates dominate from western Morocco to eastern Iran. Within these zones, subtropical high-pressure systems offer only limited opportunities for precipitation. Elsewhere, mild midlatitude climates with wet winters are found near the Mediterranean basin and Black Sea. To the south, tropical savanna climates provide summer moisture to southern Sudan.

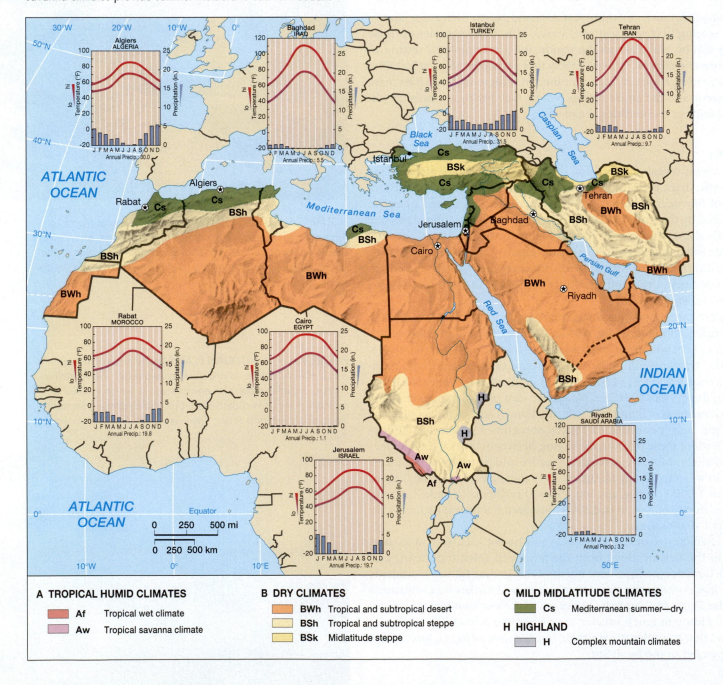

A TROPICAL HUMID CLIMATES

Af	Tropical wet climate
Aw	Tropical savanna climate

B DRY CLIMATES

BWh	Tropical and subtropical desert
BSh	Tropical and subtropical steppe
BSk	Midlatitude steppe

C MILD MIDLATITUDE CLIMATES

Cs	Mediterranean summer—dry

H HIGHLAND

H	Complex mountain climates

in which dry summers alternate with cooler, wet winters. In these areas the landscape resembles that found in nearby southern Spain or Italy (Figure 7.9). A second zone of Mediterranean climate extends along the Levant coastline into the nearby mountains and northward across sizable portions of northern Syria, Turkey, and northwestern Iran.

Legacies of a Vulnerable Landscape

The environmental history of Southwest Asia and North Africa illustrates both the short-sighted and resourceful practices of its human occupants. Littered with examples of environmental problems, the region reveals the hazards of lengthy human settlement in a marginal land (Figure 7.10).

Deforestation and Overgrazing   Deforestation is an ancient problem in Southwest Asia and North Africa. Although much of the region is too dry for trees, the more humid and elevated lands that border the Mediterranean once supported heavy forests. Included in these woodlands are the cedars of Lebanon, cut down in ancient times and now reduced to a few scattered groves that survive in a largely unforested landscape.

Human activities have generally combined with natural conditions to reduce most of the region's forests to grass and scrub. Mediterranean forests often grow slowly, are highly vulnerable to fire, and usually fare poorly if subjected to heavy grazing. Browsing by sheep and goats in particular has often been blamed for much of the region's forest loss. Deforestation has resulted in a long, slow deterioration of the region's water supplies and in accelerated soil erosion. Several governments have launched reforestation drives and forest preservation efforts. Israel and Syria have expanded their forested lands since the 1980s, and more than 5 percent of Lebanon's total area is now part of the Chouf Cedar Reserve, formed in 1996 to protect old-growth cedars.

Salinization   Salinization, or the buildup of toxic salts in the soil, is another ancient environmental issue in a region where irrigation has been practiced for centuries (see Figure 7.10). The accumulation of salt in the topsoil is a common problem wherever desert lands are subjected to extensive irrigation. When freshwater is diverted from streams into fields, dissolved salts remains in the soil after the moisture has been absorbed by the plants and evaporated by the sun. In dry regions, salt concentrations build up over time, leading to lower crop yields and land abandonment.

Hundreds of thousands of acres of once-fertile farmland within the region have been destroyed or degraded by salinization. The problem has been particularly severe in Iraq, where centuries of canal irrigation along the Tigris and Euphrates rivers have seriously degraded land quality. Similar conditions affect central Iran, Egypt, and irrigated portions of the Maghreb.

Managing Water   Residents of the region are continually challenged by many other problems related to managing water in one of the driest portions of Earth. As technological change has accelerated, the scale and impact of water management schemes have had a growing effect on the region's environment.

The people of the region have been modifying drainage systems and water flows for thousands of years. The Iranian **qanat system** of tapping into groundwater through a series of gently sloping tunnels was widely replicated on the Arabian Peninsula and in North Africa. With simple technology, farmers directed the underground flow to fields and villages where it could be efficiently utilized. In the past half century, however, the scope of environmental change has been greatly magnified. One remarkable example is Egypt's Aswan High Dam, completed in 1970 on the Nile River south of Cairo (see Figure 7.10). Many benefits came with the dam's completion. Increased storage capacity in the upstream reservoir made more water available for year-round agriculture and an expansion of cultivated lands along the Nile. The dam also generates large amounts of clean electricity for the region. But environmental costs have

FIGURE 7.8 • ARID IRAN   Large portions of Iran are arid. Settlements are found near zones of higher moisture or opportunities for irrigated agriculture. This view of remote desert lands to the north of the Zagros Mountains is near the city of Shiraz. *(Isabella Tree/Hutchison Library)*

FIGURE 7.9 • ALGERIAN ORANGE HARVEST   The Mediterranean moisture in northern Algeria produces an agricultural landscape similar to that of southern Spain or Italy. Winter rains create a scene that contrasts sharply with deserts found elsewhere in the region. *(Dott. Giorgio Gualco/Bruce Coleman Inc.)*

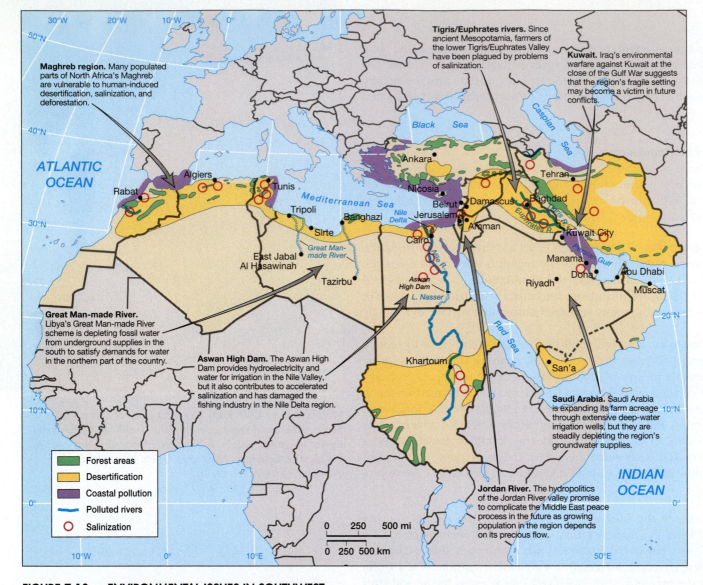

Maghreb region. Many populated parts of North Africa's Maghreb are vulnerable to human-induced desertification, salinization, and deforestation.

Tigris/Euphrates rivers. Since ancient Mesopotamia, farmers of the lower Tigris/Euphrates Valley have been plagued by problems of salinization.

Kuwait. Iraq's environmental warfare against Kuwait at the close of the Gulf War suggests that the region's fragile setting may become a victim in future conflicts.

Great Man-made River. Libya's Great Man-made River scheme is depleting fossil water from underground supplies in the south to satisfy demands for water in the northern part of the country.

Aswan High Dam. The Aswan High Dam provides hydroelectricity and water for irrigation in the Nile Valley, but it also contributes to accelerated salinization and has damaged the fishing industry in the Nile Delta region.

Saudi Arabia. Saudi Arabia is expanding its farm acreage through extensive deep-water irrigation wells, but they are steadily depleting the region's groundwater supplies.

Jordan River. The hydropolitics of the Jordan River valley promise to complicate the Middle East peace process in the future as growing population in the region depends on its precious flow.

Legend:
- Forest areas
- Desertification
- Coastal pollution
- Polluted rivers
- ○ Salinization

0 250 500 mi
0 250 500 km

FIGURE 7.10 • ENVIRONMENTAL ISSUES IN SOUTHWEST ASIA AND NORTH AFRICA Growing populations, pressures for economic development, and widespread aridity combine to create environmental hazards across the region. A long history of human activities have contributed to deforestation, irrigation-induced salinization, and expanding desertification. Saudi Arabia's deep-water wells, Egypt's Aswan High Dam, and Libya's Great Man-made River are all recent technological attempts to expand settlement, but they may carry a high long-term environmental price tag.

been high. More controlled methods of irrigation have increased salinization because water is not rapidly flushed from the fields. The new system also has meant greatly increased inputs of costly fertilizers and the infilling of Lake Nasser behind the dam with accumulating sediments. Other problems with the dam include an increased incidence of *schistosomiasis* (a harmful parasitic disease spread by waterborne snails in irrigation canals). In addition, the Mediterranean fishing industry near the Nile Delta has collapsed because nutrient-rich sediments that once supported large fish populations are now captured behind upstream dams.

Elsewhere, **fossil water**, or water supplies stored underground during earlier and wetter climatic periods, has also been put to use by modern technology. Libya's "Great Man-made River" scheme taps underground water in the southern part of the country, transports it 600 miles (965 kilometers) to northern coastal zones, and uses the precious resource to expand agricultural production in one of North Africa's driest countries (see Figure 7.10). Similarly, Saudi Arabia has invested huge sums to develop deep-water wells, allowing it to greatly expand its food output (Figure 7.11). Unfortunately, these underground supplies are being depleted much more rapidly than they are recharged, thus limiting the long-term sustainability of such water projects.

FIGURE 7.11 • SAUDI ARABIAN IRRIGATION One of Saudi Arabia's largest deep-water irrigation sites is at the oasis of Al Hofuf east of Riyadh. While significantly expanding the country's food production, such efforts are rapidly depleting underground supplies of fossil water. *(Minosa/Scorpio/Corbis/Sygma)*

Most dramatically, **hydropolitics**, or the interplay of water resource issues and politics, has raised tensions between countries that share drainage basins. For example, Sudan's plans to expand its irrigation networks along the upper Nile and Ethiopia's Blue Nile Dam project are both causes of concern in Egypt, where the availability of Nile River water is critical. To the north, Turkey's growing development of the upper Tigris and Euphrates rivers (the Southeast Anatolian Project) has raised similar issues with Iraq and Syria. Hydropolitics has also played into negotiations between Israel, the Palestinians, and other neighboring states, particularly in the valuable Jordan River drainage, which runs through the center of the area's most hotly disputed lands (see Figure 7.10). Israelis fear Palestinian and Syrian pollution; nearby Jordanians argue for more water from Syria; and all regional residents must deal with the uncomfortable reality that, regardless of their political differences, they must drink from the same limited supplies of freshwater (Figure 7.12).

Population and Settlement: Patterns in an Arid Land

The region's human geography demonstrates the intimate tie between water and life in this part of the world. The pattern is complex: Large areas of the population map remain almost devoid of permanent settlement, while lands with available moisture suffer increasingly from problems of crowding and overpopulation (Figure 7.13).

The Geography of Population

Today, more than 400 million people live in Southwest Asia and North Africa. The distribution of that population is strikingly varied: In Egypt, large zones of almost empty desert land stand in sharp contrast to crowded, well-watered locations, such as those along the Nile River. While overall population densities in such countries appear modest, **physiological densities**, a statistic that relates the number of people to the amount of arable land, are among the highest on Earth. Patterns of urban geography are also highly uneven: Although less than two-thirds of the overall population is urban, many nations are dominated by huge cities that produce the same problems of urban crowding found elsewhere in the developing world.

Across North Africa, two dominant zones of settlement, both shaped by the availability of water, account for most of the region's population (see Figure 7.13). In the Maghreb, the moist slopes of the Atlas Mountains and nearby better-watered coastal districts have supported denser populations for centuries, a stark contrast to the almost empty lands south and east of the mountains. Egypt's Nile Valley is home to the other great North African population cluster. The vast majority of Egypt's 70 million people live within 10 miles of the river. In the twenty-first century, optimistic politicians and planners hope to create a second corridor of denser settlement in Egypt's "New Valley" west of the Nile by diverting water from Lake Nasser into the Western Desert.

Most Southwest Asian residents live in well-watered coastal zones, moister highland settings, and desert localities where water is available from nearby rivers or subsurface aquifers. High population densities are found in better-watered portions of the eastern Mediterranean (Israel, Lebanon, and Syria) and Turkey. Nearby Iran is home to more than 65 million residents. Turkey's Istanbul (formerly Constantinople) and Iran's Tehran are Southwest Asia's largest urban areas, and both have grown in recent years as rural populations migrate toward economic opportunities in these cities.

Water and Life: Rural Settlement Patterns

Water and life are closely linked across the rural settlement landscapes of Southwest Asia and North Africa. Indeed, Southwest Asia is home to one of the world's earliest hearths of **domestication**, where plants and animals were purposefully selected and bred for their desirable characteristics. Beginning around 10,000 years ago, increased experimentation with wild varieties of wheat and barley led to agricultural settlements that later included domesticated animals, such as cattle, sheep,

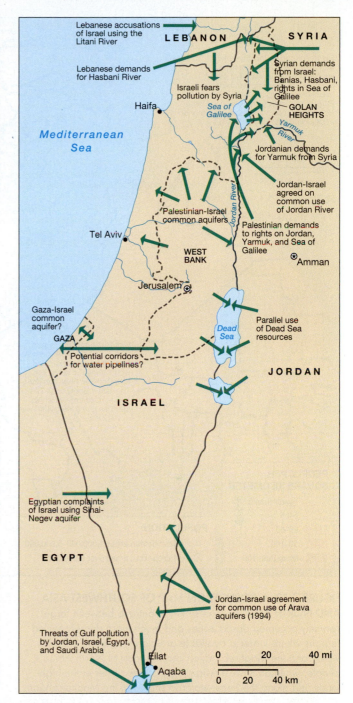

FIGURE 7.12 • HYDROPOLITICS IN THE JORDAN RIVER BASIN *Many water-related issues complicate the geopolitical setting in the Middle East. The Jordan River system has been a particular focus of conflict. (Modified from Soffer, Rivers of Fire: The Conflict over Water in the Middle East, 1999, p. 180. Reprinted by permission of Rowman and Littlefield Publishers, Inc.)*

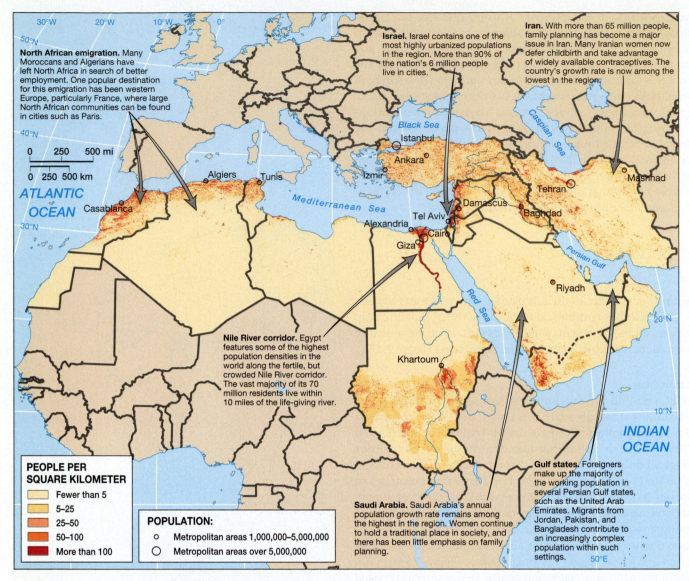

North African emigration. Many Moroccans and Algerians have left North Africa in search of better employment. One popular destination for this emigration has been western Europe, particularly France, where large North African communities can be found in cities such as Paris.

Israel. Israel contains one of the most highly urbanized populations in the region. More than 90% of the nation's 6 million people live in cities.

Iran. With more than 65 million people, family planning has become a major issue in Iran. Many Iranian women now defer childbirth and take advantage of widely available contraceptives. The country's growth rate is now among the lowest in the region.

Nile River corridor. Egypt features some of the highest population densities in the world along the fertile, but crowded Nile River corridor. The vast majority of its 70 million residents live within 10 miles of the life-giving river.

Saudi Arabia. Saudi Arabia's annual population growth rate remains among the highest in the region. Women continue to hold a traditional place in society, and there has been little emphasis on family planning.

Gulf states. Foreigners make up the majority of the working population in several Persian Gulf states, such as the United Arab Emirates. Migrants from Jordan, Pakistan, and Bangladesh contribute to an increasingly complex population within such settings.

PEOPLE PER SQUARE KILOMETER

	Fewer than 5
	5–25
	25–50
	50–100
	More than 100

POPULATION:

○ Metropolitan areas 1,000,000–5,000,000

◯ Metropolitan areas over 5,000,000

FIGURE 7.13 • POPULATION MAP OF SOUTHWEST ASIA AND NORTH AFRICA The striking contrasts between large, sparsely occupied desert zones and much more densely settled regions where water is available are clearly evident. The Nile Valley and the Maghreb region contain most of North Africa's people, while Southwest Asian populations cluster in the highlands and along the better-watered shores of the Mediterranean.

and goats. Much of the early agricultural activity focused on the **Fertile Crescent**, an ecologically diverse zone that stretches from the Levant inland through the fertile hill country of northern Syria into Iraq (see Figure 7.4). Between 5,000 and 6,000 years ago, better knowledge of irrigation techniques and increasingly powerful political states encouraged the spread of agriculture into nearby valleys such as the Tigris and Euphrates (Mesopotamia) and North Africa's Nile Valley. Throughout the region, varied types of rural settlement reflect the complex interrelationship between water and life (Figure 7.14).

Pastoral Nomadism In the drier portions of the region, **pastoral nomadism** is a traditional form of subsistence agriculture in which people depend on the seasonal movement of livestock for a large part of their livelihood. The settlement landscape of pastoral nomads reflects their need for mobility and flexibility as they seasonally move camels, sheep, and goats from place to place. Near highland zones such as the Atlas Mountains or the Anatolian Plateau, nomads practice **transhumance** by seasonally moving their livestock to cooler, greener high country pastures in the summer and then returning them to valley and lowland settings for fall and winter grazing. Elsewhere, seasonal movements often involve huge areas of desert that support small groups of a few dozen families. Fewer than 10 million pastoral nomads remain in the region today.

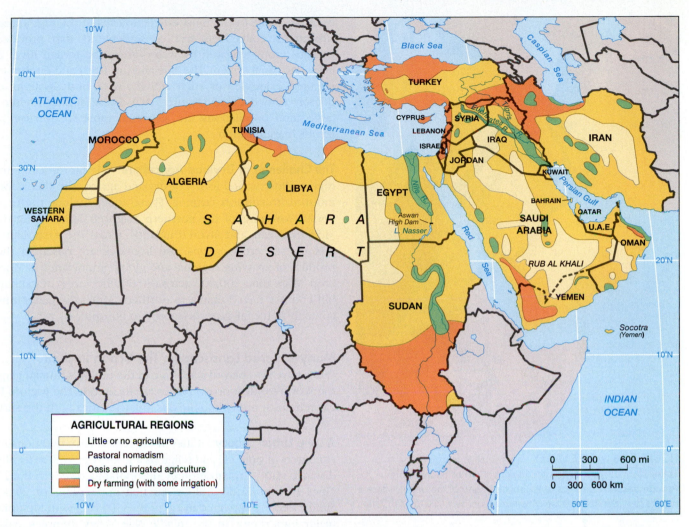

FIGURE 7.14 • AGRICULTURAL REGIONS OF SOUTHWEST ASIA AND NORTH AFRICA   Important agricultural zones include oases and irrigated farming where water is available. Elsewhere, dry farming supplemented with irrigation is practiced in midlatitude settings. *(Modified from Clawson and Fisher, 1998, World Regional Geography, Upper Saddle River, NJ: Prentice Hall; and Bergman and Renwick, 1999, Introduction to Geography, Upper Saddle River, NJ: Prentice Hall)*

Oasis Life   Permanent oasis settlements dot the arid landscape where high groundwater levels or modern deep-water wells provide reliable moisture (Figures 7.11 and 7.15). Tightly clustered, often walled villages sit next to small intensely utilized fields where underground water is carefully applied to tree and cereal crops. In more recently created oasis settlements, concrete blocks and prefabricated housing add a modern look.

Traditional oasis settlements are composed of close-knit families who work their own irrigated plots or, more commonly, work for absentee landowners. Although oasis settlements are usually small, eastern Saudi Arabia's Hofuf oasis covers more than 30,000 acres (12,150 hectares). While some crops are raised for local consumption, commercial trade is important as well. Recently, the expanding world demand for products such as figs and dates has included even these remote locations in the global economy, and many products end up on the tables of hungry Europeans or North Americans.

Exotic Rivers   For centuries, the region's densest rural settlement has been tied to its great river valleys and their seasonal floods of water and fertile nutrients. In such settings, **exotic rivers** transport precious water and nutrients from more humid lands into drier regions, where the resources are utilized for irrigated farming. For

FIGURE 7.15 • OASIS AGRICULTURE   Date palms dominate this Tunisian oasis. Often, these commercial farm products find their way to consumers in distant European and North American markets. *(Alain Le Garsmeur/Panos Pictures)*

example, some of the most efficient farms in the region are associated with Israeli **kibbutzes**, collectively worked settlements that produce grain, vegetable, and orchard crops irrigated by waters from the Jordan River and from the country's elaborate canal system. The Nile and the Tigris and Euphrates rivers offer the largest regional examples of such activity, with both systems characterized by large, densely settled deltas (Figure 7.16). These settings, while capable of supporting large rural populations, are also vulnerable to overuse, particularly if irrigation results in salinization.

The Challenge of Dryland Agriculture   Mediterranean climates in portions of the region allow varied forms of dryland agriculture that depend largely on seasonal moisture to support farming. These zones include the better-watered valleys and coastal lowlands of the northern Maghreb, lands along the shore of the eastern Mediterranean, and favored uplands across the Anatolian and Iranian plateaus. A mix of tree crops, grains, and livestock are raised in these settings. One farm product of growing regional and global importance is Morocco's thriving hashish crop. More than 200,000 acres (80,000 hectares) of cannabis are cultivated in the hill country near Ketama in northern Morocco, generating more than $2 billion annually in illegal exports (mostly to Europe).

Many-Layered Landscapes: The Urban Imprint

Cities have also played a key role in the region's human geography. Indeed, some of the world's oldest urban places are located in the region. Today, continuing political, religious, and economic ties link the cities with the surrounding countryside.

A Long Urban Legacy   Cities in the region have traditionally played important roles as centers of political and religious authority, as well as focal points of local and long-distance trade. Urbanization in Mesopotamia (modern Iraq) began by 3500 B.C., and cities such as Eridu and Ur reached populations of 25,000 to 35,000 residents. Similar centers appeared in Egypt by 3000 B.C., with Memphis and Thebes assuming major importance in the middle Nile Valley. Temples, palaces, tombs, and public buildings dominated the urban landscape, and surrounding walls (particularly in Mesopotamia) offered protection from outside invasion. By 2000 B.C., however, a different kind of city was emerging along the eastern Mediterranean and along important overland trade routes. Centers such as Beirut, Tyre, and Sidon, all in modern Lebanon, as well as Damascus in nearby Syria, exemplified the growing role of trade in creating the urban landscapes of selected cities. Expanding port facilities, warehouse districts, and commercial neighborhoods suggested how trade and commerce shaped these urban settlements, and many of these early Middle Eastern trading towns have survived to the present.

Islam also left an enduring mark on cities because urban centers traditionally served as places of Islamic religious power and education. By the eighth century, Baghdad emerged as a religious center, followed soon thereafter by the appearance of Cairo as a seat of religious authority. Urban settlements from North Africa to Turkey felt the influences of Islam. Indeed, the Moors carried Islam's influence to Spain, where it shaped the architecture and culture in centers such as Córdoba and Málaga. A characteristic Islamic cityscape exists to this day (Figure 7.17). Its traditional features include a walled urban core, or **medina**, dominated by the central mosque and its associated religious, educational, and administrative functions. A nearby bazaar, or *suq*, serves as a marketplace where products from city and countryside are traded. Housing districts feature a maze of narrow, twisting streets that maximize shade and emphasize the privacy of residents, particularly women. Houses have small windows, frequently are situated on dead-end streets, and typically open inward to private courtyards.

More recently, European colonialism added another layer of urban landscape features in selected cities. Particularly in North Africa, colonial cities added many architectural features from Great Britain and France. Victorian building blocks, French

FIGURE 7.16 • NILE VALLEY   This satellite image of the Nile Valley dramatically reveals the impact of water on the North African desert. Cairo lies at the southern end of the delta, where it begins to widen toward the Mediterranean Sea. *(NASA/MODIS)*

mansard roofs, suburban housing districts, and wide European-style commercial boulevards complicated the settlement landscapes of cities such as Algiers and Cairo and many of these characteristics remain today.

Signatures of Globalization Since 1950, cities in the region have become the gateways to the global economy. As the region has been opened to new investment, industrialization, and tourism, the urban landscape reflects the fundamental changes taking place. Expanded airports, commercial and financial districts, industrial parks, and luxury tourist facilities all reveal the influence of the global economy. Surrounding rural populations are also drawn to the new employment opportunities. The results are problematic. Many cities, such as Algiers and Istanbul, have more than doubled in population in recent years. Crowded Cairo now has more than 15 million residents. The increasing demand for homes has produced ugly, cramped high-rise apartment houses, while elsewhere extensive squatter settlements provide little in the way of quality housing or public services.

Undoubtedly, the oil-rich states of the Persian Gulf display the greatest changes in the urban landscape (Figure 7.18). Before the twentieth century, urban traditions were relatively weak in the area, and even as late as 1950 only 18 percent of Saudi Arabia's population lived in cities. All that changed, however, as the global economy's demand for petroleum increased. Today, the Saudi Arabian population is more urban (83 percent) than many industrialized nations, including the United States, and the capital city of Riyadh has grown to 3 million people. Particularly after 1970, other cities, such as Abu Dhabi (United Arab Emirates), Doha (Qatar), and Kuwait City (Kuwait), also adopted many features of modern Western urban design and futuristic architecture. The result is an urban settlement landscape where traditional and global influences combine, producing cityscapes where domed mosques, mirrored bank buildings, and oil refineries sit beside one another beneath the dusty skies and desert sun.

A Region on the Move
While pastoral nomads have crisscrossed the region for ages, new patterns of migration have been created by the global economy and recent political events. The rural-to-urban shift seen so widely in many parts of the less-developed world is at work here too. The Saudi Arabian example is repeated in many other countries within the region. In addition, large numbers of workers are migrating within the region to areas with growing job opportunities. Labor-poor countries such as Saudi Arabia, Kuwait, and the United Arab Emirates (UAE), for example, have attracted thousands of Jordanians, Palestinians, Yemenis, and Iranians to construction and service jobs within their borders. Today, more than 70 percent of Saudi Arabia's labor force are foreigners who annually export more than $18 billion in wages.

Some regional residents are migrating elsewhere. Because of its strong economy and its close location, Europe has been a particularly powerful draw. More than 2 million Turkish guest workers live in Germany. Both Algeria and Morocco also have seen large out-migrations to western Europe, particularly France. Only eight miles separates Africa from Europe, and illegal northbound boat traffic has reached record levels. Thousands more have journeyed to North America, particularly more educated professionals who seek high-paying jobs. Many wealthier residents also left Lebanon and Iran during the political turmoil of the 1980s and are today living in cities such as Toronto, Los Angeles, and Paris.

Shifting Demographic Patterns
While high population growth remains a critical issue throughout the region, the demographic picture is shifting. Uniformly high rates of population growth in the 1950s and 1960s have been replaced by more varied regional patterns, and many nations are seeing birthrates fall fairly rapidly. For example, women in Tunisia, Iran, and Turkey are now averaging fewer than three births, representing a large decline in the total fertility rate. Indeed, some experts now predict that fertility rates on the European and North African sides of the Mediterranean will be very similar (at just

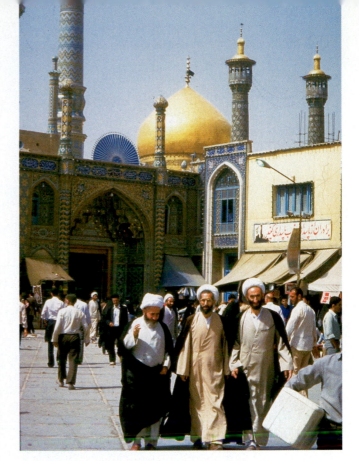

FIGURE 7.17 • THE ISLAMIC LANDSCAPE Iranian mullahs discuss the religious questions of the day beneath the minarets of the Moussavi Mosque in Qom. Islam has left a widespread mark on the region's cultural landscapes. *(S. Franklin/Corbis/Sygma)*

FIGURE 7.18 • ABU DHABI CITYSCAPE The capital city of the oil-rich United Arab Emirates vividly reveals the impact of Western wealth and architecture on the modern Arab world. *(M. Attar/Corbis/Sygma)*

over two births per woman) by 2030. Many factors help explain the changes. More urban, consumer-oriented populations have opted for fewer children. Many Arab women are now delaying marriage into the middle 20s and even early 30s. Family planning initiatives are expanding in many countries. For example, programs in Tunisia, Egypt, and even Iran have greatly increased access to contraceptive pills, IUDs, and condoms since 1980.

Still, the region faces demographic challenges. Areas such as the West Bank, Gaza, and Libya experience some of the highest rates of natural increase on the planet. In some places, continuing patterns of poverty and traditional ways of rural life contribute to the large rates of population increase, and even in more urban and industrialized Saudi Arabia, annual growth rates remain around 3 percent. Such growth rates result from the combination of high birthrates and very low death rates. In Egypt, even though birthrates are likely to decline, the labor market will need to absorb more than 500,000 new workers annually over the next 10 to 15 years just to keep up with the country's large, youthful population. As that population ages in the mid-twenty-first century, it will demand jobs, housing, and social services. Given recent trends in job growth and migration, the region's large cities probably will bear the brunt of future population increases. Growing populations will also increase demands on the region's already limited water resources.

Cultural Coherence and Diversity: Signatures of Complexity

While Southwest Asia and North Africa clearly define the heart of the Islamic and Arab worlds, a surprising degree of cultural diversity characterizes the region. Muslims practice their religion in varied ways, often disagreeing strongly on basic religious views. Religious minorities complicate the region's present-day cultural geography. Linguistically, Arabic languages form an important cultural core, but non-Arab peoples, including Persians, Kurds, and Turks, also dominate important sections of the region. These varied patterns of cultural geography can help us understand the region's political tensions, as well as help us appreciate why many of its residents resist processes of globalization.

Patterns of Religion

Religion is an important part of the lives of most people within the region. Its centrality stands in sharp contrast to less religious cultural impulses that dominate life in many other parts of the world. Whether it is the quiet ritual of morning prayers or discussions about current political and social issues, religion is a central part of the daily routine of most regional residents from Casablanca to Tehran.

Hearth of the Judeo-Christian Tradition   Both Jews and Christians trace their religious roots to an eastern Mediterranean hearth. The roots of Judaism lie deep in the past: Abraham, an early leader in the Jewish tradition, lived some 4,000 years ago and led his people from Mesopotamia to Canaan (modern-day Israel), near the shores of the Mediterranean. Jewish history, recounted in the Old Testament of the Holy Bible, focused on a belief in one God (or **monotheism**), a strong code of ethical conduct, and a powerful ethnic identity that continues to the present. During the Roman Empire, many Jews left the eastern Mediterranean to escape from Roman authority and persecution. The resulting forced migration, or diaspora, took Jews to the far corners of Europe and North Africa. Only in the past century have many of the world's far-flung Jewish populations returned to the religion's place of origin, a process that gathered speed after the formation of the Jewish state of Israel in 1948.

Christianity also emerged in the vicinity of modern-day Israel. An outgrowth of Judaism, Christianity was based on the teachings of Jesus and his disciples, who lived and traveled in the eastern Mediterranean about 2,000 years ago. While many

Christian traditions became associated with European history, some forms of early Christianity remained dominant near the religion's hearth. One stream of Christian influences associated with the Coptic Church diffused into northern Africa, shaping the cultural geographies of places such as Egypt. In the Levant, another group of early Christians, the Maronites, retained a separate cultural identity that survives today.

The Emergence of Islam Islam originated in Southwest Asia in A.D. 622, forming another cultural hearth of global significance. While Muslims can be found today from North America to the southern Philippines, the Islamic world is still centered on Southwest Asia. Most Southwest Asian and North African peoples still follow its religious teachings. Muhammad, the founder of Islam, was born in Makkah (Mecca) in A.D. 570 and taught in nearby Medinah (Medina) (Figure 7.19). His beliefs parallel the Judeo-Christian tradition. Muslims believe that both Moses and Jesus were prophets and that the Hebrew Bible (or Old Testament) and the Christian New Testament, while incomplete, are basically accurate. Ultimately, however, Muslims hold that the **Quran** (or Koran), a book of teachings received by Muhammad from Allah (God), represents God's highest religious and moral revelations to mankind.

The basic beliefs of Islam offer a detailed blueprint for leading an ethical and religious life. Islam literally means "submission to the will of God," and the practice of the religion rests on five essential activities: (1) repeating the basic creed ("There is no God but God, and Muhammad is his prophet"); (2) praying facing Makkah five times daily; (3) giving charitable contributions; (4) fasting between sunup and sundown during the month of Ramadan; and (5) making at least one religious pilgrimage, or **Hajj**, to Muhammad's birthplace of Makkah (Figure 7.20). Islam is a more austere religion than most forms of Christianity. Many Islamic fundamentalists still argue for a **theocratic state**, such as modern-day Iran, in which religious leaders (ayatollahs) guide policy.

A major religious division split Islam early on and endures today. The breakup occurred almost immediately after the death of Muhammad in A.D. 632. Questions surrounded who would get religious power. One group, now called **Shiites**, favored passing power on within Muhammad's own family, specifically to Ali, his son-in-law. Most Muslims, later known as **Sunnis**, advocated passing power down through the established clergy. This group was victorious. Ali was killed, and his Shiite supporters went underground. Ever since, Sunni Islam has formed the mainstream branch of the religion, to which Shiite Islam has presented a recurring, and sometimes powerful, challenge.

Islam quickly spread from the western Arabian peninsula, following camel caravan routes and Arab military campaigns as it expanded its geographical range and converted thousands to its beliefs (see Figure 7.19). By Muhammad's death in A.D. 632, the peoples of the Arabian peninsula were united under its banner. Shortly thereafter, the Persian Empire fell to Muslim forces and the Eastern Roman (or Byzantine) Empire lost most of its territory to Islamic influences. By A.D. 750, Arab armies had swept across North Africa, conquered most of Spain and Portugal, and established footholds in Central and South Asia. By the thirteenth century, most people in the region were Muslims, while older religions such as Christianity and Judaism became minority faiths or disappeared altogether.

Between 1200 and 1500, Islamic influences expanded in some areas and contracted in others. The Iberian Peninsula (Spain and Portugal) returned to Christianity in 1492,

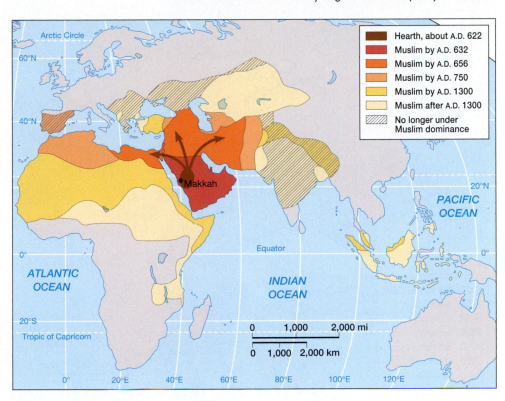

FIGURE 7.19 • DIFFUSION OF ISLAM The rapid expansion of Islam that followed its birth is shown here. From Spain to Southeast Asia, Islam's legacy remains strongest nearest its Southwest Asian hearth. In some settings, its influence has faded or has come into conflict with other religions, such as Christianity, Judaism, and Hinduism. *(Modified from Rubenstein, 2002, An Introduction to Human Geography, Upper Saddle River, NJ: Prentice Hall)*

FIGURE 7.20 • MAKKAH Millions of the faithful make the pilgrimage to Makkah, where they visit the holy al-Ka'ba, the black, cubelike structure in the center of the picture. In Muslim theology, Abraham and Ishmael constructed the al-Ka'ba, and the site was later honored by Muhammad. *(Elkoussy/Corbis/ Sygma)*

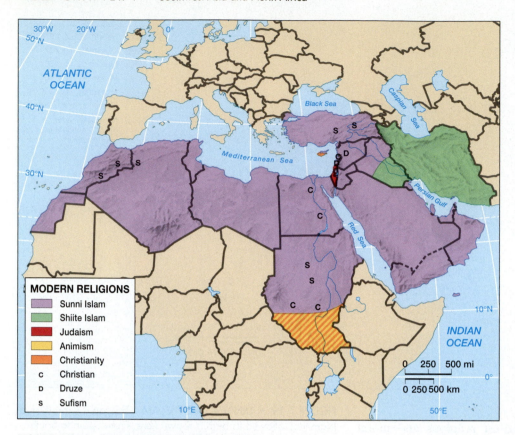

FIGURE 7.21 • MODERN RELIGIONS Islam continues to be the dominant religion across the region. Most Muslims are tied to the Sunni branch, while Shiites are found in places such as Iran and southern Iraq. In some locales, however, Christianity and Judaism remain important. African animism is found in southern portions of Sudan. *(Modified from Rubenstein, 2002, An Introduction to Human Geography, Upper Saddle River, NJ: Prentice Hall)*

FIGURE 7.22 • OLD JERUSALEM Jerusalem's historic center reflects its varied religious legacy. Sacred sites for Jews, Christians, and Muslims are all located within the Old City. The Western Wall, a remnant of the ancient Jewish temple, stands at the base of the Dome of the Rock and Islam's Al Aqsa Mosque. *(Bernard Boutrit/Woodfin Camp & Associates)*

although many Moorish (Islamic) cultural and architectural features still shape the region today. At the same time, Muslims expanded their influence southward and eastward into Africa. In addition, Muslim Turks largely replaced Christian Greek influences in Southwest Asia after 1100. One group of Turks moved into the Anatolian Plateau and finally conquered the Byzantine Empire in 1453. These Turks soon created the huge **Ottoman Empire** (named after one of its leaders, Osman), which included southeastern Europe (including modern-day Albania, Bosnia, and Kosovo) and most of Southwest Asia and North Africa. It provided a focus of Muslim political power within the region until the Empire's disintegration in the late nineteenth and early twentieth centuries.

Modern Religious Diversity Today, Muslims form the majority population in all of the countries of Southwest Asia and North Africa except Israel, where Judaism is the dominant religion, and Cyprus, where Greek Orthodox outnumber Turkish Muslims (Figure 7.21). Still, divisions within Islam create cultural differences. While most (73 percent) of the region is dominated by Sunni Muslims, the Shiites (23 percent) remain an important element in the contemporary cultural mix. Particularly dominant in Iran, southern Iraq, and Lebanon, the Shiites also form a major religious minority in many other countries, including Algeria, Egypt, and Yemen. Strongly associated with the recent flowering of Islamic fundamentalism, the Shiites also have benefited from rapid growth rates because their views are particularly appealing to the poorer, powerless, and more rural populations of the region.

While the Sunni–Shiite split is the great divide within the Muslim world, other variations of Islam can also be found in the region. One division separates the mystically inclined form of Islam—known as *Sufism*—from the more mainstream tradition. Sufism is especially prominent in the peripheries of the Islamic world, including the Atlas Mountains of Morocco and Algeria. The Druze of Lebanon practice yet another variant of Islam.

Southwest Asia is also home to many non-Islamic communities. Israel has a Muslim minority (14 percent) that is dominated by that nation's Jewish population (80 percent). Even within Israel's Jewish community, increasing cultural differences divide Jewish fundamentalists from more reform-minded Jews. In neighboring Lebanon, there was a slight Christian (Maronite and Orthodox) majority as recently as 1950. Christian out-migration and higher Islamic birthrates, however, have created a nation that today is 60 to 70 percent Muslim.

Jerusalem (now the Israeli capital) holds special religious significance for several groups and also stands at the core of the region's political problems (Figure 7.22). Indeed, the sacred space of this ancient Middle Eastern city remains deeply scarred and divided as different groups argue for more control of contested neighborhoods and nearby suburbs. Historically, Jews particularly honor the city's old Western Wall (the site of a Roman-era temple); Christians honor the Church of the Holy Sepulchre (the burial site of Jesus); and Muslims hold sacred religious sites in the city's eastern quarter (including the place from which the prophet Muhammad supposedly ascended to heaven). Further complicating the traditional religious divisions of the city, Israel's victory in the 1967 war encouraged it to establish many new Jewish settlements within the city's redefined and expanded eastern suburbs. The move angered Islamic Palestinian residents of the area, who saw the new settlements intruding on their homeland.

Geographies of Language

Although the region is often referred to as the "Arab World," linguistic complexity creates many important cultural divisions across Southwest Asia and North Africa (Figure 7.23).

Semites and Berbers Afro-Asiatic languages dominate much of the region. Within that family, Arabic-speaking Semitic peoples are found from the Persian Gulf to the Atlantic and southward into Sudan. The Arabic language has a special religious significance for all Muslims because it was the sacred language in which God delivered his message to Muhammad. While most of the world's Muslims do not speak Arabic, the faithful often memorize prayers in the language, and many Arabic words have entered the other important languages of the Islamic world.

Hebrew, another Semitic language, was reintroduced into the region with the creation of Israel. This Semitic language originated in the Levant and was used by the ancient Israelites 3,000 years ago. Today its modern version survives as the sacred tongue of the Jewish people and is the official language of Israel, although the country's non-Jewish population largely speaks Arabic.

While Arabic spread across North Africa, older languages survive in more remote areas. Older Afro-Asiatic languages are still spoken in the Atlas Mountains and in certain parts of the Sahara. Collectively known as Berber, these languages are related to each other but are not mutually intelligible. Most Berber languages have never been written, and none has generated a significant literature. Indeed, a Berber-language version of the Quran was not completed until 1999.

Persians and Kurds Although Arabic spread readily through portions of Southwest Asia, much of the Iranian Plateau and nearby mountains are dominated by older Indo-European languages. Here the principal tongue remains Persian, although, since the tenth century, the language has been enriched with Arabic words and written in the Arabic script.

Kurdish speakers of northern Iraq, northwest Iran, and eastern Turkey add further complexity to the regional pattern of languages. Kurdish, also an Indo-European language, is spoken by 10 to 15 million people in the region. The Kurds have a strong sense of shared cultural identity. Indeed, "Kurdistan" has sometimes been called the world's largest nation without its own political state because the group remains a threatened minority in several countries, particularly Turkey (Figure 7.24).

The Turkish Imprint Turkish languages provide more variety across much of modern Turkey, in portions of far northern Iran, and on the northern third of Cyprus. The Turkish languages are a part of the larger Altaic language family that originated in Central Asia. Turkey remains the largest nation in Southwest Asia dominated by the family. Tens of millions of people in other countries of Southwest and Central Asia speak related Altaic languages, such as Azeri, Uzbek, and Uighur.

Regional Cultures in Global Context

Southwest Asian and North African peoples increasingly find themselves intertwined with cultural complexities and conflicts external to the region. Islam links the region

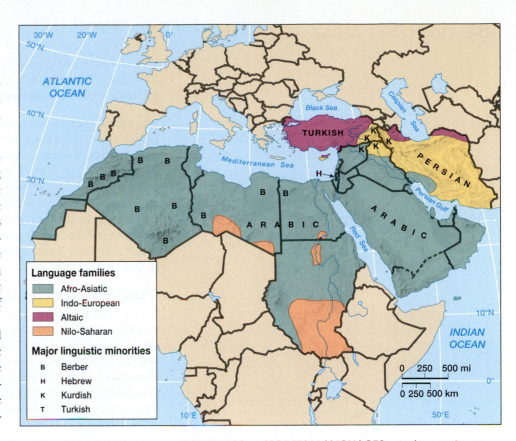

Language families
- Afro-Asiatic
- Indo-European
- Altaic
- Nilo-Saharan

Major linguistic minorities
- B Berber
- H Hebrew
- K Kurdish
- T Turkish

FIGURE 7.23 • MODERN LANGUAGES Arabic is a Semitic Afro-Asiatic language, and it dominates the region's cultural geography. Turkish, Persian, and Kurdish, however, remain important exceptions, and such differences within the region have often had long-lasting political consequences. Israel's more recent reintroduction of Hebrew further complicates the region's linguistic geography. *(Modified from Rubenstein, 2002, An Introduction to Human Geography, Upper Saddle River, NJ: Prentice Hall)*

FIGURE 7.24 • KURDISH FAMILY As an ethnic minority, these Kurdish-speaking settlers from eastern Turkey face cultural and political discrimination in their own country, as do their Kurdish neighbors in nearby nations. *(J. Egan/Hutchison Library)*

with a Muslim population that now lives in many different settings around the world. In addition, European and American cultural influences have increased greatly across the region since the mid-nineteenth century.

Islamic Internationalism   Islamic communities are well established in such distant places as central China, European Russia, central Africa, and the southern Philippines. Today, Muslim congregations are also expanding rapidly in the major urban areas of western Europe and North America, largely through migration but also through conversions. Islam is thus emerging as a truly global religion. The religion's tradition of pilgrimage ensures that Makkah will become a city of increasing global significance in the twenty-first century. The global growth of Islamic fundamentalism also focuses attention on the region. In addition, the oil wealth accumulated by many Islamic nations is being used to fund the promotion and spread of the religion.

Globalization and Cultural Change   The people of the region are also struggling to retain their traditional cultural values and at the same time enjoy the benefits of global economic growth. Islamic fundamentalism is in many ways a reaction to the threat posed by external cultural influences, particularly those of western Europe and the United States. In oil-rich countries, huge capital investments also have had important cultural implications as the number of foreign workers living in these countries has increased. In addition, many affluent young people within the region have embraced elements of Western-style music, literature, and clothing.

Technology also contributes to cultural change. Educated Turks, Tunisians, and Egyptians embrace the Internet and its power to access a global wealth of information and entertainment. In Morocco, the use of cellular phones more than quadrupled in 2000. While less than 6 percent of the country's population is served by fixed phone lines, more than 85 percent is within the reach of cellular phones. Even in conservative Iran, where satellite dishes are officially banned, millions of people have access to them, beaming in multicultural programming from around the globe.

Geopolitical Framework: A Region of Persisting Tensions

Geopolitical tensions remain high in Southwest Asia and North Africa (Figure 7.25). Some tensions relate to age-old patterns of cultural geography in which different ethnic, religious, and linguistic groups still struggle to live with one another in a rapidly changing world of new nation-states and political relationships. The history of European colonialism also contributes to present difficulties, since the modern boundaries of many countries were formed by colonial powers. Geographies of wealth and poverty enter the geopolitical mix: Some residents profit from petroleum resources and industrial expansion, while others struggle just to feed their families. Other conflicts, such as those in Iraq, have grown from specific clashes between the United States and anti-American regimes within the region. The result is a political climate charged with tension, a region in which the sounds of bomb blasts and gunfire have been an all-too-common characteristic of everyday life.

The Colonial Legacy

European colonialism arrived relatively late in Southwest Asia and North Africa, but the era left an important imprint upon the region's modern political geography. The Turks were one reason for Europe's delayed colonial presence in the region. Between 1550 and 1850, the region was dominated by the Ottoman Empire, which expanded from its Anatolian hearth to engulf much of North Africa as well as nearby areas of the Levant, the western Arabian Peninsula, and modern-day Iraq. It took a century for Ottoman influences to be replaced with largely European colonial dominance after World War I (1918). While much of that direct European control ended by the 1950s, colonial influences remain in varied settings. Indeed, it is still common to encounter British English on the streets of Cairo, while French can still be heard in Algiers and Beirut.

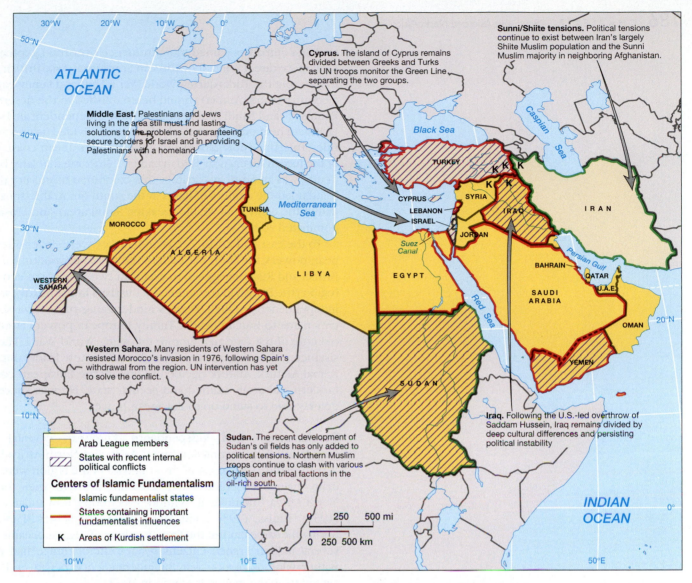

Middle East. Palestinians and Jews living in the area still must find lasting solutions to the problems of guaranteeing secure borders for Israel and in providing Palestinians with a homeland.

Cyprus. The island of Cyprus remains divided between Greeks and Turks as UN troops monitor the Green Line separating the two groups.

Sunni/Shiite tensions. Political tensions continue to exist between Iran's largely Shiite Muslim population and the Sunni Muslim majority in neighboring Afghanistan.

Western Sahara. Many residents of Western Sahara resisted Morocco's invasion in 1976, following Spain's withdrawal from the region. UN intervention has yet to solve the conflict.

Sudan. The recent development of Sudan's oil fields has only added to political tensions. Northern Muslim troops continue to clash with various Christian and tribal factions in the oil-rich south.

Iraq. Following the U.S.-led overthrow of Saddam Hussein, Iraq remains divided by deep cultural differences and persisting political instability

Legend:
- Arab League members
- States with recent internal political conflicts

Centers of Islamic Fundamentalism
- Islamic fundamentalist states
- States containing important fundamentalist influences
- K   Areas of Kurdish settlement

0   250   500 mi
0   250   500 km

FIGURE 7.25 • GEOPOLITICAL ISSUES IN SOUTHWEST ASIA AND NORTH AFRICA Political tensions continue across much of the region. While the central conflict remains oriented around Israel, its neighboring states, and the rights of resident Palestinians, other regional trouble spots periodically erupt in violence. Islamic fundamentalism challenges political stability in settings from Algeria to Sudan. Elsewhere, continuing conflicts shape daily life in Iraq, Sudan, Turkey, and Cyprus.

Imposing European Power   French colonial ties have long been a part of the region's history. Beginning around 1800, France expanded its colonial influence within North Africa. Later in the nineteenth century, Algeria ended up under French rule. The French government expected this territory to become an integral part of France, dominated by a growing French-speaking immigrant population (Figure 7.26). France also established colonial possessions (termed *protectorates*) in Tunisia (1881) and Morocco (1912). France later added more colonial territories in the Levant (Syria and Lebanon) after World War I and the defeat of the Ottoman Empire.

Great Britain's colonial influence also grew within the region before 1900. To control its sea-lanes to India, Britain established a series of nineteenth-century protectorates over the small coastal states of the southern Arabian Peninsula and the Persian Gulf. In this manner, such places as Kuwait, Bahrain, Qatar, the United Arab Emirates, and Aden (in southern Yemen) were loosely incorporated into the British Empire. Nearby Egypt also caught Britain's attention. Once the British-engineered **Suez Canal** linked the Mediterranean and Red seas in 1869, European banks and

FIGURE 7.26 • ALGIERS   European colonial influences are still plentiful in the old French capital of Algiers in northern Algeria. The city's modern bustle continues despite the increasing political and religious tensions that have torn the country apart since 1992. (Francoise Perri/Woodfin Camp & Associates)

trading companies gained more influence over the Egyptian economy. The British took more direct control in 1883. In Southwest Asia, British and Arab forces joined to force out the Turks during World War I. The Saud family convinced the British that a country (Saudi Arabia) should be established in the desert wastes of the Arabian Peninsula, and Saudi Arabia became fully independent in 1932. Elsewhere, however, Britain held onto several colonies. Britain divided its new territories into three entities: Palestine (now Israel) along the Mediterranean coast; Transjordan to the east of the Jordan River (now Jordan); and a third zone that later became Iraq. Iraq, in particular, was an artificial territory that combined three very different cultural zones. It included the centers of Basra in the south (an Arabic-speaking Shiite area), Baghdad in the center (an Arabic-speaking Sunni area), and Mosul in the north (a Kurd-dominated zone).

To the east and north, Persia and Turkey were never directly occupied by European powers, but both played into the global geopolitics of the time. In Persia, the British and Russians agreed to establish two spheres of economic influence in the region (the British in the south, the Russians in the north), while respecting Persian independence. In 1935, Persia's modernizing ruler, Reza Shah, changed the country's name to Iran. In nearby Turkey, European powers attempted to divide up the old core of the Ottoman Empire following World War I. The successful Turkish resistance to European control was based on new leadership that was provided by Kemal Ataturk. Ataturk decided to imitate the European countries and establish a modern, culturally unified and secular state. He was successful, and Turkey was quickly able to stand up to European power.

Decolonization and Independence European colonial powers began their withdrawal from several Southwest Asian and North African colonies before World War II. By the 1950s, most of the countries in the region were independent, although many maintained political and economic ties with their former colonial rulers. In North Africa, Britain finally withdrew its troops from Sudan and Egypt in 1956. Libya (1951), Tunisia (1956), and Morocco (1956) achieved independence peacefully during the same era, but the French colony of Algeria became a major problem. Since several million French citizens resided there, France had no intention of simply withdrawing. A bloody war for independence began in 1954, and France finally agreed to an independent Algeria in 1962.

Southwest Asia also lost its colonial status between 1930 and 1960. While Iraq became independent from Britain in 1932, its later instability in part resulted from its imposed borders, which never recognized much of its cultural diversity. Similarly, the French division of its Levant territories into the two independent states of Syria and Lebanon (1946) greatly angered local Arab populations and set the stage for future political instability in the region. As a favor to its small Maronite Christian majority, France carved out a separate Lebanese state from largely Arab Syria, even guaranteeing the Maronites constitutional control of the national government. The action created a culturally divided Lebanon as well as a Syrian state that repeatedly has asserted its influence over its Lebanese neighbors.

Modern Geopolitical Issues

The geopolitical instability in Southwest Asia and North Africa will continue in the twenty-first century. It remains difficult to predict political boundaries that seem certain to change as a result of negotiated settlements or political conflict. Several key problems continue to affect the region.

The Arab–Israeli Conflict The 1948 creation of the Jewish state of Israel produced a long-lasting zone of cultural and political tensions within the eastern Mediterranean. Jewish migration to the area increased rapidly after the British took over the rule of Palestine from the defeated Ottoman Empire after World War I. In 1917 Britain issued the Balfour Declaration, considered by many Jews to be a pledge to encourage the "establishment of Palestine as a home for the Jewish people." After World War II, the British agreed to withdraw from the area, and the United Nations divided the region into two states, one to be predominantly Jewish and the other primarily Muslim (Figure 7.27). Local Arab Palestinians rejected the division, and war erupted as soon as the British departed. Jewish forces proved victorious, and by 1949 Israel had gained a larger share of land than it had originally been allotted. The remainder of Palestine, including the West Bank and the Gaza Strip, passed to Jordan and Egypt, respectively. Hundreds of thousands of Palestinian refugees fled from Israel to neighboring countries, such as Egypt, Jordan, and Lebanon, where many of them remained in makeshift camps. Under these difficult conditions, the Palestinians developed the idea of creating their own state in the land that had become Israel. The Israelis, not surprisingly, remained adamant that the country was theirs.

Israel and its Arabic-speaking neighbors fought major wars in 1956, 1967, and 1973. In territorial terms, the Six-Day War of 1967 was the most important conflict. In this struggle against Egypt, Syria, and Jordan, Israel gained substantial new territories in the Sinai Peninsula, the Gaza Strip, the West Bank, and the Golan Heights. Israel occupied the eastern part of the formerly divided city of Jerusalem, arousing particular bitterness among the Palestinians, since Jerusalem is a sacred city in the Muslim tradition (it contains the Dome of the Rock and the holy Al-Aqsa Mosque). The old city is sacred in Judaism as well (it contains the Temple Mount and the holy Western Wall of the ancient Jewish temple), and Israel remains unmoved in its claims to the entire city. A peace treaty with Egypt later resulted in the return of the Sinai Peninsula in 1982, but tensions then focused on the other occupied territories that remained under Israeli control. To strengthen its geopolitical claims, Israel also built additional Jewish settlements on the West Bank and in the Golan Heights, further angering Palestinian residents and nearby Syrians.

A cycle of violent actions and equally violent reactions increased with the Palestinian uprising in 1987 that became known as the *Intifada*. The conflict continued until 1993, when the two sides began to negotiate a settlement. Preliminary agreements called for the construction of a quasi-independent Palestinian state in the Gaza Strip and across much of the West Bank. Radical forces on both sides denounced the settlement as a sellout, and political extremists among both the Palestinians and the Israelis heightened tensions across the region. Israel also has reacted militarily to ongoing Palestinian terrorism, while Palestinian leaders such as Yasser Arafat have criticized continuing construction of new Jewish settlements in the West Bank. A tentative agreement late in 1998 strengthened the potential control of the ruling **Palestinian Authority (PA)** in the Gaza Strip and West Bank (Figure 7.28). A new cycle of heightened violence erupted late in 2000, however, as the Israelis started new settlements in occupied lands. The Israelis have also begun construction of a huge wall along the borders of the West Bank region to act as a security barrier and to separate them from the Palestinians.

One thing is certain: Geographical issues will remain at the center of the regional conflict. Palestinians hope for a land they can call their own, and Israelis continue their search for more secure borders that guarantee their own political stability in

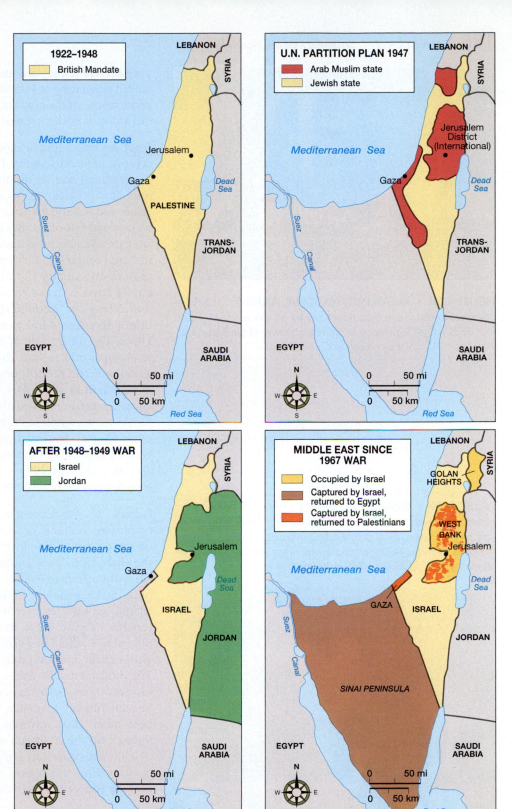

FIGURE 7.27 • EVOLUTION OF ISRAEL Modern Israel's complex evolution began with an earlier British colonial presence and a United Nations partition plan in the late 1940s. Thereafter, multiple wars with nearby Arab states produced Israeli territorial victories in settings such as Gaza, the West Bank, and the Golan Heights. Each of these regions continues to be important in the country's recent relations with nearby states and with resident Palestinian populations. *(Modified from Rubenstein, 2002, An Introduction to Human Geography, Upper Saddle River, NJ: Prentice Hall)*

FIGURE 7.28 • RAFAH INTERNATIONAL AIRPORT The Palestinian-controlled Gaza Strip now features an international airport that adds further legitimacy to the ruling power of the Palestinian Authority (PA). *(Nadav Neuhaus/Corbis/Sygma)*

FIGURE 7.29 • KHARTOUM, SUDAN The politics of fundamentalism spread to the streets of Khartoum in 1989 as a military regime overthrew a democratic government. Today, signs of the city's religious heritage mingle with office buildings and commercial streets. *(J. Hartley/Panos Pictures)*

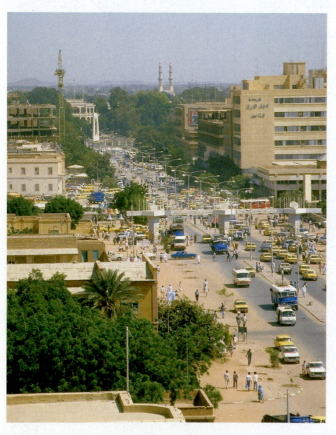

a region where they are surrounded by potentially hostile neighbors. Ultimately, the sacred geography of Jerusalem, a mere 220 acres of land within the Old City, stands at the center of the conflict. Imaginative compromises will need to recognize its special value to both its Jewish and Palestinian residents.

Troubled Iraq   Iraq is another nation born during the colonial era that has yet to achieve political stability. The recent American-led invasion of the country (2003) has replaced one set of uncertainties with another. With Saddam Hussein's departure, new political tensions have emerged, both within the country and between Iraqis and Americans. Iraq remains a culturally complex place. When the country was carved out of the British Empire in 1932, it contained the cultural seeds of its later troubles because many different groups lived within its borders. One distinctive area centers on the lower Tigris and Euphrates valley south of Baghdad, where most of the country's 11 million Shiites live. Indeed, the region focused around the city of Basra contains some of the holiest Shiite sites in the world. Its Marsh Arab population has periodically resisted centralized control from Baghdad, and it briefly threatened independence just after the first Persian Gulf War in the early 1990s. The future role of this Shiite population in Iraq remains uncertain. Meanwhile in northern Iraq, the culturally distinctive Kurds have their own political ambitions. The Kurdish population engaged in periodic revolts and raised its flags of rebellion in the 1980s, only to face the full force of the Iraqi military. Entire villages were bombed with poison gases. Now that Saddam Hussein is gone, some Kurds want complete independence from a central government in Bagdad while other Kurdish leaders argue for participation in a new Iraqi government. Northwest of Bagdad, in the so-called Sunni Triangle near Tikrit, anti-American feelings also rose after the 2003 War, making a peaceful transition to an independent Iraqi regime even more difficult.

Politics of Fundamentalism   Islamic fundamentalism is another cause for continuing regional geopolitical tensions. Fundamentalism dramatically appeared on the political scene in 1978 and 1979 as Shiite Muslim clerics in Iran overthrew the shah, a pro-Western dictator friendly to U.S. political and economic interests. Since then, Iranian leaders have continued to be hostile to many Western cultural ideas. Sudan's Sunni Muslims also followed the fundamentalist path. In 1989, an Islamic-led military overthrew a democratic government and imposed an Islamic republic (Figure 7.29). Immediately, the Sudanese fundamentalists strengthened official ties with Iran. Further, elements of the new government became associated with radical political groups in the region, provoking U.S. military attacks in 1998. The fundamentalist regime also imposed Islamic law across the country and in the process upset both moderate Sunni Muslims as well as the nation's large non-Muslim (mostly Christian and animist) population in the south. Following the September 2001 terrorist attacks in the United States, new attention has also focused on the Al Qaeda network which has often used fundamentalist language to rally anti-American support in the region. Many Al Qaeda cells (groups) are based within the region, and the area's poor, youthful population has been a popular source of new recruits into the organization.

Conflicts Within States   Two examples illustrate the internal challenges faced by nations within the region. Lebanon has seen some of the most destructive fighting over the past several decades. From its birth in the 1940s, Lebanon faced potential conflict among its many distinct Christian and Muslim communities. By 1975 civil war raged within the country. Fighting generally pitted Christians against Muslims, but the existence of numerous groups on both sides made the actual struggle far more complicated. By the late 1970s, the capital city of Beirut was a scarred and divided war zone. Both Syria and Israel sent in armies, and an international peacekeeping force, backed by the United States, tried unsuccessfully to pacify Beirut. Although the conflict flared throughout the 1980s, the nation's various armed factions finally reached a peace agreement in 1995. Syria still maintains a large military force in the country, but the fighting has ended and rebuilding has begun.

Divided Cyprus in the eastern Mediterranean also offers a classic example of a country torn in two by its complex cultural geography. The island's ancient connections with the Greek world were disrupted in the sixteenth century when Ottoman Turks occupied the area. When Cyprus passed into British hands three centuries later, an important Turkish minority population was well established on the island. When Cyprus gained independence in 1960, the country's national leaders hoped that the island's mixed Greek Orthodox and Islamic Turkish populations could find common ground in their newly won freedom. Unfortunately, conflicts increased in the 1970s, and civil war erupted. The conflict intensified when Turkey sent troops to the island in 1974, and war with neighboring Greece became a real possibility. One dramatic geographical consequence of the conflict was the spatial division of the population: Turks concentrated on the northern portion of the island, while Greeks settled to the south. As a result, UN peacekeepers established a **Green Line** that separates the two groups and that runs through the heart of the country's divided capital of Nicosia (Figure 7.30). In 1983 the Turks (20 percent of the population) declared their own Turkish Republic of Northern Cyprus, a state that has gone unrecognized by the rest of the world (with the exception of Turkey) and that has stagnated economically as the Greek Cypriots enjoy more rapid development.

An Uncertain Political Future

Few areas of the world pose more geopolitical questions than Southwest Asia and North Africa. Twenty years from now, the region's political map could look quite different than it does today. Political relations between the United States and the various countries of the region remain complex. The ultimate future of an independent Iraq and its evolving relationship with the United States are unclear. Israel and Turkey are close U.S. allies and receive large amounts of military aid, but Iran, Syria, and Libya remain firmly opposed to the United States. Relations with the oil-rich states of the Arabian Peninsula are more uncertain. Most countries in the region, especially Saudi Arabia, have been major purchasers of U.S. military hardware, and all have relied to some extent on U.S. forces for their protection. Still, questions have been raised about Saudi Arabian financial support for anti-American groups such as Al Qaeda. Conversely, many people in the region view the United States as an intrusive superpower allied with Israel and too prone to interfere directly in the region's affairs. The realm's petroleum riches also promise to keep it in the global spotlight. In the long run, a lasting settlement of the Palestinian issue and the securing of stable borders for Israel seem essential before the region can move toward greater political stability.

FIGURE 7.30 • NICOSIA, CYPRUS The United Nations-enforced Green Line separating Greeks and Turks on the island runs through the major city of Nicosia. The result is an unusual urban landscape that resembles scenes in once-divided East and West Berlin. *(David Rubinger/TimePix)*

Economic and Social Development: Lands of Wealth and Poverty

Southwest Asia and North Africa is a region of both incredible wealth and discouraging poverty. While a few of its countries enjoy great prosperity, due mainly to rich reserves of petroleum and natural gas, other nations within the region are among the least developed in the world. Overall, recent economic growth rates have fallen behind those of the more-developed world. Continuing political instability has also contributed to the region's struggling economy. Petroleum will no doubt figure significantly into the region's future economy, but many countries in the area also have focused on increasing agricultural output, investing in new industries, and promoting tourism to broaden the regional economic base.

The Geography of Fossil Fuels

The global geographies of oil and natural gas show the region's importance in the world oil economy, as well as the extremely uneven distribution of these resources within the region (Figure 7.31). Saudi Arabia remains one of the major producers of petroleum in the world, and Iran, the United Arab Emirates, Libya, and Algeria also contribute significantly. The region plays an important though less dominant role

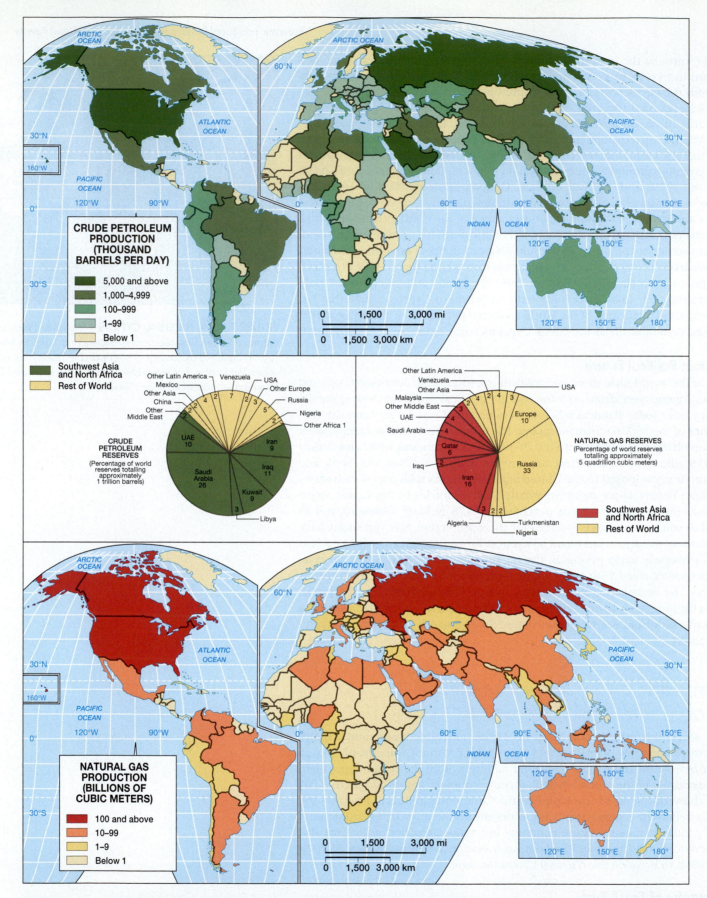

FIGURE 7.31 • CRUDE PETROLEUM AND NATURAL GAS PRODUCTION AND RESERVES The region plays a pivotal role in the global geography of fossil fuels. Abundant regional reserves suggest that the pattern will continue. *(Modified from Rubenstein, 2002,* An Introduction to Human Geography, *Upper Saddle River, NJ: Prentice Hall)*

in natural gas production. Overall, with only 7 percent of the world's population, the region holds an impressive 68 percent of the world's proven oil reserves. Saudi Arabia's key position, both regionally and globally, is also clear: Its 21 million residents live above 26 percent of the planet's known oil supplies.

Two major geological zones supply much of the region's output of fossil fuels. The world's largest concentration of petroleum lies within the Arabian-Iranian sedimentary basin, a geological formation that extends from northern Iraq and western Iran to Oman and the lower Persian Gulf (Figure 7.32). A second important zone of oil and gas deposits includes eastern Algeria, northern and central Libya, and scattered developments in northern Egypt.

Even with all these riches, the geography of fossil fuels is extremely uneven. Some nations—even those with tiny populations (Bahrain, Qatar, and Kuwait, for example)—contain very large fossil fuel reserves, especially when considered on a per capita basis. Many other countries, however, and millions of regional residents, receive relatively few benefits from the oil and gas economy. Israel, Jordan, and Lebanon all lie outside favored geological zones for either petroleum or natural gas. While Turkey has some developed fields in the far southeast, it must import substantial supplies to meet the needs of its large and industrializing population.

Regional Economic Patterns

Remarkable economic differences characterize the region. Some oil-rich countries have prospered greatly since the early 1970s, but in many cases fluctuating oil prices, political disruptions, and rapidly growing populations have reduced the likelihood of future economic growth.

Higher-Income Oil Exporters The richest countries of Southwest Asia and North Africa owe their wealth to massive oil reserves. Nations such as Saudi Arabia, Kuwait (after recovering from the First Gulf War), Qatar, Bahrain, and the United Arab Emirates benefit from fossil fuel production, as well as from their relatively small populations. Since the oil-price hikes of the 1970s, these countries have collected billions of dollars in revenues that have significantly changed their economies (Figure 7.33). Particularly in the case of Saudi Arabia, huge investments in transportation networks, in urban commercial and financial centers, and in other petroleum-related industries have reshaped the cultural landscape. The Saudi petroleum-processing and shipping centers of Jubail (on the Persian Gulf) and Yanbu (on the Red Sea) are examples of this commitment to expand their economic base beyond the simple extraction of crude oil. While much of the oil wealth in these nations remains in the hands of a small ruling elite, petrodollars also have provided real improvements for the larger population. Billions of dollars have poured into new schools, medical facilities, low-cost housing, and modernized agriculture, significantly raising the standard of living in the past 40 years.

Still, problems remain, even in these centers of relative wealth. Poverty exists in Saudi Arabia and elsewhere in the oil-rich region. For example, the Shiite minority in eastern Saudi Arabia has a standard of living far below that of urban populations elsewhere in the country. In addition, large foreign workforces are common across the oil-rich zone, and they typically are paid far less than local workers.

Another challenge relates to fluctuations of the oil economy itself. When prices rise, as they did in the 1970s, early 1980s, and in 2003, these oil-rich nations gain an economic windfall. But dependence on oil and gas revenues clearly has a negative side: Falling prices, such as those seen in the mid-1980s, the late 1990s, and in 2001

FIGURE 7.32 • PERSIAN GULF This satellite view of the Persian Gulf reveals one of the world's richest sources of petroleum. Sedimentary rocks, both on land and offshore, contain additional reserves that can sustain production for decades. *(Earth Satellite Corporation/Science Photo Library/Photo Researchers, Inc.)*

FIGURE 7.33 • SAUDI ARABIAN OIL REFINERY Eastern Saudi Arabia's Ras Tanuna Oil Refinery links the oil-rich country to the world beyond. Huge foreign and domestic investments since 1960 have dramatically transformed many other settings in the region. *(Minosa/Scorpio/Corbis/Sygma)*

created immediate economic pain for the major Middle East producers. In fact, the UAE, Qatar, and Oman all experienced real declines in income during the 1990s. While OPEC continues to be a significant global economic force, many non-OPEC producers also compete in the world marketplace (including Europe, Mexico, and Russia), thus diminishing OPEC's power to set prices.

Lower-Income Oil Exporters   Other states in the region are less dominant players in the oil trade. In these countries, different political and economic variables have slowed economic growth. Algeria illustrates this scenario. Algerian oil and natural gas dominate its exports, but the 1990s brought lower fossil-fuel prices, political instability, and increasing shortages of consumer goods. While the country contains some excellent agricultural lands in the north, the overall amount of arable land has increased little over the past 25 years, even as the country's population has grown by more than 50 percent. Problems in the rural economy have accelerated migration to the crowded cities, and many of the nation's upwardly mobile workers have left the country for employment opportunities in western Europe.

The situation in Iran is more complex. Iran's oil reserves are huge and have seen active commercial development since 1912. The country also has a large industrial base, much of it built in the 20 years prior to the fundamentalist revolution of 1979. In addition, considerable capital flowed into modernizing Iran's agricultural economy during the same period, bringing the benefits of increased mechanization and new irrigation projects. Still, most observers agree that today Iran is relatively poor, burdened with a declining standard of living. Since 1980, the country's fundamentalist leaders have downplayed the role of international trade in consumer goods and services, fearing they would import unwanted cultural influences from abroad. Making matters worse were a costly and bloody war with Iraq in the 1980s followed by the struggling oil economy of the 1990s. Recently, however, the country's economic prospects have brightened somewhat with new economic links to Central Asia, a partial recovery in oil prices, and the potential for greater economic interaction with the West.

Prospering Without Oil   Some countries, while lacking petroleum resources, have nevertheless found paths to increasing economic prosperity. Israel, for example, supports the highest standard of living in the region, even with its political challenges. Israelis have invested large amounts of capital to create a highly productive agricultural and industrial base. The country is also emerging as a global center for high-tech computer and telecommunications products. Even so, Israel has many economic problems. Its continuing struggles with the Palestinians and with neighboring states have cost a great deal of money. Defense spending absorbs a large share of total gross national income (GNI), necessitating high tax rates. Poverty among the Palestinians is also widespread, and the gap between rich and poor within the country has widened considerably.

Turkey also has a diversified economy. While its per capita income is modest even by regional standards, it has shown far greater economic growth in recent years than have many of the major oil producers in the region. Lacking petroleum, Turkey produces varied agricultural and industrial goods for export. The country's main commercial products include cotton, tobacco, wheat, and fruit. The industrial economy has grown since 1980, including exports of textiles and chemicals. Turkey remains the most important tourist destination in the region as well, attracting more than 6 million visitors annually during the 1990s. Many Turkish leaders hope to boost the economy by eventually joining the European Union. Turkey already has closer links to the West than any other country of the region except Israel (it is, for example, a member of NATO), and its trade potential with Europe is great.

Other oil-poor countries in the region are actively promoting more rapid economic development. In North Africa, Tunisia has achieved relative economic stability without the benefit of large oil reserves. Although the country possesses a few important oil fields south of Tunis, most of its workforce is in agriculture, services (particularly tourism), and manufacturing. Recent government policies have favored economic reforms, increased private investment, and encouraged participation in global markets, such as those that bring the region's citrus fruits to grocery stores in the United States. A similar situation benefits Cyprus. Although the island is divided along ethnic lines, the Greek-dominated southern portion has invested in technology, education, and tourism, while favorable tax and trade policies have attracted foreign capital. Even war-torn Lebanon has the potential for prosperity. Tourism is on the rise, and the country's new telecommunications industry signals the potential for growth in that sector of the economy.

Regional Patterns of Poverty   Poorer countries of the region share the problems of much of the less-developed world. For example, Sudan, Morocco, and Egypt each face unique economic challenges. For Sudan, continuing political problems have stood in the way of progress. Civil war has resulted in major food shortages, particularly in the south. The country's transportation and communications systems have seen little new investment, settlement remains mostly rural, and secondary school enrollments stand at less than 25 percent of the school-age population (Figure 7.34). On the other hand, Sudan's fertile soils could support more farming, and its new oil pipeline suggests petroleum's expanding role in the economy. Still, the country's sustained economic development appears delayed by continuing political instability.

Morocco's economic picture is brighter, but the country remains poorer than either Algeria or Tunisia. Poverty is especially widespread in the Atlas Mountains, where many Berber communities have little access to modern services or infrastructure. Elsewhere in the country, economic reforms have increased privatization, encouraged investment by foreign banks, and slowed inflation. Still, illiteracy is widespread, and the country suffers from the **brain drain** phenomenon as some of its brightest young people leave for better jobs in western Europe.

Egypt's economic prospects are also unclear. On the one hand, the country experienced real economic growth during the 1990s as President Hosni Mubarak pushed for smaller government deficits and a multibillion-dollar privatization program to put government-controlled assets under more efficient management. Egypt also actively invites foreign investment in its expanding industrial sector, and the streets of Cairo are increasingly sprinkled with the flash of Rolls-Royces, Porsches, and Lamborghinis. Even so, many Egyptians still live in poverty, and the gap between rich and poor continues to widen. Furthermore, the demographic clock is ticking: Egypt's 70 million people already make it the region's most heavily populated state, and recent efforts to expand the nation's farmland have met with numerous environmental, economic, and political problems.

Issues of Social Development

Measures of social development vary widely across the realm and are broadly associated with overall levels of economic development. Religion and cultural traditions also shape social geographies across the region.

Varied Regional Patterns   Israel's population benefits from its relative wealth within the region. Israelis enjoy high-quality health care and education. In fact, Israel does better in many social measures than might be expected on the basis of its per capita GNI. Average lifespans and rates of infant mortality parallel those of the United States, and the vast majority of the population is literate. Social conditions are significantly better, however, for Israel's Jewish majority than for its Muslim minority. Other relatively wealthy countries—particularly high-income oil exporters such as Saudi Arabia—have lower figures of social well-being than might be expected. In Saudi Arabia, for example, almost one-quarter of the population remains illiterate, and infant mortality is much higher than in Israel. Recent Saudi investments in health care and education may improve these statistics in the future.

A Woman's Changing World   The role of women in the largely Islamic region also remains a major social issue. Female labor participation rates in the workforce are the lowest in the world, and large gaps typically exist between male and female

FIGURE 7.34 • SUDANESE VILLAGE   Poverty and political instability remain a part of life in many Sudanese villages. The traditionally cultivated fields around the village center contrast greatly with the skyscrapers and industrial complexes found in oil-rich portions of the region. *(Liba Taylor/Hutchison Library)*

FIGURE 7.35 • THE CHANGING ROLE OF WOMEN
Traditional and Western clothing styles mingle on the streets of Tehran, Iran, suggesting slow but steady shifts in the roles played by women in many Islamic societies within the region. *(Corbis)*

literacy. In the most conservative parts of the region, few women are allowed to work outside of the home. More orthodox Islamic states impose legal restrictions on the activities of women. In Saudi Arabia, for example, women are not allowed to drive. In Iran, full facial veiling of women in public remains mandatory in more conservative parts of the country. Generally, Islamic women lead more private lives than men: Much of their domestic space is shielded from the world by walls and shuttered windows, and even their public appearances are filtered through the use of the face veil or chador (full-body veil).

Yet even in a fundamentalist state such as Iran, women's roles are changing. Young working women in Tehran, for example, are much more likely to be wearing Western-style fashions than was the case 10 years ago (Figure 7.35). Educational opportunities are also increasingly open to girls across much of the region. In Sudan and Saudi Arabia, a growing number of women pursue high-level careers. Education may be segregated by gender, but it is available. Libya has singled out the modernization of women as a high priority, and today more women than men graduate from the nation's university system.

Global Economic Relationships

Southwest Asia and North Africa share close economic ties with the world. While oil and gas remain critical commodities that dominate international economic linkages, the growth of manufacturing and tourism are redefining the region's role in the world. New regional and global ties promise to further change the ways in which this dynamic part of the world functions in the twenty-first century economy.

Changing Global Linkages OPEC certainly remains an important economic force. The region's enduring role in oil and gas production ensures that fossil fuels will continue to be a major international export for many nations in Southwest Asia and North Africa (see Figure 7.31). While OPEC can no longer control oil and gas prices globally, it still influences the cost and availability of these essential products within the developed and less-developed worlds. Western Europe, the United States, Japan, and many less industrialized countries depend on the region's fossil fuels. In the case of Saudi Arabia, for example, crude oil shipments make up more than 70 percent of its exports, with refined oil and petrochemicals constituting another 20 percent. Petrodollars also shape the global economy. Recent estimates still value investments made by Gulf Arabs at more than $800 billion. These investments influence everything from the cost of office buildings in downtown Singapore to the yields on Brazilian bonds.

Beyond the key OPEC producers, other countries within the region are less dependent on oil-related exports. Turkey, for example, ships textiles, food products, and manufactured goods to its principal trading partners, Germany, Russia, Italy, and the United States. Similarly, Israeli exports emphasize the country's highly skilled workforce: Products such as cut diamonds, electronics, and machinery parts are exported to the United States, western Europe, and Japan.

Tourists are another link to the global economy. Traditional magnets such as ancient historical sites and globally significant religious sites (for a multitude of faiths) draw millions of visitors annually. As the developed world becomes wealthier, there is also a growing global demand for recreational spots that can offer beaches, sunshine, and novel entertainment. Some travelers seek rugged, adventurous ecotourist activities such as snorkeling in Naama Bay on Egypt's Sinai Coast or four-wheeling among the Berbers in the Moroccan backcountry. Endangered wildlife also attracts photographers and poachers hoping to catch a glimpse of a South Arabian grey wolf, Cyprian mouflon (wild sheep), Nubian ibex, or a darting Persian squirrel.

Tourists also have a large impact on the economic geographies of settings such as northern Morocco and Tunisia, coastal Cyprus (in the south), and the holy Saudi Arabian cities of Makkah and Medinah. In some cases tourism remains much less developed, but the small existing inflows in such places as Libya, Jordan, and Iran may be the beginning of a much larger economic trend in future years, depending on the region's political stability.

Regional Connections   Future interconnections with the global economy depend on developing new linkages. Relations with the European Union (EU) are critical. Since 1996, Turkey has enjoyed closer economic ties with the EU, but recent attempts at full membership in the organization have failed. Other so-called Euro-Med agreements also have been signed between the EU and Morocco, Tunisia, Jordan, and Israel. Supporters argue that these agreements will bring more export-oriented industries to the region. Most Arab countries, however, are wary of too much European dominance. They formed a regional political organization known as the Arab League in 1945, and 18 League members established the Arab Free-Trade Area (AFTA) in 1998, designed to eliminate all trade barriers within the region by 2008. In addition, Saudi Arabia has played a particularly central role in regional economic development through organizations such as the Islamic Development Bank and the Arab Fund for Economic and Social Development.

CONCLUSION

Positioned at the meeting ground of Earth's largest landmasses, Southwest Asia and North Africa have played a critical role in world history and in processes of globalization that bind the planet ever more tightly together. In ancient times, the region's inhabitants were early contributors to the Agricultural Revolution, a long process of plant and animal domestication destined to reshape the world's cultural landscapes. The realm also served as a home for urban civilization, offering in the process a new type of human settlement that continues to reshape the distribution of global populations today. Three of the world's great religions—Judaism, Christianity, and Islam—emerged beneath its desert skies. More recently, from Ottoman rule to the unpredictable decisions of OPEC oil ministers, regional forces have reached beyond its boundaries, changing global political, economic, and cultural relationships.

Despite this rich legacy of global influence and power, the peoples of Southwest Asia and North Africa are struggling at the beginning of the twenty-first century. Indeed, most of the countries within the region suffer from significant economic problems and political uncertainties. Twentieth-century population growth across the region was dramatic. At the same time, it has been difficult and costly to expand the region's limited supplies of agricultural land and water resources. Political conflicts have also disrupted economic development across the region. Civil wars, conflicts between states, and regional tensions have worked against plans for greater cooperation and trade. Perhaps most important, the region must deal with the basic inconsistencies between modern ways and the more fundamentalist interpretations of Islam. One thing is certain: Future cultural change will be guided by a complex response to Western influences, a mix of fascination and suspicion that will produce its own unique regional geography. Southwest Asia and North Africa will retain its distinctive regional identity, a character defined by its environmental setting, the rich cultural legacy of its history, the selective abundance of its natural resources, and its continuing political problems.

KEY TERMS

brain drain (page 193)
culture hearth (page 169)
domestication (page 175)
exotic rivers (page 177)
Fertile Crescent (page 176)
fossil water (page 174)
Green Line (page 189)
Hajj (page 181)
hydropolitics (page 175)
Islamic fundamentalism (page 170)
kibbutzes (page 178)
Levant (page 171)
Maghreb (page 171)
medina (page 178)
monotheism (page 180)
OPEC (Organization of Petroleum Exporting Countries) (page 170)
Ottoman Empire (page 182)
Palestinian Authority (PA) (page 187)
pastoral nomadism (page 176)
physiological densities (page 175)
qanat system (page 172)
Quran (page 181)
Shiites (page 181)
Suez Canal (page 185)
Sunnis (page 181)
theocratic state (page 181)
transhumance (page 176)

Europe

FIGURE 8.1 Europe

Stretching from Iceland in the Atlantic to the Black Sea, Europe includes 37 countries, ranging in size from large states, such as France and Germany, to the microstates of Liechtenstein, Andorra, San Marino, and Monaco. Currently the population of the region is about 523 million. Europe is highly urbanized and, for the most part, relatively wealthy, particularly the western portion. However, economic and social differences between eastern and western Europe remain a problem.

SETTING THE BOUNDARIES

The European region is small compared to the United States. In fact, Europe from Iceland to the Black Sea would fit easily into the eastern two-thirds of North America. A more apt comparison would be Canada, as Europe, too, is a northern region. More than half of Europe lies north of the forty-ninth parallel, the line of latitude forming the western border between the United States and Canada (see Figure 8.3).

Europe currently contains 37 countries, and that number may grow in the near future. These states range in size from large countries, such as France and Germany, to microstates, such as Liechtenstein, Andorra, Monaco, and San Marino. Currently the population for these combined states totals about 523 million people.

The notion that Europe is a continent with clearly defined boundaries is a mistaken belief with historical roots. The Greeks and Romans divided their worlds into the three continents of Europe, Asia, and Africa separated by the Mediterranean Sea, the Red Sea, and the Bosporus Strait. A northward extension of the Black Sea was thought to separate Europe from Asia, and only in the sixteenth century was this proven false. Instead, explorers and cartographers discovered that the "continent" of Europe was firmly attached to the western portion of Asia.

Since that time, geographers have not agreed on the eastern boundary of Europe. During the existence of the Soviet Union, some took the "continent's" border all the way east to the Ural Mountains, while others drew the line at the western boundary of the Soviet Union. However, with the disintegration of the Soviet Union in 1990, the eastern boundary of Europe became even more problematic. Now some geography textbooks extend Europe to the border with Russia, which places the two countries of Ukraine and Belarus, former Soviet republics, in eastern Europe. Though an argument can be made for that expanded definition of Europe, recent events, along with a bit of crystal-ball gazing into the near future, lead us to draw our eastern border at Poland, Romania, and Moldova. To the north, the three Baltic republics of Estonia, Lithuania, and Latvia are also included in Europe. Our justification is this: These six countries are currently clearly engaged with Europe and aspire to become even more closely linked to the West in the future. This is not the case with Ukraine and Belarus, which show a decidedly eastern orientation toward Russia.

Europe is one of the most diverse regions in the world, encompassing a wide assortment of people and places in an area considerably smaller than North America. More than half a billion people reside in this region, living in 37 countries that range in size from giant Germany to microstates such as Andorra and Monaco (Figure 8.1).

The region's remarkable cultural diversity produces a geographical mosaic of different languages, religions, and landscapes. Commonly, a day's journey finds a traveler speaking two or three languages, possibly changing money several times, and sampling distinct regional food and drink.

Though the traveler may revel in Europe's cultural and environmental diversity, these regional differences are also entangled with Europe's troubled past (see "Setting the Boundaries"). Throughout history, European neighbors have fought with each other, often because of the same cultural and political differences that travelers so enjoy today. It is often said that Europe invented the nation-state. But this nationalism has also been Europe's downfall, leading it into devastating wars and destructive regional rivalries. In the twentieth century alone, Europe was the principal battleground of two world wars, followed by a 40-year **Cold War** that divided the continent and the world into two hostile, highly armed camps—Europe and the United States against the former Soviet Union. Today, however, a spirit of cooperation prevails as Europe sets aside nationalistic pride and works toward regional economic, political, and cultural integration through the **European Union (EU)**. This supranational organization is made up of 25 countries, anchored by the western European states of Germany, France, Italy, and the United Kingdom, but also including eastern European countries. Undoubtedly, the geographical reach and economic policies of the EU will continue to transform the region during the twenty-first century (Figure 8.2)

Europe, like most world regions, is caught up in the tension between globalization and national and local diversity. Given Europe's considerable impact on the rest of the world as the hearth of Western civilization, the cradle of the Industrial Revolution, and the home of global colonialism, many would argue that Europe actually invented globalization. But while all world regions struggle with this problem of trading off global convergence against national interests, Europe finds itself with an added third layer of complexity as it moves into the new and untested waters of economic, political, and cultural integration through the EU.

FIGURE 8.2 • ANTI-EU PROTEST Farmers in western England burn a European Union flag during an anti-EU demonstration. These farmers are protesting the EU ban on British beef imports into continental Europe because of BSE/Mad Cow disease. Protests against EU agricultural policies are fairly common throughout Europe as farmers struggle against the submergence of local ways to European integration. *(AP/Wide World Photos)*

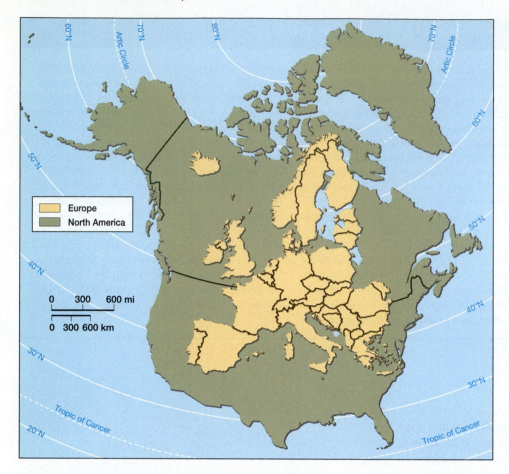

Environmental Geography: Human Transformation of a Diverse Landscape

Despite its small size, Europe's environmental diversity is extraordinary. Within its borders are found a startling range of landscapes from the Arctic tundra of northern Scandinavia to the barren hillsides of the Mediterranean islands, and from the explosive volcanoes of southern Italy to the glaciated seacoasts of western Norway.

Four factors explain this environmental diversity. First, the complex geology of this western extension of the Eurasian landmass has produced some of the newest, as well as the oldest, landscapes in the world. Second, Europe's latitudinal extent creates opportunities for diversity, because the region extends from the Arctic to the Mediterranean subtropics (Figure 8.3). Third, these latitudinal controls are further modified by the moderating influence of the Atlantic Ocean and Black, Baltic, and Mediterranean seas. Last, the long history of human settlement has transformed and modified Europe's natural landscapes in fundamental ways over thousands of years.

Environmental Issues, Local and Global, East and West

Because of its long history of agriculture, resource-extraction, industrial manufacturing, and urbanization, Europe has its share of serious environmental problems. Compounding the situation is the fact that pollution rarely stays within political boundaries. Air pollution from England, for example, creates serious acid-rain problems in Sweden, and water pollution of the upper Rhine River by factories in Switzerland creates major problems for the Netherlands, where Rhine River water is used for urban drinking supplies. When environmental problems cross national boundaries, solutions must come from intergovernmental cooperation (Figure 8.4).

Since the 1970s when the European Union (EU) added environmental issues to its economic and political agenda, western Europe has been increasingly effective in addressing its environmental problems with regional solutions. Besides focusing on the more obvious environmental problems of air and water pollution, the EU is a world leader in recycling, waste management, reduced energy usage, and sustainable resource use. As a result, western Europe is probably the "greenest" of the major world regions. Further, support for environmental protection is also expressed on the international scene. The European Union, for example, has become an aggressive advocate and world leader for the reduction of atmospheric pollutants responsible for global climate change. More specifically, as an energy-efficient group of countries, the EU has pushed other industrialized countries—including the United States—to reduce greenhouse emissions and attain the same level of energy efficiency. Given the Bush administration's cautious and conservative position on greenhouse gas reductions, political tensions have increased between the EU and the United States over emission reduction agreements.

While western Europe is successfully addressing environmental issues in that part of the region, the situation is more grim in eastern Europe (Figure 8.5). During the period of Soviet economic planning (1945–91), little attention was paid to environmental issues because of a clear emphasis on short-term industrial output. Since communist economics did not take into account environmental costs, they were therefore not a factor of concern. As a result, the environment was ignored as industry grew; there were few controls on air and water pollution, environmental safety and health, and the dumping of toxic and hazardous wastes.

FIGURE 8.3 • EUROPE: SIZE AND NORTHERLY LOCATION Europe is about two-thirds the size of North America, as shown in this cartographic comparison. Another important characteristic is the northerly location of the region, which affects its climate, vegetation, and agriculture. Much of Europe lies at the same latitude as Canada; even the Mediterranean lands are farther north than the U.S.–Mexico border.

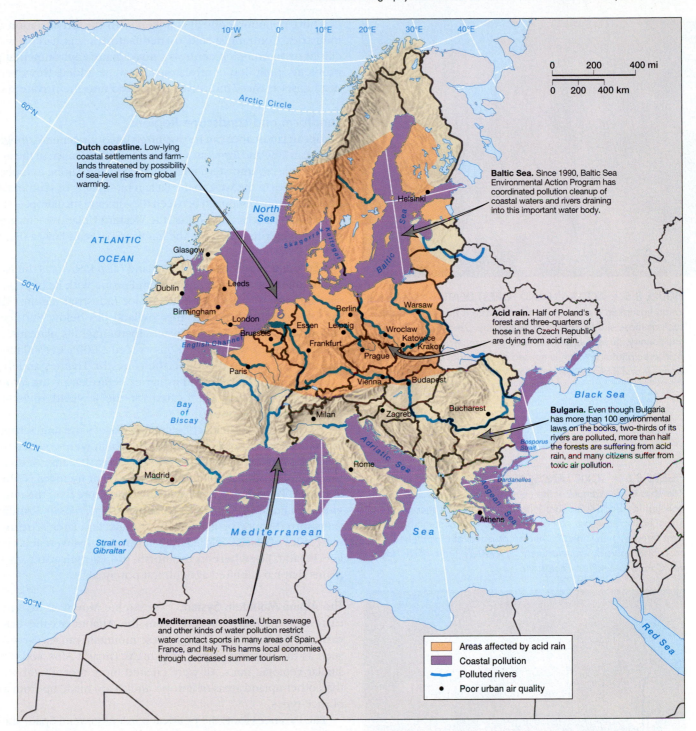

Dutch coastline. Low-lying coastal settlements and farmlands threatened by possibility of sea-level rise from global warming.

Baltic Sea. Since 1990, Baltic Sea Environmental Action Program has coordinated pollution cleanup of coastal waters and rivers draining into this important water body.

Acid rain. Half of Poland's forest and three-quarters of those in the Czech Republic are dying from acid rain.

Bulgaria. Even though Bulgaria has more than 100 environmental laws on the books, two-thirds of its rivers are polluted, more than half the forests are suffering from acid rain, and many citizens suffer from toxic air pollution.

Mediterranean coastline. Urban sewage and other kinds of water pollution restrict water contact sports in many areas of Spain, France, and Italy. This harms local economies through decreased summer tourism.

Areas affected by acid rain
Coastal pollution
Polluted rivers
• Poor urban air quality

FIGURE 8.4 • ENVIRONMENTAL ISSUES IN EUROPE In terms of environmental protection, there is a major gap between western and eastern Europe. While the West has worked energetically over the last 30 years to solve problems such as air and water pollution, those same problems are still widespread in the East because of the long environmental neglect during the communist period. Because Europe is made up of relatively small nation-states, most environmental problems must be solved at the regional level, rather than by each country alone.

Solutions to eastern Europe's environmental problems will not come easily because of several complicating factors. First, most eastern European states are still struggling with their post-1990 economic and political transitions, and it is unclear whether these new governments will either enact or enforce the necessary environmental controls. Instead, there is widespread concern that environmental regulations will inhibit economic recovery by burdening industry with higher operating costs.

Second, there is increasing evidence that capital is not flowing into certain areas or specific industrial sites because western businesses fear they will be charged for cleaning up toxic dumps or repairing environmental damage. Until eastern European governments show a willingness to share those cleanup costs, foreign investors will not put capital into these polluted areas but will instead look at less-damaged regions (Figure 8.6).

FIGURE 8.5 • ACID RAIN AND FOREST DEATH Acid precipitation has taken a devastating toll on eastern European forests, such as those shown here in Bohemia, the Czech Republic. In this country, three-quarters of the forests are dead or injured from acid precipitation, which is caused by industrial and auto emissions. *(Karol Kallay/Bilderberg Archiv der Fotografen)*

FIGURE 8.6 • TOXIC LANDSCAPE IN ROMANIA This site in Romania is an example of the numerous toxic dump sites and polluted landscapes that are leftover from the Soviet communist era in eastern Europe. Not only do these environmental problems threaten public health, but the prospect of high cleanup costs also slows down investment by Western firms. *(Filip Horvat/Corbis/SABA Press Photos, Inc.)*

Third, pollution control is linked with the privatization of former state-owned industries. Communist state-owned factories that had no wastewater treatment facilities or air emission controls caused many environmental problems. To solve the problems, pollution controls must now be added. However, this will add to overhead costs as these industries are moved into the private sector.

Landform and Landscape Regions

European landscapes can be organized into four general topographic regions. First, the European Lowland forms an arc from southwest France to the northeast plains of Poland and includes southeastern England. Second, the Alpine mountain systems extend from the Pyrenees in the west to the Balkan mountains of southeast Europe. Third, the Central Uplands are positioned between the Alps and the European Lowland, stretching from France into eastern Europe. Last, the Western Uplands include mountains in Spain, portions of the British Isles, and the highlands of Scandinavia (Figure 8.7).

The European Lowland This lowland (also known as the North European Plain) is the unquestionable focus of western Europe with its high population density, intensive agriculture, large cities, and major industrial regions. Though not completely flat by any means, most of this lowland lies below 500 feet (150 meters) in elevation, though it is broken in places by rolling hills, plateaus, and uplands (such as in Brittany, France), where elevations exceed 1,000 feet (300 meters). Many of Europe's major rivers, such as the Rhine, the Loire, the Thames, and the Elbe, meander across this lowland and form broad estuaries before emptying into the Atlantic. Several of Europe's great ports are located on the lowland, including London, Le Havre, Rotterdam, and Hamburg.

The Rhine River delta conveniently divides the unglaciated lowland to the south from the glaciated plains to the north, which were covered by a Pleistocene (or "Ice Age") ice sheet until about 15,000 years ago. Because of these continental glaciers, the area of the North European Lowland that includes Netherlands, Germany, Denmark, and Poland is far less fertile for agriculture than the unglaciated portion in Belgium and France (Figure 8.8). Rocky clay materials in Scandinavia were eroded and transported south by glaciers. As the glaciers later retreated with a warming climate, piles of glacial debris known as **moraines** were left on the plains of Germany and Poland. Elsewhere in the north, glacial meltwater created infertile outwash plains that have limited agricultural potential.

The Alpine Mountain System The Alpine Mountain System consists of a series of east–west-running mountains from the Atlantic to the Black Sea and the southeastern Mediterranean. Though these mountain ranges carry distinct regional names, such as the Pyrenees, Alps, Carpathians, Dinaric Alps, and Balkan Ranges, they have similar geologic traits. All were created more recently (about 20 million years ago) than other upland areas of Europe, and all are made up from a complex arrangement of rock types.

The Pyrenees form the political border between Spain and France (including the microstate of Andorra). This rugged range extends almost 300 miles (480 kilometers), stretching from the Atlantic to the Mediterranean. Within the mountains, glaciated peaks reaching to 11,000 feet (3,350 meters) alternate with broad glacier-carved valleys. The centerpiece of this larger geologic system is the Alpine range itself, the Alps, reaching more than 500 miles (800 kilometers) from France to eastern Austria. These impressive mountains are highest in the west, reaching more than 15,000 feet (4,575 meters) in Mt. Blanc on the French-Italian border; in Austria, to the east, few peaks exceed 10,000 feet (3,050 meters). The Appenine Mountains, located to the south, are physically connected to the Alps by the hilly coastline of the French and Italian Riviera. Forming the mountainous spine of Italy, the Appenines are generally lower and lack the scenic glaciated peaks and valleys of the true Alps. To the east, the Carpathian Mountains define the limits of the Alpine system in eastern Europe. They are a plow-shaped upland area that extends from eastern Austria to where the borders of Romania and Yugoslavia intersect. About the same length as the main

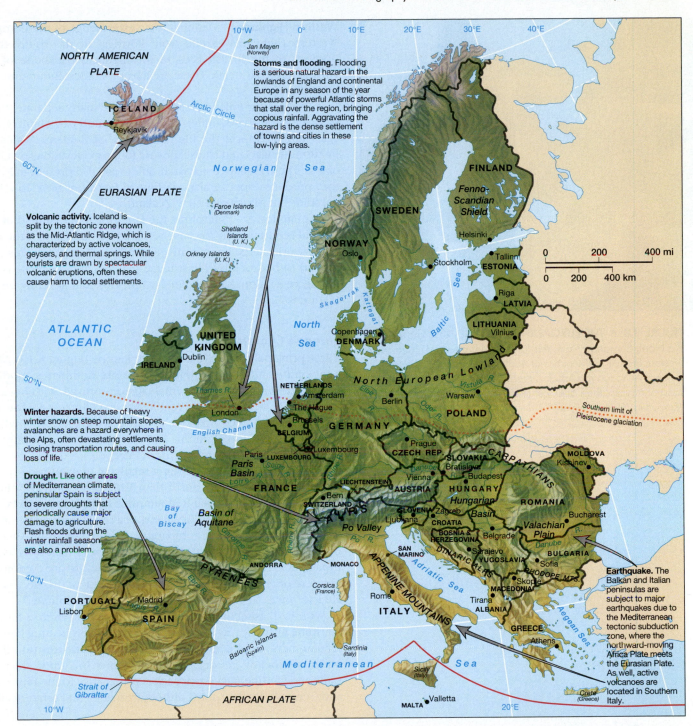

Volcanic activity. Iceland is split by the tectonic zone known as the Mid-Atlantic Ridge, which is characterized by active volcanoes, geysers, and thermal springs. While tourists are drawn by spectacular volcanic eruptions, often these cause harm to local settlements.

Storms and flooding. Flooding is a serious natural hazard in the lowlands of England and continental Europe in any season of the year because of powerful Atlantic storms that stall over the region, bringing copious rainfall. Aggravating the hazard is the dense settlement of towns and cities in these low-lying areas.

Winter hazards. Because of heavy winter snow on steep mountain slopes, avalanches are a hazard everywhere in the Alps, often devastating settlements, closing transportation routes, and causing loss of life.

Drought. Like other areas of Mediterranean climate, peninsular Spain is subject to severe droughts that periodically cause major damage to agriculture. Flash floods during the winter rainfall season are also a problem.

Earthquake. The Balkan and Italian peninsulas are subject to major earthquakes due to the Mediterranean tectonic subduction zone, where the northward-moving Africa Plate meets the Eurasian Plate. As well, active volcanoes are located in Southern Italy.

FIGURE 8.7 • PHYSICAL GEOGRAPHY OF EUROPE Much of the mountain and upland topography of Europe is a function of the gradual northward movement of the African tectonic plate into the Eurasian Plate. Besides these tectonic forces, Pleistocene glaciation has also shaped the European region. Until about 15,000 years ago, much of the region was covered by continental glaciers that extended south to the mouth of the Rhine River. The southern extent of this glaciated area is shown by the dotted line on the map.

Alpine chain, the Carpathians are not nearly as high. The highest summits in Slovakia and southern Poland are less than 9,000 feet (2,780 meters).

Central Uplands In western Europe, a much older highland region occupies an arc between the Alps and the European Lowland in France and Germany. These mountains are much lower in elevation than the Alpine system, with their highest peaks at 6,000 feet (1,830 meters). Their importance to western Europe is great because they contain the raw materials for Europe's industrial areas. In both Germany and France, for example, these uplands have provided the iron and coal necessary for each country's steel industry and in the eastern part of this upland area, mineral resources have also fueled major industrial areas in Germany, Poland, and the Czech Republic.

FIGURE 8.8 • THE EUROPEAN LOWLAND   Also known as the North European Plain, this large lowland extends from south-western France to the plains of northern Germany and into Poland. Although this landform region has some rolling hills, most of it is less than 500 feet (150 meters) in elevation. (P. Vauthey/Corbis/Sygma)

Western Highlands   Defining the western edge of the European subcontinent, the Western Highlands extend from Portugal in the south, through the northwest portions of the British Isles, to the highland backbone of Norway, Sweden, and Finland in the far north. These are Europe's oldest mountains, formed about 300 million years ago.

As with other upland areas that traverse many separate countries, specific place-names for these mountains differ from country to country. A portion of the Western Highlands form the highland spine of England, Wales, and Scotland, where picturesque glaciated landscapes are found at elevations of 4,000 feet (1,220 meters) or less. These U-shaped glaciated valleys are also present in Norway's uplands, where they produce a spectacular coastline of **fjords**, or flooded valley inlets similar to the coastlines of Alaska and New Zealand.

Though lower in elevation, the Fenno-Scandian Shield of Sweden and northern Finland is noteworthy because it is made up of some of the oldest rock formations in the world, dated conservatively at 600 million years. This **shield landscape** was eroded to bedrock by Pleistocene glaciers and, because of the cold climate and sparse vegetation, has extremely thin soils that severely limit agricultural activity (Figure 8.9).

Europe's Climates

Three principal climates characterize Europe (Figure 8.10). Along the Atlantic coast, a moderate and moist **maritime climate** dominates, modified by oceanic influences. Farther inland, continental climates prevail, with hotter summers and colder winters. Finally, dry-summer Mediterranean climates are found in southern Europe, from Spain to Greece. An extensive area of warm-season high pressure inhibits summer storms and rainfall in this region, creating the seemingly endless blue skies so attractive to tourists from northern Europe.

Though most of Europe is at a relatively high latitude (London, England, for example, is slightly farther north than Vancouver, British Columbia), the oceanic influence moderates coastal temperatures from Norway to Portugal and even inland to the western reaches of Germany. As a result, Europe has a climate 5 to 10 °F (2.8 to 5.7 °C) warmer than comparable latitudes without this oceanic effect. In the **marine west coast climate** region, no winter months average below freezing, though cold rain, sleet, and an occasional blizzard are common winter visitors. Summers are often cloudy and overcast with frequent drizzle and rain. Ireland, the Emerald Isle, offers a fitting picture of this maritime climate.

With increasing distance from the ocean (or where a mountain chain limits the maritime influence, as in Scandinavia), landmass heating and cooling produces hotter summers and colder winters. Indeed, all **continental climates** average at least one month below freezing during the winter. In Europe, the transition between maritime and continental climates takes place close to the Rhine River border of France and Germany. Farther north, although Sweden and other nearby countries are close

FIGURE 8.9 • NORTHERN LANDSCAPES   Northern Europe is a harsh land characterized by landscapes with little soil, expanses of bare rock, sparse vegetation, and thousands of lakes. This results from the glaciers that sculpted this region until about 15,000 years ago. (Macduff Everton/Corbis)

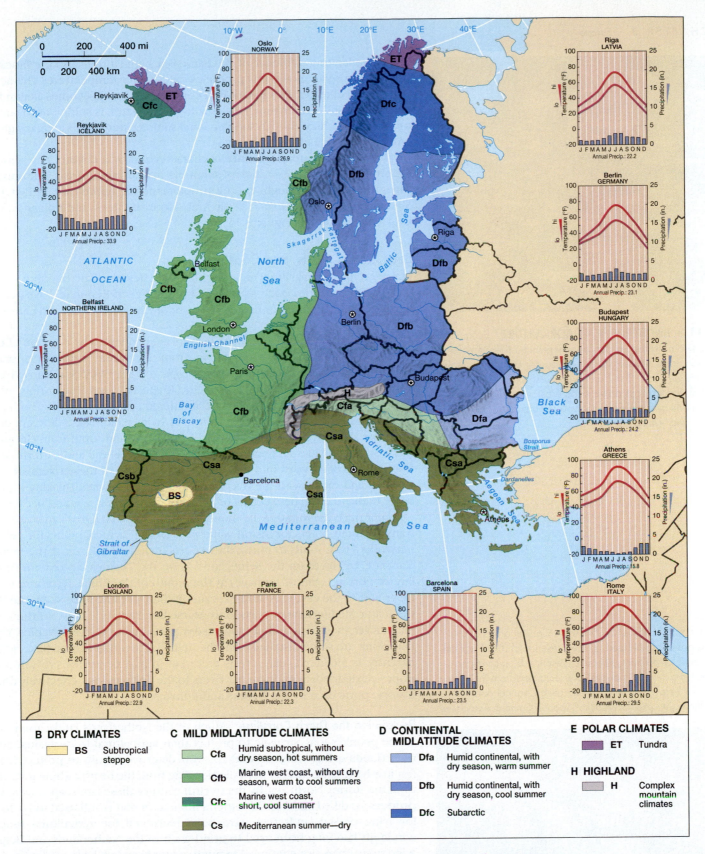

B DRY CLIMATES

| | BS | Subtropical steppe |

C MILD MIDLATITUDE CLIMATES

	Cfa	Humid subtropical, without dry season, hot summers
	Cfb	Marine west coast, without dry season, warm to cool summers
	Cfc	Marine west coast, short, cool summer
	Cs	Mediterranean summer—dry

D CONTINENTAL MIDLATITUDE CLIMATES

	Dfa	Humid continental, with dry season, warm summer
	Dfb	Humid continental, with dry season, cool summer
	Dfc	Subarctic

E POLAR CLIMATES

| | ET | Tundra |

H HIGHLAND

| | H | Complex mountain climates |

FIGURE 8.10 • CLIMATE MAP OF EUROPE Three major climate zones dominate Europe. The marine west coast climate is found close to the Atlantic Ocean, with cool seasons and steady rainfall throughout the year. Farther inland, continental climates are found. They have at least one month averaging below freezing, and they have hot summers, with a precipitation maximum falling during the summer season. The dry summer Mediterranean climate is found in southern Europe. Contrasted to continental climates, most precipitation falls during the cool winter period in the Mediterranean region.

FIGURE 8.11 • MEDITERRANEAN AGRICULTURE Because of water scarcity during the hot, dry summers in the Mediterranean climate region, local farmers have evolved agricultural strategies for making the best use of soil and water resources. In this photo from Portugal, the terraces prevent soil erosion on steep slopes, while trees shade ground crops to reduce water evaporation. *(Getty Images, Inc.—Image Bank)*

FIGURE 8.12 • POLDER LANDSCAPE Coastal areas of the Netherlands are characterized by the presence of diked agricultural settlements, or polders. Because these lands, such as the fields on the left side of this photo, are reclaimed from the sea, many are at or below sea level. The English Channel and North Sea are to the right. *(Adam Woolfitt/Woodfin Camp & Associates)*

to the moderating influence of the Baltic Sea, high latitude and the blocking effect of the Norwegian mountains produces cold winter temperatures characteristic of continental climates. Precipitation in continental climates comes as rain from summer storms and winter snowfall. Usually this moisture is sufficient to support non-irrigated agriculture, though supplemental summer watering is increasingly common where high-value crops are grown.

The **Mediterranean climate** is characterized by a distinct dry season during the summer. While these rainless summers may attract tourists from northern Europe, the seasonal drought can be problematic for agriculture. In fact, traditional Mediterranean cultures, such as the Arab, Moorish, Greek, and Roman, have all used irrigated agriculture (Figure 8.11).

Seas, Rivers, Ports, and Coastline

Europe remains a maritime region with strong ties to its surrounding seas. Even its landlocked countries, such as Austria and the Czech Republic, have access to the ocean through an interconnected network of navigable rivers and canals.

Rivers and Ports Europe's navigable rivers are connected by a system of canals and locks that allow inland barge travel from the Baltic and North seas to the Mediterranean, and between western Europe and the Black Sea. Many rivers on the European Lowland, such as the Loire, Seine, Rhine, Elbe, and Vistula, flow into Atlantic or Baltic waters. However, the Danube and the Rhône, although they both begin near the headwaters of the Rhine in Germany and Switzerland, have a different direction of flow. The Danube, Europe's longest river, flows east and south from Germany to the Black Sea. It provides a connecting artery between central and eastern Europe. The Rhône flows southward into the Mediterranean. Both of these rivers are connected by locks and canals with the rivers of the European Lowland, making it possible for barge traffic to travel between all of Europe's surrounding seas and oceans.

Major ports are found at the mouths of most western European rivers, serving as transshipment points for inland waterways as well as focal points for rail and truck networks. From south to north, these ports include Bordeaux at the mouth of the Garonne, Le Havre on the Seine, London on the Thames, Rotterdam (the world's largest port in terms of tonnage) at the mouth of the Rhine, Hamburg on the Elbe River, and, to the east in Poland, Szcezin on the Oder, and Gdansk on the Vistula. Of the major Mediterranean ports, only Marseilles, France, is close to the mouth of a major river, the Rhône. Other modern-day ports, such as Genoa, Naples, Venice, and Barcelona, are some distance from the delta harbors that served historic trade. Apparently, as Mediterranean forests were cut in past centuries, erosion on the hillslopes carried sediment down the rivers to the delta regions, effectively filling in the historic ports used by the Greeks and Romans.

Reclaiming the Dutch Coastline Much of the Netherlands landscape is a product of the people's long struggle to protect their agricultural lands against coastal and river flooding. Beginning around A.D. 900, dikes were built to protect these fertile but low-lying lands against periodic flooding from the nearby Rhine River, as well as from the stormy North Sea. By the twelfth century, these landscapes were known as **polders**, or diked agricultural settlements, a term that is still used today to describe low-lying coastal lands that have been reclaimed for agricultural usage. While windmills had long been used to grind grain in the Netherlands and Belgium, this wind power was also employed to pump water from low-lying wetlands. As a result, windmills became increasingly common on the Dutch landscape to drain marshes. This technology worked so well that the Dutch government began a widespread coastal reclamation plan in the seventeenth century that converted an 18,000-acre (7,275-hectare) lake into agricultural land (Figure 8.12).

With steam- (and later, electric-) powered pumps, even more ambitious polder reclamation was possible. The last of these plans was the massive Zuider Zee project of the twentieth century, in which the large bay north of Amsterdam was dammed, drained, and converted to agricultural lands over the course of a half-century. This project improved a major flooding hazard and also opened up new lands for farming and settlement in the heart of the Netherlands. New suburbs of Amsterdam are now spilling onto these reclaimed lands.

Settlement and Population: Slow Growth and Rapid Migration

The map of Europe's population distribution (Figure 8.13) shows that, in general, population densities are higher in the historical industrial core areas of western Europe (England, the Netherlands, northern France, northern Italy, and western Germany) than in the periphery to the east and north. While this generalization overlooks important urban clusters in Mediterranean Europe, it does express a sense of a densely settled European core set apart from a more rural, agricultural periphery. Much of this distinctive population pattern is linked to areas of early industrialization, yet there are many modern consequences of this core-periphery distribution. For example, economic subsidies from the wealthy, highly urbanized core to the less affluent, agricultural periphery have been an important part of the EU's development policies for several decades. Further, while the urban-industrial

FIGURE 8.13 • POPULATION MAP OF EUROPE The European region includes more than 523 million people, many of them clustered in large cities in both western and eastern Europe. As can be seen on this map, the most densely populated areas are in England, the Netherlands, Belgium, western Germany, northern France, and south across the Alps to northern Italy.

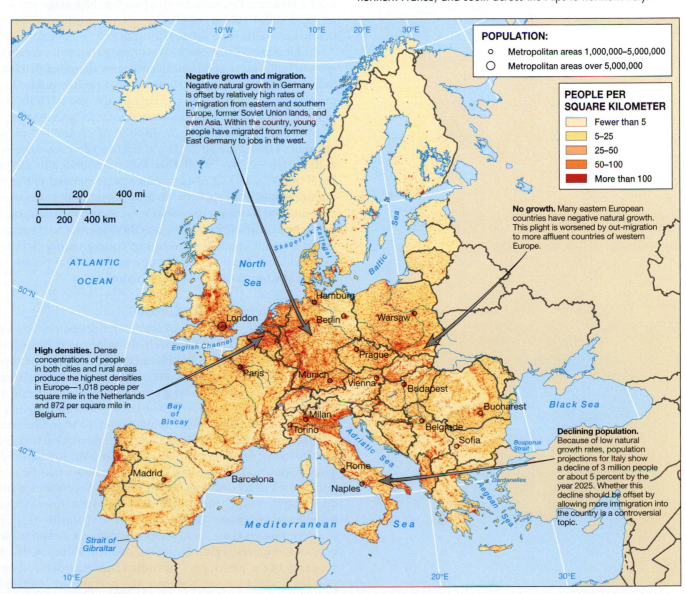

POPULATION:
- ○ Metropolitan areas 1,000,000–5,000,000
- ◯ Metropolitan areas over 5,000,000

PEOPLE PER SQUARE KILOMETER
- Fewer than 5
- 5–25
- 25–50
- 50–100
- More than 100

Negative growth and migration. Negative natural growth in Germany is offset by relatively high rates of in-migration from eastern and southern Europe, former Soviet Union lands, and even Asia. Within the country, young people have migrated from former East Germany to jobs in the west.

No growth. Many eastern European countries have negative natural growth. This plight is worsened by out-migration to more affluent countries of western Europe.

High densities. Dense concentrations of people in both cities and rural areas produce the highest densities in Europe—1,018 people per square mile in the Netherlands and 872 per square mile in Belgium.

Declining population. Because of low natural growth rates, population projections for Italy show a decline of 3 million people or about 5 percent by the year 2025. Whether this decline should be offset by allowing more immigration into the country is a controversial topic.

core is characterized by extremely low natural growth rates, it is also the target area for migrants—both legal and illegal—from Europe's peripheral countries as well as from outside Europe.

Natural Growth: Beyond the Demographic Transition

Probably the most striking characteristic of Europe's population is its slow natural growth. More to the point, in many European countries the death rate exceeds the birthrate, meaning that there is simply no growth at all. Instead, many countries are experiencing negative growth rates; were it not for in-migration from other countries and other world regions, these countries would record a decline in population over the next few decades. Italy, for example, currently has a population of 57.8 million. Yet if current natural growth holds true for 20 years and is not offset by immigration, the population will decrease to 55 million by the year 2025.

There seem to be several reasons for zero population growth in western Europe. First of all, recall from Chapter 1 that the concept of the demographic transition was based upon from the historical change in European growth rates as the population moved from rural settings to more urban and industrial locations. What we see today is an extension of that model, namely, the continued expression of the low fertility–low mortality fourth stage of the demographic transformation. Some demographers suggest adding a fifth stage to the model—a "postindustrial" phase in which population falls below replacement levels, which means parents have less than 2 children. Evidence for this possible fifth stage comes from the highly urbanized and industrialized populations of Germany, France, and England, all of which are below zero natural population growth.

Additional explanations, however, are needed to explain the negative population growth in the eastern European countries, the most striking examples of which are found in Bulgaria, Romania, Hungary, the Czech Republic, Latvia, Estonia, and Lithuania. The explanation for this current negative growth traces back to the development of these countries while they were under communist rule.

Following World War II, the growth of centralized planning and industrial development under the Soviet Union led to a labor shortage in eastern Europe, as well as in the Soviet Union itself. This was compounded by the huge losses suffered during World War II. As a result, women were needed in the workforce. So that child rearing would not conflict with jobs, Soviet policy was to make all forms of family planning and birth control available, including abortion. As a result, small families became the norm. Another factor was the widespread housing shortage that prevailed in most Soviet countries, a result of concentrating people in cities for industrial jobs. Evidence from eastern Europe suggests that people experiencing a housing crunch tend to have few (if any) children.

Migration to and Within Europe

Migration is one of the most challenging population issues facing Europe today because the region is caught in a web of conflicting policies and values. Currently there is widespread resistance to unlimited migration into Europe, primarily because of high unemployment in the western European industrial countries. Many Europeans argue that scarce jobs should go first to European citizens, not to "foreigners." However, with an improving economy, there could be a significant labor shortage that could be solved only through immigration. To illustrate, recent studies suggest that Germany would need about 400,000 in-migrants each year to provide an adequate labor force for its industry. Additionally, given the low rates of natural growth and the aging population in most of Europe, immigration may be needed to provide tax revenue for social security and other programs necessary for an older population. Many demographers (and some politicians) believe an open-door immigration policy will be necessary to ensure Europe's economic vitality in the next decade. Complicating the issue, though, is the social and political unease linked to the presence of large numbers of foreigners within European countries that are used to cultural and ethnic similarity. In the recent past, individual countries decided on their own immigration policies. Today, however, the topic has become so controversial that the European

Union (EU) has committed itself to forging a common immigration policy in the next several years.

During the 1960s postwar recovery when western Europe's economies were booming, many countries looked to migrant workers to ease labor shortages. Germany, for example, depended on workers from Europe's periphery, namely Italy, Yugoslavia, Greece, and Turkey, for industrial and service jobs. These *gastarbeiter*, or **guest workers**, arrived by the thousands. As a result, large ethnic enclaves of foreign workers became a common part of the German urban landscape. Today, there are 2.5 million Turks in Germany, most of whom are there because of the open-door foreign worker policies of past decades.

However, these foreign workers are now the target of considerable ill will. Western Europe has suffered from economic recession and stagnation for the last decade, resulting in unemployment rates approaching 25 percent for young people. As noted, many native Europeans protest that guest workers are taking jobs that could be filled by them. Additionally, the region has witnessed a massive flood of migrants from former European colonies in Asia, Africa, and the Caribbean. The former colonial powers of England, France, and the Netherlands have been the major recipients of this immigration. England, for example, has inherited large numbers of former colonial citizens from India, Pakistan, Jamaica, and, most recently, Hong Kong. Indonesians (from the former Dutch East Indies) are common in the Netherlands, while migrants in France are often from former colonies in both northern and Sub-Saharan Africa (Figure 8.14).

More recently, political and economic troubles in eastern Europe and the former Soviet Union have generated a new wave of migrants to Europe. Within Germany, for example, thousands of former East Germans have taken advantage of the Berlin Wall's removal by moving to the more prosperous and dynamic western parts of unified Germany. With the total collapse of Soviet border controls in 1990, emigrants from Poland, Bulgaria, Romania, Ukraine, and other former Soviet satellite countries poured into western Europe, looking for a better life. This flight from the post-1989 economic and political chaos of eastern Europe has also included refugees from war-torn regions of what was Yugoslavia, particularly from Bosnia and Kosovo.

As a result of these different migration streams, Germany has become a reluctant land of migrants that now receives 400,000 newcomers each year. Currently, about 7.5 million foreigners live in Germany, making up about 9 percent of the population. This is about the same percentage as the foreign-born share of the U.S. population. However, while the U.S. celebrates its history of immigration, Germany, along with other European countries, struggles with this new kind of cultural diversity.

The Landscapes of Urban Europe

One of the major characteristics of Europe's population is its high level of urbanization. All but three countries (Albania, Portugal, and Yugoslavia) have more than half their population in cities, and several countries, such as the United Kingdom and Belgium, are more than 90 percent urbanized. Once again, the gradient from the highly urbanized European heartland to the less-urbanized periphery reinforces the distinctions between the affluent, industrial core area and the more rural, less well-developed periphery.

Three historical eras dominate most European city landscapes. The medieval (roughly A.D. 900–1500), Renaissance–Baroque (1500–1800), and industrial (1800–present) periods each left characteristic marks on the European urban scene. Learning to recognize these stages of historical growth provides visitors to Europe's cities with fascinating insights into both past and present landscapes (Figure 8.15).

The Medieval Period   The **medieval landscape** is one of narrow, winding streets, crowded with three- or four-story masonry buildings with little setback from the street. This is a dense landscape with few open spaces, except around churches or public buildings. Here and there, public squares or parks are clues to medieval open-air marketplaces where commerce was transacted. As picturesque as we find medieval-era districts today, they nevertheless present challenges to modernization because of their narrow, congested streets and old housing. Often

FIGURE 8.14 • IMMIGRANT GHETTOS IN EUROPE   A large number of Africans from France's former colonies have migrated to large cities where they form distinct ethnic colonies—even ghettos—in high-rise apartment buildings on the outskirts of Paris, Lyon, and Marseille. *(S. Elbaz/Corbis/Sygma)*

FIGURE 8.15 • URBAN LANDSCAPES   This aerial view of the Marais district of Paris shows how the present-day landscape expresses different historical periods. The narrow streets and small buildings represent the medieval period, while the broad boulevards and large public buildings are examples of later Renaissance–Baroque city planning. *(Corbis)*

modern plumbing and heating are lacking, and rooms and hallways are small and cramped compared to our present-day standards. Because these medieval districts usually lack modern facilities, the majority of people living in them have low or fixed incomes. In many areas this majority is made up of the elderly, university students, and ethnic migrants.

The Rennaissance–Baroque Period   In contrast to the cramped and dense medieval landscape, those areas of the city built during the **Renaissance–Baroque** period produce a landscape that is much more open and spacious, with expansive ceremonial buildings and squares, monuments, ornamental gardens, and wide boulevards lined with palatial residences. During this period (1500–1800), a new artistic sense of urban planning arose in Europe that resulted in the restructuring of many European cities, particularly the large capitals. These changes were primarily for the benefit of the new urban elite, such as royalty and successful merchants. City dwellers of lesser means remained in the medieval quarters, which became increasingly crowded and cramped as more of the city space was devoted to the ruling classes.

During this period, defensive structures such as city walls limited the outward spread of these growing cities, thus increasing density and crowding within. With the advent of assault artillery, European cities were forced to build an extensive system of defensive walls. Once encircled by these walls, the cities could not expand outward. Instead, as the demand for space increased within the cities, a common solution was to add several new stories to the medieval houses.

The Industrial Period   Industrialization dramatically altered the landscape of European cities. Historically, factories clustered together in cities beginning in the early nineteenth century, drawn by their large markets and labor force and supplied by raw materials shipped by barge and railroad. Industrial districts of factories and worker tenements grew up around these transportation lines. In continental Europe, where many cities retained their defensive walls until the late nineteenth century, the new industrial districts were often located outside the former city walls, removed from the historic central city. In Paris, for example, when the railroad was constructed in the 1850s, it was not allowed to enter the city walls. As a result, terminals and train stations for the network of tracks were located beyond the original fortifications. Although the walls of Paris are long gone, this pattern of outlying train stations persists today. As noted, the historical industrial areas were also along these tracks, outside the walls in what were essentially suburban locations. This contrasts with the geography of industrial areas in North American cities, where factories and working-class housing often occupied central city locations.

Cultural Coherence and Diversity: A Mosaic of Differences

The rich cultural geography of Europe demands our attention for several reasons. First, the highly varied and fascinating mosaic of languages, customs, religions, ways of life, and landscapes that characterize Europe has also shaped strong local and regional identities that have all too often stoked the fires of conflict. Second, European cultures have played leading roles in processes of globalization. Through European colonialism, regional languages, religion, economies, and values have been changed in every corner of the globe. If you doubt this, consider cricket games in Pakistan, high tea in India, Dutch architecture in South Africa, and the millions of French-speaking inhabitants of equatorial Africa. Even without such modern technologies as satellite TV, the Internet, and Hollywood films and video, European culture spread across the world, changing the speech, religion, belief systems, dress, and habits of millions of people on every continent.

Today, though, many European countries resist the varied expressions of global culture (Figure 8.16). France, for example, struggles against both U.S.-dominated popular culture and the multicultural influences of its large migrant population. In

FIGURE 8.16 • GLOBAL CULTURE IN EUROPE   The Europeans have mixed feelings about the influx of U.S. popular culture. While many people welcome everything from fast food to Hollywood movies, at the same time they protest the loss of Europe's traditional regional cultures. *(David R. Frazier Photolibrary, Inc.)*

many ways the same is true of Germany and England. The cultural geography of contemporary Europe is also complicated by the fact that culture operates at many different scales and in many different ways. For example, while Europe fends off global culture with one hand, with the other it creates its own unique culture through economic and political integration. But as individual countries come together in political alliances and unions that transcend national sovereignty, like the EU, local and regional cultures—such as the Basques in Spain—demand autonomy and independence. Underlying the complexity of this ever-changing cultural mosaic are the fundamentals of language and religion, traits that form the basis of so many of Europe's cultural patterns.

Geographies of Language

Language has always been an important component of nationalism and group identity in Europe (Figure 8.17). Today, while some small ethnic groups, such as the Irish or the Bretons, work hard to preserve their local language in order to reinforce their cultural identity, millions of Europeans are busy learning multiple languages so they can communicate across cultural and national boundaries. In this age of globalization and world culture, the European who does not speak at least two languages is rare.

FIGURE 8.17 • LANGUAGE MAP OF EUROPE Ninety percent of Europeans speak an Indo-European language, grouped into the major categories of Germanic, Romance, and Slavic languages. As a first language, 90 million Europeans speak German, which places it ahead of the 60 million who list English as their native language. However, given the large number of Europeans who speak fluent English as a second language, one could make the case that English is the dominant language of modern Europe.

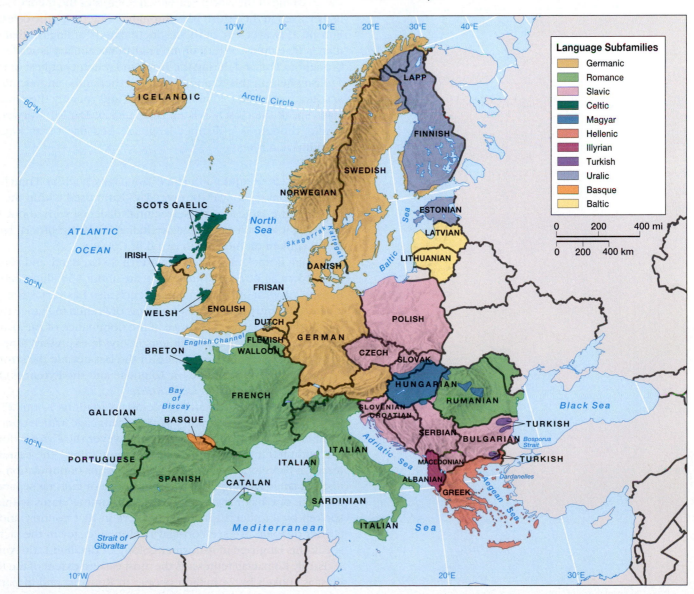

As their first language, 90 percent of Europe's population speaks Germanic, Romance, or Slavic languages belonging to western linguistic groups of the Indo-European family. Germanic and Romance speakers each number almost 200 million in the European region. There are far fewer Slavic speakers (about 80 million) when Europe's boundaries are drawn to exclude Russia, Belarus, and Ukraine.

Germanic Languages Germanic languages dominate Europe north of the Alps. Today about 90 million people speak German as their first language. This is the dominant language of Germany, Austria, Liechtenstein, Luxembourg, eastern Switzerland, and several small areas in Alpine Italy. Until recently, there were also large German-speaking minorities in Romania, Hungary, and Poland, but many of these people left eastern Europe when the Iron Curtain was lifted in 1989. As "ethnic Germans," they were given automatic citizenship in Germany itself. As is true of most languages, there are very strong regional dialects in German that set apart German-speaking Swiss, for example, from the *Letzeburgish* German spoken in Luxembourg. Nevertheless, speakers of these different dialects can usually understand each other.

English is the second-largest Germanic language, with about 60 million speakers using it as their first language. Additionally, a large number learn English as a second language, particularly in the Netherlands and Scandinavia, where many are as fluent as native speakers. Linguistically, English is closest to the Low German spoken along the coastline of the North Sea, which reinforces the theory that an early form of English evolved in the British Isles through contact with the coastal peoples of northern Europe. However, one of the distinctive traits of English that sets it apart from German is that almost a third of the English vocabulary is made up of Romance words brought to England during the Norman French conquest of the eleventh century.

Elsewhere in this region, Dutch (Netherlands) and Flemish (northern Belgium) account for another 20 million people, and roughly the same number of Scandinavians speak the closely related languages of Danish, Norwegian, and Swedish. Icelandic, though, is a distinct language because of its long separation from its Scandinavian roots.

Romance Languages Romance languages, such as French, Spanish, and Italian, evolved from the vulgar (or everyday) Latin used within the Roman Empire. Today Italian is the largest of these regional dialects, with about 60 million Europeans speaking it as their first language. Italian is also an official language of Switzerland and is spoken on the French island of Corsica.

French is spoken in France, western Switzerland, and southern Belgium, where it is known as *Walloon*. Today there are about 55 million native French speakers in Europe. As with other languages, French also has very strong regional dialects. Linguists differentiate between two forms of French in France itself, that spoken in the north (the official form because of the dominance of Paris) and the language of the south, or *langue d'oc*. This linguistic divide expresses long-standing tensions between Paris and southern France. In the last decade the strong regional awareness of the southwest (centered on Toulouse and the Pyrenees) has led to a rebirth of its own distinct language, *Occitanian*.

Spanish also has very strong regional variations. About 25 million people speak Castillian Spanish, the country's official language, which dominates the interior and northern areas of that large country. However, the *Catalan* form, which some argue is a completely separate language, is found along the eastern coastal fringe, centered on Barcelona, Spain's major city in terms of population and the economy. This distinct language reinforces a strong sense of cultural separateness that has led to the state of Catalonia's being given autonomous status within Spain. Portuguese is spoken by another 12 million speakers in that country and in the northwestern corner of Spain. We see Portugal's colonial legacy in the fact that far more people speak this language in its former colony of Brazil in Latin America than in Europe.

Finally, Romanian represents the most eastern extent of the Romance language family; it is spoken by 24 million people in Romania and in parts of the new state of

Moldova. Though unquestionably a Romance language, Romanian also contains many Slavic words. Further, in Moldava, the language is written in the **Cyrillic alphabet**, which is based on Greek letters and is used by Russian and other Slavic languages.

The Slavic Language Family   Slavic is the largest European subfamily of the Indo-European languages. Traditionally, Slavic speakers are separated into northern and southern groups, divided by the non-Slavic speakers of Hungary and Romania.

To the north, Polish has 35 million speakers, and Czech and Slovakian about 14 million each. These numbers pale in comparison, however, with the number of northern Slav speakers in nearby Ukraine, Belarus, and Russia, where one can easily count more than 150 million. Southern Slav languages include three groups: 14 million Serbo-Croatian speakers (now considered separate languages because of the political troubles between Serbs and Croats), 11 million Bulgarian-Macedonian, and 2 million Slovenian.

The use of two alphabets further complicates the geography of Slavic languages (Figure 8.18). In countries with a strong Roman Catholic heritage, such as Poland and the Czech Republic, the Latin alphabet is used in writing. In contrast, countries with close ties to the Orthodox church use the Greek-derived Cyrillic alphabet, as is the case in Bulgaria, Macedonia, Croatia, and the Serbian areas of Yugoslavia.

Geographies of Religion, Past and Present

Religion is an important component of the geography of cultural coherence and diversity in Europe because so many of today's ethnic tensions result from historical religious events. To illustrate, strong cultural borders in the Balkans and eastern Europe are drawn based upon the eleventh-century split of Christianity into eastern and western churches; in Northern Ireland blood is still shed over the tensions between the seventeenth-century division of Christianity into Catholicism and Protestantism; and much of the terrorism of ethnic cleansing in Yugoslavia comes from the historical struggle between Christianity and Islam in that part of Europe. Additionally, there is considerable tension regarding the large Muslim migrant populations in England, France, and Germany. Understanding these important contemporary issues thus involves a brief look back at the historical geography of Europe's religions (Figure 8.19).

The Split Between Western and Eastern Christianity   In southeastern Europe, early Greek missionaries spread Christianity through the Balkans and into the lower reaches of the Danube. Progress was slower than in western Europe, perhaps because of continued invasions by peoples from the Asian steppes. There were other problems as well, primarily the refusal of these Greek missionaries to accept the control of Roman bishops from western Europe.

This tension with western Christianity led in 1054 to an official split of the eastern church from Rome. This eastern church subsequently splintered into Orthodox sects closely linked to specific nations and states. Today, for example, we find Greek Orthodox, Bulgarian Orthodox, and Russian Orthodox churches, all of which have different rites and rituals, yet share cultural traits. The current political ties between Russians and Serbian Yugoslavs, for example, grow from their shared language and religion.

Another factor that distinguished eastern Christianity from western was the Orthodox use of the Cyrillic alphabet instead of the Latin. Because Greek missionaries were primarily responsible for the spread of early Christianity in southeastern Europe, it is not surprising that they used an alphabet based on Greek characters. More precisely, this alphabet is attributed to the missionary work of St. Cyril in the ninth century. As a result, the division between western and eastern churches, and between the two alphabets, remains one of the most problematic cultural boundaries in Europe.

The Protestant Revolt   Besides the division between western and eastern churches, the other great split within Christianity occurred between Catholics and Protestants. This division arose in Europe during the sixteenth century and has

FIGURE 8.18 • THE ALPHABET OF ETHNIC TENSION   After the war with Kosovo's Serbs, signs in the Serbian Cyrillic alphabet were removed from many stores, shops, and restaurants as Kosovars of Albanian ethnicity restated their claims to the region. Here the Albanian owner of a restaurant in the capital city of Pristina scrapes off Cyrillic letters following the recent war. (David Brauchll/AP/Wide World Photos)

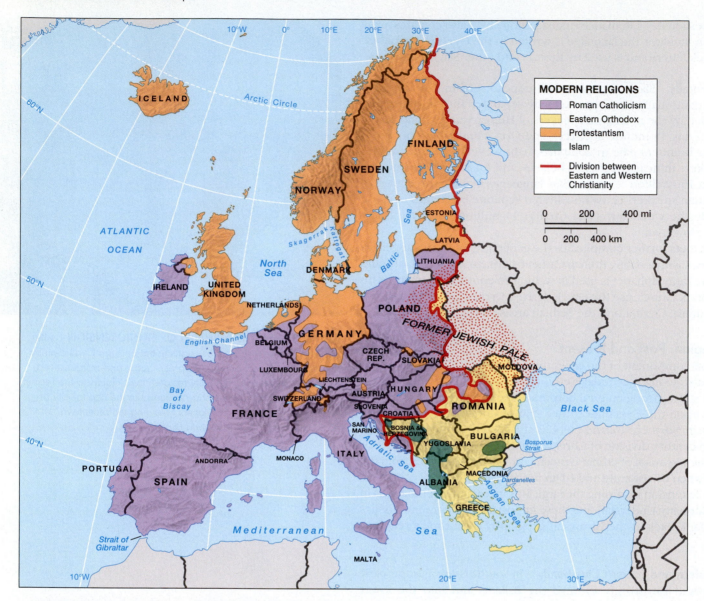

FIGURE 8.19 • RELIGIONS OF EUROPE This map shows the divide in western Europe between the Protestant north and the Roman Catholic south. Historically this distinction was much more important than it is today. Note the location of the former Jewish Pale, which was devastated by the Nazis during World War II. Today ethnic tensions with religious overtones are found primarily in the Balkans, where adherents to Roman Catholicism, Eastern Orthodoxy, and Islam live in close proximity.

divided the region ever since. With the exception of the troubles in Northern Ireland, tensions today between these two major groups are far less damaging than in the past.

Conflicts with Islam Both eastern and western Christian churches also struggled with challenges from Islamic empires to Europe's south and east. Even though historical Islam was reasonably tolerant of Christianity in its conquered lands, Christian Europe was not accepting of Muslim imperialism. The first crusade to reclaim Jerusalem from the Turks took place in 1095. After the Ottoman Turks took Constantinople in 1453 and gained control over the Bosporus strait and the Black Sea, they moved rapidly to spread their Muslim empire throughout the Balkans and arrived at the gates of Vienna in the middle of the sixteenth century. There, Christian Europe turned back the Turks. However, Ottoman control of southeastern Europe lasted until the empire's end in the early twentieth century. This historical presence of Islam explains the current mosaic of religions in the Balkans, with intermixed areas of Muslims, Orthodox, and Roman Catholics. Commonly, ethnic boundaries in that part of Europe are drawn along religious lines (Figure 8.20).

A Geography of Judaism   Europe has long been a difficult homeland for Jews after they were forced to leave Palestine during the Roman Empire. At that time, small Jewish settlements were found in cities throughout the Mediterranean. Later, by A.D. 900, about 20 percent of the Jewish population was clustered in the Muslim lands of the Iberian Peninsula, where Islam showed greater tolerance than Christianity had for Judaism. Furthermore, Jews played an important role in trade activities both within and outside of the Islamic lands. After the Christian reconquest of Iberia, however, Jews once more faced severe persecution and fled from Spain to more tolerant countries in western and central Europe.

One focus for migration was the area in eastern Europe that became known as the Jewish Pale. In the late Middle Ages, at the invitation of the Kingdom of Poland, Jews settled in cities and small villages in what is now eastern Poland, Belarus, western Ukraine, and northern Romania (see Figure 8.19). Jews collected in this region for several centuries in the hope of establishing a true European homeland, despite the poor natural resources of this marshy, marginal agricultural landscape.

Until emigration to North America began in the 1890s, 90 percent of the world's Jewish population lived in Europe. Most were clustered in the Pale. Even though many emigrants to the United States and Canada came from this area, the Pale remained the largest grouping of Jews in Europe until World War II. Tragically, Nazi Germany was able to use this ethnic clustering to its advantage by focusing extermination activities on this area.

In 1939, on the eve of World War II, there were 9.5 million Jews in Europe, or about 60 percent of the world's Jewish population. During the war German Nazis murdered some 6 million Jews in the horror of the Holocaust. Today fewer than 2 million Jews live in Europe. Since 1990 and the lifting of quotas on Jewish emigration from Russia, Belarus, and Ukraine, more than 100,000 Jews have emigrated to Germany, giving it the fastest-growing Jewish population outside of Israel.

The Patterns of Contemporary Religion   In Europe today there are about 250 million Roman Catholics and fewer than 100 million Protestants. Generally, Catholics are found in the southern half of the region, except for significant numbers in Ireland and Poland, while Protestants dominate in the north. Additionally, since World War II there has been a noticeable loss of interest in organized religion, mainly in western Europe, which has led to declining church attendance in many areas. This trend is so marked that the term **secularization** is used, referring to the widespread movement away from the historically important organized religions of Europe.

Catholicism dominates the religious geography and cultural landscapes of Italy, Spain, France, Austria, Ireland, and southern Germany. In these areas large cathedrals, monasteries, monuments to Christian saints, and religious place-names draw heavily on pre-Reformation Christian culture. Since visible beauty is part of the Catholic tradition, elaborate religious structures and monuments are more common than in Protestant lands.

Protestantism is most widespread in northern Germany, the Scandinavian countries, and England, and is intermixed with Catholicism in the Netherlands, Belgium, and Switzerland. Because of its reaction against ornate cathedrals and statues of the Catholic Church, the landscape of Protestantism is much more sedate and subdued. Large cathedrals and religious monuments in Protestant countries are associated primarily with the Church of England, which has strong historical ties to Catholicism; St. Paul's Cathedral and Westminster Abbey in London are examples.

Tragically, another sort of religious landscape has emerged in Northern Ireland, where barbed-wire fences and concrete barriers separate Protestant and Catholic neighborhoods in an attempt to reduce violence between warring populations (Figure 8.21). Religious affiliation is a major social force that influences where people live, work, attend school, shop, and so on. Of the 1.6 million inhabitants of North Ireland, 54 percent are Protestant and 42 percent Roman Catholic. The Catholics feel they have been discriminated against by the Protestant majority and have been treated as second-class citizens. As a result, they want a closer relationship with the Republic of Ireland across the border to the south. The Protestant reaction is to forge

FIGURE 8.20 • BALKAN LANDSCAPE   The urban landscape of Mostar, Bosnia-Herzegovina, shows the different religious affiliations of the local people. Muslim mosques, with their distinctive minarets, are found side by side with Roman Catholic and Eastern Orthodox churches. *(A. Ramey/Woodfin Camp & Associates)*

FIGURE 8.21 • RELIGIOUS TENSIONS IN NORTHERN IRELAND   Despite ongoing efforts to forge peace between Protestants and Catholics in Northern Ireland, outbreaks of violence are still common. Here, in Belfast, fighting between Catholics and Protestants broke out only hours after groups from both religions met to discuss ways to end the fighting. *(Cathal McNaughton/Getty Images, Inc.—Liaison)*

FIGURE 8.22 • TURKISH STORE IN GERMANY This Turkish store in the Kreuzberg district of Berlin, sometimes referred to as "Little Istanbul," serves not only the large Turkish population, but also German shoppers who enjoy this aspect of the country's increasing ethnicity. Unfortunately, to others, Europe's migrant cultures are less appealing. *(Stefano Pavasi/Matrix International, Inc.)*

even stronger ties with the United Kingdom. Unfortunately, these differences have been expressed in prolonged violence that has led to about 4,000 deaths since the 1960s, largely resulting from paramilitary groups on both sides who promote their political and social agendas through terrorist activities.

This religious and social strife goes back to the seventeenth century when Britain attempted to colonize Ireland and dilute Irish nationalism by subsidizing the migration and settlement of Protestants from England and Scotland. This strategy was most successful in the north, as that population chose to remain part of the United Kingdom when the rest of Ireland declared its independence from Britain in 1921. However, peace between the two religious groups was short-lived in Northern Ireland. The breakdown of law and order was so bad that British troops were ordered onto the streets of Belfast in 1969, leading to a military occupation that lasted for decades. Current peace talks have centered on three difficult issues: the Protestant insistence that their desire for a continued relationship with the United Kingdom be honored; the Catholic insistence on a closer relationship with Ireland that might lead to eventual unification with that country; and the need for some form of political power-sharing by these two groups. While mainstream paramilitary groups have declared a series of cease-fires, splinter groups from both sides continue to engage in terrorism.

European Culture in a Global Context

Europe, like all world regions, is currently caught up in a period of profound cultural change; in fact, many would argue that the pace of cultural change in Europe is accelerated because of the complicated interactions between globalization and Europe's internal agenda of political and economic integration. While newspaper analysts celebrate the "New Europe" of integration and unification, other critics refer to a more tension-filled New Europe of foreign migrants and guest workers troubled by ethnic discrimination and racism (Figure 8.22).

Globalization and Cultural Nationalism   Since World War II, Europe has been trying to control cultural contamination from North America. Some countries are more outspoken than others. While a few large countries, such as England and Italy, seem to accept the onrush of U.S. popular culture, other countries have expressed outright indignation over the corrupting impact of U.S. popular culture on speech, music, food, and fashion. France, for example, is often thought of as the poster child for the antiglobalization struggle as it fights to preserve local foods and culture against the homogenizing onslaught of fast food outlets and genetically engineered crops. Currently, the outspoken leader of French antiglobalization is José Bové, a sheep farmer who became internationally famous for trashing a McDonald's restaurant to protest a 100 percent tax placed on Roquefort cheese by the United States, which was the U.S. response to the EU's banning U.S. hormone-treated beef.

France, in fact, has taken cultural nationalism to new levels with laws protecting other cultural elements, such as music, films, and language. Radio stations, for example, must devote at least 40 percent of their air play to French songs and musicians, and French filmmakers are subsidized in the hope of preventing a complete takeover by Hollywood. The official body of the French Academy, whose job it is to legislate proper usage of the French language, has a long list of English and American words and phrases that are banned from use in official publications and from highly visible advertisements or billboards. Though the French commonly use "le weekend" or "le software" in their everyday speech, these words will never be found in official governmental speeches or publications.

Migrants and Culture   Migration patterns are also influencing the cultural mix in Europe. Historically, Europe spread its cultures worldwide through aggressive colonialization. Today, however, the region is experiencing a reverse flow as millions of migrants move into Europe, bringing their own distinct cultures from the far-flung countries of Africa, Asia, and Latin America. Unfortunately, in some areas of Europe, the products of this cultural exchange are highly problematic.

Ethnic clustering leading to the formation of ghettos is now common in the cities and towns of western Europe. The high-density apartment buildings of suburban Paris, for example, are home to large numbers of French-speaking Africans and Arab Muslims caught in the crossfire of high unemployment, poverty, and racial discrimination. Recent estimates are that there are 4.5 million Muslims in France, constituting 7.5 percent of the total population. Similarly, more than 2.5 million Muslim Turks live in Germany (see Figure 8.22). Added to that are increasing numbers of emigrants from eastern Europe and the former Soviet Union. As a result, cultural battles have emerged in many European countries. For example, French leaders, unsettled by the country's large Muslim migrant population, attempted to speed assimilation of female high school students into French mainstream culture by banning a key symbol of conservative Muslim life, the head scarf. This rule triggered riots, demonstrations, and counterdemonstrations. As a result of these kinds of conflicts, the political landscape of many European countries now contains far-right, nationalistic parties with thinly veiled agendas of excluding migrants from their countries

Geopolitical Framework: A Dynamic Map

One of Europe's unique characteristics is its dense fabric of 37 independent states within a relatively small area. No other world region demonstrates the same kind of geopolitical division. Europe invented the nation-state. Later though, these political ideas founded in Europe fueled the flames of political independence and democracy worldwide that replaced Europe's colonial rule in Asia, Africa, and the Americas.

Europe's diverse geopolitical landscape has been as much problem as promise. Twice in the last century Europe shed blood to redraw its political borders. Within the last decade alone, seven new states have appeared in eastern Europe, more than half through war. Further, today's map of geopolitical troubles suggests that still more political fragmentation may take place in the near future (Figure 8.23).

Most of Europe, however, sees a brighter geopolitical future. For many, this is based on a widespread spirit of cooperation through unification rather than continued fragmentation. This results mainly from the European Union's recent successes at political and economic integration since the sudden and unanticipated end of the Cold War in 1989. Many argue that the disasters of the twentieth century were of Europe's own making. If true, this region seems determined to avoid those mistakes in the twenty-first century by giving the world a new geopolitical model for peace and prosperity. The first decade of the new century should provide signs as to whether this optimism is well founded.

Redrawing the Map of Europe Through War

Two world wars redrew the geopolitical maps of twentieth-century Europe (Figure 8.24). Because of these conflicts, empires and nation-states have appeared and disappeared within the last 100 years. By the early twentieth century, Europe was divided into two opposing and highly armed camps that tested each other for a decade before the outbreak of World War I in 1914. France, Britain, and Russia were allied against the new nation-states of Italy and Germany, along with the Austro-Hungarian or Hapsburg Empire, which controlled a complex assortment of ethnic groups in central Europe and the Balkans. Though at the time World War I was referred to as the "war to end all wars," it fell far short of solving Europe's geopolitical problems. Instead, according to many experts, it made a further European land war unavoidable.

When Germany and Austria-Hungary surrendered in 1918, the Treaty of Versailles peace process set about redrawing the map of Europe with two goals in mind: to punish the losers through loss of territory and severe financial reparations and, second, to recognize the nationalistic aspirations of unrepresented peoples by creating new nation-states. As a result, the new states of Czechoslovakia and Yugoslavia were born. Additionally, Poland was reestablished, as were the Baltic states of Finland, Estonia, Latvia, and Lithuania.

Though the goals of the Treaty were admirable, few European states were satisfied with the resulting map. New states were resentful when their ethnic citizens

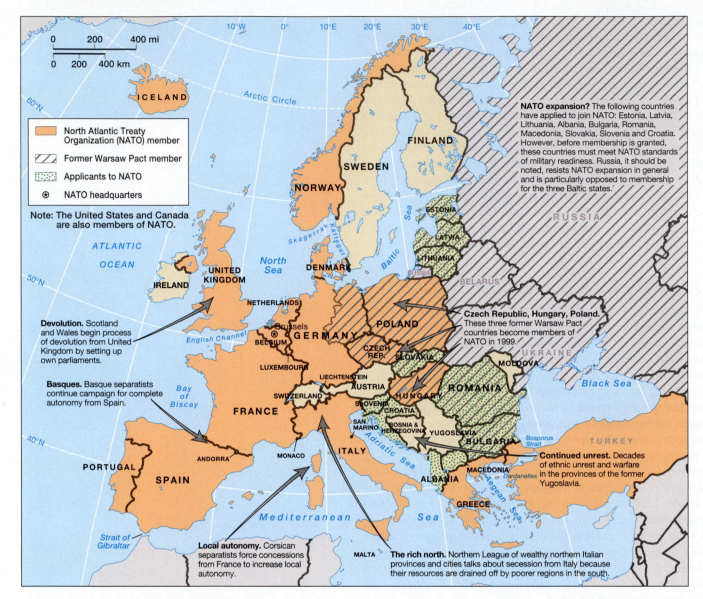

FIGURE 8.23 • GEOPOLITICAL ISSUES IN EUROPE   While the major geopolitical issue of the early twenty-first century remains the integration of eastern and western Europe in the European Union, numerous issues of ethnic nationalism could cause geopolitical fragmentation. In other parts of Europe, such as Spain, France, and Great Britain, issues of local ethnic autonomy challenge central governments.

were left outside the new borders and became minorities in other new states. This created an epidemic of **irredentism**, or state policies for reclaiming lost territory and peoples. Examples include the large German population in the western portion of the newly created state of Czechoslovakia, and the Hungarians stranded in western Romania by border changes.

This imperfect geopolitical solution was aggravated greatly by the global economic depression of the 1930s, which brought high unemployment, food shortages, and political unrest to Europe. Three competing ideologies promoted their own solutions to Europe's pressing problems: Western democracy (and capitalism), communism from the Soviet revolution to the east, and a fascist totalitarianism promoted by Mussolini in Italy and Hitler in Germany. With industrial unemployment commonly approaching 25 percent in western Europe, public opinion fluctuated wildly between extremist solutions of facisim and communism. In 1936 Italy and Germany once again joined forces through the Rome–Berlin "axis" agreement. As in World War I, this alignment was countered with mutual protection treaties between France, Britain, and the Soviet Union. When an imperialist Japan signed a pact with Germany, the scene was set for a second global war.

Nazi Germany tested Western resolve in 1938 by first annexing Austria, the country of Hitler's birth, and then Czechoslovakia under the pretense of providing protection for ethnic Germans located there. After signing a nonaggression pact with

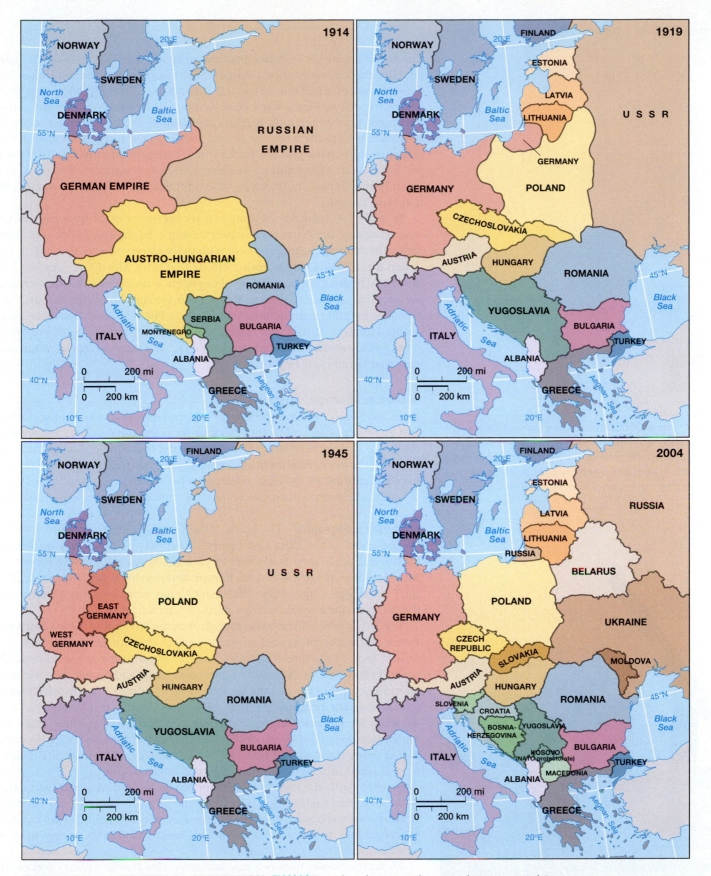

FIGURE 8.24 • A CENTURY OF GEOPOLITICAL CHANGE When the twentieth century began, central Europe was dominated by the German, Austro-Hungarian (or Hapsburg), and Russian empires. Following World War I, these empires were largely replaced by a mosaic of nation-states. More border changes followed World War II, largely as a result of the Soviet Union's turning that area into a buffer zone between itself and western Europe. With the fall of Soviet power in 1989, additional political division took place.

the Soviet Union, Hitler invaded Poland on September 1, 1939. Two days later France and Britain declared war on Germany. Within a month the Soviet Union moved into eastern Poland, the Baltic states, and Finland to reclaim territories lost through the peace treaties of World War I. Nazi Germany then moved westward and occupied Denmark, the Netherlands, Belgium, and France and began preparations to invade England. In 1941 the war took several startling new turns. In June, Hitler broke the nonaggression pact with the Soviet Union and, catching the Red Army by surprise, took the Baltic states and drove deep into Soviet territory. When Japan attacked Pearl Harbor, Hawaii, in December, the United States entered the war in both the Pacific and Europe.

By early 1944 the Soviet army had recovered most of its territorial losses and moved against the Germans in eastern Europe, beginning a long communist domination in that region. By agreement with the Western powers, the Red Army stopped when it reached Berlin in April 1945. At that time, Allied forces crossed the Rhine River and began their occupation of Germany. With Hitler's suicide, Germany signed an unconditional surrender on May 8, 1945, ending the war in Europe. But with Soviet forces firmly entrenched in the Baltics, Poland, Czechoslovakia, Bulgaria, Romania, Hungary, Austria, and eastern Germany, the military battles of World War II were quickly replaced by the ideological Cold War between communism and democracy that lasted until 1989.

A Divided Europe, East and West

From 1945 until 1989, Europe was divided into two geopolitical and economic blocs, east and west, separated by the infamous **Iron Curtain** that descended shortly after the peace agreement of World War II. East of the Iron Curtain border, the Soviet Union imposed the heavy imprint of communism on all activities—political, economic, military, and cultural. To the west, as Europe rebuilt from the destruction of the war, new alliances and institutions were created to counter the Soviet presence in Europe.

Cold War Geography   The seeds of the Cold War are commonly thought to have been planted at the Yalta Conference of February 1945, when Britain, the Soviet Union, and the United States met to plan the shape of postwar Europe. Since the Red Army was already in eastern Europe and moving quickly on Berlin, Britain and the United States agreed that the Soviet Union would occupy eastern Europe and the Western allies would occupy parts of Germany.

The larger geopolitical issue, though, was the Soviet desire for a **buffer zone** between its own territory and western Europe. This buffer zone consisted of an extensive bloc of satellite countries, dominated politically and economically by the Soviet Union, that would cushion the Soviet heartland against possible attack from western Europe. In the east the Soviet Union took control of the Baltic states, Poland, Czechoslovakia, Hungary, Bulgaria, Romania, Albania, and Yugoslavia. Austria and Germany were divided into occupied sectors by the four (former) allied powers. In both cases the Soviet Union dominated the eastern portion of each country, areas that contained the capital cities of Berlin and Vienna. Both capital cities, in turn, were divided into French, British, U.S., and Soviet sectors.

In 1955, with the creation of an independent and neutral Austria, the Soviets withdrew from their sector, effectively moving the Iron Curtain eastward to the Hungary–Austria border. This was not the case with Germany, however, which quickly evolved into two separate states (West Germany and East Germany) that remained apart until 1990.

Along the border between east and west, two hostile military forces faced each other for almost half a century. Both sides prepared for and expected an invasion by the other across the barbed wire of a divided Europe (Figure 8.25). In the west, NATO (the North Atlantic Treaty Organization) forces, including the United States, were stationed from West Germany south to Turkey. To the east, Warsaw Pact forces were anchored by the Soviets but also included small military units from satellite countries. Both NATO and the Warsaw Pact countries were armed with nuclear weapons, making Europe a tinderbox for another world war.

FIGURE 8.25 • THE IRON CURTAIN   The former Czechoslovakian–Austrian border was marked by a barbed-wire fence as part of an extensive border zone that included mine fields, tank barricades, watch towers, and a security zone in which all trespassers were shot on sight. *(James Blair/NGS Image Collection)*

(a)

(b)

FIGURE 8.26 • THE BERLIN WALL   In August 1961, the East German and Soviet armies built a concrete and barbed-wire structure, known simply as "the Wall," to stem the flow of East Germans leaving the Soviet zone. It was the most visible symbol of the Cold War until November 1989, when Berliners physically broke it apart after the Soviet Union gave up its control over eastern Europe. *[(a) Bettmann/Corbis; (b) Anthony Suau/Getty Images, Inc.—Liaison]*

Berlin was the flashpoint that brought these forces close to a fighting war on two occasions. In winter 1948 the Soviets imposed a blockade on the city by denying Western powers access to Berlin across its East German military sector. This attempt to starve the city into submission by blocking food shipments from western Europe was thwarted by a nonstop airlift of food and coal by NATO. Then, in August 1961 the Soviets built the Berlin Wall to curb the flow of East Germans seeking political refuge in the west. The Wall became the concrete-and-mortar symbol of a firmly divided postwar Europe. For several days while the Wall was being built and the West agonized over destroying it, NATO and Warsaw Pact tanks and soldiers faced each other with loaded weapons at point-blank range. Though war was avoided, the Wall stood for 28 years, until November 1989.

The Cold War Thaw   The symbolic end of the Cold War in Europe came on November 9, 1989, when East and West Berliners joined forces to rip apart the Wall with jackhammers and hand tools (Figure 8.26). By October 1990 East and West Germany were officially reunified into a single nation-state. During this period, all other Soviet satellite states, from the Baltic to the Black Sea, also underwent major geopolitical changes that have resulted in a mixed bag of benefits and problems. While some, like the Czech Republic, appear to have made a successful transformation to democracy and capitalism, others, such as Romania, appear stalled politically and economically in their search for new directions.

The Cold War's end came as much from a combination of problems within the Soviet Union (discussed in Chapter 9) as from rebellion in eastern Europe. By the mid-1980s, the Soviet leadership was advocating an internal economic restructuring and also recognizing the need for a more open dialogue with the West. Financial problems from supporting a huge military establishment, along with heavy losses from an unsuccessful war in Afghanistan, lessened the Soviet appetite for occupying other countries.

In August 1989 Poland elected the first noncommunist government to lead an eastern European state since World War II. Following this, with just one exception, peaceful revolutions with free elections spread throughout eastern Europe as communist governments renamed themselves and broke with doctrines of the past. In Romania, though, street fighting between citizens and military resulted in the violent overthrow and execution of the communist dictator, Nicolae Ceaucescu.

In October 1989, as Europeans nervously awaited a Soviet response to developments in eastern Europe, President Mikhail Gorbachev said that the Soviet Union had no moral or political right to interfere in the domestic affairs of eastern Europe. Following this statement, Hungary opened its borders to the west, and eastern Europeans freely visited and migrated to western Europe for the first time in 50 years. Estimates are that more than 100,000 East Germans fled their country during this time.

As a result of the Cold War thaw, the map of Europe began changing once again. Germany reunified in 1990. Elsewhere, political separatism and ethnic nationalism, long suppressed by the Soviets, were unleashed in southeastern Europe. Yugoslavia broke into the independent states of Slovenia, Croatia, and Bosnia. On January 1, 1993, Czechoslovakia was replaced by two separate states, the Czech Republic and Slovakia. Although the twentieth century ended with a largely united Europe from a geopolitical perspective, the challenge of the early twenty-first century is lessening the economic and social disparities between east and west.

Economic and Social Development: Integration and Transition

As the acknowledged birthplace of the Industrial Revolution, Europe in many ways invented the modern economic system of industrial capitalism. Though Europe was the world's industrial leader in the early twentieth century, it was soon eclipsed by Japan and the United States while Europe struggled to cope with the effects of two world wars, a decade of global depression, and the Cold War.

In the last 40 years, however, economic integration guided by the European Union has been increasingly successful. In fact, Western Europe's success at blending national economies has given the world a new model for regional cooperation, an approach that may be imitated in Latin America and Asia in the twenty-first century. Eastern Europe, however, has not fared so well. The results of four decades of Soviet economic planning were, at best, mixed. The total collapse of that system in 1990 cast eastern Europe into a period of chaotic economic, political, and social transition that may result in a highly differentiated pattern of rich and poor regions. While some regions like the Czech Republic prospers, others like Bulgaria and Romania stagnate.

Accompanying western Europe's economic boom has been an unprecedented level of social development as measured by worker benefits, health services, education, literacy, and gender equality. Though the improved social services set an admirable standard for the world, cost-cutting politicians argue that these services now increase the cost of business so that European goods cannot compete in the global marketplace.

The early twenty-first century brings Europe renewed hope that it will once again be a world economic superpower. Yet this region also faces new and difficult challenges. In terms of economic and social development, it is at a major crossroads.

Europe's Industrial Revolution

Europe is the cradle of modern industrialism. Two fundamental changes were associated with this industrial revolution. First, machines replaced human labor in many manufacturing processes, and, second, inanimate energy sources, such as water, steam, electricity, and petroleum, powered the new machines. Though we commonly apply the term *Industrial Revolution* to this transformation, implying rapid change, in reality it took more than a century for the interdependent pieces of the industrial system to come together. This new system emerged first in England between 1730 and 1850. Later, this new industrialism spread to other parts of Europe and the world.

Centers of Change England's textile industry, located on the flanks of the Pennine Mountains, was the center of early industrial innovation, which took place in small towns and villages away from the rigid control of the urban guilds. The town of Yorkshire, on the eastern side of the Pennines, had been a center of woolen textile making since medieval times, drawing raw materials from the

FIGURE 8.27 • WATER POWER AND TEXTILES Europe's industrial revolution began on the slopes of England's Pennine Mountains, where swift-running streams and rivers were used to power large cotton and wool looms. Later many of these textile plants switched to coal power. *(James Marshall/Corbis)*

extensive sheep herds of that region and using the clean mountain waters to wash the wool before it was spun in rural cottages. By the 1730s, water wheels were used to power mechanized looms at the rapids and waterfalls of the Pennine streams (Figure 8.27). By the 1790s, the steam engine had become the preferred source of energy to drive the new looms. However, steam engines needed fuel, and only with the building of railroads after 1820 could coal be moved long distances at a reasonable cost.

Development of Industrial Regions in Continental Europe
By the 1820s, the first industrial districts had begun appearing in continental Europe. These hearth areas were, and still are, near coalfields (Figure 8.28). The first area outside of Britain was the Sambre-Meuse region, named for the two river valleys straddling the French-Belgian border. Like the English Midlands, it also had a long history of cottage-based wool textile manufacturing that quickly converted to the new technology of

FIGURE 8.28 • INDUSTRIAL REGIONS OF EUROPE From England, the Industrial Revolution spread to continental Europe, starting with the Sambre-Meuse region on the French-Belgian border, then diffusing to the Ruhr area in Germany. Readily accessible surface coal deposits powered these new industrial areas. Early on, iron ore for steel manufacture came from local deposits, but later it was imported from Sweden and other areas in the shield country of Scandinavia. Most of the newer industrial areas are closely linked to urban areas.

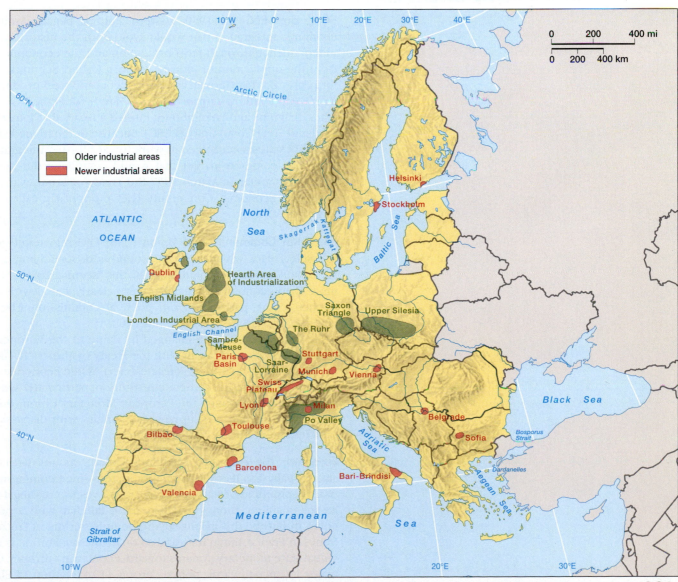

221

FIGURE 8.29 • THE RUHR INDUSTRIAL LANDSCAPE Long the dominant industrial region in Europe, the Ruhr region was bombed heavily during World War II because of its central role in providing heavy arms to the German military. It has been rebuilt, has modernized, and remains competitive with newer industrial regions. *(Andrej Reiser/Bilderberg Archiv der Fotografen)*

steam-powered mechanized looms. Additionally, a metalworking tradition drew on charcoal-based iron foundaries in the nearby forests of the Ardenne Mountains. Coal was also found in these mountains and in 1823 the first blast furnace outside of Britain started operation in Liege, Belgium.

By the second half of the nineteenth century, the dominant industrial area in all Europe (including England) was the Ruhr district in northwestern Germany, near the Rhine River. Rich coal deposits close to the surface fueled the Ruhr's transformation from a small textile region to one oriented around heavy industry, particularly iron and steel manufacturing. By the early 1900s, the Ruhr had used up its modest iron ore deposits and was importing ore from Sweden, Spain, and France. Several decades later the Ruhr industrial region became synonymous with the industrial strength behind Nazi Germany's war machine and thus was bombed heavily in World War II (Figure 8.29).

Rebuilding Postwar Europe: Economic Integration in the West

Europe was unquestionably the leader of the industrial world in the early twentieth century. More specifically, before World War I, European industry was estimated to produce 90 percent of the world's manufactured output. However, four decades of political and economic chaos and two world wars left Europe divided and in shambles. By mid-century, its cities were in ruins; industrial areas were destroyed; vast populations were dispirited, hungry, and homeless; and millions of refugees moved about Europe looking for safety and stability.

ECSC and EEC In 1950 western Europe began discussing a new form of economic integration that would avoid both the historical pattern of nationalistic independence through tariff protection and the economic inefficiencies resulting from the duplication of industrial effort. Robert Schuman, France's foreign minister, proposed that German and French coal and steel production be coordinated by a new supranational organization. In May 1952, France, Germany, Italy, the Netherlands, Belgium, and Luxembourg ratified a treaty that joined them in the European Coal and Steel Community (ECSC). Because of the immediate success of the ECSC, these six states soon agreed to work toward further integration by creating a larger European common market that would encourage the free movement of goods, labor, and capital. In March 1957 the Treaty of Rome was signed, establishing the European Economic Community (EEC), popularly called the *Common Market* (Figure 8.30).

European Community and Union The EEC reinvented itself in 1965 with the Brussels Treaty, which laid the groundwork for adding a political union to the successful economic community. In this "second Treaty of Rome," aspirations for more than economic integration were clearly stated with the creation of an EEC council, court, parliament, and political commission. The EEC also changed its name to the European Community (EC).

In 1991, the EC again changed its name to the European Union (EU) and expanded its goals once again with the Treaty of Maastricht (named after the town in the Netherlands in which delegates met). While economic integration remains an underlying theme in the new constitution, particularly with its commitment to a single currency through the European Monetary Union (discussed below), the EU has moved further into supranational affairs with discussion of common foreign policies and mutual security agreements.

Euroland: The European Monetary Union As any world traveler knows, each state usually has its own monetary system since coining money has long been a basic component of state sovereignty. As a result, crossing a political border usually means changing money and becoming familiar with a new system of bills and coins. In Europe this meant buying pounds sterling in England, marks in Germany, francs in France, lira in Italy, and so on. However, Europe is now moving from individual state monetary systems toward a common currency. On January 1, 1999, in a major advance toward a united Europe, 11 of the then 15 EU member states joined

Legend:
- Current members of the European Union (EU)
- 1995 Date of entry into the European Union
- Current applicants to the European Union
- Current members of the European Monetary Union ('Euroland')
- Not a European Union (EU) member

FIGURE 8.30 • THE EUROPEAN UNION The driving force behind Europe's economic and political integration has been the European Union (EU), which began in the 1950s as an organization focused solely on rebuilding the region's coal and steel industries. After that proved successful, the Common Market (as it was then called) expanded into a wider range of economic activities. Today the EU has 25 members.

in the European Monetary Union (EMU). As of that day, cross-border business and trade transactions began taking place in the new monetary unit, the *euro*. Then, on January 1, 2002, new euro coins and bills became available for everyday use. This currency completely replaced the different national currencies of **Euroland** (countries belonging to the EMU) member states in July 2002.

By adopting a common currency, Euroland members expect to increase the efficiency and competitiveness of both domestic and international business. Formerly, when products were traded across borders, there were transaction costs associated with payments made in different currencies. These are now eliminated within Euroland. Germany, for example, exports two-thirds of its products to other EU members. With a common currency, this business now becomes essentially domestic trade, protected from the fluctuations of different currencies and without transaction costs.

Economic Integration, Disintegration, and Transition in Eastern Europe

Through history, the economy of eastern Europe has been less developed than its western counterpart. This is explained in part by the fact that eastern Europe is not particularly rich in natural resources. It has only modest amounts of coal, less iron ore, and little in terms of oil and natural gas. Furthermore, what few resources have been found in the region have been historically exploited by outside interests, including the Ottomans, Hapsburgs, Germans, and, more recently, the Soviet Russians. In many ways eastern Europe has been an economic colony of the more-developed states to its west, south, and east.

FIGURE 8.31 • POLISH AGRICULTURE One of the most problematic issues facing eastern European countries is the modernization of the agricultural sector. This problem is particularly pressing now that eastern European countries are in the European Union and must compete with western European farmers. (Hans Madej/Bilderberg Archiv der Fotografen)

The Soviet-dominated economic planning of the postwar period (1945–91) was an attempt to develop eastern Europe by coordinating regional economies in a way that also served Soviet interests. When the Soviets took control of eastern Europe, their goals were complete economic, political, and social integration through a **command economy**, one that was centrally planned and controlled. However, the collapse of that centralized system in 1991 threw many eastern European countries into deep economic, political, and social chaos. Thus, the present-day problems of the region are closely tied to economic activities during the Soviet era.

The Results of Soviet Economic Planning After 40 years of communist economic planning, the results were mixed and varied widely within eastern Europe. In Poland and Yugoslavia, for example, many farmers strongly resisted the national ownership of agriculture; thus, most productive land remained in private hands. However, in Romania, Bulgaria, Hungary, and Czechoslovakia, 80 to 90 percent of the agricultural sector was converted to state-owned communal farms. Across eastern Europe, despite these changes in agricultural structure, food production did not increase dramatically, and, in fact, food shortages became commonplace (Figure 8.31).

Perhaps most notable during the Soviet period were dramatic changes to the industrial landscape, with many new factories built in both rural and urban areas that were fueled by cheap energy and raw materials imported from the Soviet Union. As a means to economic development in eastern Europe, Soviet planners chose heavy industry, such as steel plants and truck manufacturing, over consumer goods. As in the Soviet Union, the bare shelves of retail outlets in eastern Europe became both a sign of communist shortcomings and, perhaps more important, a source of considerable public tension. As western Europeans enjoyed an increasingly high standard of living with an abundance of consumer goods, eastern Europeans struggled to make ends meet as the utopian vision promised by Soviet communists became increasingly elusive.

Transition and Turmoil Since 1991 As Soviet domination over eastern Europe collapsed in the late 1980s, so did the forced economic integration of the region. In place of Soviet coordination and subsidy has come a painful period of economic transition as eastern European countries rebuild and reorient their national economies.

Most countries are redirecting their economic and political attention to western Europe with the clear goal of joining the EU sometime in the twenty-first century. To link up with the West, many have moved away from a socialist-based economy of state ownership and control toward a capitalist economy of private ownership and free markets. To achieve this, states such as the Czech Republic, Hungary, and Poland have gone through a period of **privatization**, which is the transfer to private ownership of those firms and industries previously owned and run by state governments.

Eastern Europeans have witnessed many changes as a result of this shift in economic policy. Price supports, tariff protection against imports, and subsidies have been removed from consumer goods as countries have made the transition to a free market. For the first time, goods from western Europe and other parts of the world are plentiful and common in eastern European retail stores. However, this prolonged period of economic change has taken its toll on eastern European consumers, most of who cannot afford these long-dreamed-of products. Unemployment has often risen to more than 15 percent, and underemployment is common. Financial security is difficult, with irregular paychecks for those with jobs and uncertain welfare benefits for those without. Furthermore, for most people the basic costs of food, rent, and utilities are higher under a free market system than under the subsidized economies of communism.

The causes of this economic pain are complex. As the Soviet Union turned its attention to its own economic and political turmoil in 1991, it stopped exporting cheap natural gas and petroleum to eastern Europe. Instead, Russia sold these fuels on the open global market to gain hard currency. Without cheap energy, many eastern European industries were unable to operate and shut down operations, laying off thousands of workers. In the first two years of the transition (1990–92), industrial production fell 35 percent in Poland and 45 percent in Bulgaria (Figure 8.32). In addition, markets guaranteed under a command economy, many of them in the Soviet Union, evaporated. Consequently, many factories and services closed because they lacked a market for their goods. In some cases products from the West, particularly beverages, clothes, and cigarettes, moved in quickly to fill this void.

With the privatization of industry, services, and property in eastern Europe, state subsidies ended. As a result, the cost of living increased because of higher prices for goods, services, and rent. Free-market advocates point out that this should be a temporary stage and that increased competition should drive down prices. High inflation rates have also troubled eastern European populations in recent years, adding to the cost-of-living increases. Food supplies remain problematic since agricultural production decreased during the transition. A number of factors explain this decline, including shortages of fuel to run farm equipment, uncertainty about land ownership and access to fields in former state or communal farms, and the inefficiency of small farms where privatization has taken place. In Bulgaria, for example, which previously had 90 percent of its productive land in state or communal farms, most of the new privatized farms are smaller than 2.5 acres (1 hectare).

FIGURE 8.32 • POST-1989 HARDSHIP With the fall of communism in eastern Europe after 1989, economic subsidies and support from the Soviet Union ended. Accompanying this transition has been a high unemployment rate resulting from the closure of many industries, such as this plant in Bulgaria. *(Rob Crandall/Rob Crandall, Photographer)*

CONCLUSION

The twentieth century brought numerous challenges to Europe. Disruption, even chaos, characterized many decades as a result of two world wars, a handful of minor ones, and almost half a century of political and military division during the Cold War. The scars from these hardships are still apparent, particularly in the east and southeast. Most of Europe, however, met the twenty-first century with optimism and confidence.

In terms of environmental issues, western Europe has made great progress in the last several decades. Not only have individual countries enacted strong environmental laws, but the EU has played an important role with its strong commitment to regional environmental solutions. Additionally, the EU has become an aggressive environmental advocate at the global level. This is illustrated by its participation in global climate conferences and treaties where it has demonstrated a willingness to cut back its atmospheric emissions far more than any other major world region.

The region also faces ongoing challenges related to population and migration. For example, slow and no growth is being addressed by individual countries. The most pressing problem is how Europe deals with in-migration from Asia, Africa, Latin America, former Soviet lands, and its own underdeveloped regions. If Europe's economy booms once again, this migration could be the solution to a new labor shortage. But if it stagnates, migration will continue to complicate a large number of issues throughout the region.

Cultural tension will remain one of the problems caused by migration. As the bulk of Europeans embrace a common culture of sorts, the cultural differences between Europeans and outsiders will increase. While there is little doubt the cultural geography of Europe will change profoundly in the next few years, how Europeans will react to the social changes brought about by migration is unclear.

With the end of the Cold War, Europe's geopolitical issues are continually being redefined.

More specifically, in the last two decades several new countries have appeared on the European map. Another theme is the integration of the former Soviet satellites in eastern Europe in western European—and even international—geopolitics. Twenty-first-century Europeans must also address issues of economic and social development. Much uncertainty exists as the EU implements its common monetary policy and expands into eastern Europe. While it is easy to envision a positive future for the newly enlarged EU that includes member states in eastern Europe like Poland, Hungary, and the Czech Republic, it is much more difficult to think of a positive economic, political, and social future for Moldova, Albania, Bulgaria, and Romania. Sharing the EU's twenty-first-century dream with these eastern European countries will constitute political and economic challenge for Europe. If this challenge is met, the future seems bright for this dynamic world region.

KEY TERMS

The Russian Domain

FIGURE 9.1 The Russian domain

Russia and its neighboring states of Belarus, Ukraine, Georgia, and Armenia make up a dynamic and unpredictable world region that has experienced tremendous change in the past decade. Sprawling from the Baltic Sea to the Pacific, the region includes huge industrial centers, vast farmlands, and almost-empty stretches of tundra. New global connections are on the rise, but it remains unclear how this region will participate in the evolving world economy.

Elevation in meters

4000+
2000–4000
500–2000
200–500
0–200
Below sea level
Sea Level

RUSSIAN DOMAIN
Political Map

⊗ ● Over 1,000,000
○ • 500,000–1,000,000 (selected cities)
* · Selected smaller cities

0   250   500 mi
0   250   500 km

SETTING THE BOUNDARIES

The boundaries of the Russian domain have shifted over time. For decades the regional definitions were relatively easy because the highly centralized Soviet Union (or Union of Soviet Socialist Republics) dominated the region's political geography. Most geographers agreed that the Soviet Union could thus be viewed as a single world region. Although the nation contained many cultural minorities, the Soviet Union wielded remarkable political and economic power. After World War II, some geographers even included much of Soviet-dominated eastern Europe within the region in response to the country's expanded military role in nations such as East Germany, Poland, and Hungary.

The maps were suddenly redrawn late in 1991. The once-powerful Soviet state was officially dissolved, and in its place stood 15 former "republics" that had once been united under the Soviet Union. Now independent, each of these republics has tried to make its own way in a post-Soviet world. While some geographers initially treated the region as the "Former Soviet Union," it quickly became clear that diverse cultural forces, economic trends, and political orientations were taking the republics in different directions. Even so, the Russian Republic remained dominant in size and area and thus came to form the core of a new Russian domain that was considerably smaller than the Soviet Union and yet included some characteristics shared by Russia and several of its neighboring states.

The new regional definition reflects the changing political and cultural map since the breakup of the Soviet Union. The term *domain* suggests persisting Russian influence within the four other nations included in the region. Russia, Ukraine, and Belarus make up the core of the new region. Enduring cultural and economic ties closely connect these three countries. In addition, Georgia and Armenia, while more culturally distinctive, are still best classified within a zone of Russian influence. In fact, Armenia and Russia recently signed a treaty of "friendship, cooperation, and mutual understanding" that will maintain Russian military bases in the country for another 25 years. While Georgia has had more stressful relations with its giant neighbor, Russian troops also remain stationed in that former Soviet republic. Two significant areas that were once a part of the Soviet Union have been eliminated from the domain. The mostly Muslim republics of Central Asia and the Caucasus (Kazakstan, Uzbekistan, Kyrgyzstan, Turkmenistan, Tajikistan, and Azerbaijan) have become aligned with a Central Asia world region (Chapter 10), while the Baltic republics (Estonia, Latvia, and Lithuania) and Moldova are best grouped with Europe (Chapter 8).

The Russian domain sprawls across the vast northern half of Eurasia and includes not only Russia itself, but also the nations of Ukraine, Belarus, Georgia, and Armenia (see Figure 9.1; see also "Setting the Boundaries"). The land is rich with superlatives: endless Siberian spaces, unlimited natural resources, legends of ruthless Cossack warriors, and tales of epic wars and revolutions are all part of the region's geographical and historical mythology. Indeed, the rise of Russian civilization remarkably parallels the story of the United States. Both cultures grew from small beginnings to become imperial powers that benefited from the fur trade, gold rushes, and transcontinental railroads during the nineteenth century. In addition, both countries experienced dramatic change resulting from industrialization in the twentieth century.

Recently, however, the parallels have ended: Incredible political and economic changes have rocked the Russian-dominated region, and its immediate future remains uncertain. The challenges are indeed overwhelmingly daunting. Economic collapse across much of the region in the late 1990s produced steep declines in living standards. The region is politically unstable as a result of tensions between neighboring states, as well as within the countries themselves. The region also faces some of the most severe environmental challenges anywhere in the developed world.

Globalization is shaping the Russian domain in complex ways. The region's relationship with the rest of the world shifted dramatically during the last 10 years of the twentieth century. Until the end of 1991, all five countries belonged to the Soviet Union, the world's most powerful communist state. Under Soviet control, the region's twentieth-century economy saw large increases in industrial output that made the nation a major global producer of steel, weaponry, and petroleum products. Its communist system offered a powerful set of ideas that promised economic prosperity and hope to residents within the region and beyond. Indeed, the political and military reach of the Soviet Union spanned the globe, making it a superpower on par with the United States. The Soviet presence dominated many eastern European countries, and nations from Cuba to Vietnam enjoyed close strategic ties with the country.

Suddenly, as the old communist order evaporated early in the 1990s, the now-independent republics of Russia, Ukraine, Belarus, Georgia, and Armenia had to carve out new regional and global relationships. With the breakdown of Soviet control, the region also felt the growing presence of western European and American influences. Westernized popular culture, as well as both modest and radical economic changes, accompanied the political transformations within the realm. The region's economic stability has been threatened by its exposure to the competitive pressures of the global economy. The result is a world region that has seen its global linkages radically redefined in the recent past.

Slavic Russia (population 145 million) dominates the region. Although only about three-quarters the size of the former Soviet Union, Russia's dimensions still make it the largest state on Earth. West of Moscow, the country's European front borders Finland and Poland, while far to the east the nations of Mongolia and China share a thinly peopled boundary with sprawling Russian Siberia. Its area of 6.6 million square miles (17 million square kilometers) dwarfs even Canada, and its 11 time zones are a reminder that dawn in Vladivostok on the Pacific Ocean is still only dinnertime in Moscow. With the fall of the Soviet Union, the Russians ended almost 75 years of Marxist rule. What comes next is difficult to predict. Will Russia, spurred by its numerous ethnic minorities, further fragment into smaller political units, or might it consolidate its power and reassert direct control over neighboring states such as Belarus and Ukraine?

Arguably, stability in the region may be a long way off, but two things are certain. First, because of its size and location, change within Russia will inevitably shape political and economic geographies far beyond its borders. Second, even with its present problems, Russians share a long history and common cultural heritage. They also possess a rich storehouse of natural resources (particularly in energy) that can bind the nation together and provide the foundation for future political stability and economic development.

The bordering states of Ukraine, Belarus, Georgia, and Armenia will inevitably be linked to the evolution of their giant neighbor, even as they attempt to make their own way as independent nations. Emerging from the shadows of Soviet dominance has been difficult. Ukraine, in particular, has the size, population, and resource base to become a major European nation, but it has struggled to create real political and economic change since independence. With about 50 million people and a rich storehouse of resources, Ukraine's size of 233,000 square miles (604,000 square kilometers) is similar to that of France. Nearby Belarus is smaller (80,000 square miles or 208,000 square kilometers), and its population of 10 million is likely to remain more closely tied economically and politically to Russia. South of Russia and beyond the bordering Caucasus Mountains, the Transcaucasian countries of Armenia and Georgia are smaller still. Their populations depart culturally from their Slavic neighbor to the north. In addition, these two nations face significant political challenges: Armenia shares a hostile border with Azerbaijan (see Chapter 10), and Georgia's ethnic diversity threatens its political stability.

Environmental Geography: A Vast and Challenging Land

The Russian domain faces a number of serious and somewhat frightening environmental issues, one of which concerns the Caspian and Black Sea sturgeon. Recently the World Wildlife Federation warned that several varieties of this valuable caviar-producing fish were in danger of extinction. Indeed, the annual sturgeon harvest has fallen to barely 3 percent of the levels seen in the late 1970s. The culprit has been the growing global black market for the exotic Russian delicacy, combined with few domestic regulations that effectively protect the caviar beds of the Caspian Sea, the Volga River Delta, and the Sea of Azov (a part of the Black Sea). Criminals in the Russian mafia, looking for a quick profit on the illegal sale of sturgeon, have also been central players in this ecological disaster. Mafia-supported poaching operations and transport connections have resulted in the rapid depletion of the sturgeon as the illegal goods flow north to Moscow and onward to even larger international markets. The vanishing fish exemplifies the larger predicament of the Russian domain's devastated environment. Caught by inadequate legal protections, a global economic setting that encourages its continued harvest, and the powerful presence of the mafia, the luckless fish is being sacrificed to turn a quick profit. Sadly, the same can be said for much of the region's natural resource base.

Still, the Russian domain possesses a unique physical setting, characterized by its huge size, northern latitude, and varied topography (Figure 9.2). The fact that the region occupies a major portion of the world's largest landmass means that huge distances separate people and resources. But size is also a blessing: The domain's varied natural resources are widely distributed across the realm, and the region is still home to some of Eurasia's best farmlands, metals resources, and petroleum reserves.

The region's northern latitudinal position is critical to understanding basic geographies of climate, vegetation, and agriculture. Indeed, the Russian domain provides the world's largest example of a high-latitude continental climate where seasonal temperature extremes and short growing seasons greatly limit opportunities for human settlement. In terms of latitude, Moscow is positioned as far north as Ketchikan, Alaska, and even the Ukrainian capital of Kiev (Kyiv) would sit north of the Great Lakes in Canada. Thus, apart from a subtropical zone near the Black Sea, much of the region experiences a classic continental climate with hard, cold winters and marginal agricultural potential (Figure 9.3).

The European West

An airplane flight over the western portions of the Russian domain would reveal a vast, barely changing landscape below. European Russia, Belarus, and Ukraine cover the eastern portions of the vast European Plain, which runs from southwest France

FIGURE 9.2 • PHYSICAL GEOGRAPHY OF THE RUSSIAN DOMAIN   In the west, the European Plain stretches across Belarus and the Russian heartland to the Urals. In the far south, the rugged Caucasus Mountains impose significant barriers to movement. East of the Urals, more large plains and plateaus alternate with mountains, particularly in eastern Siberia, where higher peaks and volcanic uplifts further isolate the region from the world beyond.

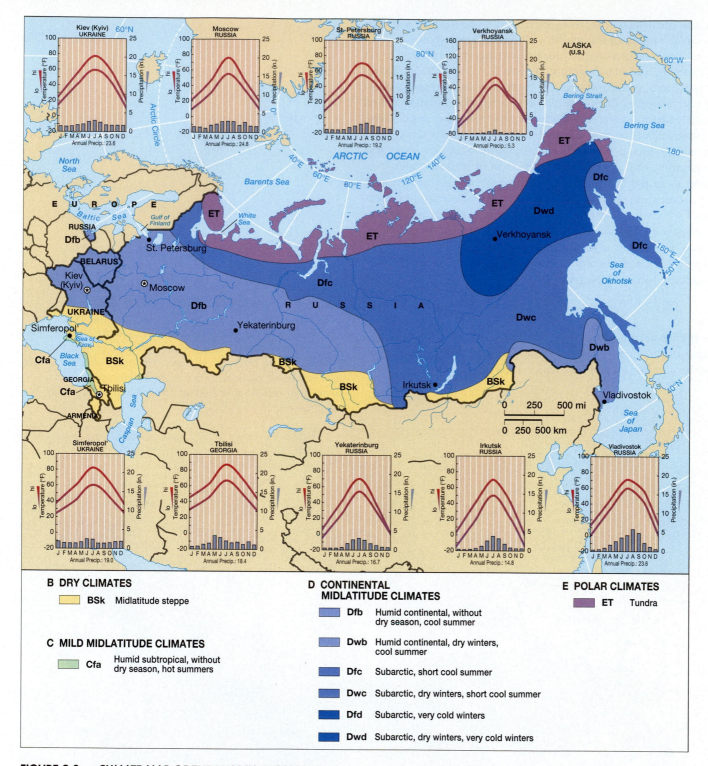

FIGURE 9.3 • CLIMATE MAP OF THE RUSSIAN DOMAIN
The region's northern latitude and large landmass suggest that continental climates dominate. Indeed, farming is greatly limited by short growing seasons across much of the region. Aridity imposes limits elsewhere. Only a few small zones of mild midlatitude climates are found on the warming shores of the Black Sea in the far southwestern corner of the region, producing subtropical conditions in western Georgia.

to the Ural Mountains. One of the major geographical advantages of European Russia is the fact that different river systems, all now linked by canals, flow into four separate drainages. The result is that trade goods can easily flow in many directions within and beyond the region. The Dnieper and Don rivers flow into the Black Sea; the West and North Dvina rivers drain into the Baltic and White seas, respectively; and the Volga River runs to the Caspian Sea (Figures 9.2 and 9.4).

Even though European Russia has a milder climate than Siberia, most of it experiences cold winters and cool summers by North American standards. Moscow, for example, is about as cold as Minneapolis in January, yet not nearly as warm in

FIGURE 9.4 • VOLGA VALLEY The city of Nizhniy Novgorod lies along the shores of the Volga River east of Moscow. The huge Volga Basin remains a center of Russian settlement and economic development. It drains much of western Russia and empties into the Caspian Sea. *(Tass/Sovfoto/Eastfoto)*

July. In Ukraine, Kiev is milder, however, and Simferopol', near the Black Sea, offers wintertime temperatures that average more than 20°F warmer than those of Moscow.

Three distinctive environments shape agricultural potential in the European West (Figure 9.5). North of Moscow and St. Petersburg, poor soils and cold temperatures severely limit farming. Belarus and central portions of European Russia possess longer growing seasons, but acidic **podzol soils**, typical of northern forest environments, limit agricultural output and the ability of the region to support a highly productive farm economy. Still, diversified agriculture includes grain (rye, oats, and wheat) and potato cultivation, swine and meat production, and dairying. South of 50° latitude, agricultural conditions improve across much of southern Russia and Ukraine. Forests gradually give way to steppe environments dominated

FIGURE 9.5 • AGRICULTURAL REGIONS Harsh climate and poor soils combine to limit agriculture across much of the Russian domain. Better farmlands are found in Ukraine and in European Russia south of Moscow. Portions of southern Siberia support wheat production, but yield marginal results. In the Russian Far East, warmer climates and better soils translate into higher agricultural productivity. *(Modified from Clawson and Fisher, 1998, World Regional Geography, Upper Saddle River, NJ: Prentice Hall)*

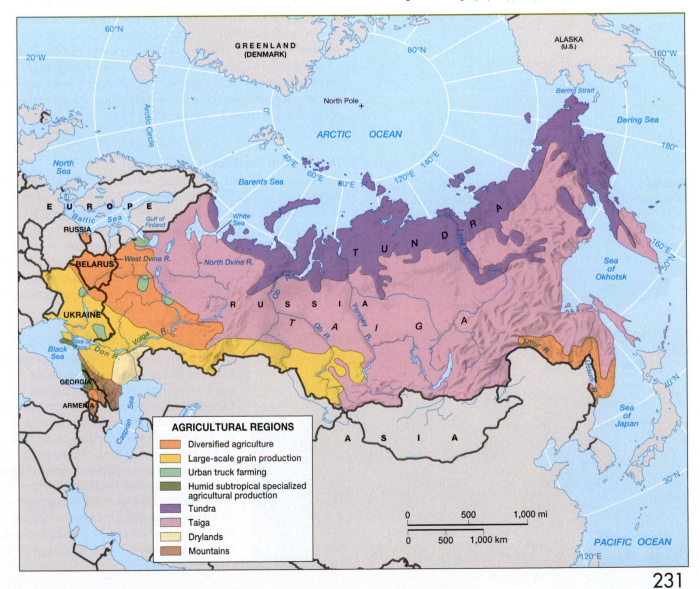

231

FIGURE 9.6 • COMMERCIAL WHEAT PRODUCTION Ukraine possesses some of the region's best cropland, including a sizable zone of commercial wheat production. Highly mechanized operations improve farm productivity on both state-controlled and privately managed acreage. *(Tass/Sovfoto/Eastfoto)*

by grasslands and by fertile "black earth" **chernozem soils**, which have proven valuable for commercial wheat, corn, and sugar beet cultivation and for commercial meat production (Figure 9.6).

The Ural Mountains and Siberia

The Ural Mountains (see Figure 9.2) mark European Russia's eastern edge, separating it from Siberia, or Asian Russia. Despite their geographical significance as the traditional division between continents, the Urals are not a particularly impressive range; several of their southern passes are less than 1,000 feet (305 meters) high. Although not agriculturally productive, the range is still significant for two reasons: Its ancient rocks contain many valuable mineral resources, and it once marked Russia's eastern cultural boundary.

East of the Urals, the vastness of Russian Siberia unfolds across the landscape for thousands of miles. The great Arctic-bound Ob, Yenisey, and Lena rivers (see Figure 9.5) drain millions of square miles of northern country that includes the flat West Siberian Plain, the hills and plateaus of the Central Siberian Uplands, and the rugged and isolated Northeast Highlands. Wintertime climatic conditions are severe across the region.

Siberian vegetation and agriculture reflect the climatic setting. The northern portion of the region is too cold for tree growth and instead supports tundra vegetation, which is characterized by mosses, lichens, and a few ground-hugging flowering plants. South of the tundra, the Siberian **taiga**, or coniferous forest zone, dominates a large portion of the Russian interior (Figure 9.7). In the east, much of the tundra and taiga regions are associated with **permafrost**, a cold-climate condition of unstable, seasonally frozen ground that limits the growth of vegetation and causes problems for railroad construction.

The Russian Far East

The Russian Far East is a distinctive subregion characterized by proximity to the Pacific Ocean, a more southerly latitude, and a pair of fertile river valleys. Located at about the same latitude as North America's New England, the region features longer growing seasons and milder climates than those found to the west or north. Here, the continental climates of the Siberian interior meet the seasonal monsoon rains of East Asia. It is also a fascinating zone of ecological mixing: Conifers of the taiga mingle with Asian hardwoods, and reindeer, Siberian tigers, and leopards find common ground.

FIGURE 9.7 • SIBERIAN TAIGA Siberia's vast coniferous forests stretch from the Urals to the Pacific. Known as the taiga, these fir, spruce, and larch forests lie in a zone too cold for commercial agriculture. Lumbering and industrial pollution increasingly threaten this national resource. *(Andrey Zvoznikov/The Hutchison Library)*

The Caucasus and Transcaucasia

In European Russia's extreme south, flat terrain gives way first to hills and then to the Caucasus Mountains, a large range stretching between the Black and Caspian seas (see Figure 9.2). The Caucasus mark Russia's southern boundary and are characterized by major earthquakes. Farther south lies Transcaucasia and the distinctive natural setting of Georgia and Armenia.

Patterns of both climate and terrain in the Caucasus and Transcaucasia are very complex. Rainfall is generally high in the western zone, with some slopes supporting dense forests. The area's eastern valleys, on the other hand, are semiarid or arid. In areas of adequate rainfall or where irrigation is possible, agriculture can be quite productive. Georgia in particular has long been an important producer of fruits, vegetables, flowers, and wines (Figure 9.8).

FIGURE 9.8 • SUBTROPICAL GEORGIA The modifying influences of the Black Sea and a more southern latitude produce a small zone of humid subtropical agriculture in Georgia. The fertile landscapes of these tea plantations offer a sharp contrast to the colder country found north of the Caucasus. *(Sovfoto/Eastfoto)*

A Devastated Environment

As the story of the struggling sturgeon suggests, the Russian domain faces many environmental challenges. The breakup of the Soviet Union and subsequent opening of the region to international public examination revealed some of the world's most severe environmental degradation (Figure 9.9). Seven decades of intense and rapid Soviet industrialization caused environmental problems that extend across the entire region.

Air and Water Pollution Poor air quality affects hundreds of cities and industrial complexes throughout the region. The Soviet policy of building large clusters of industrial processing and manufacturing plants in concentrated areas, often with minimal environmental controls, has produced a collection of polluted cities that

FIGURE 9.9 • ENVIRONMENTAL ISSUES IN THE RUSSIAN DOMAIN Varied environmental hazards have left a devastating legacy across the region. The landscape has been littered with nuclear waste, heavy metals, and air pollution. Fouled lakes and rivers pose additional problems in many localities. Present economic difficulties and political uncertainties only add to the costly challenge of improving the region's environmental quality in the twenty-first century.

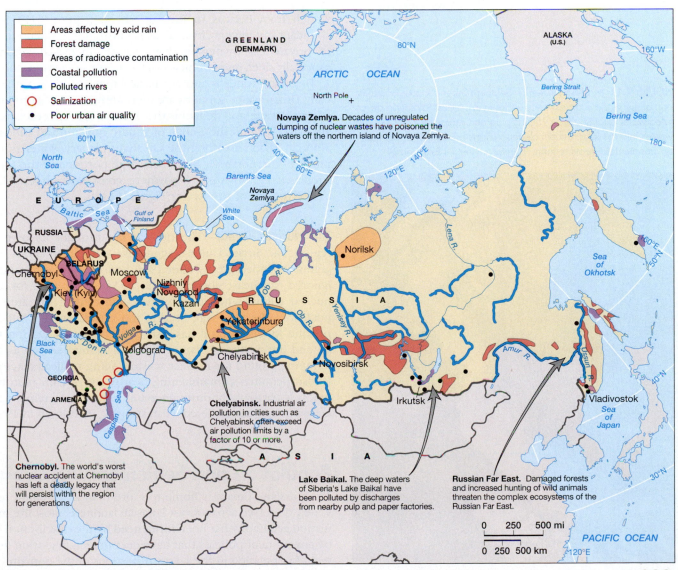

233

FIGURE 9.10 • RUSSIAN AIR POLLUTION Siberia's northern mining and smelting city of Norilsk was thought during the Soviet period to be one of the dirtiest cities in Siberia, if not the entire Northern Hemisphere. Toxic air, damaged forests, and water pollution are continuing problems in the area today. *(Bilderberg Archiv der Fotografen)*

FIGURE 9.11 • LAKE BAIKAL Southern Siberia's Lake Baikal is one of the world's largest deep-water lakes. Industrialization devastated water quality after 1950 as pulp and paper factories poured wastes into the lake. Recent cleanup efforts have helped, but many environmental threats remain. *(Digital Image © 1996 Corbis)*

stretch from Belarus to Russian Siberia (Figure 9.10). A traditional reliance on abundant, but low-quality coal also contributes to pollution problems. Growing rates of private car ownership have greatly increased automobile-related pollution, especially since many vehicles lack catalytic converters and still use leaded gasoline.

Degraded water is another hazard that residents of the region must cope with daily. Urban water supplies are constantly vulnerable to industrial pollution, flows of raw sewage, and demands that increasingly exceed capacity. Oil spills and seepage have harmed thousands of square miles in the tundra and taiga of the West Siberian Plain and along the Ob River estuary. Water pollution has also affected much of the northern Black Sea, large portions of the Caspian Sea shoreline, the Arctic waters off Russia's northern coast, and Siberia's Lake Baikal (the world's largest reserve of freshwater; Figure 9.11).

The Nuclear Threat The nuclear era brought its own deadly dangers to the region. The Soviet Union's aggressive nuclear weapons and nuclear energy programs expanded greatly after 1950, and issues of environmental safety were often ignored. Northeast Siberia's Sakha (or Yakutia) region, for example, suffered regular nuclear fallout when tests were conducted in the atmosphere. In other areas, nuclear explosions were used to move earth in dam-building projects. The once-pristine Russian Arctic has also been poisoned. During the Soviet era, the area around the northern island of Novaya Zemlya served as a huge and unregulated dumping ground for nuclear wastes. Aging nuclear reactors also dot the region's landscape, often contaminating nearby rivers with plutonium leaks. Nuclear pollution is particularly pronounced in northern Ukraine, where the Chernobyl nuclear power plant suffered a catastrophic meltdown in 1986.

Population and Settlement: An Urban Domain

The Russian domain is home to more than 200 million residents. While they are widely dispersed across a vast Eurasian landmass, most live in cities. The region's population geography has been influenced by the distribution of natural resources and by government policies that have encouraged migration out of the traditional centers of population in the western portions of the domain.

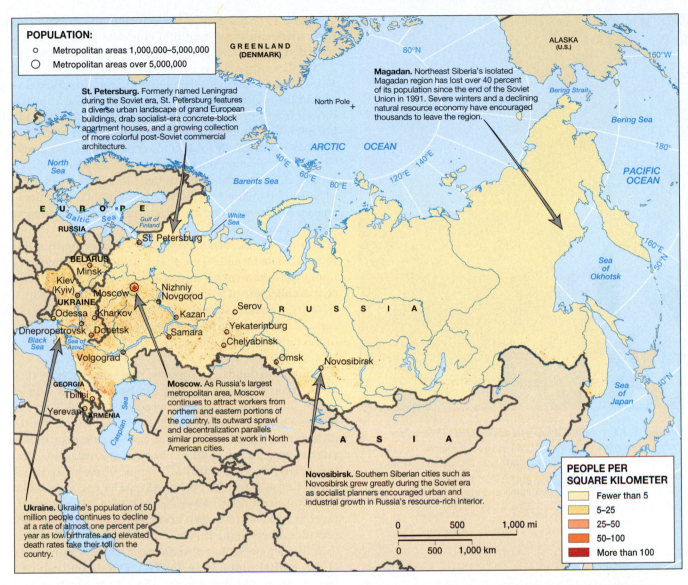

FIGURE 9.12 • POPULATION MAP OF THE RUSSIAN DOMAIN Population within the region is strongly clustered west of the Ural Mountains. Dense agricultural settlements, extensive industrialization, and large urban centers are found in Ukraine, much of Belarus, and across western Russia south of St. Petersburg and Moscow. A narrower chain of settlements follows the better lands and transportation corridors of southern Siberia, but most of Russia east of the Urals remains sparsely settled.

Population Distribution

The more favorable agricultural setting of the European West offered a home to more people than did the more inhospitable conditions found across central and northern Siberia. Although Russian efforts over the past century have encouraged a wider dispersal of the population, it remains heavily concentrated in the west (Figure 9.12). European Russia is home to more than 110 million persons, while Siberia, although far larger, holds only some 35 million. When one adds the 60 million inhabitants of Belarus and Ukraine, the imbalance between east and west becomes even more striking.

The European Core The region's largest cities, biggest industrial complexes, and most productive farms are located in the European Core, a subregion that includes Belarus, much of Ukraine, and Russia west of the Urals. The sprawling city of Moscow and its nearby urbanized region clearly dominate the settlement landscape with a metropolitan area of more than 8.5 million people (Figure 9.13). Within 250 miles, a series of other major urban centers are closely linked to Moscow.

On the shores of the Baltic Sea, St. Petersburg (Leningrad in the Soviet period) (4.8 million) has traditionally had a great deal of contact with western Europe. Between

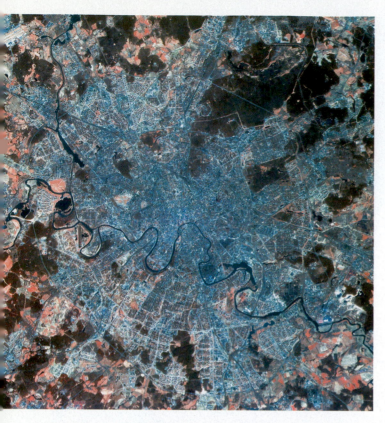

FIGURE 9.13 • METROPOLITAN MOSCOW Sprawling Moscow extends more than 50 miles (80 kilometers) beyond the city center. The city is home to more than 8.5 million people, and the relative strength of its urban economy continues to attract migrants from elsewhere in the country, thus putting more pressure on its infrastructure. *(CNES/Spot Image/Photo Researchers, Inc.)*

FIGURE 9.14 • ST. PETERSBURG Picturesque Canal Street captures some of the architectural character that makes St. Petersburg one of Russia's most beautiful cities. Despite the construction of many concrete buildings in the Soviet period, St. Petersburg retains a charm that echoes the urban landscapes of western Europe. *(Nick Nicholson/Getty Images, Inc.—Image Bank)*

1712 and 1917 it served as the capital of the Russian Empire. Its rich skyline of ornate architecture and beautiful churches gave the city an urban landscape many have compared to the great cities of western Europe (Figure 9.14).

Another urban focus is oriented along the lower and middle stretches of the Volga River and includes the cities of Kazan, Samara, and Volgograd. Industrialization within the region accelerated during World War II, as the region lay somewhat removed from German advances in the west. Today, the highly commercialized river corridor, also containing important petroleum reserves, supports a diverse industrial base strategically located to serve the large populations of the European Core.

A third collection of key population centers on the eastern edge of the Core is located along the resource-rich Ural Mountains. From the gritty industrial landscapes of Serov (1.0 million) to Chelyabinsk (1.1 million) in the south, the Urals region specializes in iron and steel manufacturing, metals smelting and refining, and heavy machinery construction.

Beyond Russia, major population clusters within the European Core are also found in Belarus and Ukraine. The Belorussian capital of Minsk (1.7 million) is the dominant urban center in that country, and its landscape recalls the drab Soviet-style architecture of an earlier era (Figure 9.15). In nearby Ukraine, the capital of Kiev (Kyiv) (2.6 million) straddles the Dnieper River, and the city's old and beautiful buildings are a visual reminder of its historical role as a cultural and trading center within the European interior.

Siberian Hinterlands Leaving the southern Urals city of Yekaterinburg on a Siberia-bound train, one is aware that the land ahead is ever more sparsely settled (see Figure 9.12). The distance between cities grows, and the intervening countryside reveals a landscape shifting gradually from farms to forest. The Siberian hinterland is divided into two characteristic zones of settlement, each of which can be linked to railroad lines. To the south, a collection of isolated, but sizable, urban centers follows the **Trans-Siberian Railroad** (Figure 9.16), a key railroad passage to the Pacific completed in 1904. The eastbound traveler encounters Omsk (1.3 million) as the rail line crosses the Irtysh River, Novosibirsk (1.3 million) at its junction with the Ob River, and Irkutsk (600,000) near Lake Baikal. The port city of Vladivostok (730,000) provides access to the Pacific.

A thinner sprinkling of settlement appears along the more recently completed (1984) **Baikal-Amur Mainline (BAM) Railroad**, which parallels the older line but runs north of Lake Baikal to the Amur River. North of the BAM line, however, the almost empty spaces of central and northern Siberia dominate the scene. Settlements are few and far between, and larger urban areas are either regional administrative capitals, such as Yakutsk (200,000), or resource extraction and processing centers, such as Norilsk (200,000).

Regional Migration Patterns

Over the past 150 years, millions of people within the Russian domain have been on the move. These major migrations, both forced and voluntary, reveal sweeping examples of human mobility that rival the great movements from Europe and Africa or the transcontinental spread of settlement across North America.

Eastward Movement Just as settlers of European descent moved west across North America, exploiting natural resources and displacing native peoples, European Russians moved east across the Siberian frontier to extend their influence within the Eurasian interior. Although these migrations into Siberia began several centuries earlier, the pace and volume of the eastward drift accelerated in the late nineteenth century once the Trans-Siberian Railroad was completed. Peasants were attracted to the region by its agricultural opportunities (in the south) and by greater political freedoms than they traditionally enjoyed under the **tsars** (or czars; Russian for *Caesar*), the authoritarian leaders who dominated politics during the pre-1917 Russian Empire. Almost 1 million Russian settlers moved into the Siberian hinterland between 1860 and 1914. The eastward migration continued during the

Soviet period, and by the end of the communist era, 95 percent of Siberia's population was classified as Russian (including Ukrainians and other western immigrants).

Political Motives Political motives also shaped migration patterns. Particularly in the case of Russia, leaders from both the imperial and Soviet eras saw advantages in moving people to new locations. Clearly, for example, the infilling of the southern Siberian hinterland was useful for both political and economic reasons. Both the tsars and the Soviet leaders saw their political power grow as Russians moved into the Eurasian interior and developed its resources. Political dissidents and troublemakers were also sent there. Especially in the Soviet period, uncounted millions were forcibly relocated to the region's well-known **Gulag Archipelago**, a vast collection of political prisons in which inmates often disappeared or spent years far removed from their families and home communities.

Russification, the Soviet policy of resettling Russians into non-Russian portions of the Soviet Union, also changed the region's human geography. Millions of Russians were given economic and political incentives to move elsewhere in the Soviet Union in order to increase Russian dominance in many of the outlying portions of the country. By the end of the Soviet period, Russians made up significant minorities within former Soviet republics such as Kazakstan (38 percent Russian), Latvia (34 percent), and Estonia (30 percent). Among its Slavic neighbors, Belarus remains 13 percent Russian and Ukraine more than 22 percent Russian, with concentrations particularly high in the eastern portions of the countries.

New International Movements In the post-Soviet era, Russification has often been reversed (Figure 9.17). Several of the newly independent non-Russian countries have imposed rigid language and citizenship requirements, which encouraged many Russian residents to leave. In other settings, ethnic Russians have simply experienced discrimination when it came to economic and social opportunities in these non-Russian countries. By 2000, about 6 million Russians had left former Soviet republics and had returned to their homeland.

The domain's more open borders have also made it easier for other residents to leave the region (see Figure 9.17). The "brain drain" of young, well-educated, upwardly mobile Russians has been considerable. Many have left for better job opportunities outside of the region. Sometimes, ethnic links play a part in migration patterns. For example, many Russian-born ethnic Finns have moved to nearby Finland. Russia's Jewish population also continues to fall. Furthermore, Russians have become one of the largest new immigrant groups in the United States. In fact, a recent U.S. Immigration Service report suggests that young Ukrainian women have been a favorite choice for American men searching for marriage partners on the Web.

The Urban Attraction Regional residents have also been bound for the cities. The Marxist philosophy followed by Soviet planners encouraged urbanization. In 1917 the Russian Empire was still overwhelmingly rural and agrarian; 50 years later, the Soviet Union was primarily urban. Planners saw great economic and political advantages in efficiently clustering the population, and Soviet policies dedicated to large-scale industrialization obviously favored the growth of cities.

Soviet cities grew according to strict governmental plans. Planners selected different cities for different purposes. Some were designed for specific industries, while others were given administrative roles. All cities were assigned set population levels. A system of internal passports prohibited people from moving freely from city to city. Instead, people generally went where the government assigned them jobs. Moscow, the country's leading administrative city, thrived under the Soviet regime. It formed the core of Soviet bureaucratic power, as well as the center of education, research, and the media.

With the end of the Soviet Union, people gained basic freedoms of mobility. In addition, with a more open economy, especially in Russia, employment shifts related to how well an area could compete in the global economy. This new economic

FIGURE 9.15 • MINSK Almost 2 million people reside in the Belorussian capital of Minsk. Its drab apartments and office buildings show the legacy of the Soviet period on the urban scene. Indeed, Belarus retains many Soviet-style characteristics, trailing far behind Russia in the pace of economic reforms. *(Novosti/Sovfoto/Eastfoto)*

FIGURE 9.16 • TRANS-SIBERIAN RAILROAD Snaking its way through winter and the vastness of the southern Siberian landscape, the Trans-Siberian Railroad provides a critical connection between European and Asian portions of the Russian domain. *(Tass/Sovfoto/Eastfoto)*

FIGURE 9.17 • RECENT MIGRATION FLOWS IN THE RUSSIAN DOMAIN Recent events are encouraging the return of ethnic Russians from former Soviet republics, while other Russians are emigrating from the domain for economic, cultural, and political reasons. Within Russia, both political and economic forces are also at work encouraging people to be on the move.

reality, for example, has led to decline of many older industrial areas, as they simply cannot produce raw materials or finished industrial goods at competitive global prices. Since 1991 more than 12 percent of the population of resource-rich northern Russia has emigrated.

Inside the Russian City

Large Russian cities possess a core area, or center, that features superior transportation connections; the best-stocked, upscale department stores and shops; the most desirable housing; and the most important offices (both governmental and private) (Figure 9.18). In the largest urban centers, cities such as Moscow and St. Petersburg also feature extensive public spaces and examples of monumental architecture at the city center.

Within the city, there is usually a distinctive pattern of circular land-use zones, each of which was built at a later date as one moves outward from the center. Such a ringlike urban morphology is not unique. However, as a result of the extensive power of government planners during the Soviet period, this urban form is probably more highly developed here than in most parts of the world.

At their very center, the cores of many older cities predate the Soviet Union. Pre-1900 stone buildings often dominate older city centers. Some of these are former private mansions that were turned into government offices or subdivided into apartments during the communist period but are now being privatized again.

Beyond this pre-communist urban core, one can often find a ring of public housing projects and fully planned *sotzgorods*, or socialist neighborhoods. Sotzgorods are based on a close connection between workplace and home and typically include simple dormitory-style housing.

Another common urban zone, removed some distance from city cores, is the *chermoyuski*. Chermoyuski are large, uniform apartment blocks built during the 1950s and 1960s. Most of these nearly uniform apartment blocks are five stories or less in height, since they typically are not equipped with elevators. Today, many of these apartment complexes remain, but they are now rundown, with little money available for repair.

Farther out from the city centers are the **mikrorayons**, the much-larger housing projects of the 1970s and 1980s (Figure 9.19). Mikrorayons are typically composed of massed blocks of standardized apartment buildings, ranging from 9 to 24 stories in height. Each mikrorayon was to form a self-contained community, with grocery stores and other basic services located within walking distance of each apartment building. The largest of these supercomplexes contain up to 100,000 residents. While planners hoped that mikrorayons would foster a sense of community, most now serve largely as anonymous bedroom communities for larger metropolitan areas.

Some of Russia's most rapid urban growth in recent years has occurred on the metropolitan periphery, paralleling the North American experience. Moscow, for example, has seen its urban reach expand far beyond the city center. The surrounding administrative district (the Moscow Oblast) contains about 7 million residents, and local officials estimate that foreigners invested more than $1 billion in the fast-growing suburban fringe in 2001.

The Demographic Crisis

Russia's Ministry of Labor and Social Development recently acknowledged that the country faced a crisis of persistently declining populations. Government studies predicted that Russia's population could fall by 3 million people by 2005 and by a startling 25 million by 2030. Similar conditions are affecting the other countries within the region. Declining populations, low birthrates, and rising mortality, particularly among middle-aged males, are all troubling symptoms of the region's population problems. Death rates began exceeding birthrates in the early 1990s.

The region's demographic crisis appears related to the fraying social fabric and uncertain economic times. Very low birthrates within the region may reflect a lack of optimism about the future. In Russia, for example, total live births fell from a peak of 2.5 million in 1987 to fewer than 1.4 million in the late 1990s, a period of tremendous economic and political disruption. The health of women of childbearing age has also declined, while problem pregnancies, maternal childbirth death rates, and birth defects are on the rise. Similarly, sharp increases in death rates, especially among Russian men, appear related to many stress-related conditions, such as alcoholism and heart disease. Murder and suicide rates have climbed rapidly in the past 15 years. In addition, perhaps 20 to 30 percent of the increase in death rates may be attributed to the region's increasingly toxic environment.

Cultural Coherence and Diversity: The Legacy of Slavic Dominance

For hundreds of years, Slavic peoples speaking the Russian language expanded their influence from an early homeland in central European Russia. Eventually, Slavic cultures spread north to the Arctic Sea, south to the Black Sea and Caucasus, west to the shores of the Baltic, and east to the Pacific Ocean. Russian cultural patterns and social institutions spread widely during this Slavic expansion, influencing many non-Russian ethnic groups that continued to live under the rule of the Russian empire. The legacy of this cultural spread continues today. It offers Russians a rich historical identity and sense of nationhood. It also helps us understand how present-day Russians are dealing with forces of globalization and how non-Russian cultures have evolved within the region.

FIGURE 9.18 • DOWNTOWN MOSCOW   The bustling traffic along central Moscow's Novy Arbat parallels urban scenes elsewhere in Europe and North America. The city's landscape remains a complex and fascinating mix of imperial, Soviet, and post-Soviet influences. *(Itar-Tass/Sovfoto/Eastfoto)*

FIGURE 9.19 • MOSCOW HOUSING   For many residents in larger Russian cities, home is a high-rise apartment house. Most of these satellite centers were built in the Soviet era. Poor construction and a lack of landscaping often yield a bleak suburban scene, but nearby stores, entertainment, and public transportation offer important services. *(Itar-Tass/Sovfoto/Eastfoto)*

The Heritage of the Russian Empire

The expansion of the Russian Empire paralleled similar events in western Europe. As Spain, Portugal, France, and Britain carved out empires in the Americas, Africa, and Asia, the Russians expanded eastward and southward across Eurasia.

Growth of the Russian Empire The origin of the Russian Empire lies in the early history of the **Slavic peoples**, defined linguistically as a distinctive northern branch of the Indo-European language family. Slavic political power grew by A.D. 900 as they intermarried with southward-moving warriors from Sweden known as *Varangians*, or *Rus*. Within a century, the state of Rus extended from Kiev (the capital) to near St. Petersburg. The new Kiev-Rus state interacted with the Byzantine Empire of the Greeks, and this influence brought Christianity to the Russian realm by A.D. 1000. Along with the new religion came many other aspects of Greek culture, including the Cyrillic alphabet. Even as the Russians converted to **Eastern Orthodox Christianity**, a form of Christianity historically linked to eastern Europe and church leaders in Constantinople (modern Istanbul), their Slavic neighbors to the west (the Poles, Czechs, Slovaks, Slovenians, and Croatians) accepted Catholicism. This early Russian state soon faltered and split into several principalities that were then ruled by invading Mongols and Tatars (a group of Turkish-speaking peoples).

By the fourteenth century, however, northern Slavic peoples overthrew Tatar rule and established a new and expanding Slavic state (Figure 9.20). The core of the

FIGURE 9.20 • GROWTH OF THE RUSSIAN EMPIRE
Beginning as a small principality in the vicinity of modern Moscow, the Russian Empire took shape between the fourteenth and sixteenth centuries. After 1600, Russian influence stretched from eastern Europe to the Pacific Ocean. Later, portions of the empire were added in the Far East, Central Asia, and near the Baltic and Black seas. *(Modified from Bergman and Renwick, 1999, Introduction to Geography, Upper Saddle River, NJ: Prentice Hall)*

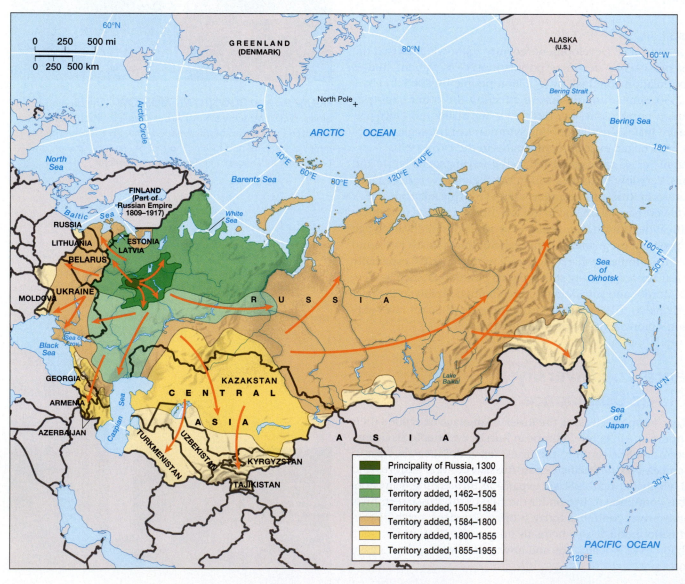

new Russian Empire lay near the eastern fringe of the old state of Rus. Gradually this area's language diverged from that spoken in the new Russian core, and *Ukrainians* and Russians developed into two separate peoples. A similar development took place among the northwestern Russians, who experienced several centuries of Polish rule and over time were transformed into a distinctive group known as the *Belorussians*.

The Russian Empire expanded remarkably in the sixteenth and seventeenth centuries. Former Tatar territories in the Volga Valley (near Kazan) were incorporated into the Russian state in the mid-1500s. The Russians also allied with the seminomadic **Cossacks**, Slavic-speaking Christians who had earlier migrated to the region to seek freedom in the ungoverned steppes (Figure 9.21). The Russian empire granted them a number of privileges in exchange for their military service, an alliance that smoothed the way for Russian expansion into Siberia during the seventeenth century. Premium Siberian furs were the chief attraction of this immense northern territory.

While the Russian Empire rapidly expanded to the east, its westward expansion was slow and halting. When Tsar Peter the Great (1682–1725) defeated Sweden in the early 1700s, he obtained a territorial foothold on the Baltic Sea. There, he built the new capital city of St. Petersburg, designed to give the Russian Empire better access to western Europe. Later in the eighteenth century, Russia defeated both the Poles and the Turks and gained all of modern-day Belarus and Ukraine. Tsarina Catherine the Great (1762–1796) was especially important in colonizing Ukraine and bringing the Russian Empire to the warm-water shores of the Black Sea.

The nineteenth century witnessed the Russian Empire's final expansion. Large gains were made in Central Asia, where a group of once-powerful Muslim states was no longer able to resist the Russian army. The mountainous Caucasus region proved a greater challenge, as the peoples of this area had the advantage of rugged terrain in defending their lands. South of the Caucasus, however, the Christian Armenians and Georgians accepted Russian power with little struggle, since they found it preferable to rule by the Persian or Ottoman empires.

The Significance of Empire   The expansion of the Russian people was one of the greatest human movements Earth has ever witnessed. By 1900 a traveler going from St. Petersburg on the Baltic to Vladivostok on the Sea of Japan would everywhere encounter Russian peoples speaking the same language, following the same religion, and living under the rule of the same government. Nowhere else in the world did such a tightly integrated cultural region cover such a vast space.

Geographies of Language

Slavic languages dominate the region (Figure 9.22). The geographic pattern of the Belorussian people is relatively simple. The vast majority of Belorussians reside in Belarus, and most people in Belarus are Belorussians. The Ukrainian situation is more complex. Russians historically have dominated large portions of eastern Ukraine, while they make up a much smaller portion of western Ukraine's population. Similarly, the Crimean Peninsula, now a part of Ukraine, has long ethnic and political roots within Russia. Conversely, ethnic Ukrainians live in scattered communities in southern Russia and southwestern Siberia. During the Soviet period, such geographical mixing had little consequence, since the distinction between Russians and Ukrainians was not viewed as important in official circles. Now that Russia and Ukraine are separate countries, this issue has emerged as a significant source of tension across the region.

Approximately 80 percent of Russia's population claims a Russian linguistic identity. Russians inhabit most of European Russia, but there are large groups of other peoples within the region. The Russian zone extends across southern Siberia to the Sea of Japan. In sparsely settled lands of central and northern Siberia, Russians are more numerous in many areas, but they share territory with varied native peoples.

Finno-Ugric (Finnish-speaking peoples), though small in number, dominate sizable portions of the non-Russian north. Altaic speakers also complicate the country's linguistic geography. This language family includes the Volga Tatars, whose territory is

FIGURE 9.21 • COSSACK   Natives of the Ukrainian and Russian steppe, the highly mobile Cossacks played a pivotal role in aiding Russian expansion into Siberia during the sixteenth century. Many modern descendents retain their skills of horsemanship and are proud of their distinctive ethnic heritage. *(Julien Chatelin/ Getty Images, Inc.—Liaison)*

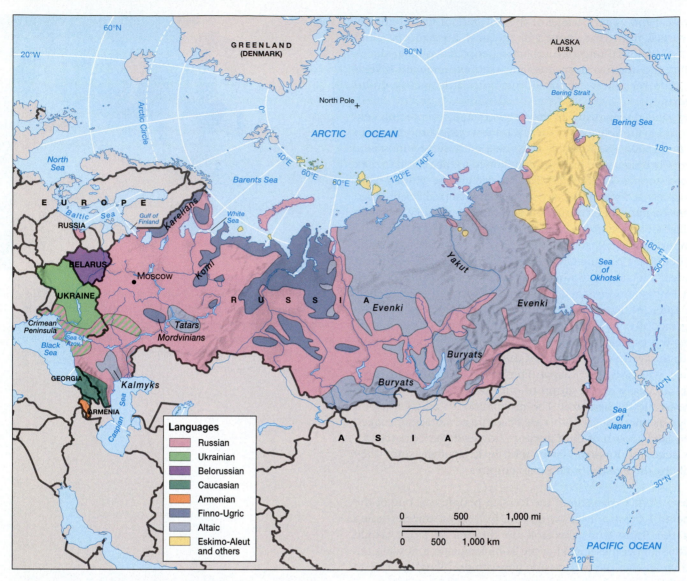

FIGURE 9.22 • LANGUAGES OF THE RUSSIAN DOMAIN
Slavic Russians dominate the realm, although many linguistic minorities are present. Siberia's diverse native peoples add cultural variety in that area. To the southwest, the Caucasus Mountains and the lands beyond contain the region's most complex linguistic geography. Ukrainians and Belorussians, while sharing a Slavic heritage with their Russian neighbors, add further variety in the west.

centered on the city of Kazan in the middle Volga Valley. While retaining their ethnic identity, the Turkish-speaking Tatars have extensively intermarried with and borrowed from their Russian neighbors. Yakut peoples of northeast Siberia also represent Turkish speakers within the Altaic family. In the east, the Buryats live in the vicinity of Lake Baikal and represent a local Siberian group closely tied to the cultures and history of Central Asia (Figure 9.23).

The plight of many native peoples in central and northern Siberia parallels the situation in the United States, Canada, and Australia. Rural indigenous peoples in each of these settings remain distinct from dominant European cultures. These peoples are also internally diverse and are often divided into a number of unrelated linguistic groups. One entire linguistic grouping of Eskimo-Aleut speakers is limited to approximately 25,000 people who are widely dispersed through northeast Siberia. Many of these Siberian peoples have seen their traditional ways challenged by the pressures of Russification, just as native peoples elsewhere in the world have been subjected to similar pressures of cultural and political assimilation. Unfortunately, other common traits seen within such settings are low levels of education, high rates of alcoholism, and widespread poverty.

Although small in size, Transcaucasia offers a bewildering variety of languages. From Russia, along the north slopes of the Caucasus, and to Georgia and Armenia east of the Black Sea, a complex history and a complicated physical setting have combined to produce some of the most complex language patterns in the world. Several language families are spoken within a region smaller than Ohio, and many individual languages are represented by small, isolated cultural groups.

Geographies of Religion

Most Russians, Belorussians, and Ukrainians share a religious heritage of Eastern Orthodox Christianity. For hundreds of years Eastern Orthodoxy served as a central cultural presence within the Russian Empire (Figure 9.24). Indeed, church and state were tightly fused, until the demise of the empire in 1917. Under the Soviet Union, however, religion in all forms was severely discouraged and actively persecuted. Most monasteries and many churches were converted into museums or other kinds of public buildings, and schools rejected traditional religious ideas.

With the downfall of the Soviet Union, however, a religious revival has swept much of the Russian domain. In the past 15 years, more than 11,000 Orthodox churches have been returned to religious uses. Now, an estimated 75 million Russians are members of the Orthodox Church, including almost 400 monasteries dispersed across the country.

Other forms of Western Christianity are also present in the region. For example, the people of western Ukraine, who experienced several hundred years of Polish rule, eventually joined the Catholic Church. Eastern Ukraine, on the other hand, remained fully within the Orthodox framework. This religious split reinforces the cultural differences between eastern and western Ukrainians. Elsewhere, Christianity came early to the Caucasus, but modern Armenian forms—their roots dating to the fourth century A.D.—differ somewhat from both Eastern Orthodox and Catholic religious traditions. Evangelical Protestantism has also been on the rise since the fall of the Soviet Union.

Non-Christian religions also appear and, along with language, shape ethnic identities and tensions within the region. Islam is the largest non-Christian religion and claims between 20 and 25 million followers. Most are Sunni Muslims, and they include peoples in the North Caucasus, the Volga Tatars, and Central Asian peoples near the Kazakstan border. To date, Islamic fundamentalism has not become a major cultural or political issue in the region. Still, a growing Islamic political consciousness is present, particularly in Russia. In mid-2001, the first large-scale meeting of the new Muslim Prosperity Party took place. Russia, Belarus, and Ukraine are also home to more than 1 million Jews, who are especially numerous in the larger cities of the European West. Jews suffered severe persecution under both the tsars and the communists. Recent outmigrations, encouraged by new political freedoms, have further reduced their numbers in Russia, Belarus, and Ukraine. Buddhists are also represented in the region, associated with the Kalmyk and Buryat peoples of the Russian interior.

Russian Culture in Global Context

Russian culture has developed its own distinctive traditions and symbols, but it has also been influenced greatly by western Europe. By the nineteenth century, even as Russian peasants interacted rarely with the outside world, Russian high culture had become thoroughly Westernized, and Russian composers, novelists, and playwrights gained considerable fame in Europe and the United States.

Soviet Days   During the Soviet period, new cultural influences shaped the socialist state. Initially, European-style modern art flourished in the Soviet Union, encouraged by the radical Marxist rhetoric of the new rulers. By the late 1920s, however, Soviet leaders turned against modernism, which they viewed as the decadent expression of a declining capitalist world. Many Soviet artists fled to the West, and others were exiled to Siberian labor camps. Increasingly, state-sponsored Soviet artistic productions centered on **socialist realism**, a style devoted to the realistic depiction of workers heroically challenging nature or struggling against capitalism. Still, traditional high arts, such as classical music and ballet, continued to receive generous state subsidies, and to this day Russian artists regularly achieve worldwide fame.

FIGURE 9.23 • MINORITY BURYATS   Closely related to residents of Mongolia, Russia's Buryats live in the vicinity of Lake Baikal. They enjoy some degree of political autonomy in recognition of their distinctive ethnic background. *(Hans-Jurgen Burkard/ Bilderberg Archiv der Fotografen)*

FIGURE 9.24 • RUSSIAN ORTHODOX CHURCH, SIBERIA   A revival of interest in the Russian Orthodox Church followed the collapse of the Soviet Union in the early 1990s. The newly built Znamensky Cathedral in Kemerovo displays many of the faith's characteristic cultural landscape signatures. *(Novosti/Sovfoto/ Eastfoto)*

FIGURE 9.25 • ALSOU   Symbol of a new Russia and the globalizing power of popular music, Russian singer Alsou recorded English-language songs in 2001 for young audiences in western Europe and North America. *(Hy/Newsmakers/Getty Images, Inc.—Liaison)*

Turn to the West   By the 1980s it was clear that the attempt had failed to fashion a new Soviet culture based on Communist ideals. The younger generation instead adopted a rebellious attitude, turning for inspiration to fashion and rock music from the West. The mass-consumer culture of the United States proved very popular, symbolized above all by brand-name chewing gum, jeans, and cigarettes.

After the fall of the Soviet Union in 1991, increased information and movement into the region brought a rush of global cultural influences, particularly to the region's larger urban areas such as Moscow. Shops were quickly flooded with Western books and magazines; people sought advice about home mortgages and condominium purchases, and they enjoyed the newfound pleasures of fake Chanel handbags and McDonald's hamburgers. English-language classes became even more popular in cities such as Moscow, where Russians hurried to embrace the world their former leaders had warned them about for generations. Cultural influences streaming into the country were not all Western in inspiration. Films from Hong Kong and Mumbai (Bombay), as well as the televised romance novels (*telenovelas*) of Latin America, for example, proved far more popular in the Russian domain than in the United States.

The Music Scene   Younger residents of the region have also embraced the world of popular music, and their enthusiasm for U.S. and European performers, as well as their support of a budding home-grown music industry, symbolizes the changing values of an increasingly post-Soviet generation. MTV Russia went on the air at midnight on September 26, 1998. By 2001 Russian MTV reached more than 60 million viewers. Perhaps even more important, Sony Music Entertainment established its Russian operations in December 1999. Sony, BMG, and other labels have signed multiple Russian acts for domestic markets, helping to boost a native pop-music culture. Universal has also opened the way for Russian performers to go global. For example, it sponsored the English-language debut of Russian pop singer Alsou in 2001 (Figure 9.25).

Geopolitical Framework: The Remnants of a Global Superpower

The geopolitical legacy of the former Soviet Union still weighs upon the Russian domain. After all, the bold lettering of the "Union of Soviet Socialist Republics" dominated the Eurasian map for much of the twentieth century, and the country's global political influence affected every part of the world. Many of the present political uncertainties that affect the region stem from the Soviet period. Former Soviet republics continue to struggle to stand on their own as independent countries. Present demands for more local political control within countries such as Russia, Ukraine, and Georgia can still be understood in the context of the Soviet era, when a strong central government did not tolerate any regional dissent.

Geopolitical Structure of the Former Soviet Union

The Soviet Union rose from the ashes of the Russian Empire, which collapsed abruptly in 1917. The Russian tsars did little to modernize the country or improve the life of its people. The Russian peasants, who formed the majority of the population, had always resisted the powerful land-owning aristocracy. After the fall of the tsar and the aristocracy, a broad-based government that represented several political groups assumed authority. Several months later, however, the **Bolsheviks**, a faction of Russian communists representing the interests of the industrial workers, seized power within the country. The leader of these Russian communists was Vladimir Ilyich Ulyanov, usually known by his self-selected name, Lenin. Lenin became the main architect of the Soviet Union, the state that replaced the failed Russian Empire.

The new socialist state reconfigured Eurasian political geography. Although it resembled the territory of the Russian Empire and centralized authority continued to

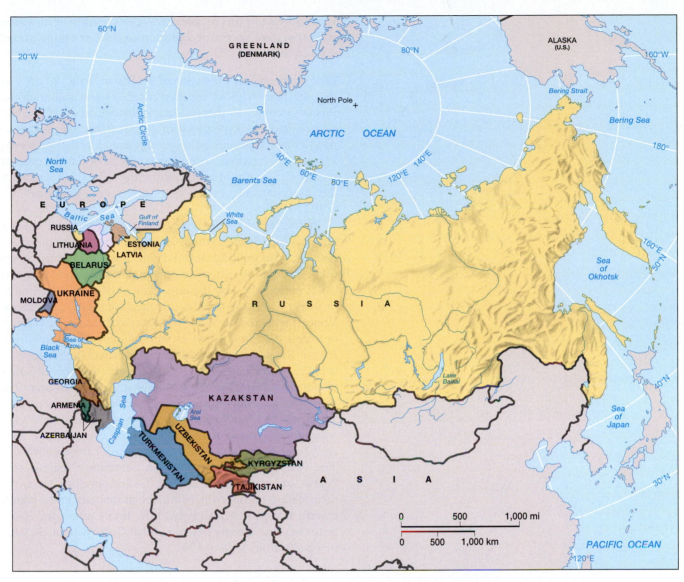

FIGURE 9.26 • SOVIET GEOPOLITICAL SYSTEM   During the Soviet period, the boundaries of the country's 15 internal republics often reflected major ethnic divisions. Ultimately, however, many of the ethnically non-Russian republics pressured the Soviet government for more political power. As the Soviet empire disintegrated, the former republics became politically independent states and now form an uneasy ring of satellite nations around Russia. *(Modified from Rubenstein, 1999, Introduction to Human Geography, Upper Saddle River, NJ: Prentice Hall)*

be concentrated in European Russia, the basic economic geography of the country was radically transformed. When the Soviet Union emerged in 1917, Lenin and the other communist leaders were aware that they faced a major challenge in organizing the new state.

The Soviet Republics and Autonomous Areas
Soviet leaders designed a geopolitical solution that maintained their country's territorial boundaries and recognized, at least theoretically, the rights of its many non-Russian citizens. Each major nationality was to receive its own "union republic," provided it was situated on one of the nation's external borders (Figure 9.26). Eventually 15 such republics were established, thus creating the Soviet Union. The massive Russian Republic sprawled over roughly three-quarters of the Soviet terrain. Each republic was to have considerable political freedom, even the right to withdraw from the union if it so desired. In practice, however, the Soviet Union remained a centralized state, with important decisions made in the capital of Moscow.

The Soviets came up with different geopolitical solutions to acknowledge smaller ethnic groups and nationalities that were not situated on the country's external borders. Indeed, dozens of significant minority groups pressed for recognition.

One solution to this problem was the creation of **autonomous areas** of varying sizes that gave special recognition to smaller ethnic homelands within the different republics.

Centralization and Expansion of the Soviet State　In the early Soviet era, it appeared that the framework of separate republics and autonomous areas might allow non-Russian peoples to protect their own cultures and establish their own social and economic policies (provided, of course, that such policies embodied Marxist principles). From the beginning, however, it was clear that such self-determination would be a temporary measure. According to official beliefs the gradual development of a communist society would see the withering away of all significant ethnic differences and the disappearance of religion. In the future, a new classless Soviet society was supposed to emerge.

By the 1930s it was clear that local and regional political units would not have any real significance within an increasingly centralized Soviet state. The chief architect of this political consolidation was Soviet leader Joseph Stalin, who did everything he could to centralize power in Moscow and to assert Russian authority. The Stalin period also saw the enlargement of the Soviet Union. As a victorious power in World War II, the country acquired southern Sakhalin and the Kuril Islands from Japan. It regained the Baltic republics (Lithuania, Latvia, and Estonia—independent between 1917 and 1940), as well as substantial territories formerly belonging to Poland, Romania, and Czechoslovakia. A small but strategic addition on the Baltic Sea was the northern portion of East Prussia (the port of Kaliningrad), previously part of Germany. It still forms a small but strategic Russian **exclave**, which is defined as a portion of a country's territory that lies outside of its contiguous land area.

After World War II, the Soviet Union also gained significant authority, although not actual sovereignty, over a broad swath of eastern Europe. As they pushed the German army west toward the end of the war, Soviet troops advanced across much of the region, actively working to establish communist regimes thereafter. In the words of British leader Winston Churchill, the Soviets extended an **"Iron Curtain"** between their eastern European allies and the more democratic nations of western Europe. As eastern Europe retreated behind the Iron Curtain, the Soviet Union and the United States became antagonists in a global **Cold War** of escalating military competition that lasted from 1948 to 1991.

End of the Soviet System　Ironically, Lenin's system of republics based on cultural differences sowed the seeds of the Soviet Union's demise. Even though the republics were never allowed real freedom, they provided a political framework that encouraged the survival of distinct cultural identities. Indeed, contrary to the expectations of Soviet leaders, ethnic nationalism intensified in the post-World War II era as the Soviet system grew less repressive. When Soviet President Mikhail Gorbachev initiated his policy of **glasnost**, or greater openness, during the 1980s, several republics—most notably the Baltic states of Lithuania, Latvia, and Estonia—demanded outright independence.

Meanwhile, other forces worked toward the political end of the Soviet regime. A failed war in Afghanistan in the early 1980s frustrated both the Soviet leaders and population. In eastern Europe, growing protests over Soviet dominance emerged from Czechoslovakia and Poland. Worsening domestic economic conditions, increasing food shortages, and the declining quality of life led to fundamental questions concerning the value of centralized planning within the country. In response, President Gorbachev introduced **perestroika**, or planned economic restructuring, aimed at making production more efficient and more responsive to the needs of Soviet citizens.

In 1991, however, Gorbachev saw his authority slip away amid rising pressures for political decentralization and more dramatic economic reforms. During the summer, Gorbachev's regime was further endangered by the popular election of reform-minded Boris Yeltsin as the head of the Russian Republic and by a failed military takeover by communist hard-liners. By late December, all of the country's 15 constituent republics had become independent states, and the Soviet Union ceased to exist.

Current Geopolitical Setting

The political geography of post-Soviet Russia and the nearby independent republics has changed dramatically since the collapse of the Soviet Union in 1991. All of the former republics still struggle to establish stable political relations with their neighbors, and increasing calls for further political decentralization in many settings threaten the stability of Russia and nearby states (Figure 9.27).

Russia and the Former Soviet Republics   For a time it seemed that a looser political union of most of the former republics, called the **Commonwealth of Independent States (CIS)**, would emerge from the ruins of the Soviet Union. All the former republics, with the exception of the three Baltic states, joined the CIS soon after the end of the old union (Figure 9.27). By the early twenty-first century, however, the CIS had developed into little more than a forum for discussion, without real economic or political power. Among former Soviet republics, Russia, Belarus, and Ukraine have enjoyed the closest political and economic relations.

Another complicating factor in the post-Soviet period has been the ongoing military relationships between Russia and its former republics. **Denuclearization**, the return of nuclear weapons from outlying republics to Russian control and their partial dismantling, was completed during the 1990s. The Soviet-era nuclear arsenals of Kazakstan, Ukraine, and Belarus were removed in the process. Naval

FIGURE 9.27 • GEOPOLITICAL ISSUES IN THE RUSSIAN DOMAIN   The Russian Federation Treaty of 1992 created a new internal political framework that acknowledged many of the country's ethnic minorities. Political stability, however, continues to elude the country as many of these groups press for complete independence. Elsewhere, Armenia struggles with neighboring Azerbaijan, and Russia's relations with nearby states remain in flux. *(Modified from Bergman and Renwick, 1999,* Introduction to Geography, *Upper Saddle River, NJ: Prentice Hall)*

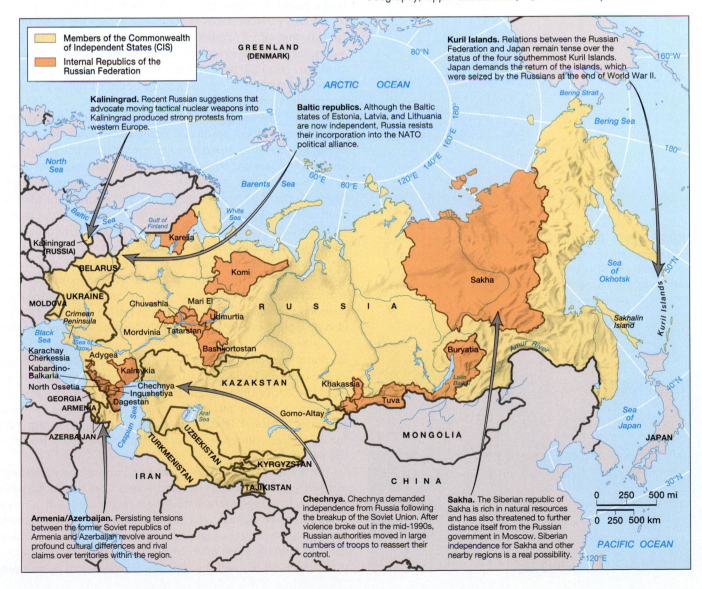

FIGURE 9.28 • KAZAN Capital of the internal Tatarstan Republic, Kazan is home to many Islamic residents who press for even more freedom from distant Moscow. *(Novosti/Sovfoto/Eastfoto)*

FIGURE 9.29 • DEVASTATED GROZNYY The recent war for independence in the Chechnyan Republic met stiff resistance from Russia's central government. Much of the capital city has been caught in the crossfire and continues to reveal the scars of the conflict. *(Tass/Sovfoto/Eastfoto)*

bases in the former Soviet Union present another dilemma. The Soviet-era Black Sea fleet was traditionally based in the Crimean port of Sevastopol, now in Ukraine. After lengthy negotiations, Russia and Ukraine agreed in 1997 to share the naval base, but the longer-term political control of the entire Crimean Peninsula remains in doubt.

The former Transcaucasian republics have witnessed political tensions in the post-Soviet era. The territories of the Christian Armenians and the Muslim Azeris, for example, interpenetrate one other in a complex fashion. The far southwestern portion of Azerbaijan (Nakhichevan) is separated from the rest of the country by Armenia, while the important Armenian-speaking district of Nagorno-Karabakh is officially a self-governing part of Azerbaijan. Almost immediately after independence in 1991, Armenia and Azerbaijan went to war over these territories. After Armenia successfully occupied much of Nagorno-Karabakh, the fighting between the countries diminished. No peace treaty has been signed, however, and Azerbaijan demands the return of the territory.

Devolution and the Russian Federation Within Russia, further pressures for devolution, or more localized political control, produced the March 1992 signing of a new Russian Federation Treaty. The treaty granted Russia's internal autonomous republics and its local political units greater political, economic, and cultural freedoms, including more control of their natural resources and foreign trade. On the other hand, it weakened Moscow's centralized authority to collect taxes and to shape policies within its varied hinterlands. Defined essentially along ethnic lines, 21 regions possess status as republics within the federation and now have constitutions that often challenge national policies and laws. Tatarstan in the Volga Valley, for example, even extracted agreements from Moscow to develop their own "foreign economic policy" (Figure 9.28). In addition, dozens of smaller regions and cities, including Moscow itself, have declared varying degrees of political independence from federal rule.

Internal Russian geopolitics remain caught between the forces of devolution, aimed at more local political control, and a renewed effort on the part of the central government in Moscow to assert its national presence. The outcome of that struggle remains uncertain. Local political control has its advantages. After a long era of Soviet dominance, it allows people to have a more direct role in running their own political affairs. They also can claim fuller control of their economic resources and more effectively manage public expenditures on economic development.

Devolution, however, comes with a price. Central leaders in Moscow fear that Russia could break apart, resulting in further political instability. Indeed, many of the country's local and regional governments have experienced incompetence and widespread corruption. Such political division makes it more difficult to introduce national reforms, to make large-scale investments in infrastructure, and to engage in comprehensive long-term planning. A more centralized state, supporters argue, can also deal more effectively with global political and economic affairs.

Regional Tensions Regional geopolitical trouble spots within Russia reveal some of the tensions between local and national demands for power (see Figure 9.27). After the Soviet Union disintegrated, several of Russia's internal autonomous areas threatened to become independent. While Moscow has recently controlled powerful regional leaders in such settings, enduring political and economic forces will continue to encourage more autonomy for these distant portions of the country.

The Caucasus Mountains have been an explosive setting for internal political tensions. In 1994 leaders of the internal Chechnyan Republic vowed to establish a genuinely independent state. Russia responded with a massive military invasion. After intensive fighting, the Russian army took control of Chechnya's capital of Groznyy, located in the plains just to the north of the Caucasus (Figure 9.29). A new Russian military force of around 100,000 soldiers was sent into the small republic late in 1999 to quell the Chechnyan rebels and rally domestic political support around the struggling central government in Moscow. Since their return, the Russians have been accused of looting Chechnyan natural resources (especially metals and oil) and trying to grab a share of the profitable global drug trade that flows through the remote mountainous region.

The Shifting Global Setting

Since the fall of the Soviet Union, regional political tensions continue to challenge the Russians in both the east and west. In East Asia, the boundary between Russia and China was imposed by the Russian Empire in 1858 and has never been fully accepted by Beijing. The two countries battled over the area in 1969, but since then relations have improved. Territorial disagreements also complicate Russia's relationship with Japan in a dispute over the Kuril Islands. To the west, Russia worries about the expansion of the North Atlantic Treaty Organization (NATO). Although most Russian leaders accept the inclusion of Poland, Hungary, and the Czech Republic into NATO, they oppose the entry of former Soviet territories.

Today, Russian leaders are eager to increase the country's status as a global power. Its nuclear arsenal, while reduced in size, remains a powerful counterpoint to U.S., European, and Chinese interests. While Russia can no longer directly challenge the United States as it did in the days of the Soviet Union, it can act as a partial counterweight to the United States in international maneuverings. Russia also retains a permanent seat on the United Nations Security Council, arguably the world's most important geopolitical body. Russia's recent inclusion in some of the Group of Seven (G-7) economic meetings (which include the United States, Canada, Japan, Germany, Great Britain, France, and Italy) also signifies its remaining international clout. Indeed, the long geopolitical history of the domain suggests that, despite the region's recent weakness, it is likely to play a major role in future global political affairs.

Economic and Social Development: An Era of Ongoing Adjustment

The economic future of the Russian domain remains difficult to predict. Economic declines devastated the region for much of the 1990s, but the situation stabilized between 2000 and 2004. Russia's gross national product declined by more than 40 percent between 1990 and 1999, and similar economic disasters unfolded in nearby Belarus, Ukraine, Armenia, and Georgia. These declines marked the most abrupt economic collapse within the industrialized world since the Great Depression of the early 1930s. Recently, in the case of Russia, higher oil and gas prices brought some economic improvement, but significant challenges remain for the region.

The Legacy of the Soviet Economy

The birth of the Soviet Union in 1917 initiated a radical change within the region's economy. Under the Russian Empire, most people were peasants, farming the land much as they had done for centuries. Following the revolution, however, the Soviet Union quickly emerged to rival, and even surpass, many of the most powerful economies on Earth. During that era of unmatched growth, much of the region's present economic infrastructure was established, including new urban centers and industrial developments, as well as a modern network of transportation and communication linkages. Thus, even though the Soviet Union has departed from the scene, much of its economic legacy remains to shape the progress and the challenges of its contemporary economy.

As communist leaders such as Stalin consolidated power in the 1920s and 1930s, they nationalized Russian industries and agriculture, creating a system of **centralized economic planning** in which the state controlled production targets and industrial output. The Soviets emphasized heavy, basic industries (steel, machinery, chemicals, and electricity generation), postponing demand for consumer goods to the future. Huge increases in production were realized, but at a significant human cost as Soviet citizens endured poor-quality consumer goods (shoes, cars, appliances) for decades. The Soviets also nationalized agriculture. By the late 1920s, communist leader Stalin was shifting agricultural land into large-scale collectives and state farms that organized production around state-authorized production goals. Most peasants initially resisted collectivization, and many slaughtered their animals rather than turn them over to state ownership. Later efforts to put new lands into production also failed to meet the high expectations of Soviet planners.

Soviet-era industrial expansion was far more successful than any gains seen in agriculture. The nation's development of heavy industries during the 1930s proved vitally important in World War II. Although the German army advanced to within miles of Moscow, new centers of heavy industry in the resource-rich Urals and in Siberia's Kuznetsk Basin in the vicinity of Novosibirsk remained untouched, supplying the Russian army with the materials necessary to conduct the war (Figure 9.30). The industrial economy boomed in the 1950s and 1960s, and Soviet leaders confidently predicted that their country would be the world's most highly developed nation by 1980.

Much of the Russian domain's basic infrastructure—its roads, rail lines, canals, dams, and communications networks—also originated during the Soviet period. Dam and canal construction, for example, turned the main rivers of European Russia into a virtual network of interconnected reservoirs. Invaluable links such as the Volga-Don Canal (completed in 1952), which connected those two key river systems, have greatly eased the movement of industrial raw materials and manufactured goods within the country (Figure 9.31). The Soviets also improved the country's railroad network. Thousands of miles of new track were added in the European west, and the Trans-Siberian Line was modernized and then complemented by the

FIGURE 9.30 • MAJOR NATURAL RESOURCES AND INDUS-TRIAL ZONES   The region's varied natural resources and chief industrial zones are widely distributed. Fossil fuels are in abundance, although their distance from markets often imposes special costs. In southern Siberia, rail corridors offer access to many mineral resources. In the mineral-rich Urals and eastern Ukraine, proximity to natural resources sparked industrial expansion, while Moscow's industrial might is related to its proximity to markets and capital. *(Modified from Bergman and Renwick, 1999, Introduction to Geography, Upper Saddle River, NJ: Prentice Hall, and Rubenstein, 1999, Introduction to Human Geography, Upper Saddle River, NJ: Prentice Hall)*

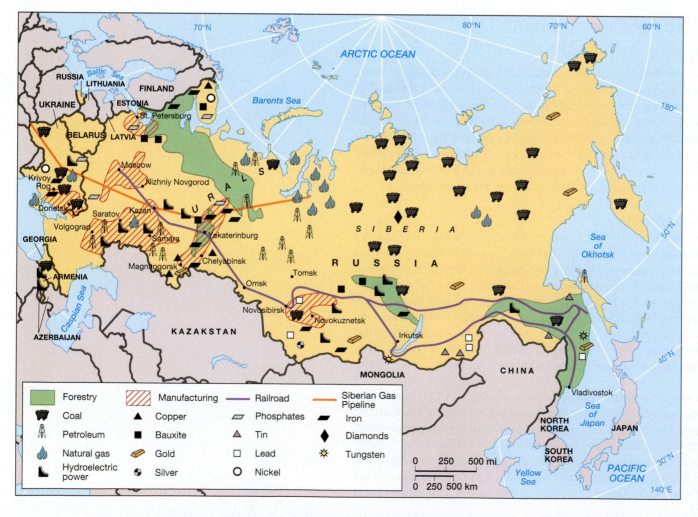

addition of the BAM link across central Siberia. Farther north, the Siberian Gas Pipeline was built to link the energy-rich fields of the Soviet arctic with growing demand in Europe.

Overall, the postwar period saw real economic improvements for the Soviet people. No longer a peasant nation, Russia, like the other Soviet republics, was now industrialized and largely urbanized. A massive effort to build new, more modern housing units in the 1950s and 1960s ensured most families a private apartment. Consumer goods, such as televisions, telephones, and even private automobiles, began to spread more widely through society. Literacy was virtually universal, quality health care was readily available, and every capable person was guaranteed employment.

Despite these successes, however, economic and social problems increased during the 1970s and 1980s. Soviet agriculture was still inefficient, and the country increasingly relied on grain imports. Manufacturing efficiency and quality failed to match the standards of the West, particularly in regard to consumer goods. Equally troubling was the fact that the Soviet Union was failing to participate fully in the technological revolutions that were transforming the United States, Europe, and Japan, a failure that had serious military consequences. Disparities also visibly grew between the Soviet elite and an everyday population that still enjoyed few personal freedoms. By the late 1980s, the Soviet Union had reached both an economic and a political impasse.

FIGURE 9.31 • VOLGA-DON CANAL   This view near Volgograd suggests the enduring economic importance of the Volga-Don Canal. Built during the Soviet era, the canal remains a key commercial link that facilitates the economic integration of southern Russia. *(Carl Wolinsky/Stock Boston)*

The Post-Soviet Economy

Fundamental economic changes have shaped the Russian domain since the demise of the Soviet Union. Particularly within Russia itself, much of the highly centralized state-controlled economy has been replaced by a mixed economy of state-run operations and private enterprise. The changeover has been very difficult. Fundamental problems of unstable currencies, corruption, and changing government policies plague the system. Indeed, the region experienced an unprecedented economic decline for much of the 1990s. While selective portions of the region's economy stabilized early this century, no one can confidently predict how the economy will evolve in the future.

Redefining Regional Economic Ties   Economic ties still link the five nations of the Russian domain, but the region has seen many of those relationships redefined in the post-Soviet era. The collapse of the communist state has meant that economic relationships between the former Soviet republics can no longer be controlled by a single, centralized government. Now, trade between Russia and Ukraine, for example, is a more complex, unpredictable exchange between two foreign countries. Russia still retains an overwhelming economic presence within the region. Belarus, for example, is still strongly dependent upon Russian economic assistance, particularly in the form of cheap energy exports. Recently, cash-starved Ukrainian companies have also looked to Russia for help, and the result has been a Russian-led buying binge of cheap Ukrainian corporations and real estate.

Privatization and Economic Uncertainty   Even as hopes initially ran high for a smooth restructuring of the post-Soviet economy, recent years have been very difficult for most residents of the region. In 1992 the Russian government suddenly freed prices from state control, encouraging more production and filling store shelves. As prices rose, however, inflation soared, wages failed to keep pace, and savings were wiped out as the value of the Russian currency (ruble) declined. Another dramatic move came in October 1993 when the government initiated a program to privatize the Russian economy. It took dozens of state-run companies and sold them to private investors. Unfortunately, however, a lack of legal and financial safeguards created many problems during this privatization process and many companies experienced mismanagement and corruption. Elsewhere in the region, the Belorussian and Ukrainian economies remain crippled by rigid, corrupt state controlled systems.

The agricultural sector continues to struggle. No matter its economic system, much of the region will always be challenged by short growing seasons, poor soils, and moisture deficiencies. Russia's best farmlands remain limited to a small slice of southern territory wedged between the Black and Caspian seas, leaving most of the rest of the country on the agricultural margins. About two-thirds of the country's farmland was privatized by 2000, with many farmers now jointly owning and working the same acreage they did under the Soviet system.

Russia, in particular, has encouraged the rapid privatization of the service and industrial sectors. Thousands of privatized retailing establishments have appeared, and they now dominate that portion of the economy. In addition, the long-established "informal economy" continues to flourish. Even during the Soviet era, millions of citizens earned extra money by informally selling Western consumer goods, manufacturing food and vodka, and providing skilled services such as computer and automobile repairs. Substantial portions of the natural resource and heavy industrial sectors of the economy have also been privatized.

Privatization, on scales both great and small, has also brought tremendous challenges. The privatization process has often been corrupt, unfairly benefiting company insiders and trade union officials who obtained controlling interest of many corporations. Special state rulings permitting privatization also benefited corrupt politicians who were paid off by the newly private companies. Many workers have seen little real economic gain. Complex and awkward tax policies also prevent efficient revenue collections from most private companies and individuals involved in the informal economy, and thus the government sees little benefit from such operations.

The Russian Mafia   Organized crime is everywhere in Russia and controls many aspects of the economy. The government's own interior ministry estimates that the Russian mafia controls about 40 percent of the private economy and 60 percent of state-run enterprises. Unofficially, sources also suggest that 80 percent of the country's banks are under mafia influence. While organized crime certainly existed in the Soviet era, the more liberal economic and political environment that followed has allowed it to thrive even more widely and openly. The mafia provided critical funds and jobs to many unemployed young men in the unstable months immediately following the collapse of communism. Today, bribery and protection are part of the cost of doing business across much of the Russian domain. Many businesses must turn to the mafia to obtain loans in an economy where willing banks and foreign lending institutions are in short supply. More than 8,000 syndicates, or mafia organizations, now exist in Russia alone. While one group might control the construction business in a Moscow suburb, another syndicate oversees drug dealing and prostitution, and still another helps to funnel illegal CDs and DVDs to eager consumers.

The Russian mafia has also gone global. In 2001, it was implicated in a huge money-laundering scheme that involved Russian, British, and U.S. banks as well as the flow of International Monetary Fund investments into the region. Banks in Switzerland, Liechtenstein, and Cyprus have been involved in similar schemes. Mafia money has also been invested in every corner of the world economy, including Sri Lankan gambling syndicates, Colombian drug enterprises, and legitimate Israeli high-technology companies. Russian emigrants help spread mafia influence and act as key local contacts in cities such as Tel Aviv, Paris, London, and New York.

Social Problems   Tough economic times and political uncertainties have contributed to a growing number of social problems within the Russian domain. Rates of violent crime increased late in the Soviet period and have risen further since the fall of communism. High unemployment, rising housing costs, and declining social welfare expenditures have hit many families hard. Often, both husband and wife work multiple low-paying jobs with few benefits and long hours.

Women are paying an especially heavy price amid the current economic and social problems. Increasing domestic violence has particularly impacted women throughout the region. Beatings and rapes are common. A survey in Moscow suggested that one-third of divorced women had experienced domestic violence, while a women's rights group in Ukraine reported that rape was an all-too-common crime in many villages. Divorce rates also continue to rise as families struggle to make ends meet. Economically, women often have suffered first as unemployment has risen. Forced out of traditional industrial and service-sector jobs, many have been compelled to sell homemade products or services on city streets for extremely low wages (Figure 9.32). Many housewives and schoolteachers are forced into prostitution, simply as a way to survive.

Russia's health-care crisis also illustrates the relationship between the region's troubled economy, limited government expenditures, and larger social issues. Soviet spending on health care averaged 6.6 percent of the country's gross national product in 1960 but declined to 2.3 percent by the end of the Soviet period. A shortage of vaccines and medical services has contributed to the return of such diseases as cholera, typhus, and the bubonic plague. Chronic illnesses, often related to lifestyle, produce the highest mortality rates in the developed world for cardiovascular disease, alcoholism, and smoking, particularly for Russian men. AIDS is rapidly on the rise, spreading especially among the region's young drug-using population. In addition, toxic environmental conditions have affected both human health and the overall quality of life in the region.

FIGURE 9.32 • LIFE ON THE STREET This young street vendor at the Moscow Food Market hopes to sell her pickled vegetables to passing consumers. Economic instability and high unemployment, however, make even modest purchases more difficult for many Russians. (*John Egan/The Hutchison Library*)

Growing Economic Globalization

The relationship between the Russian domain and the world beyond has shifted greatly since the end of communism. During much of the Soviet era, the region was relatively isolated from the world economic system. By the 1970s, however, the Soviet Union began to export large quantities of fossil fuels to the West while importing more food products. Connections with the global economy became stronger after the downfall of the Soviet Union.

A New Day for the Consumer Today, residents of the Russian domain, particularly those living in its larger cities, are bombarded by new consumer imports. McDonald's hamburgers, Calvin Klein jeans, and many other symbols of global capitalism are visible in the heart of Moscow and, increasingly, in many other settings throughout the region. Luxury goods from the West have also found a small but enthusiastic market among the newly emergent Russian upper class; a group noted for its devotion to BMW automobiles, Rolex watches, and other status emblems. Most important, a consumer consciousness seems more apparent today within the Russian domain than it was during the Soviet era.

Attracting Foreign Investment The Russian domain (particularly Russia itself) has seen some growth in foreign investment in the post-Soviet era, but the region struggles to attract enough outside capital. Still, money has flowed into Russian-controlled enterprises (many Russian companies are now listed on European and U.S. stock exchanges) and has also been directly invested into the region's economy. As measured by total foreign investment, the strongest global ties by far have been with the United States and western Europe, particularly Germany and Great Britain. Large investments have been made in Russia's oil and gas economy, as well as in the food, telecommunications, and consumer goods industries. In 2000 foreign investment in Russia totaled more than $9.5 billion and was growing at more than 14 percent annually.

Globalization and Russia's Petroleum Economy Russia's oil and gas industry remains one of the strongest economic links between the region and the global economy. The diverse international connections it has built suggest the importance of the energy sector to the region's future. The statistics are impressive: Russia has 35 percent of the world's natural gas reserves (mostly in Siberia), and, even with its

FIGURE 9.33 • SIBERIAN GAS PIPELINE Snaking its way across the remote Russian taiga, the Siberian Gas Pipeline connects arctic gas fields to needy consumers in far-off Europe. The costly pipeline requires frequent maintenance in the harsh northern environment. *(Tass/Sovfoto/Eastfoto)*

FIGURE 9.34 • VLADIVOSTOK The busy harbor of Vladivostok remains Russia's leading trade center in the Far East. With easy access to markets in Japan, China, and even the United States, this Pacific port is poised to grow as Russia's economy recovers. *(Itar-Tass/Sovfoto/Eastfoto)*

recent economic turmoil, it is the world's largest gas exporter. As for oil, Russia is by far the world's largest non-OPEC producer: It outpaces even the United States in annual output (major oilfields are in Siberia, the Volga Valley, the Far East, and the Caspian Sea region), and it possesses more than twice the proven petroleum reserves of the United States.

The changing geography of the Russian oil and gas business reveals the global character of the energy economy. Prior to the breakup of the Soviet Union, about half of Russia's oil and gas exports went to other Soviet republics, such as Ukraine and Belarus. While these two nations still depend on Russian supplies, the primary destination for Russian petroleum products has shifted to western Europe. Russia now supplies that region with more than 25 percent of its natural gas and 16 percent of its crude oil. An agreement between Russia and the European Union in 2000 aimed at the rapid expansion of these East–West linkages. The Siberian Gas Pipeline already links distant Asian fields with western Europe via Ukraine, and those connections are being supplemented by new lines through Belarus (the Yamal-Europe Pipeline) and Turkey (the Blue Stream Pipeline) (Figure 9.33). Similar expansions are dramatically changing the geography of oil exports. The huge Druzhba Pipeline already serves western Europe. Large new export terminals are opening on the Baltic and Black seas, including a major new line from the vast Caspian Sea reserves that opened in 2001.

There are several indicators that the world is making a large bet that Russia's petroleum products will help integrate that troubled region with the rest of the global economy. For example, British, U.S., Dutch, and Norwegian oil and gas companies are spearheading many technical aspects of these huge initiatives. Similarly, Italian and Japanese banks are financing much of the Blue Stream Pipeline, and a broad assortment of other global financial loan guarantees and assistance are backing other exploration and pipeline projects. While Russia has much to gain in this process, the region will also have to bear the human and environmental costs associated with growth in the petroleum industry.

Local Impacts of Globalization As the geography of the Russian petroleum industry illustrates, the local impacts of globalization vary a great deal from place to place. For example, portions of the region that are close to oil and gas wells, pipeline and refinery infrastructure, and key petroleum shipping points are greatly impacted (both economically and environmentally) by the changing global oil economy. The same is true more broadly: Globalization has affected different locations within the Russian domain in very distinctive ways. Within Russia, capitalism has brought its most dramatic, though selective, benefits to Moscow. Indeed, more than 32 percent of new foreign investment in the country has recently gone to the Moscow area.

Beyond Moscow, St. Petersburg and the Siberian city of Omsk have seen growing global investment, while new oil and gas prospects in Siberia, the Far East (Sakhalin Island), and near the Caspian Sea have attracted other global capital. Cities such as Nizhniy Novgorod and Samara have also succeeded in attracting global capital by building a probusiness economic environment. In addition, port cities such as Vladivostok are well positioned to trade with nearby markets, although political red tape and corruption have hampered growth in the region (Figure 9.34).

Elsewhere, globalization has clearly produced local economic declines. Older, less competitive industrial centers in the Urals, Siberia, and eastern Ukraine have been hit hard. Aging steel plants, for example, no longer have guaranteed markets for their high-cost, low-quality products as they did in the days of the planned Soviet economy. Instead, they must compete on the global market, a market that is increasingly prone to lower prices and weakening demand for many of the traditional industrial goods the region produces.

CONCLUSION

Many influences have shaped the human geographies of the Russian domain. Much of its underlying cultural geography was formed centuries ago, the complex product of Slavic languages, Orthodox Christianity, and numerous ethnic minorities that continue to complicate the scene today. Further changing the country are new global influences, a set of products, technologies, and attitudes that often clash with traditional cultural values. Much of the region's political legacy is rooted in the Russian Empire, a land-based system of colonial expansion that greatly enlarged Russian influence after 1600 and then reappeared as the Soviet Union expanded its own influence. Only large remnants of that empire survive on the modern map, yet it has stamped the geopolitical character of the region in lasting ways. Much of the region's economic geography remains defined by its varied natural setting, as well as by the changes wrought by twentieth-century industrialization and urbanization during the Soviet period. The result is a modern human geography rooted in many different historical influences ranging from Tatar nationalism to twenty-first century consumerism.

Looking toward the future, the region's basic political geography remains in doubt. Will Belarus or eastern Ukraine rejoin Russia? Northern Kazakstan, still home to a large Russian population, is another target for possible Russian expansion. The most nationalistic Russian leaders would like to go several steps further and recreate the Soviet Union. While partial reunification of the Russian domain remains a possibility, there is also the potential for more division. Many non-Russian peoples within Russia might prefer independence. The Russian government appears determined to prevent political fragmentation, but its financial resources and even its centralized military power remain limited.

Equally uncertain are the region's future political and economic systems, which will significantly influence its international relations. Recent Russian leaders promised a quick transition to democracy and capitalism, as well as rapid integration into the global economic system. The economic and social difficulties that followed, however, have largely discredited this approach in the eyes of the Russian population.

Despair remains the greatest challenge facing the people of the region. Recent disappointments have been monumental. Communism spectacularly failed in its social and economic promises to bring about a better life. Unfortunately, Russia's experimentation with capitalism has brought not the freedom and prosperity its supporters promised, but rather crime, poverty, and uncertainty. Longer term, the Russian domain's links to the world economy will be crucial, but those connections have yet to provide many immediate rewards for the majority of the population. More than anything, the people of the Russian domain need a renewal of hope, a prospect that tomorrow's world might be better than today's.

KEY TERMS

autonomous areas (page 246)
Baikal-Amur Mainline (BAM) Railroad (page 236)
Bolsheviks (page 244)
centralized economic planning (page 249)
chernozem soils (page 232)

Cold War (page 246)
Commonwealth of Independent States (CIS) (page 247)
Cossacks (page 241)
denuclearization (page 247)
Eastern Orthodox Christianity (page 240)

exclave (page 246)
glasnost (page 246)
Gulag Archipelago (page 237)
Iron Curtain (page 246)
mikrorayons (page 239)
perestroika (page 246)
permafrost (page 232)

podzol soils (page 231)
Russification (page 237)
Slavic peoples (page 240)
socialist realism (page 243)
taiga (page 232)
Trans-Siberian Railroad (page 236)
tsars (page 236)

Central Asia

FIGURE 10.1 Central Asia

Central Asia, an extensive region in the center of the Eurasian continent, is dominated by arid plains and basins and high mountain ranges and plateaus. Eight independent countries— Kazakstan, Turkmenistan, Uzbekistan, Kyrgyzstan, Tajikistan, Azerbaijan, Afghanistan, and Mongolia—form Central Asia's core. The region also includes China's lightly populated far west, which has cultural and environmental similarities to the rest of Central Asia.

CENTRAL ASIA
Political Map

⊗ ● Over 1,000,000

○ ● 500,000–1,000,000

★ ● Selected smaller cities

Elevation in meters

4000+
2000–4000
500–2000
200–500
0–200
Below sea level

Sea Level

SETTING THE BOUNDARIES

The term *Central Asia* is defined differently by different writers. Most authorities agree that it includes five newly independent former-Soviet republics: Kazakstan, Kyrgyzstan, Uzbekistan, Tajikistan, and Turkmenistan. This chapter, however, adds another former Soviet state, Azerbaijan, in addition to Mongolia and Afghanistan, as well as the autonomous regions of western China (Tibet and Xinjiang). Several other provinces and regions of western China, such as Nei Mongol (Inner Mongolia) and Qinghai, are occasionally discussed.

The inclusion of these additional territories within Central Asia is controversial. Azerbaijan is often classified with its neighbors in the Caucasus region (Georgia and Armenia); western China is obviously part of East Asia by political criteria; and Mongolia is also often placed within East Asia because of both its location and its historical connections with China. Afghanistan is just as often located within either South Asia or Southwest Asia and the Middle East.

But considering Central Asia's historical unity, its similar environmental settings, and its recent reentry onto the stage of global geopolitics, we think that it deserves consideration in its own right. It also makes sense to define its limits rather broadly. Azerbaijan, for example, is linked by both cultural (language and religion) and economic (oil) factors more to Central Asia than it is to Armenia and Georgia.

Central Asia is not a very well-defined region, nor is it politically stable. Continuing Chinese political control over, and Han Chinese migration into, southeastern Central Asia threatens whatever claims may be made for regional coherence. Central Asia itself remains deeply divided along cultural lines. Most of the people of the region speak Turkish languages and practice Islam, but residents in both the northeastern and southeastern sections (Mongolia and Tibet) are firmly Buddhist in their religious beliefs.

Central Asia does not appear in most books on world regional geography. Although it covers a larger expanse than the United States, it is a remote and lightly populated area dominated by high mountains, barren deserts, and semiarid steppes (grasslands). Central Asia is unique among world regions because, until recently, most of its countries have been under foreign political control. Until 1991 it contained only two independent states, Mongolia and Afghanistan. The rest of the region was at that time divided between the Soviet Union and China (Figure 10.1).

Central Asia began to reappear in discussions of global geography following the breakup of the Soviet Union. Suddenly a handful of new countries appeared on the international scene, prompting scholars to pay more attention to this rapidly changing region (see "Setting the Boundaries"). Central Asia was more firmly established on the world map after September 11, 2001, when it became evident that the attack on the World Trade Center and the Pentagon had been planned and organized by Osama bin Laden and his Al Qaeda organization operating out of Afghanistan.

One reason for Central Asia's insignificance as a world region until recently is that it was poorly integrated into international trade networks. This began to change, too, in the 1980s and 1990s as large oil and gas reserves were found, especially in Kazakstan, Turkmenistan, and Azerbaijan. Estimates of the total oil reserves in the region run between 70 and 200 billion barrels, second only to the 600 billion barrels in the Persian Gulf area. As a result, Western oil companies are showing increasing interest in Central Asia. A number of important countries, moreover, are seeking to exert influence over Central Asia, including Iran, Pakistan, Turkey, the United States, and Russia. International concerns over China's strict control and periodic repression of its Central Asian lands also highlight the significance of the region.

Central Asia forms a large, compact region in the center of the Eurasian landmass. Alone among all the world regions, it lacks ocean access. Owing to its continental position in the center of the world's largest landmass, Central Asia is noted for its severe climate. High mountains, deep basins, and extensive plateaus increase its climatic extremes. The aridity of the region, as we shall see, has also contributed to some of the most severe environmental problems in the world.

Environmental Geography: Steppes, Deserts, and Threatened Lakes

One of the great environmental tragedies of the twentieth century was the near destruction of the Aral Sea, a large, saline lake (until recently, larger than Lake Michigan) located on the boundary of Kazakstan and Uzbekistan in western Central Asia. The Aral's only sources of water are the Amu Darya and Syr Darya rivers, which

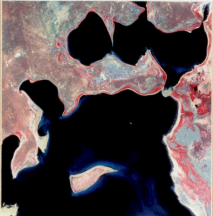
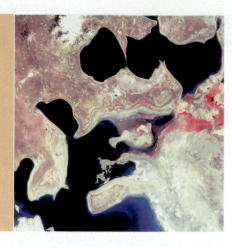

FIGURE 10.2 • THE SHRINKING OF THE ARAL SEA These three satellite images—the first taken in May 1973, the second in August 1987, and the third in July 2000—show the dramatic shrinkage of the Aral Sea. More than 60 percent of the lake's water has been lost since the 1960s, resulting in severe economic damage and environmental degradation. *(EROS Data Center, U.S. Geological Survey)*

FIGURE 10.3 • A DYING LAKE Due to the shrinkage of the Aral Sea, former lakeside villages are now located far inland, as evidenced by these two beached ships. Not only have fishing economies been destroyed, but the desiccated lake bed itself is now a source of pollution, as desert winds deposit salt and agricultural chemicals on fields. *(© David Turnley/Corbis)*

flow out of the Pamir Mountains, some 600 miles to the southeast. Both of these rivers have been intensively used for irrigation for thousands of years, but the size of diversion greatly expanded after 1950. The valleys of the two rivers formed the southernmost farming districts of the Soviet Union and thus became critical suppliers of warm-season crops, such as rice and especially cotton. Soviet agricultural planners favored huge engineering projects that used irrigation canals from the Amu Darya and Syr Darya rivers to deliver water to arid lands and thus "make the deserts bloom."

Unfortunately, more water delivered to produce crops meant less freshwater available for the Aral Sea. As the inflow of water was reduced, the shallow waters of the Aral began to recede. By the 1970s the sea began to shrink at a rapid rate; eventually a number of "seaside" villages found themselves up to 40 miles (64 kilometers) inland. With less freshwater flowing into the lake, the water of the Aral Sea grew increasingly salty. As a result, an estimated 135—out of a total of 173—animal species in the lake disappeared. New islands began to emerge, and by the 1990s the Aral Sea had been divided into two separate lakes (Figure 10.2).

The destruction of the Aral Sea resulted in economic and cultural damage, as well as ecological devastation. Fisheries that were once large enough to support a canning industry began to close as the lake grew increasingly salty. Even agriculture has suffered. The retreating lake has left large salt flats on its exposed beds. Windstorms pick up the salt, along with the agricultural chemicals that have accumulated in the lake's shallows, and deposit it in nearby fields. As a result, agricultural yields have declined, desertification has accelerated, and public health has been threatened (Figure 10.3).

Major Environmental Issues

Despite the tragedy of the Aral Sea, much of Central Asia has a relatively clean environment, owing largely to its generally low population density. Industrial pollution is a serious problem only in the larger cities, such as Tashkent (in Uzbekistan) and Baku (in Azerbaijan). Some parts of Central Asia, such as northwestern Tibet, remain unspoiled, with little human impact of any kind. Elsewhere, however, the typical environmental problems of arid environments plague the region: desertification (the spread of deserts resulting from poor land-use practices), salinization (the accumulation of salt in the soil), and desiccation (the drying up of lakes and wetlands).

Desertification Desertification, caused by over-grazing and poor farming practices, is a major concern in Central Asia (Figure 10.4). In the eastern part of the region, the Gobi Desert has gradually spread southward, encroaching on densely settled lands in northeastern China proper. The Chinese have tried to prevent the march of desert with massive tree- and grass-planting campaigns, as the roots of such plants stabilize the soil and help to keep sand dunes from moving. Such efforts, however, have been only partially successful. Northern Kazakhstan has also seen extensive

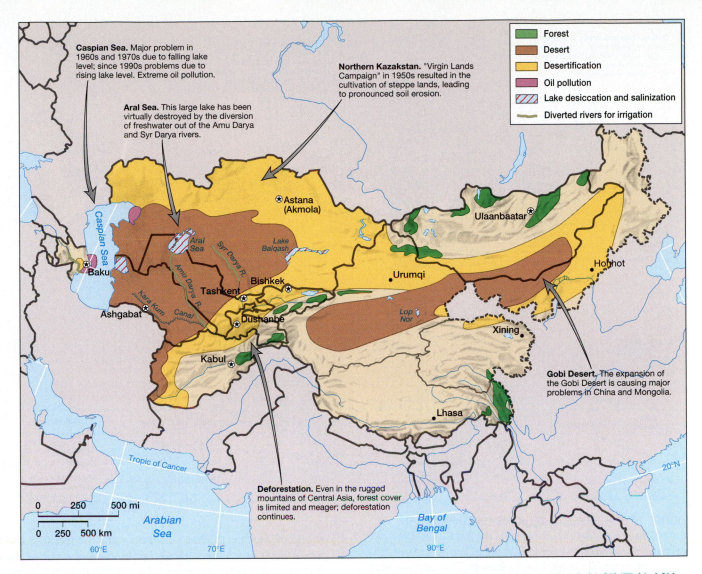

Caspian Sea. Major problem in 1960s and 1970s due to falling lake level; since 1990s problems due to rising lake level. Extreme oil pollution.

Aral Sea. This large lake has been virtually destroyed by the diversion of freshwater out of the Amu Darya and Syr Darya rivers.

Northern Kazakstan. "Virgin Lands Campaign" in 1950s resulted in the cultivation of steppe lands, leading to pronounced soil erosion.

■ (green)	Forest
■ (brown)	Desert
■ (yellow)	Desertification
■ (pink)	Oil pollution
▨	Lake desiccation and salinization
∿	Diverted rivers for irrigation

Gobi Desert. The expansion of the Gobi Desert is causing major problems in China and Mongolia.

Deforestation. Even in the rugged mountains of Central Asia, forest cover is limited and meager; deforestation continues.

FIGURE 10.4 • ENVIRONMENTAL ISSUES IN CENTRAL ASIA
Desertification is perhaps more widespread in Central Asia than in any other world region. Soil erosion and overgrazing have led to the advance of desertlike conditions in much of western China and Kazakstan. In western Central Asia, the most serious environmental problems are associated with the diversion of rivers for irrigation and the corresponding desiccation of lakes. Oil pollution is a particularly serious issue in the Caspian Sea area.

desertification. This area was one of the main sites of the ambitious Soviet "Virgin Lands Campaign" of the 1950s, in which semiarid grasslands were plowed and planted with wheat. As a result of plowing, wind erosion stripped away much of the area's soil. Many of these lands have since returned to native grass cover.

Shrinking and Expanding Lakes Western Central Asia contains large lakes because it forms a low-lying basin, without drainage to the ocean, that is surrounded by mountains and other more humid areas. The world's largest lake, by a huge margin, is the Caspian Sea, located along the region's western boundary; the fourth largest is (or more precisely, was) the Aral Sea, situated some 300 miles (480 kilometers) to the east, and the fifteenth largest is Lake Balqash, found 600 miles (960 kilometers) farther east. Like the Aral Sea, Lake Balqash has become smaller and saltier over the past several decades.

The story of the Caspian, however, is more complicated. The Caspian Sea receives most of its water from the large rivers of the north, the Ural and the Volga, which drain much of European Russia. As a result of the construction of large reservoirs and the development of extensive irrigation in the lower Volga basin, the volume of freshwater reaching the Caspian began to decline in the second half of the twentieth century. With a reduced flow of water, the level of the great lake dropped, exposing as much as 15,000 square miles (39,000 square kilometers) of former lake bed. Decreased water volume and increased salinity disrupted the ecosystem and devastated fisheries. The Russian caviar industry, centered in the northern Caspian, was particularly damaged by these changes.

The water level of the Caspian Sea reached a low point in the late 1970s. At that point it began to rise, probably because of higher than normal precipitation in its drainage

259

Caspian Sea and Basin. The Caspian Sea is the world's largest lake by a wide margin. It lies within the Caspian Basin, containing the world's largest area of dry land below sea level.

Turfan Depression. At 505 feet (154 meters) below sea level, the Turfan fault trough depression is one of the lowest places in the world. Intensively farmed, the basin is famous for its fruits and melons.

The gorge country of eastern Tibet. Several extremely steep canyons alternate with lofty ridges, making eastern Tibet one of the most topographically forbidding places in the world.

The Pamir Knot. A complex tangle of east–west and north–south trending ranges, the Pamir Knot forms the Asian highland core. Peaks reach up to 24,584 feet (7,495 meters).

Legend:
— Tibetan Plateau
— Highlands
— Steppe Zone
- - - Desert
— Important rivers of desert zone

FIGURE 10.5 • PHYSICAL REGIONS OF CENTRAL ASIA

Central Asia is divided into three main regions based on physical geography. Relatively flat, grassy plains known as the *steppes* dominate the north. Desert plains and basins cover most of the central area. Finally, mountain ranges and plateaus of the highland zone, although scattered throughout Central Asia, are particularly common in the south. Since most of Central Asia is arid, rivers running out of the highlands have special significance for the region's human geography.

basin, and by the late 1990s it had risen some 8.2 feet (2.5 meters). This enlargement, too, has caused problems, such as the flooding of some of the newly reclaimed farmlands in the Volga Delta. At present, the most serious environmental threat to the Caspian is probably pollution from the oil industry, rather than fluctuation in size.

Central Asia's Physical Regions

To understand why Central Asia suffers from these environmental problems, it is necessary to examine the region's physical geography in greater detail. In general, Central Asia is characterized by high plateaus and mountains in the south-center and southeast, grassland plains (or steppe) in the north, and desert basins in the southwestern and central areas (Figure 10.5).

The Central Asian Highlands The highlands of Central Asia originated in one of the great geological events of Earth's history: the collision of the Indian subcontinent into the Asian mainland. This ongoing collision has created the highest mountains in the world, the Himalayas, located along the boundary of South Asia and Central Asia. To the northwest, the Himalayas merge with the Karakoram Range and then the Pamir Mountains. From the so-called Pamir Knot, a complex tangle of mountains located where Pakistan, Afghanistan, China, and Tajikistan meet, other towering ranges spread outward in several directions. The Hindu Kush curves to the

FIGURE 10.6 • TIBETAN PLATEAU Alpine grasslands and tundra interspersed with rugged mountains and saline lakes dominate the Tibetan Plateau. In summer the sparse vegetation offers forage for the herds of nomadic Tibetan pastoralists. Much of northern Tibet, however, is too high to support pastoralism and is therefore uninhabited. *(Michel Peissel/SIPA Press)*

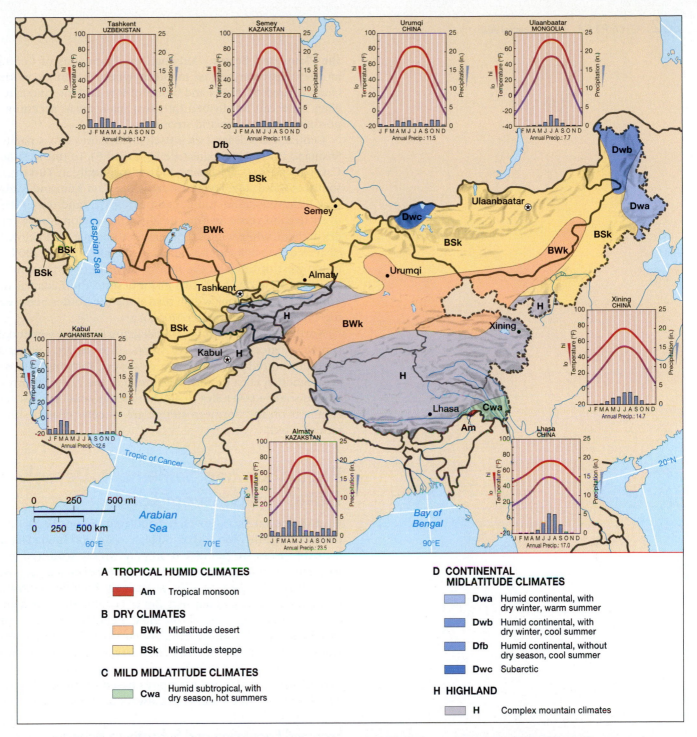

FIGURE 10.7 • CLIMATES OF CENTRAL ASIA Central Asia is a dry region dominated by desert and steppe climates. Even in most of Central Asia's highlands, marked "H" on this map, arid conditions prevail. Truly humid areas in Central Asia are found only in small portions of the far north and extreme southeast. Since Central Asia is located in the interior of a large continent, its climate is characterized by significant differences between winter and summer temperatures.

southwest through central Afghanistan, the Kunlun Shan extends to the east, and the Tien Shan swings out to the northeast into China's Xinjiang province. All of these ranges have peaks higher than 20,000 feet (6,000 meters) in elevation.

Much more extensive than these mountain ranges, however, is the Tibetan Plateau (Figure 10.6). This massive upland extends some 1,250 miles (2,000 kilometers) from east to west and 750 miles (1,200 kilometers) from north to south. Its size is more remarkable than its elevation; almost the entire area is higher than 12,000 feet (3,700 meters) above sea level, and its average height is about 15,000 feet (4,600 meters). In fact, the greater part of the Tibetan Plateau lies near the maximum elevation at which human life can exist. Rather than forming a flat surface, the plateau has numerous east–west running mountain ranges alternating with basins. Although the southeastern sections of the plateau receive plenty of rainfall, most of Tibet is arid (Figure 10.7). Winters on the Tibetan Plateau are cold; and while summer afternoons can be warm, summer nights remain chilly.

The Plains and Basins   Although the mountains of Central Asia are higher and more extensive than those found anywhere else in the world, most of the region is characterized by plains and basins of low and intermediate elevation. This lower-lying zone can be divided into two main areas: a central belt of deserts and a northern strip of semiarid steppe.

Central Asia's desert belt is itself divided into two separate segments by the Tien Shan and Pamir Mountains (see Figure 10.5). To the west lie the arid plains of the Caspian and Aral Sea basins, located primarily in Turkmenistan, Uzbekistan, and southern Kazakstan. The driest areas, the Kara Kum and Kyzyl Kum deserts, support very little vegetation and contain extensive sand dunes. Most of this area is relatively flat and very low. The climate of this region is very continental; summers are dry and hot, while winter temperatures average well below freezing. Central Asia's eastern desert belt extends for almost 2,000 miles (3,200 kilometers) from the extreme west of China at the foot of the Pamirs to the southeastern edge of Inner Mongolia. Here one finds two major deserts: the Taklamakan, in the Tarim Basin of Xinjiang, and the Gobi, which runs along the border between Mongolia proper and the Chinese region of Inner Mongolia.

North of the desert zone, rainfall gradually increases and desert eventually gives way to the great grasslands, or steppe, of northern Central Asia. Near the region's northern boundary, trees begin to appear, showing that one is approaching the great Siberian taiga (coniferous forest) of the north. Nearly continuous grasslands extend some 4,000 miles (6,400 kilometers) east to west across the entire region. Summers on the northern steppe are usually pleasant, but winters are often extremely cold.

Population and Settlement: Densely Settled Oases amid Vacant Lands

Most of Central Asia is sparsely populated (Figure 10.8). Large areas are essentially uninhabited, either too arid or too high to support human life. Even many of the more favorable areas are populated only by widely scattered groups of nomadic **pastoralists** (people who raise livestock for subsistence purposes). Mongolia, which is more than twice the size of Texas, has only 2.5 million inhabitants—fewer than live in the Dallas metropolitan area. But as is common in arid environments, those few lowland settings with good soil and dependable water supplies are thickly settled. Despite its overall aridity, Central Asia has many rivers and fertile oases (areas where wells or other sources of water allow irrigated agriculture in a desert environment). While the nomadic pastoralists of the steppe zone have dominated the history of Central Asia, creating great empires based on their military power, the sedentary peoples of the river valleys have always been more numerous.

Highland Population and Subsistence Patterns

The environment of the Tibetan Plateau is particularly harsh. Only sparse grasses and herbaceous plants—so-called *mountain tundra*—can survive in this high-altitude climate, and human subsistence is very difficult. The only feasible way of life in this area is nomadic pastoralism based on the yak, an altitude-adapted relative of the cow. Several hundred thousand people manage to make a living in such a manner, roaming with their herds over great distances.

Although most of the Tibetan Plateau can support only nomadic pastoralism, the majority of Tibetans are sedentary farmers. Farming in Tibet is possible only in the few locations that are *relatively* low in elevation and that have good soils and either adequate rainfall or a dependable irrigation system based on local streams. The main zone of sedentary settlement lies in the far south, where protected valleys offer these favorable conditions. The population of Tibet proper (the Chinese autonomous region of Xizang) is only 2.5 million, while that of China's Qinghai province (most of which lies on the Tibetan Plateau) is 4.2 million. Considering the huge size of this area, these are very small numbers.

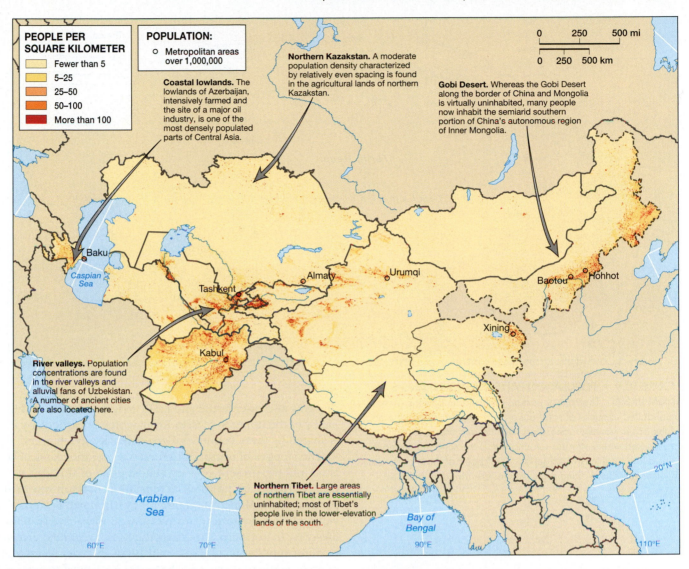

PEOPLE PER SQUARE KILOMETER
- Fewer than 5
- 5–25
- 25–50
- 50–100
- More than 100

POPULATION:
○ Metropolitan areas over 1,000,000

Coastal lowlands. The lowlands of Azerbaijan, intensively farmed and the site of a major oil industry, is one of the most densely populated parts of Central Asia.

Northern Kazakstan. A moderate population density characterized by relatively even spacing is found in the agricultural lands of northern Kazakstan.

Gobi Desert. Whereas the Gobi Desert along the border of China and Mongolia is virtually uninhabited, many people now inhabit the semiarid southern portion of China's autonomous region of Inner Mongolia.

River valleys. Population concentrations are found in the river valleys and alluvial fans of Uzbekistan. A number of ancient cities are also located here.

Northern Tibet. Large areas of northern Tibet are essentially uninhabited; most of Tibet's people live in the lower-elevation lands of the south.

FIGURE 10.8 • POPULATION DENSITY IN CENTRAL ASIA
Central Asia as a whole remains one of the world's most sparsely populated regions, although it does contain distinct clusters of higher population density. Most of Central Asia's large cities are located near the region's periphery or in its major river valleys.

Throughout Central Asia, the mountainous areas are vitally important for people living in the adjacent lowlands, whether they are migratory pastoralists or settled farmers. Many herders use the highlands for summer pasture; when the lowlands are dry and hot, the high meadows provide rich grazing. The Kyrgyz (of Kyrgyzstan) are noted for their traditional economy based on **transhumance**, moving their flocks from lowland pastures in the winter to highland meadows in the summer. The farmers of Central Asia rely on the highlands for their wood and, more important, for their water.

Lowland Population and Subsistence Patterns

Most of the inhabitants of Central Asian deserts live in the narrow belt where the mountains meet the basins and plains. Here water supplies are adequate and soils are neither salty nor alkaline, as is often the case in the basin interiors. For example, the population distribution pattern of China's Tarim Basin forms an almost perfect ringlike structure (Figure 10.9). Streams flowing out of the mountains are diverted to irrigate fields and orchards in the fertile band along the basin's edge.

The population west of the Pamir range, in former Soviet Central Asia, is also concentrated in the transitional zone between the highlands and the plains. A series of **alluvial fans** (fan-shaped deposits of sediments dropped by streams flowing out of the mountains) have long been devoted to intensive cultivation. **Loess**, a fertile, silty soil deposited by the wind, is plentiful in this region, and in a few favored areas winter precipitation is high enough to allow rain-fed agriculture. Several large valleys in

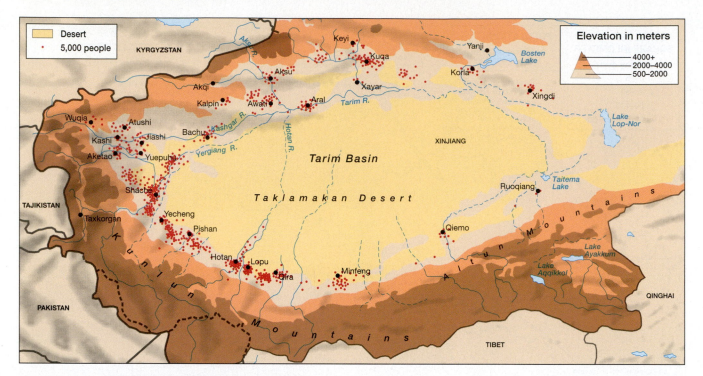

FIGURE 10.9 • POPULATION PATTERNS IN XINJIANG'S TARIM BASIN

The central portion of the Tarim Basin is a nearly uninhabited expanse of sand dunes and salt flats. Along the edge of the basin, however, dense agricultural and urban settlements are located where streams running out of the surrounding mountains allow for intensive irrigation. The largest of these oasis communities are found along the southwestern fringe of the basin.

FIGURE 10.10 • FARMLAND IN UZBEKISTAN

The fertile river valleys of Uzbekistan have been intensely cultivated for many centuries, producing large harvests of fruits and vegetables in addition to cotton and grain. Here tomatoes are harvested—a subtropical crop that has traditionally been exported to Russia and other areas. (Jeremy Nicholl/Katz/Corbis/SABA Press Photos, Inc.)

this area also offer fertile and easily irrigated farmland (Figure 10.10). The Fergana Valley of the upper Syr Darya River, which is particularly noted for its productivity, is shared by three countries: Uzbekistan, Kyrgyzstan, and Tajikistan.

The steppes of northern Central Asia are the classical land of nomadic pastoralism. Until the present century, almost none of this area had ever been plowed and farmed. To this day, pastoralism remains a common way of life across the grasslands, particularly in Mongolia (Figure 10.11). In northwestern China and the former Soviet republics, however, many pastoral peoples have been forced to adopt sedentary lifestyles. In northern Kazakstan, the Soviet regime converted the most productive pastures into farmland in the mid-1900s in order to increase the country's supply of grain. Some of these lands have since reverted to steppe, but large areas are still farmed, and Kazakstan is a major producer of spring wheat. Consequently, northern Kazakstan has the highest population density in the steppe region.

Population Issues

Although Central Asia has a low population density, some portions of it are growing at a moderately rapid pace. In western China, much of the population growth over the past 30 years has stemmed from the migration of Han Chinese into the area—a movement much resented by many of the local people. Population growth in the southern half of the former Soviet zone, on the other hand, has stemmed not from immigration but rather from relatively high levels of fertility. This area, particularly Kazakstan, has actually witnessed a substantial migration of people out of the region. These emigrants are mostly ethnic Russians returning to the Russian homeland.

Fertility patterns vary substantially from one part of Central Asia to another. Afghanistan, the least-developed and most male-dominated country of the region, has the highest birthrate by a substantial margin. Although good data are difficult to find, much evidence would suggest that Tibet's birthrate remains quite low. Kazakstan's birthrate, just slightly under the natural replacement level, is also low for the region, reflecting in part the extremely low fertility level of the country's Russian speakers.

Urbanization in Central Asia

Although the steppes of northern Central Asia had no real cities before the modern age, the river valleys and oases have been partially urbanized for thousands of years. Such cities as Samarkand and Bukhara in Uzbekistan were famous even in medieval

Europe for their riches and their lavish architecture (Figure 10.12). Cities such as these with a rich cultural past contrast sharply with the those that were built during the period of Russian/Soviet rule. The city of Tashkent in Uzbekistan is largely a Soviet creation and looks quite different from Bukhara, which still reflects the older urban patterns. Several major cities, such as Kazakstan's former capital of Almaty, did not exist before Russian colonization. In Azerbaijan, Baku emerged as a major city in the early twentieth century as the first Caspian oil fields began to be intensively exploited. In cities throughout this former Soviet region, one can see the effects of centralized Soviet urban planning and design.

Parts of Central Asia have experienced substantial urbanization in recent decades. North-central Kazakstan is currently seeing the rise of a new major city, Astana, which the government has designated as a new capital. Even Mongolia, long a land without many permanent settlements, now has more people living in cities than in the countryside. In some parts of the region, however, cities remain relatively few and far between. Only 28 percent of the people of Tajikistan, for example, are urban residents. Tibet similarly remains a predominantly rural society.

Cultural Coherence and Diversity: A Meeting Ground of Different Traditions

Although Central Asia has a certain environmental similarity, its cultural unity is more questionable. The western half of the region is largely Muslim and is often classified as part of Southwest Asia. In Mongolia and Tibet, in the eastern part of the region a distinctive form of Buddhism exists called either *Tibetan Buddhism* or *Lamaism*. Tibet is culturally linked to both South and East Asia, and Mongolia is historically associated with China, but neither fits easily within any world region.

Historical Overview: Changing Languages and Populations

The river valleys and oases of Central Asia were early sites of sedentary, agricultural communities. Archaeologists have discovered abundant evidence of farming villages dating back to the Neolithic period (beginning around 8000 B.C.) in the Amu Darya and Syr Darya valleys. After the domestication of the horse around 4000 B.C., nomadic pastoralism, based on the raising of livestock, emerged in the steppe belt. Eventually pastoral peoples gained power over the entire region.

The earliest recorded languages of Central Asia, spoken both in the oasis communities and among some of the pastoralists, belonged to the Indo-European linguistic family. These languages were replaced on the steppe more than a thousand years ago by languages in another major family: Altaic (which includes the Turkish and Mongolian languages). In the oasis communities as well, Turkish languages gradually began to replace the Indo-European tongues (which were closely related to Persian) as Turkish power spread through most of Central Asia. Protected by mountain barriers and the harsh conditions of the plateau, Tibet has taken a different course from the rest of Central Asia. Tibet developed into a strong kingdom after it was unified around A.D. 700. Tibetan power did not persist, however, as unity was difficult to maintain over such vast distances. Eventually the region reverted to its former state of semi-isolation. Interactions between Tibet and Mongolia, however, resulted in the establishment of Mongolian communities in the northeastern portion of the plateau and in the eventual conversion of the Mongolian people to Tibetan Buddhism.

Contemporary Linguistic and Ethnic Geography

Today peoples speaking Turkish and Mongolian languages inhabit most of Central Asia (Figure 10.13). A few native Indo-European languages are confined to the southwest, while Tibetan remains the main language of the plateau. Russian is also widely spoken in the west, while Chinese is increasingly important in the east. Chinese is perhaps beginning to threaten the long-term survival of several Central Asian languages, particularly Tibetan.

FIGURE 10.11 • STEPPE PASTORALISM The steppes of northern and central Mongolia offer lush pastures during the summer. Mongolians, some of the world's most skilled horse-riders, have traditionally followed their herds of sheep and cattle, living in collapsible, felt-covered yurts. Many Mongolians still follow this way of life. *(Goussard/SIPA Press)*

FIGURE 10.12 • TRADITIONAL ARCHITECTURE IN SAMARKAND Samarkand, Uzbekistan, is famous for its lavish Islamic architecture, some of it dating back to the 1400s. The city owes part of its rich architectural heritage to the fact that it was the capital of the empire created by the great medieval conqueror Tamerlane. *(Haley/SIPA Press)*

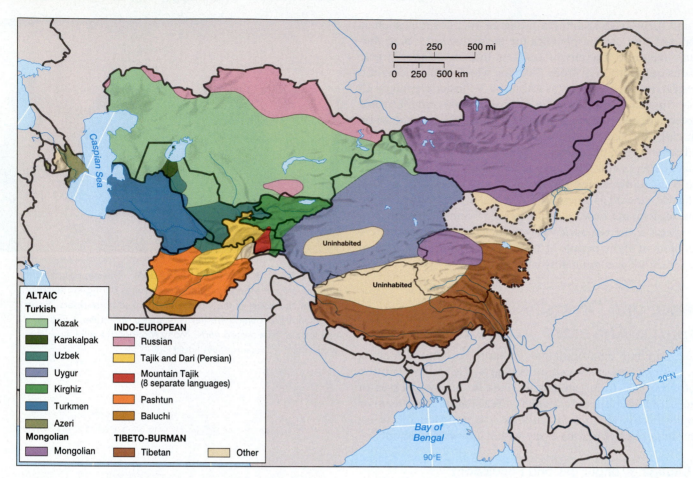

FIGURE 10.13 • LINGUISTIC GEOGRAPHY OF CENTRAL ASIA Most of Central Asia is dominated by languages in the Altaic family, which includes both the Turkish languages (found through most of the central and western parts of the region) and Mongolian (found in the northeast). Several Indo-European languages, however, are spoken in both the far northwest and the south-center, while the Tibeto-Burmese language of Tibetan covers most of the Tibetan Plateau in the southeast.

Tibetan   Tibetan is usually placed in the Sino-Tibetan language family, suggesting a shared linguistic history between the Chinese and the Tibetan peoples. Many students of Tibetan, however, argue that no definite relationship between the two languages has ever been established. The Tibetan language is divided into a number of distinct dialects that are spoken over almost the entire inhabited portion of the Tibetan Plateau. Of the 2.5 million people in Tibet itself, only about 1.5 million people speak Tibetan while most of the rest speak Chinese. Perhaps another 3 million Tibetan speakers live in China's provinces of Qinghai and Sichuan.

Mongolian   The Mongolian language includes a cluster of closely related dialects spoken by approximately 5 million people. The standard Mongolian of both the independent country of Mongolia and China's Inner Mongolia is called *Khalkha*. Mongolian has its own distinctive script, which dates back some 800 years, but Mongolia itself adopted the Cyrillic alphabet of Russia in 1941. Efforts are now being made to revive the old script.

Mongolian speakers form about 90 percent of the population of Mongolia. In China's Inner Mongolian Autonomous Province, Han Chinese who migrated into the area over the past 50 years today outnumber Mongolian speakers. Currently only about 2 million out of 22 million residents of Inner Mongolia speak Mongolian, causing concern that the influence of Mongolian culture may eventually be lost from this area.

Turkish Languages   Far more Central Asians speak Turkish languages than Mongolian and Tibetan combined. The Turkish linguistic sphere extends from Azerbaijan in the west through Xinjiang province in the east. Six main Turkish languages are found in Central Asia; five are associated with former Soviet republics of the west, while the sixth, Uygur, is the main native language of China's Xinjiang province.

Five of the six countries of the former Soviet Central Asia—Azerbaijan, Uzbekistan, Turkmenistan, Kyrgyzstan, and Kazakstan—are named after the Turkish languages of their dominant populations. In three of these countries, the native people still form a clear majority. Some 82 percent of the people of Azerbaijan speak Azeri; some 70 percent of the people of Uzbekistan speak Uzbek; and some 73 percent of the people in Turkmenistan speak Turkmen. With more than 17 million speakers, Uzbek is the most widely spoken Central Asian language. In the Amu Darya delta in the far north of Uzbekistan, however, most people speak a different Turkish language called *Karakalpak,* while Kazak speakers are found in the sparsely populated Uzbek deserts.

In the two other Turkish republics, the main nationality forms only about half of the total population. Some 52 percent of the inhabitants of Kyrgyzstan speak Kyrgyz as their native language, while approximately 42 percent of the people of Kazakstan speak Kazak. In Kazakstan the population is almost evenly split between speakers of Turkish languages and European languages (Russian, Ukrainian, and German). In general, the Kazaks and other Turks live in the center and south of the country, while the people of European descent live in the agricultural districts of the north and in the cities of the southeast.

Uygur, spoken in China's Xinjiang province, is an old language that dates back almost 2,000 years. The Uygur people number about 8 million, almost all of whom live in Xinjiang (Figure 10.14). As recently as 1953 the Uygur formed about 80 percent of the population of Xinjiang; at present, because of Han Chinese immigration, they actually form a minority in their own homeland. There are also about 1 million Kazak speakers living in the Xinjiang province.

Linguistic Complexity in Tajikistan
The sixth republic of the former Soviet Central Asia, Tajikistan, is dominated by people who speak an Indo-European rather than a Turkish language. Tajik is so closely related to Persian that it is often considered to be a Persian (or Farsi) dialect. Roughly 3.5 million people in Tajikistan, about 65 percent of the total population, speak Tajik as their first language. In the remote mountains of eastern Tajikistan, people speak a variety of distinctive Indo-European languages, sometimes collectively referred to as "Mountain Tajik."

The remainder of Tajikistan's population speaks a complex mixture of languages. For example, about a quarter of its people speak Uzbek. In addition, the highland portions of Azerbaijan—like other parts of the Caucasus Mountains—are especially noted for their ethnic and linguistic complexity.

Language and Ethnicity in Afghanistan
The linguistic geography of Afghanistan is even more complex than that of the former Soviet zone (Figure 10.15). Afghanistan became a country in the 1700s under the leadership of the Pashtun people. The Pashtuns did not attempt, however, to build a nation-state around Pashtun identity because approximately half of the Pashtun people live in Pakistan. In Afghanistan itself, estimates of the amount of people speaking this language vary from 40 percent to 60 percent.

In Afghanistan, Pashtun speakers live primarily to the south of the Hindu Kush mountain range. Almost as many people in Afghanistan speak Dari, Afghanistan's variant of Persian. Dari speakers are concentrated in the cities of the west, in the central mountains, and near the boundary with Tajikistan. Two separate ethnicities divide the Dari-speaking people; those in the west and north are considered to be Tajik, whereas those in the

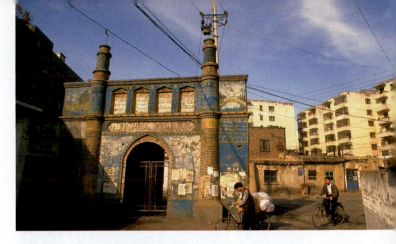

FIGURE 10.14 • UYGUR MOSQUE A small mosque, illustrating traditional Uygur architecture, survives amid blocks of modern apartments in Urumchi, Xinjiang. Traditional forms of housing and urban design can still be found in Uygur communities in northwestern China, but they are gradually disappearing. *(Chris Stowers/Panos Pictures)*

FIGURE 10.15 • AFGHANISTAN'S ETHNIC PATCHWORK
Afghanistan is one of the world's more ethnically complex countries. Its largest ethnic group is that of the Pashtuns, who live in most of the southern portion of the country as well as the adjoining borderlands of Pakistan. Northern Afghanistan, however, is mostly inhabited by Uzbeks, Tajiks, and Turkmens—whose main population centers are located in Uzbekistan, Tajikistan, and Turkmenistan, respectively. The Hazaras of Afghanistan's central mountains, like the Tajiks, speak a form of Persian. However, the Hazaras are considered to be a separate ethnic group in part because they, unlike other Afghans, follow Shiite rather than Sunni Islam.

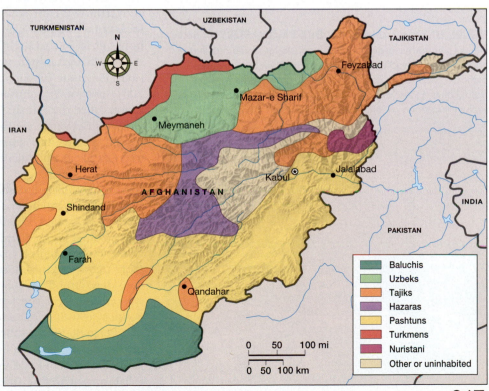

FIGURE 10.16 • AFGHAN WOMEN IN PUBLIC Especially in the Pashtun areas of Afghanistan, women have traditionally been forced to cover their entire bodies when in public areas. In places that were controlled by the Taliban, such dress codes were strictly enforced. Extremely modest dress still prevails in Afghanistan. *(Lemoyne/Getty Images, Inc.—Liaison)*

FIGURE 10.17 • LAMAIST BUDDHIST MONASTERY Tibet is well known for its large Buddhist monasteries, buildings that in earlier years served as seats of political as well as religious authority. The Potala Palace in Lhasa, traditional seat of the Dalai Lama, is the largest, most important, and most famous of such monasteries. *(Alain le Garsmeur/Panos Pictures)*

central mountains are called *Hazaras* and are said to be descendants of Mongol conquerors who arrived in the twelfth century. Finally, another 11 percent of the people of Afghanistan speak Turkish languages, mainly Uzbek.

Geography of Religion

At one time, Central Asia was noted for its religious complexity. Major overland trading routes crossed the region, giving easy access to both merchants and missionaries. By 1500, however, the region was divided into two opposed spiritual camps: Islam in the west and center, and Tibetan Buddhism in Tibet and Mongolia. Other religions are practiced in former Soviet Central Asia but on a smaller scale. For example, many Russians belong to the Russian Orthodox Church, and Uzbekistan has a small Jewish population.

Islam in Central Asia As is true elsewhere in the Muslim world, different Central Asian peoples are known for their different interpretations of Islam. The Pashtuns of Afghanistan are noted for their strict Islamic ideals—although critics contend that many Pashtun rules, such as never allowing women's faces to be seen in public, are actually based on pre-Islamic customs (Figure 10.16). The traditionally nomadic groups of the northern steppes, such as the Kazaks, on the other hand, are often considered to be rather relaxed about religious beliefs and practices. And while most of the region's Muslims are Sunnis, Shiism is dominant among both the Hazaras of central Afghanistan and the Azeris of Azerbaijan.

Under the communist rule of China, the Soviet Union, and Mongolia, all forms of religion were discouraged. Chinese authorities and student radicals attempted to suppress Islam in Xinjiang during the Cultural Revolution of the late 1960s and early 1970s. Chinese Muslims now enjoy basic freedom of worship, but the state still closely monitors religious expression out of fear that it will lead to political unrest and resistance to Chinese rule. Periodic persecution of Muslims also occurred in Soviet Central Asia, and until the 1970s it appeared that the Islamic faith might slowly disappear from the region entirely.

Religious expression was not, however, so easily repressed. Interest in Islam began to grow in former Soviet Central Asia in the 1970s and 1980s. In the post-Soviet period Islam continues to revive as people seek to return to their cultural roots. Thus far, however, there have been few signs of Islamic political fundamentalism aimed at overthrowing the state. In Xinjiang, Islam does indeed seem to be emerging as a focal point of a political movement among the Uygur people. Most Uygur leaders, however, insist that their beliefs are not fundamentalist.

Only in Afghanistan, parts of Tajikistan, and the Fergana Valley (mostly in Uzbekistan) is Islamic fundamentalism a powerful movement. From the mid-1990s until 2001, most of Afghanistan was controlled by an extremely fundamentalist organization called the Taliban. The Taliban insisted that all aspects of society conform to its own strict version of Islamic orthodoxy. The commitment to this goal was demonstrated in early 2001 when Afghanistan's religious authorities oversaw the destruction of ancient Buddhist statues in the country.

Tibetan (Lamaist) Buddhism Mongolia and Tibet stand apart from the rest of Central Asia—and indeed, from the rest of the world—in their people's practice of Tibetan, or Lamaist, Buddhism. Buddhism entered Tibet from India many centuries ago, where it merged with the native religion of the area, called *Bon*. The resulting mix is more oriented toward magic than are other forms of Buddhism, and it is more tightly organized. Standing at the head of Lamaist society is the Dalai Lama, considered to be a reincarnation of the Buddha. Until the Chinese conquest, Tibet was essentially a **theocracy** (or religious state), with the Dalai Lama enjoying political as well as religious authority (Figure 10.17).

Tibetan Buddhists suffered particularly brutal persecution after 1959 when China invaded Tibet. The Chinese hold on Tibet has never been as secure as that on Xinjiang, and Tibetan Buddhism has inspired many Tibetans to resist Chinese rule. The Dalai Lama, who fled Tibet for India in 1959, has also been a powerful advocate for the Tibetan cause in international circles. During the 1960s and 1970s, an estimated 6,000 Tibetan Buddhist monasteries were destroyed and thousands of monks were killed; the number of active monks today is only about 5 percent of what it was before the Chinese occupation. Currently, the Chinese have allowed many monasteries to reopen, but their activities remain limited.

Central Asian Culture in International and Global Context

In cultural terms, western Central Asia is most closely linked to Russia, whereas eastern Central Asia is more closely tied to China. In the east, the main cultural issue is the migration of Han Chinese into the area, which has resulted in serious ethnic and political tensions. In the west, in contrast, the cultural influence of the Russians is diminishing.

During the Soviet period, the Russian language spread widely through western Central Asia. Russian served both as a lingua franca (or common language) and as a means of instruction in higher education. One had to be fluent in Russian in order to reach any position of responsibility. Russian speakers settled in all of the major cities, and many became influential. Today, however, many Russian speakers have migrated back to Russia. The use of Russian in education, government, business, and the media is gradually declining in favor of local languages. However, because Russian served an important role as the international language of the area for many years, it is still regarded as the common language of western Central Asia.

Although Central Asia is remote and poorly integrated into global cultures, it is hardly immune to the forces of globalization. The tensions existing in the region between religious and non-religious political philosophies, and between ethnic nationalism and cultural pluralism, are found throughout the world. The increased usage of English and the influence of U.S. culture throughout Central Asia show that this part of the world is not cut off from global culture. Such influences have been especially marked in the oil cities of the Caspian Basin, such as Baku, which have seen an influx of U.S. oil workers. Although their number is low, English speakers in Central Asia, especially those with computer skills, are increasingly valued as the region strives to find a place and a voice in the global community.

Geopolitical Framework: Political Reawakening

Central Asia has played a minor role in global political affairs for the past several hundred years. Before 1991 the entire region, except Mongolia and Afghanistan, lay under direct Soviet and Chinese control. Mongolia, moreover, had been a close Soviet ally, and even Afghanistan came under Soviet domination in the late 1970s. Today, of course, the southeastern third of Central Asia is still a part of China. And although the breakup of the Soviet Union saw the emergence of six new Central Asian countries, all of them are troubled by economic problems and political instability (Figure 10.18).

Partitioning of the Steppes

Before 1500 Central Asia was a power center, a region whose mobile armies threatened the far more populous, sedentary states of Asia and Europe. The development of gunpowder and of effective hand weapons changed the balance of power, however, allowing the wealthier agricultural states to conquer the nomads. By the 1700s their armies had been defeated and their lands taken. The winners in this struggle were the two largest states bordering the steppes: Russia and China.

By the mid-1700s the Chinese empire (under the Manchu, or Qing, dynasty) stood at its greatest territorial extent, including Mongolia, Xinjiang, Tibet, and a slice of modern Kazakstan. From the height of its power in the late 1700s, Manchu-ruled China declined rapidly. By the early 1900s, Chinese power had greatly declined in Central Asia. When the Manchu dynasty fell in 1911, Mongolia became independent, although China still ruled the extensive borderlands of Inner Mongolia (Nei Mongol). Tibet had earlier gained effective independence, although this status was not recognized by China.

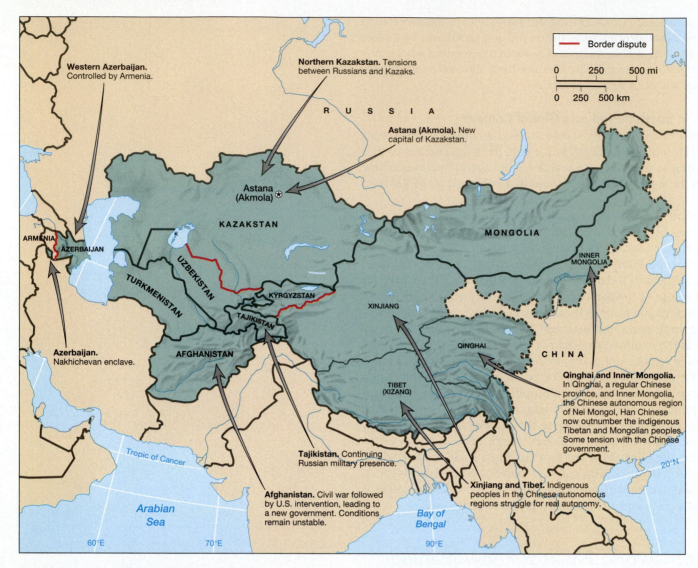

Border dispute

Western Azerbaijan. Controlled by Armenia.

Northern Kazakstan. Tensions between Russians and Kazaks.

Astana (Akmola). New capital of Kazakstan.

RUSSIA

Astana (Akmola)

KAZAKSTAN

MONGOLIA

ARMENIA

AZERBAIJAN

INNER MONGOLIA

TURKMENISTAN

UZBEKISTAN

KYRGYZSTAN

XINJIANG

TAJIKISTAN

CHINA

Azerbaijan. Nakhichevan enclave.

AFGHANISTAN

QINGHAI

TIBET (XIZANG)

Qinghai and Inner Mongolia. In Qinghai, a regular Chinese province, and Inner Mongolia, the Chinese autonomous region of Nei Mongol, Han Chinese now outnumber the indigenous Tibetan and Mongolian peoples. Some tension with the Chinese government.

Tropic of Cancer

20°N

Tajikistan. Continuing Russian military presence.

Arabian Sea

Afghanistan. Civil war followed by U.S. intervention, leading to a new government. Conditions remain unstable.

Xinjiang and Tibet. Indigenous peoples in the Chinese autonomous regions struggle for real autonomy.

Bay of Bengal

60°E   70°E   90°E

FIGURE 10.18 • CENTRAL ASIAN GEOPOLITICS Six of the eight independent states of Central Asia came into existence in 1991 with the breakup of the Soviet Union. A number of border disputes and other geopolitical problems are left over from the period of Soviet rule. In eastern Central Asia, the most serious geopolitical problems stem from China's maintenance of control over areas in which the native peoples are not Chinese. Afghanistan, scene of a prolonged and brutal civil war, has experienced the most extreme forms of geopolitical tension in the region.

Russia began to advance into Central Asia at roughly the same time as China. In the 1700s the Russian Empire conquered the Kazak steppe. Expansion farther to the south, however, was blocked by several native states in Uzbekistan. Only in the late 1800s, when European military techniques and materials became superior to those of Asia, was Russia able to conquer the Amu Darya and Syr Darya valleys. Its conquest of this area was not completed until the early 1900s, just before the Soviet Union replaced the Russian Empire.

One reason for the Russian advance into Central Asia was concern over possible British influence in the area. Britain did indeed attempt to conquer Afghanistan, but it failed to do so. Subsequently, Afghanistan's position as an independent "buffer state" between the Russian Empire (later the Soviet Union) and the British Empire in South Asia remained secure.

Central Asia Under Communist Rule

Western Central Asia came under communist rule after the formation of the Soviet Union in 1917; Mongolia followed in 1924. After the Chinese revolution of 1949, the communist system was also imposed on Xinjiang and Tibet. In all of these areas, major changes in the geopolitical order soon followed.

Soviet Central Asia The newly established Soviet Union retained all of the Central Asian territories of the old Russian Empire. The policies that it directed toward this region, however, changed. The new regime sought to create a socialist

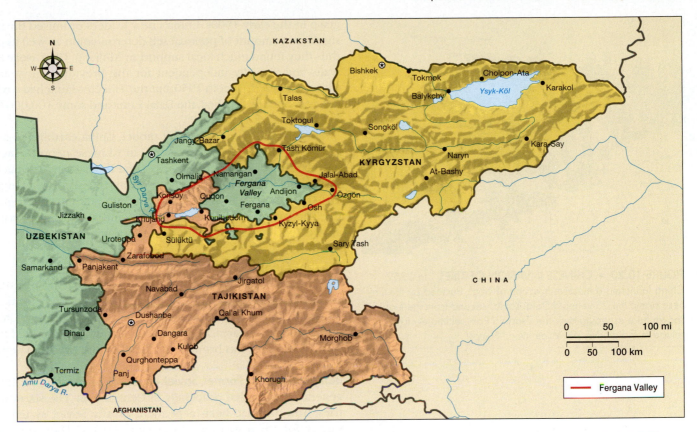

FIGURE 10.19 • POLITICAL BOUNDARIES AROUND THE FERGANA VALLEY Some of the world's most complex political boundaries can be found in the vicinity of the Fergana Valley. The central portion of the valley belongs to Uzbekistan, which is otherwise separated from it by high mountains. The lower valley, on the other hand, is part of Tajikistan, the core area of which is likewise separated from the valley by highlands. The Fergana's upper periphery belongs to Kyrgyzstan. Note also the small enclaves of Uzbekistan within Kyrgyzstan.

economy and to build a new Soviet society that would eventually knit together all of the massive territories of the Soviet Union. Central Asia's leaders were replaced by Communist Party officials loyal to the new state; Russian immigration was encouraged; and local languages could no longer be written in the Arabic script.

Although the early Soviet leaders thought that a new Soviet nationality would eventually emerge, they realized that local ethnic diversity would not disappear overnight. They therefore divided the Soviet Union into a series of nationally defined "union republics" in which a certain degree of autonomy would be allowed. The Soviet leaders were uncertain, however, about what the ethnic divisions in Central Asia should be. Was there a single Turkish-speaking nationality, or were the Turkish peoples themselves divided into a number of separate nationalities, with the Tajiks forming another? For several years boundaries shifted as new "republics" suddenly appeared on the map. Finally, in the 1920s the modern republics of Kazakhstan, Kyrgyzstan, Tajikistan, Uzbekistan, Turkmenistan, and Azerbaijan assumed their present shapes. In certain areas, such as the fertile Fergana Valley, the new political boundaries were extremely complex (Figure 10.19).

Some scholars argue that the Soviet policy backfired severely. The new republics of the Soviet Union encouraged the development of non-Russian nationalities. Identities such as Turkmen, Uzbek, and Tajik that had been somewhat vague in the pre-Soviet period were now given real political significance, weakening the Soviet system.

Another problem undercutting Soviet unity was the fact that cultural and economic gaps separating Central Asians from Russians did not decrease as much as planned. Islam remained well established in some areas and began to revive in others during the 1970s. Central Asia also remained poorer than most other parts of the Soviet Union, and by the 1980s it was becoming something of a burden on the national economy. In addition, the region's higher birthrates led many Russians to fear that the Soviet Union risked being dominated by Turkish-speaking Muslims.

The Chinese Geopolitical Order After decades of political and economic chaos, China reemerged as a united country in 1949. Its new communist government was able to reclaim most of the Central Asia territories that had slipped out of China's

FIGURE 10.20 • CHINESE INVASION OF TIBET In 1959 China launched a massive invasion of Tibet. Pursued by the Chinese army, the 23-year-old Dalai Lama (second from the lead) is shown here escaping over the Zsagola Pass, ultimately to find refuge in India. *(Hg/AP/Wide World Photos)*

grasp in the early 1900s. China's new leaders promised the non-Chinese peoples a significant amount of political self-determination as well as cultural autonomy, and thus they found much local support in Xinjiang. Tibet, isolated behind its mountain walls and virtually independent for the previous 150 years, was more difficult. China occupied Tibet in 1950, but the Tibetans launched a rebellion in 1959. When this was brutally crushed, the Dalai Lama and some 100,000 followers found refuge in India (Figure 10.20).

Loosely following the Soviet model, China established autonomous regions in areas occupied primarily by non-Han Chinese peoples, including Xinjiang, Tibet proper (called *Xizang* in Chinese), and Inner Mongolia. Such autonomy, however, often meant little, and it did not prevent the massive immigration of Han Chinese into these areas. Nor were all parts of Chinese Central Asia granted autonomous status. The large and historically Tibetan and Mongolian province of Qinghai, for example, remained an ordinary Chinese province.

Current Geopolitical Tension

The former Soviet portion of Central Asia made the post-1991 transition to independence rather smoothly, but the region still suffers from a number of conflicts. Much of China's Central Asian territory is unsettled, but China keeps a strong political hold on the area. Afghanistan, unfortunately, suffered from a particularly brutal civil war that was brought to a head by the events of September 11, 2001.

Independence in Former Soviet Lands The breakup of the Soviet Union in 1991 generally proceeded peacefully in Central Asia. The six newly independent countries, however, had been dependent on the Soviet system, and it was no simple matter for them to chart their own courses. They still had to cooperate with Russia, for example, on security issues. In most cases authoritarian rulers, rooted in the old order, retained power and sought to weaken opposition groups. All told, democracy has made less progress in Central Asia than in most other parts of the former Soviet Union.

Kazakstan and Tajikistan faced a particularly difficult transition. Kazakstan is the largest and most resource-rich Central Asian state and therefore might seem to have the best chance for success. Potential ethnic strife, however, looms over the country. Many Kazaks want to create a national state centered on Kazak identity, and they resent the presence and power of Russians, Ukrainians, Germans, and others of European background. These Europeans, for their part, are worried about the development of Kazak nationalism and the growing strength of Islam in the area.

In Tajikistan war broke out almost immediately after independence in 1991. Many members of the smaller ethnic groups living in the mountainous east resented the authority of the lowland Tajiks and thus rebelled. They were joined by several Islamic groups seeking to overthrow the secular state. Although the civil war officially ended after several years, periodic fighting continues, and much of the mountainous west remains beyond the control of the central government. The Tajikistan government remains in power in part because of its use of Russian troops. The presence of Russian military forces, however, brings into question Russia's long-term intentions in the region.

Strife in Western China Local opposition to Chinese rule in Central Asia increased during the 1990s, but without success. Although any form of protest in Tibet has been severely repressed, the Tibetans have brought the attention of the world to their struggle, through the work of the Dalai Lama as well as by global communication using the Internet. China maintains several hundred thousand troops in a region that has only 2.4 million civilian inhabitants. Such an overwhelming military presence is considered necessary because of both Tibetan resistance and the strategic importance of the border zone with India—one of China's main rivals.

China maintains that all of its Central Asian lands are essential parts of its national territory. Those who advocate independence are viewed as traitors. Chinese officials have also linked separatist elements in Xinjiang to a global movement of radical

Islamic fundamentalists. Leaders of Xinjiang's "Free Eastern Turkestan Movement," however, insist that their movement is founded on territory rather than religion or ethnicity and that it is based on an alliance of all non-Han Chinese peoples of Xinjiang. The involvement of a number of Uygur separatists in Osama bin Laden's Al Qaeda organization in Afghanistan, however, did not help their cause in the arena of international public opinion.

War in Afghanistan Before September 11, 2001   None of the conflicts in the former Soviet republics or western China compares in intensity to the struggle waged in Afghanistan. Afghanistan's troubles began in 1978 when a Soviet-supported military "revolutionary council" seized power. The new Marxist-oriented government began to suppress religion, which led almost immediately to widespread rebellion. When the government was about to collapse, the Soviet Union responded with a massive invasion. Despite its power, the Soviet military was never able to gain control of the more rugged parts of the country. Pakistan, Saudi Arabia, and the United States, moreover, ensured that the anti-Soviet forces remained well armed.

The exhausted Soviets finally withdrew their troops in 1989. The pro-Russian government that they installed remained to face the rebel fighters alone. It managed to hold the country's core around the city of Kabul for a few years, largely because the opposition forces were themselves divided. Local warlords grabbed power in the countryside, destroying central authority and committing a number of atrocities, including the mass raping of women in several conquered villages (Figure 10.21).

In 1995–1996, a new movement called the Taliban arrived in Afghanistan. Founded by young Muslim religious students disgusted with the anarchy that was consuming their country, the Taliban believed in the strict enforcement of Islamic law. The Taliban model attracted large numbers of soldiers, and by September 2001 only about a tenth of the country, mostly in the far northeast, lay outside of their power. Here the anti-Taliban group known as the Northern Alliance held control, but as it continued to lose territory its future looked increasingly dim.

By 2000, however, most of Afghanistan's people were turning against the Taliban rulers, mostly because of the severe restrictions on daily life that they imposed. These restrictions were most pronounced for women, but even men were compelled, by both whippings and imprisonment, to obey the Taliban's numerous laws. Most forms of recreation were outlawed, including television, films, music, and even kite flying.

International Dimensions of Central Asian Tension

With the collapse of the Soviet Union in 1991, Central Asia emerged as a key arena of geopolitical tension. A number of important countries, including China, Russia, Pakistan, Iran, Turkey, and the United States, competed for influence in the region. The revival of Islam has also generated international geopolitical issues.

The Roles of Russia, Iran, Pakistan, and Turkey   Russia's economic and military ties to Central Asia were not erased when the Soviet Union collapsed. Russia maintains armed forces in Tajikistan and continues to be one of the main markets for Central Asian exports. Furthermore, the main transportation line linking European Russia to central Siberia cuts across northern Kazakstan, while former Soviet Central Asia's rail links to the outside world are still largely oriented toward Russian lands.

Like Russia, Iran is also interested in the new republics of Central Asia. Iran is now a major trading partner for these countries, and it offers the best access route to sea ports on the Persian Gulf. Since the completion of a rail link between Iran and Turkmenistan in 1996, part of Central Asia's global trade has been reoriented toward Iran's ports. Iran's cultural ties with the region are old and deep, particularly in Tajikistan and northern Afghanistan, where Persian is the major language.

Pakistan has also been eager to gain influence in Central Asia. Extremely rugged topography, however, separates Pakistan from the Amu Darya and Syr Darya valleys; Afghanistan, moreover, occupies the intervening territory. During the Taliban period, Pakistan enjoyed very close relations with Afghanistan. In fact, the Taliban received tremendous support from Pakistan's secret services. The aftermath of

FIGURE 10.21 • WAR IN AFGHANISTAN   Prolonged civil war in Afghanistan created a social disaster. Many Afghans sought refuge from the extreme fundamentalist forces of the Taliban either by heading into other countries or by fleeing to areas of Afghanistan not under Taliban rule. This photo shows refugees who fled Kabul and headed north. Since the fall of the Taliban, many refugees have been heading home. *(Martin Adler/Panos Pictures)*

September 11 thus put Pakistan's government in a tight situation. Owing to both U.S pressure and promises, Pakistan joined the anti-Taliban coalition, and it supplied the U.S. military with valuable intelligence. Support for the Taliban, however, remains pronounced in Pakistan, especially in the Pashtun-speaking areas along its border with Afghanistan.

Turkey also maintains close cultural connections with Central Asia. Most people of this region speak Turkish languages. In contrast to Iran and Pakistan, Turkey offers itself as the model of the modern Muslim state. In this secular model, religion provides social and cultural glue, but politics and economics operate without religious content. Turkey itself, however, is experiencing some tensions between political secularism and religious fundamentalism, and it is not obvious that the Turkish system can easily be exported, even to other Turkish-speaking states.

Islamic Fundamentalism   Most of the governments of Central Asia were concerned in the late 1990s that an Islamic fundamentalist movement similar to that of the Taliban could disrupt their own countries. Such worries were seemingly confirmed with the formation of the IMU (Islamic Movement in Uzbekistan), a group dedicated to the creation of a new Islamic state in the Fergana Valley. Receiving assistance from the Taliban and Al Qaeda, the IMU set off bombs in Tashkent and other cities and engaged the armies of Uzbekistan and Kyrgyzstan.

The aftermath of terrorism on September 11, 2001, however, completely changed the balance of power in the region. The United States and Britain, working with Afghanistan's Northern Alliance, launched a major war against Al Qaeda and the Taliban government that had offered it refuge. By early 2002, the Taliban was in retreat and a new provisional government for Afghanistan was soon established. Spontaneous celebration erupted in many cities as music blared, men cut their beards, and carefully wrapped television sets were brought out of hiding.

The fall of the Taliban, unfortunately, did not bring any real peace or security to Afghanistan. As of early 2004, the new government—assisted by an International Security Assistance Force—held power only in Kabul and a few other cities. Elsewhere, local warlords reestablished their power, while bandit groups made travel extremely difficult by robbing vehicles passing through remote areas. Tensions between ethnic Tajiks (who dominated the interim government) and Pashtuns made Afghanistan's recovery all the more uncertain. More worrisome was the fact that the Taliban continued to operate in remote areas of southern and eastern Afghanistan, and by the end of 2003, it was apparently again growing in power.

Economic and Social Development: Abundant Resources, Devastated Economies

Central Asia is by most measures one of the poorer regions of the world. One of its countries, Afghanistan, stands near the bottom of almost every list of economic and social indicators. Central Asia does, however, contain substantial natural resources, particularly oil and natural gas. Much of the region also enjoys relatively high levels of health and education, a legacy of the social programs enacted by the communist regimes. These same regimes, however, built inefficient economic systems, and since the fall of the Soviet Union, western Central Asia has experienced a huge economic decline.

The Post-Communist Economies

Soviet economic planners sought to spread economic development widely across their country. This required building large factories even in remote areas of Central Asia, regardless of the costs involved. Such Central Asian industries relied heavily on subsidies from the Soviet government. When those subsidies ended, the industrial base of the region began to collapse, leading to a huge drop in the standard of living. As is true elsewhere in the former Soviet Union, however, certain individuals have grown very wealthy since the fall of communism.

Today, Kazakhstan stands as the most developed Central Asian country, and it may have the best economic prospects. Since Kazakhstan's agriculture is potentially productive and its population density is low, it could emerge as a major food exporter. More important, it has two of the world's largest untapped deposits of oil and natural gas, the vast Tengiz and Kashagan fields, as well as extensive deposits of other minerals.

As a result of its large population, Uzbekistan has the second biggest economy in the region. The Uzbek economy, moreover, has not declined as sharply as those of its neighbors. This is largely because it retains many aspects of the old command economy (one run, in other words, by governmental planners rather than by private firms responding to the market) and has resisted opening its economy to private firms and market forces. Uzbekistan's industries, however, will likely become increasingly uncompetitive unless they are opened to market forces.

Turkmenistan also has a substantial agricultural base, due mainly to Soviet irrigation projects, and it remains a major cotton exporter. It also retains a state-run economy and has a very repressive government. Government planners, however, hope that the development of new oil and gas fields will bring prosperity, proclaiming that Turkmenistan will soon become the "Kuwait of natural gas."

The region's best-developed fossil fuel industry is located in Azerbaijan (Figure 10.22). Azerbaijan has attracted a great deal of international interest and investment, promising to revitalize its oil industry. Thus far, however, its economy has largely failed to respond, and Azerbaijan—despite its promise—remains a very poor country.

The most economically troubled of the former Soviet republics is Tajikistan. With a per capita gross national income of only some $280, Tajikistan rates as one of the world's poorer countries. It has few natural resources. A prolonged civil war has also undercut its economy, and Tajikistan is burdened by its remote location and rugged topography.

Mongolia, although never part of the Soviet Union, was a close Soviet ally run by a communist party. It too suffered an economic collapse in the 1990s. Mongolia no longer receives Soviet subsidies, and its industries are not competitive in the global market. It has little farming and its traditional trade in livestock products is not very profitable. Isolation also plagues the country, as Mongolia is far from the ocean ports through which global trade is conducted. Mongolia does have some substantial mineral reserves, however, and its low population density keeps it from suffering from urban crowding and from rural landlessness.

The Economy of Tibet and Xinjiang in Western China
The Chinese portions of Central Asia have not suffered the same economic crash that occurred in the other parts of the region. China as a whole has one of the world's fastest-growing economies, although its centers of economic growth are all located in the coastal zone. Still, the most remote parts of the country have not experienced the economic decline that has been seen in the former Soviet zone. But as China had been poorer than the Soviet Union before 1991, it is not surprising that poverty in Chinese Central Asia remains widespread.

Tibet in particular remains one of the world's poorest places. Most of the plateau is relatively isolated. But many Tibetans are at least able to provide for most of their basic needs. Both Tibet and Xinjiang, moreover, do not suffer the overcrowding that strains the poorer parts of China proper. Then again, their harsh environments cannot support high population densities. This leads critics to argue that Han Chinese immigration threatens to upset the local balance between human numbers and the environment throughout western China.

Xinjiang does have tremendous mineral wealth, including a substantial portion of China's oil reserves. Its agricultural sector is also productive, although limited in extent. Many of the local Muslim peoples believe that the region's wealth is being taken by the Chinese state and the Han Chinese immigrants. At the turn of the millennium, China announced ambitious new plans to build road and rail links from eastern China to Tibet and Xinjiang. While these lines will bring economic benefits, they may also result in an even larger migration stream, further threatening the native cultures of the region.

FIGURE 10.22 • OIL DEVELOPMENT IN AZERBAIJAN
Although oil has brought a certain amount of wealth to Azerbaijan, it has also resulted in extensive pollution. Since most of the petroleum is located either near or under the Caspian Sea, this sea—actually the world's largest lake—is becoming increasingly polluted and is suffering from declining fisheries. *(John Spaull/ Panos Pictures)*

Economic Misery in Afghanistan

Afghanistan is unquestionably the poorest country in the region, with one of the weakest economies in the world. Reliable statistics, however, are impossible to obtain. Afghanistan suffered nearly continuous war starting in the late 1970s, which made most economic activities difficult to maintain. Even before the war it was an impoverished country with little industrial or commercial development.

By the late 1990s, however, Afghanistan began producing illegal drugs for the global market. According to the CIA, it was by 1999 the world's largest producer of opium. Most of the opium produced in the country is sold as heroin in the large cities of North America and Europe. In many areas, opium production seems to have increased after the fall of the Taliban, as the central government again lost control over most regions of the country. Although the United States and a host of other countries have promised billions of dollars in economic aid to rebuild Afghanistan's economy, little progress has yet been made.

Central Asian Economies in Global Context

Despite its poverty and relative isolation, Afghanistan is closely connected to the global economy through the export of illegal drugs. Since drugs are extremely lightweight, they can be successfully exported from even the most isolated parts of the world. The transportation of drugs, moreover, is a major issue throughout the former Soviet zone. Yet in general terms, Central Asia has not experienced extensive economic globalization (Figure 10.23).

FIGURE 10.23 • GLOBAL LINKAGES: DIRECT FOREIGN INVESTMENT Due to their oil industries, Kazakstan and Azerbaijan receive the bulk of the direct foreign investment coming into Central Asia. Turkmenistan, also a fossil-fuel exporter, and Mongolia and Kyrgyzstan, both relatively open economies, attract more foreign investment than do Uzbekistan, Tajikistan, or Afghanistan. However, with the exception of the oil and natural gas industries, Central Asia is not closely connected with the global system. This is true even in regard to net aid flows to desperately poor Central Asian countries such as Afghanistan and Tajikistan.

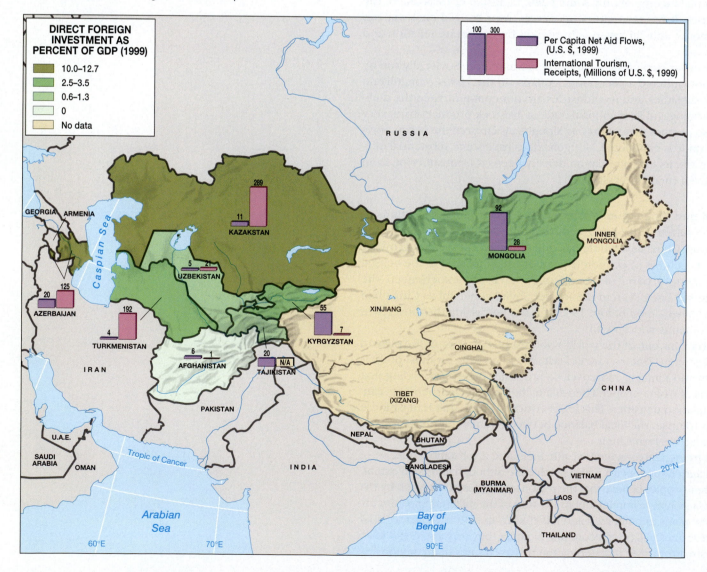

The United States and other Western countries are drawn to the area by its oil and natural gas deposits. Most large oil companies have established operations in Azerbaijan, Kazakstan and Turkmenistan, which are thought to contain the world's largest undeveloped fossil fuel deposits open to Western firms. These same fossil fuel reserves have created a very complex international economic environment. For the region's large oil fields to be economically viable, a vast pipeline system must be constructed. Existing pipelines are already inadequate and pass through Russia, which charges extremely high transit fees. International negotiations began in the late 1990s for the construction of new pipelines. The most efficient route for Turkmenistan would be through Iran, and in 1997 a small gas pipeline was opened between the two countries. The United States, however, opposes any new pipelines passing through Iran. The United States has successfully lobbied for an alternative, and much more expensive, pipeline system. This system will eventually extend from Azerbaijan into Georgia and then across Turkey to the Mediterranean port of Ceyhan.

Social Development in Central Asia

Social conditions in Central Asia vary more than economic conditions. Afghanistan falls at the bottom of the scale regarding almost every aspect of social conditions. In the former Soviet territories, levels of health and education are relatively high but are declining. Social conditions are mostly unknown in western China, though it is likely that Tibet and Xinjiang have not kept up with the progress made in China proper.

Social Conditions and the Status of Women in Afghanistan
Social conditions in Afghanistan are no better than economic circumstances. The average life expectancy in the country is a mere 45 years, one of the lowest figures in the world. Infant and childhood mortality levels remain extremely high. Not only does Afghanistan endure constant warfare, but its rugged topography hinders the provision of basic social and medical services. Illiteracy is commonplace, especially for women. Afghanistan's 15 percent adult female literacy figure is one of the lowest in the world.

Women in traditional Afghan society—and especially in Pashtun society—have very little freedom. In many areas they must completely conceal their bodies, including their faces, when venturing into public. Such restrictions intensified in the 1990s. Dress controls were strictly enforced where the Taliban gained power. In most areas, Taliban forces prevented women from working, attending school, and often even from obtaining medical care. The ban on work brought particular hardship to Afghanistan's war widows, leaving many in outright destitution.

The fall of the Taliban brought much joy to the women of Afghanistan (Figure 10.24). Many promptly uncovered their faces and began working—or seeking work. But Afghan women also remained nervous, uncertain if the new government would uphold women's rights. In rural Pashtun-speaking areas, most evidence indicates that the position of women is not much better now than it had been under the Taliban.

Social Conditions in the Former Soviet Republics
Women in the former Soviet portion of Central Asia enjoy a higher social position than those of Afghanistan. Traditionally, women had much more autonomy among the northern pastoral peoples (especially the Kazak) than among the Uzbek and Tajik oasis dwellers of the south, but under Soviet rule the position of women everywhere improved. In the former Soviet republics today, women's educational rates are comparable to those of men, and women are well represented in the workplace. Some evidence would suggest, however, that the position of women has declined in many areas since the fall of the Soviet Union.

Overall, social indicators such as literacy, average life expectancy, and levels of education remain relatively high in the former Soviet zone (Figure 10.25). The social successes of western Central Asia reflect investments made during the Soviet period. It appears unlikely, however, that health and educational facilities can be maintained considering the deep economic declines that these countries have experienced.

FIGURE 10.24 • AFGHANI WOMEN AND THE FALL OF THE TALIBAN When the Taliban government fell from power in Kabul on November 13, 2001, many residents of the city erupted into spontaneous celebrations. A number of women demonstrated their new—although still limited—freedom by exposing their faces in public. Such an "offense" would have earned them severe beatings under Taliban rule. (Scott Peterson/Getty Images, Inc.—Liaison)

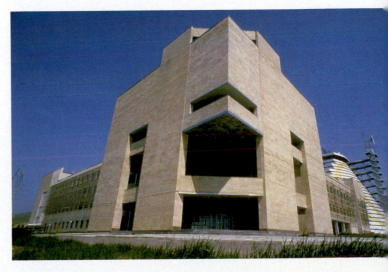

FIGURE 10.25 • TARKENT HELIOSTATION IN TASHKENT, UZBEKISTAN Despite its generally poor economic statistics, Uzbekistan boasts a relatively well-developed medical, educational, and technological infrastructure. The Tarkent heliostation focuses the sun's rays to generate the extraordinarily high temperatures needed to produce super-pure industrial minerals. (Novosti/Getty Images, Inc.—Liaison)

Social Conditions in Western China It is particularly difficult to obtain reliable information about social conditions in the Chinese portion of Central Asia. Certainly China as a whole has made significant progress in health and education, but many reports suggest that the peoples of Tibet and Xinjiang have been left behind. According to the Eastern Turkestan Web pages, some 60 percent of the non-Han people of Xinjiang are illiterate. Social conditions may be slow to improve in these more remote areas.

CONCLUSION

Central Asia, long hidden by Russian and Chinese domination, has recently reappeared on the map of the world. Environmental problems, among others, have brought the region to global attention. The destruction of the Aral Sea, after all, is one of the worst environmental disasters the world has seen. Large movements of people—Han Chinese into Western China, and Russians out of former Soviet Central Asia—have also attracted global attention.

Central Asia remains something of a question mark on the map of the world. A large part of the region recently received its independence after more than 100 years of Russian control, and it remains to be seen what economic and political directions it will follow. Much of the rest of the region—Tibet and Xinjiang—remains under the authority of what most native residents consider a foreign power. Although the Mongol, Uygur, and Tibetan peoples of this area are struggling to maintain their lands and cultural autonomy, they may not be able to withstand Han Chinese migration. The Chinese government, moreover, shows no indication that it would be willing even to discuss the possibility of granting genuine autonomy to these territories.

While China maintains a firm grip on Tibet and Xinjiang, the rest of Central Asia has emerged as a key area of geopolitical and economic competition. Russia, the United States, China, Iran, Pakistan, and Turkey all contend for influence. Kazakstan, Uzbekistan, Tajikistan, Kyrgyzstan, Turkmenistan, and Azerbaijan have attempted to play these powers off against each other in order to maintain or increase their power. And while political structures throughout the area remain largely authoritarian, the economies of the region are gradually opening up to global connections. However, Central Asia is likely to face serious economic difficulties for some time, especially since the region is not a significant participant in global trade, and has attracted little foreign investment outside of the oil industry.

Afghanistan, of course, remains Central Asia's biggest question mark. It has the region's most serious economic, social, and ethnic problems—problems that now command the attention of the global community. Global efforts will probably be necessary if Afghanistan is to rebuild itself economically and find political stability. If that were to happen, the future of all of Central Asia would be much more secure than it is today.

KEY TERMS

alluvial fans (page 263)   pastoralist (page 262)   theocracy (page 268)   transhumance (page 263)
loess (page 263)

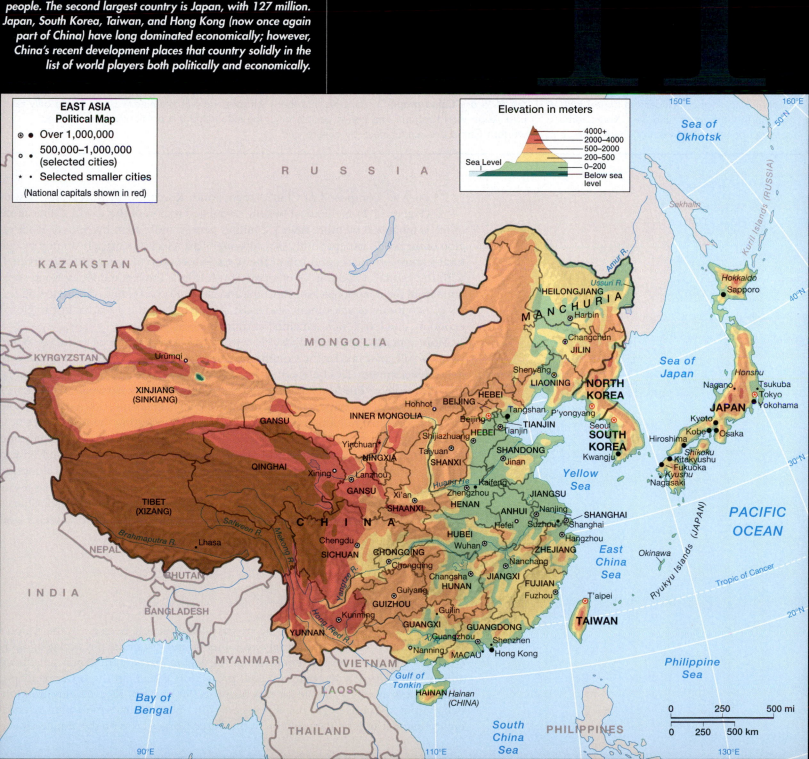

people. The second largest country is Japan, with 127 million.
Japan, South Korea, Taiwan, and Hong Kong (now once again
part of China) have long dominated economically; however,
China's recent development places that country solidly in the
list of world players both politically and economically.

EAST ASIA
Political Map

⊗ ● Over 1,000,000

○ ○ 500,000–1,000,000
      (selected cities)

★ • Selected smaller cities

(National capitals shown in red)

Elevation in meters

4000+
2000–4000
500–2000
200–500
0–200
Below sea
level

Sea Level

SETTING THE BOUNDARIES

East Asia is easily marked off on the map of the world by the territorial extent of its countries: China, Taiwan, North Korea, South Korea, and Japan (see Figure 11.1). Japan is composed of four main islands, but also includes a chain of much smaller islands (the Ryukyus) that extends almost to Taiwan. Taiwan is basically a single-island country, but its political status is confusing because China claims it as part of its own territory. China itself is a vast continental state—the third most extensive in the world—that reaches well into Central Asia. It also includes the large island of Hainan in the South China Sea. Korea forms a compact peninsula between northern China and Japan, but it is divided into two independent states: North Korea and South Korea.

In political terms, such a straightforward definition of East Asia is appropriate. If one turns to cultural considerations, however, the issue becomes more complicated. The main cultural problem arises with regard to the western half of China. This is a huge but lightly populated space; some 95 percent of the residents of China live in the east (see Figure 11.14). By certain criteria, only the eastern half of China (often referred to as **"China proper"**) really fits into the East Asian world region. The native people of western China are not Chinese by culture and language, and they have never accepted the religious and philosophical beliefs that have given historical unity to East Asian civilization. In the northwestern portion of China, called *Xinjiang* in Chinese, most native residents speak Turkish languages and are Muslim in religion. Tibet, in southwestern China, has a highly distinctive culture, and the Tibetans in general resent Chinese authority. Xinjiang and Tibet are thus perhaps more appropriately classified within Central Asia, which is covered in Chapter 10. In this chapter we will examine western China only to the extent that it is politically part of China.

East Asia, composed of China, Japan, South Korea, North Korea, and Taiwan (Figure 11.1), is the most heavily populated region of the world. China alone is inhabited by more than 1.2 billion people, more than live in any other region of the world except South Asia. Although East Asia is historically unified by cultural features, in the second half of the twentieth century it was politically divided, with the capitalist economies of Japan, South Korea, Taiwan and Hong Kong separated from the communist bloc of China and North Korea. Distinct differences in levels of economic development also exist. As Japan became a leader in the global economy, much of China remained extremely poor.

More recently, however, divisions within East Asia have been reduced. While China is still governed by the Communist Party, it is now on a path of modified capitalist development. Ties between its booming coastal zone and Japan, South Korea, and Taiwan have quickly strengthened. Although relations between North Korea and South Korea and between China and Taiwan are still hostile, East Asia as a whole has seen a gradual reduction in political tensions (see "Setting the Boundaries").

East Asia must be recognized as one of the core areas of the world economy, and it is emerging as a center of political power as well. Japan, South Korea, and Taiwan are among the world's key trading states. Tokyo, Japan's largest city, stands alongside New York and London as one of the financial centers of the globe. China has fewer international connections, but it too is becoming a global trading power. China's coastal zone is now closely tied to global networks of information, commerce, and entertainment.

Environmental Geography: Resource Pressures in a Crowded Land

Environmental problems in East Asia are particularly severe because of the region's large population, its rapid industrial development, and its unique physical geography (Figure 11.2). The wealthier countries of the region, particularly Japan, have been able to invest heavily in environmental protection. China, on the other hand, not only has less money available for this purpose, but many of its problems are more deeply rooted. Let us begin our discussion with one of China's most controversial environmental actions, the building of the Three Gorges Dam on the Yangtze River.

Flooding, Dam-Building, and Related Issues in China

The Yangtze River (also called the Chang Jiang) is one of the most important physical features of East Asia. This river, the third largest (by volume) in the world, emerges from the Tibetan highlands onto the rolling lands of the Sichuan Basin, passes

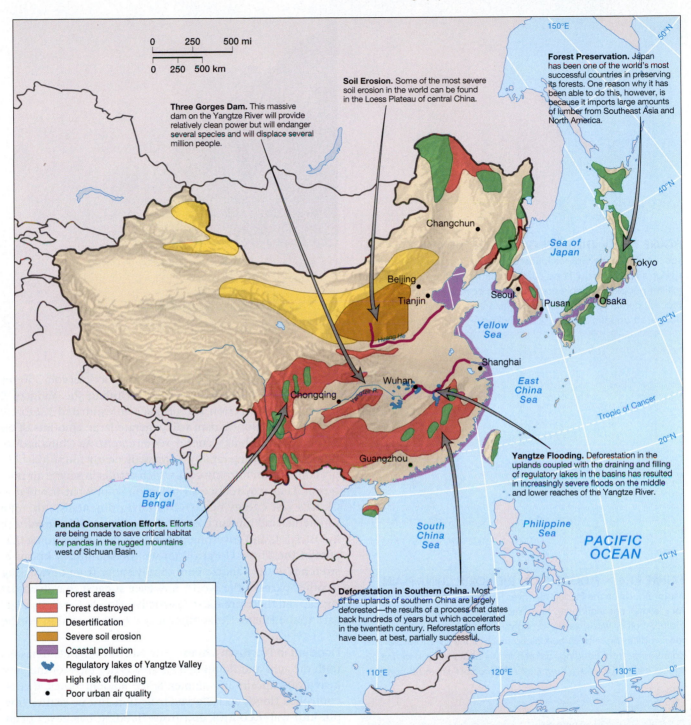

Three Gorges Dam. This massive dam on the Yangtze River will provide relatively clean power but will endanger several species and will displace several million people.

Soil Erosion. Some of the most severe soil erosion in the world can be found in the Loess Plateau of central China.

Forest Preservation. Japan has been one of the world's most successful countries in preserving its forests. One reason why it has been able to do this, however, is because it imports large amounts of lumber from Southeast Asia and North America.

Panda Conservation Efforts. Efforts are being made to save critical habitat for pandas in the rugged mountains west of Sichuan Basin.

Yangtze Flooding. Deforestation in the uplands coupled with the draining and filling of regulatory lakes in the basins has resulted in increasingly severe floods on the middle and lower reaches of the Yangtze River.

Deforestation in Southern China. Most of the uplands of southern China are largely deforested—the results of a process that dates back hundreds of years but which accelerated in the twentieth century. Reforestation efforts have been, at best, partially successful.

Legend:
- Forest areas
- Forest destroyed
- Desertification
- Severe soil erosion
- Coastal pollution
- Regulatory lakes of Yangtze Valley
- High risk of flooding
- Poor urban air quality

FIGURE 11.2 • ENVIRONMENTAL ISSUES IN EAST ASIA This huge world region has been almost completely transformed from its natural state and continues to have serious environmental problems. In China, some of the more pressing environmental issues involve deforestation, flooding, water control, and soil erosion.

through a magnificent canyon in the Three Gorges area (Figure 11.3), and then meanders across the lowlands of central China before entering into the sea in a large delta near the city of Shanghai. The Yangtze has historically been the main transportation corridor into the interior of China, and it has long been famous in Chinese literature for its beauty and power. More recently, however, it has become the focal point of an environmental controversy of global proportions.

The Three Gorges Controversy The Chinese government wants to control the Yangtze for two main reasons: to prevent flooding and to generate electricity. To do so, it must build a series of large dams, the largest of which is a massive structure in the Three Gorges area. This $25 billion structure—600 feet (180 meters) high and 1.2 miles (1.9 kilometers) long—will be the largest hydroelectric dam in the world

FIGURE 11.3 • THE THREE GORGES OF THE YANGZTE The spectacular Three Gorges landscape of the Yangzte River is the site of a controversial flood control dam that will displace several million people. Not only are the human costs high to those displaced, but there may also be significant ecological costs to endangered aquatic species. *(Greg Baker/AP/Wide World Photos)*

FIGURE 11.4 • FLOODING ON THE NORTH CHINA PLAIN
Major floods on the Huang He River have occurred throughout history, sometimes inundating large sections of the North China Plain. At other times, severe droughts affect the same region. Extensive dikes have been built along much of the river to protect the countryside from flooding, as seen in this photo taken near the historical city of Kaifeng. *(Yang Xiuyun/ChinaStock Photo Library)*

when completed in 2009 and will form a reservoir 350 miles long. It will jeopardize several endangered species (including the Yangtze River dolphin), flood a major scenic attraction, and displace an estimated 1.2 to 1.9 million persons.

The Three Gorges dam will generate large amounts of electricity, meeting up to 11 percent of China's energy requirement. As China industrializes, its demand for power is increasing rapidly. At present, nearly four-fifths of China's total energy supply comes from burning coal, which results in severe air pollution. But while this series of dams may reduce air pollution somewhat, it will not be adequate to supply China's vast energy needs. Most environmentalists therefore argue that the environmental and human costs of these dams will be greater than their benefits. Most industrialists and government planners disagree, arguing that the water-control benefits of the Three Gorges Dam will be significant because the lower and middle stretches of the Yangtze periodically suffer from devastating flooding. Experiences in other parts of the world, however, suggest that even dams cannot prevent the largest, and therefore most destructive, floods. Certainly that has been the case in regard to the Huang He, another major river located in northern China.

Flooding in Northern China The North China Plain has historically suffered from both drought and flooding. This area is dry most of the year, yet often experiences heavy downpours in summer. Since ancient times, levees and canals have both controlled floods and allowed irrigation. But no matter how much effort has been put into water control, disastrous flooding has never been completely prevented (Figure 11.4).

The worst floods in northern China are caused by the Huang He, or Yellow River, which cuts across the North China Plain. As a result of upstream erosion, the Huang He carries a huge **sediment load** (the amount of suspended clay, silt, and sand in the water), making it the muddiest major river in the world. When the river enters the low-lying plain, its velocity slows and its sediments begin to settle and accumulate in the riverbed. As a result, the level of the river gradually rises above that of the surrounding lands. Eventually it must break free of its course to find a new route to the sea over lower-lying ground.

Throughout recorded history, the Huang He has changed course 26 times, usually causing much loss of human life. The North China Plain has been densely populated for millennia and is now home to more than 300 million people. Since ancient times, the Chinese have attempted to keep the river within its banks by building ever-larger dikes. Eventually, however, the riverbed rises so high that the flow can no longer be contained. The resulting floods have been known to kill several million people in a single episode. For this reason, the Huang He has been called "the river of China's sorrow." While the river has not changed its course since the 1930s, most geographers agree that another change is inevitable.

Erosion on the Loess Plateau   The Huang He's sediment load comes from the eroding soils of the Loess Plateau, located to the west of the North China Plain. **Loess** is a fine, windblown material that was deposited on this upland area during the last Ice Age and which sometimes accumulated to depths of several hundred feet. Loess forms fertile soil, but it washes away easily when exposed to running water. Cultivation requires plowing, which leads to soil erosion. As the population of the region gradually increased, areas of woodland and grassland diminished, leading to ever-greater rates of soil loss. As the erosion process continued, great gullies—some of them hundreds of feet deep—cut across the plateau, steadily reducing the extent of productive land.

Today the Loess Plateau is one of the poorest parts of China, and thus one of the poorer parts of the world. Population is only moderately dense by Chinese standards, but good farmland is limited and drought is common. The Chinese government encourages the construction of terraces to conserve the soil, but such efforts have not been effective everywhere. Campaigns to plant woody vegetation to help stabilize the most severely eroded lands have been even less successful, as the young plants seldom survive the area's harsh climate.

Other East Asian Environmental Problems

Although the problems associated with the Yangtze and Huang He rivers are among the most well known environmental issues in East Asia, they are hardly the only ones. Pollution and deforestation also affect much of the region.

Forests and Deforestation   Most the uplands of China and South Korea support only grass, small amounts of scrub vegetation, and stunted trees (Figure 11.5). China lacks the historical tradition of forest conservation that, as we shall see, characterizes Japan. In much of southern China, sweet potatoes, maize, and other crops have been grown on steep and easily eroded hillsides for several hundred years. After centuries of exploitation, many upland areas have lost so much soil that they cannot easily support forests.

Although the Chinese government has started large-scale reforestation programs, few have been very successful. As it now stands, substantial forests are found only in China's far north, where a cool climate prevents fast growth, and along the eastern

FIGURE 11.5 • DENUDED HILLSLOPES IN CHINA
Deforestation of China's mountain slopes has occurred throughout history in order to provide wood products and agricultural lands. Without forest cover, soil erosion is a serious issue. *(Andrew Wong/Reuters/Getty Images, Inc.—Hulton Archive Photos)*

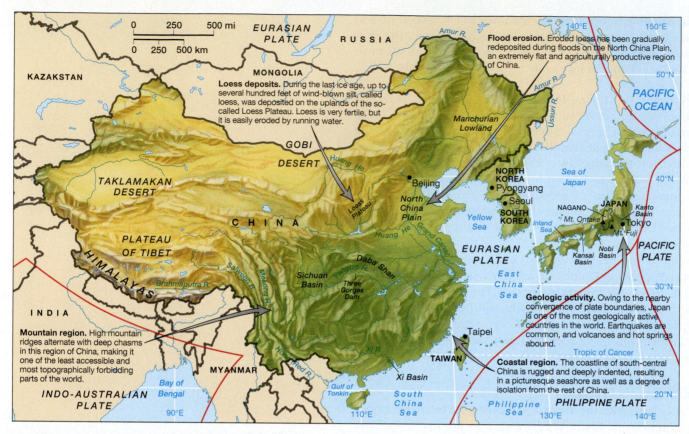

FIGURE 11.6 • PHYSICAL GEOGRAPHY OF EAST ASIA The physical geography of mainland East Asia varies widely from the high plateaus and deserts basins of western China to the broad river valleys and vast plains of eastern China. In the island region, rugged mountains and volcanoes shape the landscape in this geologically unstable area.

slopes of the Tibetan Plateau, where rugged terrain restricts commercial forestry. As a result, China suffers a severe shortage of forest resources. If its economy continues to boom, China will likely become a major importer of lumber, pulp, and paper in the near future.

Mounting Pollution As China's industrial base expands, other environmental problems, such as water pollution and toxic-waste dumping, are becoming more severe, particularly in the booming coastal areas. The burning of high-sulfur coal has resulted in serious air pollution, which is made worse by the increasing number of automobiles being driven by the Chinese. Such problems emerged earlier in other portions of East Asia, namely Taiwan and South Korea, which responded by imposing strict environmental controls as they grew wealthier. These two countries are also following in Japan's footsteps by setting up new factories in poorer countries—including China—that have looser environmental standards.

Environmental Issues in Japan Considering its large population and intensive industrialization, Japan's environment is relatively clean. Actually, the very density of its population gives certain environmental advantages. In crowded cities, for example, travel by private automobile is often slower than travel on subways and other public transportation systems. In the 1950s and 1960s, Japan's most intensive period of industrial growth, the country did suffer from some of the world's worst water and air pollution. Soon afterward the Japanese government passed strict air and water pollution laws.

Japan's environmental cleanup was aided by its location, since winds usually carry smog-forming chemicals out to sea. Equally important has been the phenomenon of **pollution exporting.** Because of Japan's high cost of production and its strict environmental laws, many Japanese companies have relocated their dirtier factories to other countries, particularly those of Southeast Asia. This practice, which is also followed by the United States and western Europe, means that Japan's pollution is thus partially displaced to poorer countries.

East Asia's Physical Geography

An examination of the physical geography of East Asia helps shed light on these environmental issues. East Asia is situated in the same general latitudinal range as the United States, although it extends considerably farther north and south (Figure 11.6). The northernmost tip of China lies as far north as central Quebec, while China's southernmost point is at the same latitude as Mexico City. The climate of southern China is thus roughly comparable to southern Florida and the Caribbean, while that of northern China is similar to south-central Canada (Figure 11.7). The area, especially the islands of Japan, is geologically active, with earthquakes and volcanic eruptions occurring frequently.

FIGURE 11.7 • CLIMATE MAP OF EAST ASIA East Asia is located in roughly the same latitudinal zone as North America, so there are climatic parallels between the two world regions. The northernmost tip of China lies at about the same latitude as Quebec and shares a similar climate, whereas southern China approximates the climate of Florida. In Japan, maritime influences produce a milder climate.

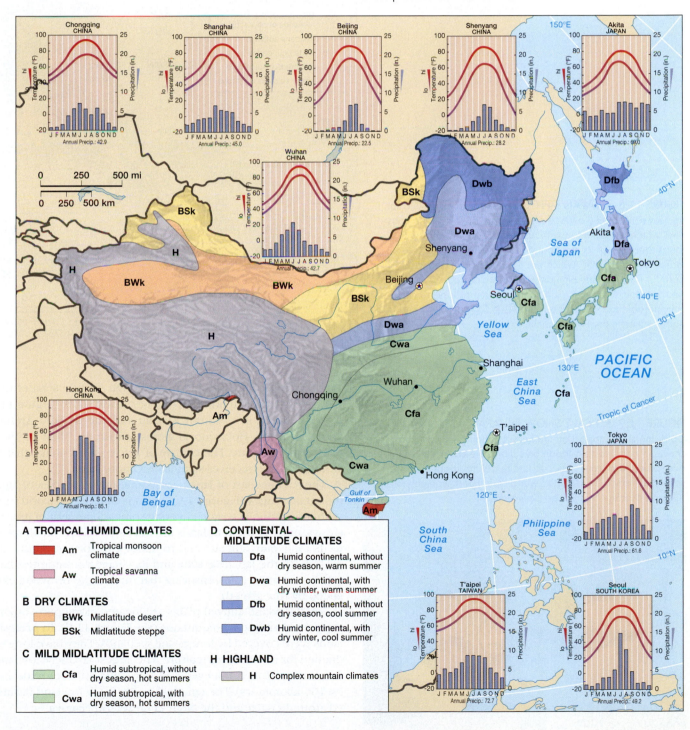

A TROPICAL HUMID CLIMATES

- **Am** Tropical monsoon climate
- **Aw** Tropical savanna climate

B DRY CLIMATES

- **BWk** Midlatitude desert
- **BSk** Midlatitude steppe

C MILD MIDLATITUDE CLIMATES

- **Cfa** Humid subtropical, without dry season, hot summers
- **Cwa** Humid subtropical, with dry season, hot summers

D CONTINENTAL MIDLATITUDE CLIMATES

- **Dfa** Humid continental, without dry season, warm summer
- **Dwa** Humid continental, with dry winter, warm summer
- **Dfb** Humid continental, without dry season, cool summer
- **Dwb** Humid continental, with dry winter, cool summer

H HIGHLAND

- **H** Complex mountain climates

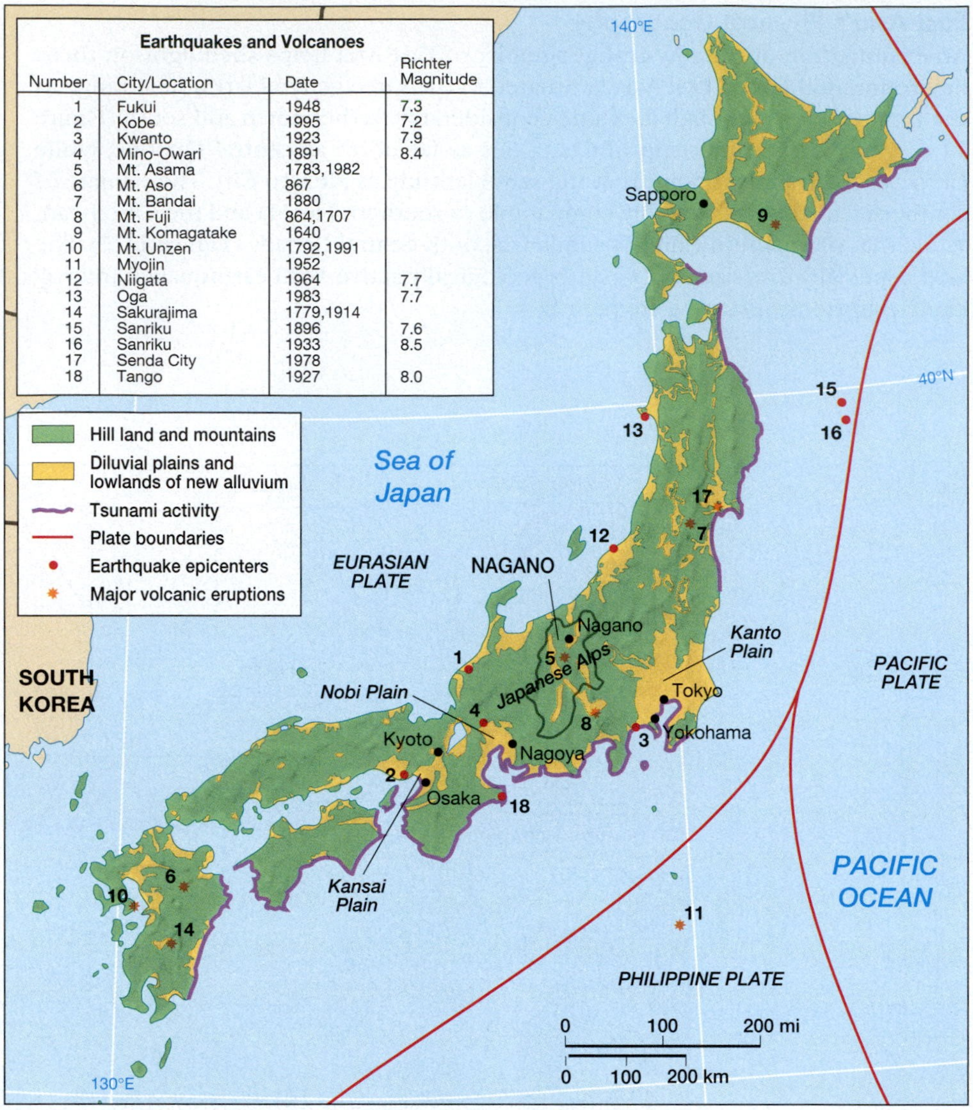

FIGURE 11.8 • JAPAN'S PHYSICAL GEOGRAPHY Japan has several sizable lowland plains, primarily along the coastline, which are mixed in with rugged mountains and uplands. Because of its location at the convergence of three major tectonic plates (which are the basic building blocks of Earth's crust), Japan commonly experiences both earthquakes and volcanic eruptions. Additionally, much of Japan's coast is vulnerable to devastating tsunamis (tidal waves) caused by earthquakes in the Pacific basin.

Earthquakes and Volcanoes

Number	City/Location	Date	Richter Magnitude
1	Fukui	1948	7.3
2	Kobe	1995	7.2
3	Kwanto	1923	7.9
4	Mino-Owari	1891	8.4
5	Mt. Asama	1783,1982	
6	Mt. Aso	867	
7	Mt. Bandai	1880	
8	Mt. Fuji	864,1707	
9	Mt. Komagatake	1640	
10	Mt. Unzen	1792,1991	
11	Myojin	1952	
12	Niigata	1964	7.7
13	Oga	1983	7.7
14	Sakurajima	1779,1914	
15	Sanriku	1896	7.6
16	Sanriku	1933	8.5
17	Senda City	1978	
18	Tango	1927	8.0

Legend:
- Hill land and mountains
- Diluvial plains and lowlands of new alluvium
- Tsunami activity
- Plate boundaries
- Earthquake epicenters
- Major volcanic eruptions

FIGURE 11.9 • HEAVY SNOW IN JAPAN'S MOUNTAINS Cold, moist air moving off the Sea of Japan produces heavy snows in northwestern Japan and along the country's mountainous spine. Numerous major ski areas dot the Japanese Alps, several of which have hosted world-class sports competitions. (George Mobley/NGS Image Collection)

Japan's Physical Environment Although slightly smaller than California, Japan is more elongated, extending farther north and south (Figure 11.8). As a result, Japan's extreme south, in southern Kyushu and the Ryukyu Archipelago, is subtropical, while northern Hokkaido is almost subarctic. Most of the country, however, is distinctly temperate. The climate of Tokyo is not unlike that of Washington, D.C.— although Tokyo receives significantly more rain.

The Pacific coast of Japan is separated from the Sea of Japan coast by a series of mountain ranges (Figure 11.9). Japan is one of the world's most rugged countries, with mountainous terrain covering some 85 percent of its territory. As Figure 11.10 shows, most of these uplands are heavily forested. Japan owes its lush forests both to its mild, rainy climate and to its long history of conservation. For hundreds of years, both the Japanese state and its village communities have enforced strict forest conservation rules, ensuring that timber and firewood extraction would be balanced by tree growth.

Small lowland alluvial plains are located along parts of Japan's coastline and interspersed among its mountains (see Figure 11.8). In prehistoric times, these lowlands were covered by forests and wetlands. They have long since been cleared and drained for intensive agriculture. The largest Japanese lowland is the Kanto Plain to the north of Tokyo, but even it is only some 80 miles wide and 100 miles long (130 by 160 kilometers). The country's other main lowland basins are the Kansai, located around Osaka, and the Nobi, centered on Nagoya.

FIGURE 11.10 • FORESTED LANDSCAPES OF JAPAN
Although much of Japan is heavily forested and supports a wood products industry, the yield is not large enough to satisfy demand. As a result, Japan imports timber extensively from North America, Southeast Asia, and Latin America. *(Robert Holmes/Corbis)*

Taiwan's Environment   Taiwan, an island about the size of Maryland, sits at the edge of the continental landmass. To the west, the Taiwan Strait is only about 200 feet (60 meters) deep; to the east, ocean depths of many thousands of feet are found 10 to 20 miles (16 to 32 kilometers) offshore.

Taiwan's central and eastern regions are rugged and mountainous, while the west is mainly a lowland alluvial plain. Bisected by the tropic of Cancer, Taiwan has a mild winter climate, but it is often hit by typhoons in the early autumn. Unlike nearby areas of China proper, Taiwan still has extensive forests, especially in its remote central and eastern uplands.

Chinese Environments   Even if one excludes China's Central Asian provinces, it is a vast country with diverse environmental regions. For the sake of convenience, it can be divided into two main areas, one lying to the north of the Yangtze River Valley, the other including the Yangtze and all areas to the south. As Figure 11.11 shows, each of these can be subdivided into a number of distinctive regions.

Southern China is a land of rugged mountains and hills interspersed with lowland basins. The lowlands of southern China are far larger than those of Japan. Large

FIGURE 11.11 • LANDSCAPE REGIONS OF CHINA   The Yangtze Valley divides China into two general areas. Immediately to the north is the large fertile area of the North China Plain, bisected by the Huang He (or Yellow) River. To the west is the Loess Plateau, an upland area of soil derived from wind-deposited silt after the prehistoric glacial period, about 15,000 years ago.

FIGURE 11.12 • THE FUJIAN COAST OF CHINA   In southeast China lies the rugged coastal province of Fujian. Here the coastal plain is narrow and the shoreline deeply indented, producing a charming landscape. Because of limited agricultural opportunities along this rugged coastline, many Fujianese people work in maritime activities. *(Dean Conger/Corbis)*

valleys and moderate-elevation plateaus are also found in the far south, where the climate is truly tropical, free of frost. The coastal areas in southeast China are rugged and offer limited agricultural opportunities. Thus people in these coastal provinces often find work in fishing and shipping industries (Figure 11.12). North of the Yangtze Valley, the climate is both colder and drier. Summer rainfall is generally abundant except along the edge of the Gobi Desert, but the other seasons are often dry. With the exception of a few low mountains in Shandong province, the entire area is a virtually flat plain. The North China Plain, a large area of fertile soil crossed by the Huang He (or Yellow) River, is cold and dry in winter and hot and humid in the summer. Overall precipitation is somewhat low and unpredictable, however, and some areas of the North China Plain are in danger of **desertification** (or the spread of desert conditions). Seasonal water shortages are growing increasingly severe through much of the region as withdrawals for irrigation and industry increase, leading to concern about an impending water crisis.

China's far northeastern region is called *Dongbei* in Chinese and *Manchuria* in English. Manchuria is dominated by a broad, fertile lowland sandwiched between mountains and uplands stretching along China's borders with North Korea, Russia, and Mongolia. Although winters here can be brutally cold, summers are usually warm and moist. Manchuria's upland areas have some of China's best-preserved forests and wildlife refuges.

Korean Landscapes   Korea forms a well-defined peninsula, partially cut off from Manchuria by rugged mountains and sizable rivers. The far north, which just touches Russia's Far East, has a climate not unlike that of Maine, whereas the southern tip is more similar to the Carolinas. Korea, like Japan, is a mountainous country with scattered alluvial basins. The lowlands of the southern portion of the peninsula are more extensive than those of the north, giving South Korea a distinct agricultural advantage over North Korea. The north, however, has much more abundant mineral deposits, forest resources, and rivers capable of producing hydroelectricity.

Population and Settlement: A Realm of Crowded Lowland Basins

East Asia, along with South Asia, is the most densely populated region of the world. The lowlands of Japan, Korea, and China are among the most intensely used portions of Earth, containing not only the major cities, but also most of the agricultural lands of these countries. Although the density of East Asia is extremely high, the region's population growth rate has declined dramatically since the 1970s. In Japan, the current concern is one of population loss. While China's population is still expanding, its rate of growth is by global standards relatively low.

Japanese Settlement and Agricultural Patterns

Japan is a highly urbanized country, supporting two of the largest metropolitan areas in the world: Tokyo and Osaka. Yet it is also one of the world's most mountainous countries, and its uplands are lightly populated. Agriculture must therefore share the limited lowlands with cities and suburbs, resulting in extremely intensive farming practices.

Japan's Agriculture Lands   Japanese agriculture is largely limited to the country's coastal plains and interior basins. Rice is Japan's major crop, and irrigated rice demands flat land. Japanese rice farming has long been one of the most productive forms of agriculture in the world, helping support a large population—some 127 million people—on a relatively small and rugged land. Although rice is grown in almost all Japanese lowlands, the country's premier rice-growing districts lie along the Sea of Japan coast of central and northern Honshu.

Vegetables are also grown intensively in all of the lowland basins, even on tiny patches within urban neighborhoods (Figure 11.13). The valleys of central and northern Honshu are famous for their temperate-climate fruit, while citrus comes from the milder southwestern reaches of the country. Crops that thrive in a cooler climate, such as potatoes, are produced mainly in Hokkaido and northern Honshu. Dairy products as well come from these northern areas.

Settlement Patterns   All Japanese cities—and the vast majority of the Japanese people—are located in the same lowlands that support the country's agriculture. Not surprisingly, the three largest metropolitan areas—Tokyo, Osaka, and Nagoya— sit near the centers of the three largest plains.

Overall Japan is a densely populated country (Figure 11.14). The fact that its settlements are largely restricted to roughly 15 percent of the country's land area means that Japan's effective population density—the actual crowding that the country experiences—is one of the highest in the world. This is especially true in the main industrial belt, which extends from Tokyo south through Nagoya and Osaka and to the northern coast of Kyushu.

Due to such space limitations, all Japanese cities are characterized by dense settlement patterns. In the major urban areas, the amount of available living space is highly restricted for all but the most affluent families. In Tokyo, an apartment the size of a large U.S. living room often shelters an entire middle-class family. Many observers—especially American ones—argue that Japan should allow its cities and suburbs to expand into nearby rural areas. However, since most uplands are too steep for residential use, such expansion would have to come at the expense of agricultural land.

Settlement and Agricultural Patterns in China, Taiwan, and Korea

Like Japan, Taiwan and Korea are essentially urban. China, however, remains largely rural, with only some 36 percent of its population living in cities. Almost all Chinese cities are growing at a rapid pace, despite governmental efforts to keep their size under control.

FIGURE 11.13 • JAPANESE URBAN FARM   Japanese landscapes often combine dense urban settlement with small patches of intensively farmed agriculture. Here a cabbage farm coexists with an urban neighborhood. *(Kyodo News International, Inc.)*

FIGURE 11.14 • POPULATION MAP OF EAST ASIA   Parts of East Asia are very densely settled, particularly in the coastal lowlands of China and Japan. This contrasts with the sparsely settled lands of western China, North Korea, and northern Japan. Although the total population of this world region is high, as is the overall density, the rate of natural population increase has slowed rather dramatically in the last several decades because of several factors, primarily China's well-known "one-child" policy.

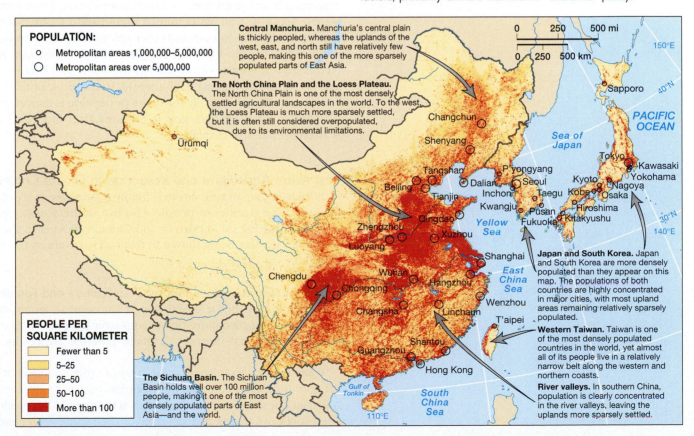

FIGURE 11.15 • LOESS SETTLEMENT This photo shows a typical subterranean dwelling carved out of the soft loess sediment in central China. This area is prone to major earthquakes that take a high toll on the local population because of dwelling collapse. *(Christopher Liu/ChinaStock Photo Library)*

China's Agricultural Regions A line drawn just to the north of the Yangtze Valley divides China into two main agricultural regions. To the south, rice is the dominant crop. To the north, wheat, millet, and sorghum are the most common.

In southern and central China, population is highly concentrated in the broad lowlands, which are famous for their fertile soil and intensive agriculture. More than 100 million people live in the Sichuan Basin, while more than 70 million reside in Jiangsu, a province smaller than Ohio. Cropping occurs year-round in most of southern and central China; summer rice alternates with winter barley or vegetables in the north, while two rice crops can be harvested in the far south. Southern China also produces a wide variety of tropical and subtropical crops, and moderate slopes throughout the area provide areas for sweet potatoes, corn, and other upland crops.

In northern China, population is more variable. The North China Plain is one of the most thoroughly **anthropogenic landscapes** in the world (an anthropogenic landscape is one that has been heavily transformed by human activities). Virtually its entire extent is either cultivated or occupied by houses, factories, and other structures of human society. Too dry in most areas for rice, agriculture on the North China Plain specializes in wheat, millet, and sorghum.

Manchuria was a lightly populated frontier zone as recently as the mid-1800s. Today, with a population of more than 100 million, its central plain is thoroughly settled. Still, Manchuria remains less crowded than many other parts of China, and it is one of the few parts of China to produce a consistent food surplus.

The Loess Plateau is more thinly settled yet, supporting only some 70 million residents (Figure 11.15). But considering its aridity and widespread soil erosion, this is a high figure indeed. Like other portions of northern China, the Loess Plateau produces wheat and millet, most of which is eaten at home rather than sold in the market.

Settlement and Agricultural Patterns in Korea and Taiwan Korea is also a densely populated area. It contains some 70 million people (22 million in the north and 48 million in the south) in an area smaller than Minnesota. South Korea's population density is actually significantly higher than that of Japan. Most of these people are crowded into the alluvial plains and basins of the west and south. The highland spine, extending from the far north to northeastern South Korea, remains relatively sparsely settled. South Korean agriculture is dominated by rice. North Korea, on the other hand, relies heavily on corn and other upland crops that do not require irrigation.

Taiwan is the most densely populated state in East Asia. Roughly the size of the Netherlands, it contains more than 22 million inhabitants. Its overall population density is one of the highest in the world. Because mountains cover most of central and eastern Taiwan, virtually the entire population is concentrated in the narrow lowland belt in the north and west. In this area, large cities and numerous factories are scattered amid lush farmlands.

East Asian Agriculture and Resources in Global Context

Although East Asian agriculture is highly productive, it is not productive enough to feed the huge number of people who live in the region. Japan, Taiwan, and South Korea are major food importers, and China is moving in the same direction. Other resources are also being drawn in from all quarters of the world by the powerful economies of East Asia.

The Global Dimensions of Japanese Agriculture and Forestry Japan may be self-sufficient in rice, but it is still one of the world's largest food importers. As the Japanese have grown more prosperous over the past 50 years, their diet has grown more diverse. That diversity is made possible largely by importing food from elsewhere.

Japan imports food from a wide array of other countries. It obtains both meat and the feed used in its domestic livestock industry from the United States, Canada, and Australia. These same countries supply wheat needed to produce bread and noodles. Even soybeans, traditionally a staple of the Japanese diet, must be purchased from

Brazil and the United States. Japan has one of the highest rates of fish consumption in the world, and the Japanese fishing fleet must scour the world's oceans to supply the demand.

Japan depends on imports to supply its demand for forest resources. While its own forests produce high-quality cedar and cypress logs, it buys most of its construction lumber and pulp (for papermaking) from western North America and Southeast Asia. Almost half of Southeast Asia's forest-product exports are shipped to Japan. As the rain forests of Malaysia, Indonesia, and the Philippines diminish, Japanese interests are beginning to turn to Latin America and Africa as sources of tropical hardwoods. Japanese and South Korean firms are also looking to the forests of Siberia (eastern Russia), which are nearby and relatively unexploited, to meet its resource needs.

Japan is able to support its large and prosperous population on such a restricted land base because it can purchase resources from abroad. Certainly all countries engage in trade, but Japan's basic resource dependence is quite distinct. Almost all of the oil, coal, and other minerals that it consumes are imported. If forced to rely on its own resources, Japan would find itself in an economic and ecological bind. But because its industries are successful in the global market, Japan has avoided such a problem.

The Global Dimensions of Chinese Agriculture   Through the 1980s, China was essentially self-sufficient in food, despite its huge population and crowded lands. But increasing wealth, combined with changing diets and the introduction of foreign foods, has brought about an increased consumption of meat, which requires large amounts of feed-grain. It has also resulted in the loss of agricultural lands to residential and industrial development. As a result, in the mid-1990s, China had to import large amounts of grain for several years.

Optimists argue that China could produce much more food than it now does by increasing its use of fertilizers and by converting its "wastelands" of grass and scrub into agricultural fields. Chinese leaders are well aware of the current grain shortage, but they promise that agricultural production will soon increase. Many concede, however, that by the year 2030 China will have to import 5 to 10 percent of its total grain demand.

Urbanization in East Asia

China has one of the world's oldest urban foundations, dating back more than 3,500 years. In medieval and early modern times, East Asia as a whole possessed a well-developed system of cities that included some of the largest settlements on the planet. In the early 1700s, Tokyo, then called *Edo*, probably overshadowed all other cities, with a population of more than 1 million.

But despite this early urbanization, East Asia was overwhelmingly rural at the end of World War II. Some 90 percent of China's people then lived in the countryside, and even Japan was only about 50 percent urbanized. But as the region's economy began to grow after the war, so did its cities. Japan, Taiwan, and South Korea are now between 70 and 80 percent urban, which is typical for advanced industrial countries. Only about 36 percent of China's people live in cities, but this figure is increasing steadily.

Chinese Cities   Traditional Chinese cities were clearly separated from the countryside by defensive walls. Most were planned in accordance with strict geometrical principles with straight streets meeting at right angles. The old-style Chinese city was dominated by low buildings and characterized by straight streets. Houses were typically built around courtyards, and narrow alleyways served both commercial and residential functions.

China's cities began to change as Europeans started to gain power in the 1800s. A group of port cities was taken over by European interests, which proceeded to build Western-style buildings and modern business districts. By far the most important of these semicolonial cities was Shanghai, built near the mouth of the Yangtze River, the main gateway to interior China.

FIGURE 11.16 • CONTEMPORARY SHANGHAI This vibrant city of more than 14 million symbolizes the new China with its massive high-rise apartments, industrial developments, and office towers. This photo shows the Nan Pu bridge and high-rise buildings in the newly developed Pudong area. *(Keren Su/Pacific Stock)*

When the communists came to power in 1949, Shanghai, with a population of more than 10 million people, was the second-largest city in the world. The new authorities considered it a foreign city dominated by capitalist exploiters. They therefore refused to invest money in the city, and as a result it began to decay. Since the late 1980s, however, Shanghai has experienced a major revival. Migrants are now pouring into Shanghai, and building cranes crowd the skyline. Official statistics now put the population of the metropolitan area at some 14 million, but the actual number, including temporary migrants, may well be significantly larger. The new Shanghai is a city of massive high-rise apartments and concentrated industrial developments (Figure 11.16).

Beijing was China's capital during the Manchu period (1644–1912), a status it regained in 1949. Under communist rule, Beijing was radically transformed; old buildings were razed and broad avenues were plowed through old neighborhoods. Crowded residential districts gave way to large blocks of apartment buildings and massive government offices. Some historically significant buildings were saved; the buildings of the Forbidden City, for example, where the Manchu rulers once lived, survived as a complex of museums.

In the 1990s Beijing and Shanghai vied for the first position among Chinese cities, with Tianjin, serving as Beijing's port, coming in a close third. All three of these cities have historically been removed from the regular provincial structure of the country and granted their own metropolitan governments. In 1997 another major city, Hong Kong, passed from British to Chinese control and was granted a unique status as a self-governing "special administrative region." While not as heavily populated as Beijing or Shanghai, Hong Kong is far wealthier. The emerging greater metropolitan area of the Xi Delta, composed of Hong Kong, Shenzhen, and Guangzhou (called *Canton* in the West), may now perhaps be regarded as China's premier urban area.

City Systems of Japan and South Korea The urban structures of Japan and South Korea are quite different from those of China. South Korea is noted for its pronounced **urban primacy** (the concentration of urban population in a single city), whereas Japan is the center of a new urban phenomenon, that of the **superconurbation** (a superconurbation, or megalopolis, is a huge zone of coalesced metropolitan areas) (Figure 11.17).

Seoul, the capital of South Korea, is by far the largest city in the country. Seoul itself is home to more than 10 million people, and its greater metropolitan area contains some 40 percent of South Korea's total population. All of South Korea's major governmental, economic, and cultural institutions are concentrated there. However, Seoul's explosive and generally unplanned growth has resulted in serious congestion. Although the government has promoted growth in other cities, it is unlikely that another urban area will ever challenge the primacy of the capital. Japan has traditionally been characterized by

FIGURE 11.17 • URBAN CONCENTRATION IN JAPAN The inset map shows the rapid expansion of Tokyo in the postwar decades. Today the Greater Tokyo metropolitan area is home to almost 30 million people. The larger map shows the cluster of urban settlements along Japan's southeastern coast. The major area of urban concentration is between Tokyo and Osaka, a distance of some 300 miles, known as the *Tokkaido corridor*. Roughly 65 percent of Japan's population lives in this area. The Osaka–Kobe metropolitan area ranks second to Tokyo, with approximately 14 million inhabitants.

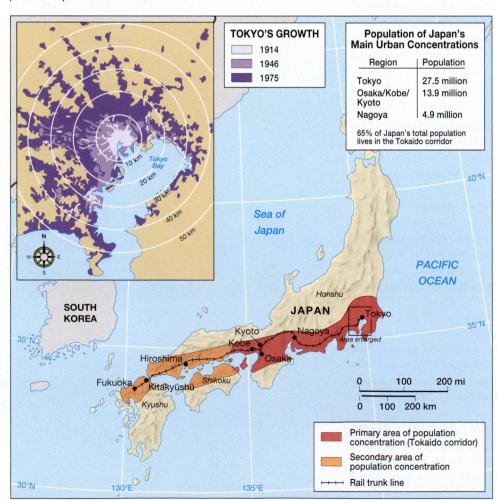

TOKYO'S GROWTH
- 1914
- 1946
- 1975

Population of Japan's Main Urban Concentrations	
Region	**Population**
Tokyo	27.5 million
Osaka/Kobe/ Kyoto	13.9 million
Nagoya	4.9 million

65% of Japan's total population lives in the Tokaido corridor

■ Primary area of population concentration (Tokaido corridor)
■ Secondary area of population concentration
┼┼┼ Rail trunk line

urban "bipolarity" rather than urban primacy. Until the 1960s, Tokyo, the capital and main business and educational center, together with the neighboring port of Yoko-hama, was balanced by the trading center of Osaka and its port of Kobe. Kyoto, the former imperial capital and the traditional center of elite culture, is also located in the Osaka region. As Japan's economy boomed in 1960s, 1970s, and 1980s, however, so did Tokyo. The capital city then outpaced all other urban areas in almost every urban function. The Greater Tokyo metropolitan area today contains 25 to 28 million persons, depending on where its borders are defined.

Japanese cities sometimes strike foreign visitors as rather gray and dull places, lacking historical interest. Little of the country's premodern architecture remains intact. Traditional Japanese buildings were made of wood, which survives earthquakes much better than stone or brick. Fires have therefore been a long-standing hazard, and in World War II the U.S. Air Force fire-bombed most Japanese cities, virtually destroying them (Hiroshima and Nagasaki were, on the other hand, completely destroyed by atomic bombs). The one exception was Kyoto, the old imperial capital, which was spared devastation. As a result, Kyoto is famous for its beautiful (wooden) Buddhist monasteries and Shinto temples, which ring the basin in which central Kyoto lies.

Cultural Coherence and Diversity: A Confucian Realm?

East Asia is in some respects one of the world's more unified cultural regions. Although different East Asian countries, as well as the regions within them, have their own unique culture, the entire region shares certain historically rooted ways of life and systems of ideas.

Most of these common features can be traced back to ancient Chinese civilization. Chinese culture emerged roughly 4,000 years ago, largely in isolation from the Eastern Hemisphere's other early centers of civilization in the valleys of the Indus, Tigris-Euphrates, and Nile rivers. As a result, East Asian civilization developed along several unique lines.

Unifying Cultural Characteristics

The most important unifying cultural characteristics of East Asia are related to religious and philosophical beliefs. Throughout the region, Buddhism and especially Confucianism have shaped both individual beliefs and social and political structures. Although the role of traditional belief systems has been seriously challenged in recent decades, especially in China, traditional cultural patterns never disappear overnight.

The Chinese Writing System The clearest distinction between East Asia and the world's other cultural regions is found in written language. Existing writing systems elsewhere in the world are based on the alphabetic principle, in which each symbol represents a distinct sound. East Asia, on the other hand, evolved an entirely different system of **ideographic writing**. In ideographic writing, each symbol (or ideograph—more commonly called *character*) represents primarily an idea rather than a sound (although the symbols can represent sounds in certain circumstances). As a result, ideographic writing requires the use of a large number of distinct symbols.

The East Asian writing system can be traced to the dawn of Chinese civilization. As the Chinese Empire expanded and Chinese civilization influenced other lands, the Chinese writing system spread. Japan, Korea, and Vietnam all came to use the same system, although in Japan it was substantially modified, while in Korea it was later largely replaced by alphabetic systems.

The Chinese ideographic writing system has one major disadvantage and one major advantage when compared with alphabetic systems, both of which stem from the fact that it is largely separated from spoken language. The disadvantage is that it is difficult to learn; to be literate, a person must memorize thousands of characters. The main benefit is that two literate persons do not have to speak the same language to be able to communicate, since the written symbols that they use to express their ideas are the same.

This advantage was tremendously important for the creation of a unified Chinese culture. When the Chinese Empire expanded south of the Yangtze River beginning in about 200 B.C., ethnic groups speaking a variety of languages were suddenly brought into the same political and cultural system. Because they could adopt the Chinese writing system, the peoples of southern China were able to integrate fully into the rest of the country without adopting its spoken language.

Korean Modifications In Korea, Chinese characters were adopted at an early date and were used exclusively for hundreds of years. In the 1400s, however, Korean officials decided that the country needed its own alphabet. They wanted to encourage more widespread literacy, hoping also to more clearly separate Korean culture from that of China. The use of the new script spread quickly through the country. Korean scholars and officials, however, continued to use Chinese characters, regarding the new script as suitable only for popular writings. Today the Korean script is used for almost all purposes, but scholarly works still contain Chinese characters.

Japanese Modifications The writing system of Japan is even more complex (Figure 11.18). Initially the Japanese simply borrowed Chinese characters, referred to in Japanese as **kanji**. Owing to the grammatical differences between spoken Japanese and Chinese, the exclusive use of kanji resulted in awkward sentences. The Japanese solved this quandary by developing **hiragana**, a writing system in which each symbol represents a distinct syllable, or combination of a consonant and a vowel sound. A different but essentially parallel system, called *katakana*, was devised for spelling words of foreign origin. Written Japanese employs a complex mixture of all three of these symbolic systems.

The Confucian Legacy Just as the use of a common writing system helped build cultural linkages throughout East Asia, so too the idea system of **Confucianism** (the philosophy developed by Confucius) came to occupy a significant position in all of the societies of the region. Indeed, so strong is the heritage of Confucius that some writers refer to East Asia as the "Confucian world." In Japan, however, Confucianism never had the influence that it did in China and Korea.

The premier philosopher of Chinese history, Confucius (or Kung Fu Zi, in Mandarin Chinese) lived during the sixth century B.C., a period of political instability. The North China Plain and Loess Plateau, then the main areas of Chinese culture, were divided into a number of hostile states. Confucius's goal was to create a philosophy that could generate social stability. While Confucianism is often considered to be a religion, Confucius himself was far more interested in how to lead a correct life and organize a proper society.

Confucius stressed obedience to authority, but he thought that those in power have a responsibility to act in a caring manner. Confucian philosophy also stresses the need for a well-rounded education focused on literature and philosophy. The most basic level of the traditional Confucian moral order is the family unit, considered the bedrock of society. The ideal family structure is patriarchal (male-dominated), and children are told to obey and respect their parents. At the highest level of the traditional moral order sat the emperor of China, who was regarded as an almost godlike father figure for the entire country.

The Modern Role of Confucian Ideas The significance of Confucianism in East Asian development has been hotly debated for most of this century. In the early 1900s many observers believed that conservative philosophy, derived from its respect for tradition and authority, was responsible for the economically backward position of China and Korea. But because East Asia has more recently enjoyed the world's fastest rates of economic growth, such a position is no longer supportable. New voices now argue that Confucianism's respect for education and the social stability that it generates give East Asia a tremendous advantage in international competition.

FIGURE 11.18 • JAPANESE WRITING The writing system of Japan was originally based on Chinese characters, known in Japan as *kanji*. Because of grammatical differences, however, the Japanese developed two unique "alphabets" of syllables, known as *katakana* and *hiragana*. Here *kanji* and *katakana* symbols are visible. *(Hiroshi Harada/DUNQ/Photo Researchers, Inc.)*

Other scholars remain skeptical of both views, preferring to credit such factors as economic policy for the region's rapid economic growth. They also note that most interior portions of China have not participated much in the current economic boom, even though they share the Confucian legacy. Confucianism has, moreover, recently lost much of the hold that it once had on public morality throughout the entire region as new ideas have flowed in from other parts of the world.

Religious Unity and Diversity in East Asia

Certain religious beliefs have worked alongside Confucianism to unite the East Asian region. The most important culturally unifying beliefs are associated with Mahayana Buddhism. Other religious practices, however, have tended to challenge the cultural unity of the region.

Mahayana Buddhism   Buddhism, a religion that stresses the quest to escape an endless cycle of rebirths and reach union with the cosmos (or nirvana), originated in India in the sixth century B.C. By the second century A.D., Buddhism had reached China, and within a few hundred years it had spread throughout East Asia. Today Buddhism remains widespread everywhere in the region, although it is far less significant here than it is in mainland Southeast Asia, Sri Lanka, and Tibet (Figure 11.19).

The variety of Buddhism practiced in East Asia—Mahayana, or Greater Vehicle—is distinct from the Therevada Buddhism of South and Southeast Asia. Most important, Mahayana Buddhism simplifies the quest for nirvana (total enlightenment), in part by putting forward the existence of beings (*boddhisatvas*) who refuse divine union for themselves in order to help others spiritually. Mahayana Buddhism, unlike other forms, is nonexclusive; in other words, one may follow it while simultaneously professing the beliefs of other faiths. Thus, most Japanese are at some level both Buddhists and followers of Shinto, while many Chinese consider themselves to be both Buddhists and Taoists (as well as Confucianists).

Shinto   The Shinto religion is so closely bound to the idea of Japanese nationality that it is questionable whether a non-Japanese person can truly follow it. Shinto began as the worship of nature spirits, but it was gradually refined into a subtle set of beliefs about the harmony of nature and its connections with human existence. Shinto is still a place- and nature-centered religion. Certain mountains, particularly Mount Fuji, are considered sacred and are thus climbed by large numbers of people (Figure 11.20). Major Shinto shrines, often located in scenic places, attract numerous religious pilgrims; the most notable of these is the Ise Shrine south of Nagoya, which is devoted to honoring the emperor.

Taoism and Other Chinese Belief Systems   The Chinese religion of Taoism (or Daosim) is similarly rooted in nature worship. Like Shinto, it stresses spiritual harmony and the pursuit of a balanced life. Taoism is indirectly associated with *feng shui*, commonly called **geomancy** in English, which is the Chinese and Korean practice of designing buildings in accordance with the spiritual powers that supposedly flow through the local topography. Even in hypermodern Hong Kong, skyscrapers worth millions of dollars have occasionally gone unoccupied because their construction failed to follow feng shui principles.

Minority Religions   Small numbers of followers of virtually all world religions can be found in the increasingly cosmopolitan cities of East Asia. Millions of Chinese and Japanese belong to Christian churches, even though they constitute less than 1 percent of the population of either country. Far more South Koreans, some 6 million in all, are Christian, mostly Protestant.

Much larger than China's Christian population is its Muslim community. Several tens of millions of Chinese-speaking Muslims, called Hui, are concentrated in Gansu and Ningxia in the northwest and in Yunnan province along the south-central border. Smaller clusters of Hui, often separated out in their own villages, live in almost every province of China.

FIGURE 11.19 • THE BUDDHIST LANDSCAPE   This Buddhist temple is located in Yunnan Province, China, and was used as the headquarters of the Yunnan provincial Buddhist association before the communist revolution of the mid-twentieth century. Today it is preserved as a historical museum. *(Brian Vikander/ Corbis)*

FIGURE 11.20 • MT. FUJI, JAPAN   This beautiful volcanic mountain, sacred to Japan's Shinto religion, is climbed by large numbers of religious pilgrims each year. In the foreground are tea fields. *(Pacific Stock)*

Secularism in East Asia   Despite all of these varied forms of religious expression, East Asia is still one of the most secular regions of the world. In Japan, while a small section of the population is very religious, most people only occasionally observe Shinto or Buddhist rituals and maintain a small shrine for their own ancestors. Japan also has a number of "new religions," a few of which are noted for their strong beliefs. But for Japanese society as a whole, religion is simply not very important.

Chinese culture was formerly dominated by Confucianism, which is more of a philosophy than a faith. After the communist regime took power in 1949, all forms of religion and traditional philosophy—including Confucianism—were discouraged and sometimes severely repressed. Under the new regime, atheistic **Marxist** philosophy (the communistic belief system developed by Karl Marx) became the official set of ideas. With the easing of Marxism during 1980s and 1990s, however, many forms of religious expression began to return. This seems to be especially true in China's more prosperous coastal areas. In North Korea, on the other hand, rigid Marxist beliefs are still strictly enforced.

Linguistic and Ethnic Diversity in East Asia

While written languages may have helped unify East Asia, the same cannot be said for spoken languages (Figure 11.21). Japanese and Mandarin Chinese may partially share a system of writing, but the two languages bear no direct relationship. Like Korean, however, Japanese has adopted many words of Chinese origin, just as English has borrowed heavily from Greek and Latin.

Language and National Identity in Japan   Japanese, according to most lingusts, is not related to any other language. Korean is also usually classified as the only member of its language family. Many linguists, however, think that Japanese and Korean should be classified together because they share many basic grammatical features.

From many perspectives, the Japanese form one of the world's more homogeneous peoples, and they tend to regard their own culture as unique. To be sure, minor cultural and linguistic distinctions are noted between the people of western Japan (centered on Osaka) and eastern Japan (centered on Tokyo), and many local areas have distinctive customs. Overall, however, such differences are of little significance.

In earlier centuries, however, the Japanese islands had been divided between two very different peoples: the Japanese living to the south and the Ainu inhabiting the north. The Ainu are completely distinct from the Japanese. They possess their own language and have a different physical appearance. Unlike Japanese men, Ainu men usually have heavy beards and were at one time mocked by their southern neighbors as "hairy barbarians." For centuries the Japanese and the Ainu competed for land, and by the tenth century A.D. the Ainu had been largely driven off the main island of Honshu. Until the 1800s, however, Hokkaido, as well as Sakhalin (now part of Russia), largely remained Ainu territory. The Japanese people subsequently began to colonize Hokkaido, putting renewed pressure on the Ainu. Today only about 24,000 Ainu remain, and most of them have mixed Japanese-Ainu ancestry.

Minority Groups in Japan   While the Japanese are relatively homogeneous as a culture, their language is divided into several dialects. But dialectal differences in the main islands are minor, and only in the Ryukyu Islands does one encounter a variant of Japanese so distinct that it might be considered a separate language. Many Ryukyu people believe that they have not been considered full members of the Japanese nation, and they have suffered a certain amount of discrimination.

Approximately 700,000 persons of Korean descent living in Japan today have also felt discrimination. Many of them were born in Japan (their parents and grandparents having left Korea early in the century) and speak Japanese rather than Korean as their primary language. Despite their deep bonds to Japan, however, such individuals are not easily able to obtain Japanese citizenship. Perhaps as a result of such treatment, many Japanese Koreans hold radical political views, and the community as a whole sends substantial amounts of money to North Korea to help out its communist government.

FIGURE 11.21 • THE LANGUAGE GEOGRAPHY OF EAST ASIA The linguistic geography of Korea and Japan is very straightforward, as the vast majority of people in those countries speak Korean and Japanese, respectively. In China, the dominant Han Chinese speak a variety of closely related *sinitic* languages, the most important of which is Mandarin Chinese. In the peripheral regions of China, a large number of languages—belonging to several different linguistic families—can be found.

Starting in the 1980s, other immigrants began to arrive in Japan, mostly from the poorer countries of Asia. Most do not have legal status. Men from China and southern Asia typically work in the construction industry; women from Thailand and the Philippines often work as entertainers or prostitutes. Almost 200,000 Brazilians of Japanese ancestry have returned to Japan for the relatively high wages they can earn. Immigration is, however, less pronounced in Japan than in most other wealthy countries, and relatively few migrants acquire permanent residency, let alone citizenship.

Language and Identity in Korea The Koreans are also a relatively homogeneous people. The vast majority of residents in both North and South Korea speak Korean and definitely consider themselves to be members of the Korean nation. There is, however, a strong sense of regional identity, some of which can be traced back to the medieval period when the peninsula was divided into three separate kingdoms.

Not all Koreans live in Korea. Several hundred thousand reside directly across the border in northern China. Substantial Korean communities can also be found in Kazakstan in Central Asia, as a result of the Soviet Union deporting Koreans from far eastern Russia in the mid-1900s. A more recent Korean **diaspora** has brought hundreds of thousands of people to the United States (a diaspora is a scattering of a particular group of people over a vast geographical area). As recently as 2001, some 15,000 Koreans were leaving annually, mostly moving to Canada, the United States, Australia, and New Zealand.

Language and Ethnicity Among the Han Chinese   The geography of language and ethnicity in China is far more complex than that of Korea or Japan. This is true even if one considers only the eastern half of the country, so-called *China proper*. The most important distinction is that separating the Han Chinese from the non-Chinese peoples. The Han, who form the vast majority, are those people who have historically been incorporated into the Chinese cultural and political systems and whose languages are expressed in the Chinese writing system. They do not, however, all speak the same language.

Northern, central, and southwestern China—a vast area extending from Manchuria through the middle and upper Yangtze Valley to the valleys of Yunnan in the far south—constitutes a single linguistic zone. The spoken language here is generally called *Mandarin Chinese*. In China today Mandarin—often called simply "the common language"— is the national tongue.

In southeastern China, from the Yangtze Delta to China's border with Vietnam, a number of separate but related languages are spoken. The most important of these languages are located along the coast. Traveling from south to north, one encounters Cantonese (or Yue) spoken in Guangdong, Fujianese (alternatively Hokkienese, or Min locally) spoken in Fujian, and Shanghaiese (or Wu), spoken in and around the city of Shanghai and in Zhejiang province. These are true languages, not dialects, because they are not mutually intelligible. They are usually called *dialects*, however, because they have no distinct written form.

Despite their many differences, all of the languages of the Han Chinese are closely related to each other, belonging to the same Sinitic language subfamily. Since their basic grammars and sound systems are similar, it is not difficult for a person speaking one of these languages to learn another. All Sinitic languages are **tonal** and monosyllabic; their words are all composed of a single syllable (although compound words can be formed from several syllables), and the meaning of each basic syllable changes completely according to the pitch in which it is uttered.

The Non-Han Peoples   Many of the more remote upland districts of China proper are inhabited by various groups of non-Han peoples speaking non-Sinitic languages. Such peoples are usually classified as **tribal**, implying that they have a traditional social order based on self-governing village communities. Such a view is not entirely accurate, however, because some of these groups once had their own kingdoms and all are now subject to the Chinese state. What they do have in common is a tradition of cultural and sometimes political struggle against the Han Chinese (Figure 11.22).

As many as 11 million Manchus live in the more remote portions of Manchuria. The Manchu languages are related to those of the tribal peoples of central and southeastern Siberia. Few Manchus, however, still speak their own language, having abandoned it for Mandarin Chinese. Thus the Manchu languages may disappear entirely. This is an ironic situation, since the Manchus ruled the entire Chinese Empire from 1644 to 1912. Until the end of this period, the Manchus prevented the Han from settling in central and northern Manchuria. Once Chinese were allowed to settle in Manchuria in the 1800, the Manchus soon found themselves vastly outnumbered. As they began to intermarry with and adopt the language and lifeways of the newcomers, their own culture began to disappear.

Much larger communities of non-Han peoples are found in the far south, especially in Guangxi. Since most of the inhabitants of Guangxi's uplands and remote valleys speak languages of the Tai family, it has been designated an **autonomous region**. Such autonomy was designed to allow non-Han peoples to experience "socialist modernization" at a different pace from that expected of the rest of the country. Critics contend that very little real autonomy has ever existed, despite the official designation. (In addition to Guangxi, there are four other autonomous regions in China. Three of these, Xizang [Tibet], Nei Monggol [Inner Mongolia], and Xinjiang, are located

FIGURE 11.22 • TRIBAL VILLAGES IN SOUTH CHINA   Non-Han people are usually classified as "tribal" in China, which assumes they have a traditional social order based upon self-governing village communities. Shown are Yi people at an open-air market in the village of Xhanghe in Yunnan province. *(Michael S. Yamashita/NGS Image Collection)*

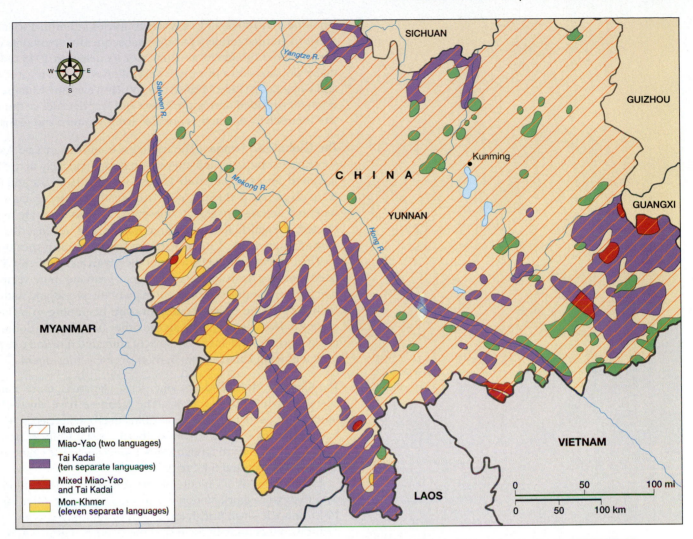

FIGURE 11.23 • LANGUAGE GROUPS IN YUNNAN
China's Yunnan Province is the most linguistically complex area in East Asia. In Yunnan's broad valleys and relatively level plateau areas and in its cities, most people speak Mandarin Chinese. In the hills, mountains, and steep-sided valleys, however, a wide variety of tribal languages, falling into several linguistic families, are spoken. In certain areas, several different languages can be found in very close proximity.

in Central Asia and are thus discussed at length in Chapter 10. The final autonomous region, Ningxia, located in northwestern China, is distinguished by its large concentration of Hui [Mandarin-speaking Muslims].)

Other areas with sizable numbers of non-Han peoples are Yunnan and Guizhou, in southwestern China, and western Sichuan. Figure 11.23 shows that in Yunnan the resulting ethnic mosaic is extremely complex.

Language and Ethnicity in Taiwan   Taiwan is also noted for its linguistic and ethnic complexity. In the island's mountainous eastern region, a few small groups of "tribal" peoples speak languages related to those of Indonesia (belonging to the Austronesian language family). These peoples resided throughout Taiwan before the sixteenth century. At that time, however, Han migrants began to arrive in large numbers. Most of the newcomers spoke Fujianese dialects, which eventually evolved into the distinctive language of Taiwanese.

Taiwan was transformed almost overnight in 1949, when China's nationalist forces, defeated by the communists, sought refuge on the island. Most of the nationalist leaders spoke Mandarin, which they made the official language. Taiwan's new leadership discouraged Taiwanese, viewing it as a simple local dialect. As a result, considerable tension developed between the Taiwanese and the Mandarin communities. Only in the 1990s did Taiwanese speakers begin to reassert their linguistic identity.

East Asian Cultures in Global Context

East Asia, like most other parts of the world, has long been torn between separating itself from the rest of the world and welcoming foreign influences and practices. This divide has both a cultural and, as we shall later see, an economic dimension.

Until the mid-1800s, all East Asian countries attempted to insulate themselves from Western cultural influences. Japan subsequently opened its doors but remained uncertain about foreign ideas. Only after its defeat in 1945 did Japan really choose to make globalization a priority. It was followed in this regard by South Korea, Taiwan, and Hong Kong (then a British colony). However, the Chinese and North Korean governments decided during the early Cold War decades of the 1950s and 1960s to isolate themselves as much as possible from Western and global culture.

The Globalized Fringe   The capitalist countries of East Asia are characterized to some extent by a cultural internationalism, especially in the large cities. Virtually all Japanese, for example, study English for 6 to 10 years, and although relatively few learn to speak it fluently, most can read and understand a good deal. Business meetings among Japanese, Chinese, and Korean firms are more often than not conducted in English. Relatively large numbers of advanced students, especially from Taiwan, study in the United States and other English-speaking countries.

The current cultural flow is not merely from a globalist West to a previously isolated East Asia. Instead, the exchange is growing more reciprocal. Hong Kong's action films are popular throughout most of the world, and with the success of director John Woo in the United States, they are beginning to influence filmmaking techniques in Hollywood. Japan, on the other hand, nearly dominates the world market in video games, and its *anime* style of animated film and television programming are now following karaoke bars in their overseas movement.

The Chinese Heartland   In one sense, Japan is more culturally predisposed to cosmopolitanism than is China. The Japanese have always borrowed heavily from other cultures (particularly from China itself), whereas the Chinese have historically been more self-sufficient. The southern coastal Chinese have, however, more often been oriented toward foreign lands, especially to the Chinese diaspora communities of Southeast Asia and the Pacific.

In most periods of Chinese history, the internal orientation of the center prevailed over the external orientation of the southern coast. After the communist victory of 1949, only the small British enclave of Hong Kong was able to pursue international cultural connections. In the rest of the country, a grim and puritanical cultural order was rigidly enforced. After China began to liberalize its economy and open its doors to foreign influences in the late 1970s and early 1980s, however, the southern coastal region suddenly assumed a new prominence. Through its doors global cultural patterns began to penetrate the rest of the country. The result has been the emergence of a vibrant but somewhat flashy urban popular culture in China that contains such global features as nightclubs, karaoke bars, fast-food franchises, and theme parks.

The Geopolitical Framework and Its Evolution: The Imperial Legacies of China and Japan

The political history of East Asia revolves around the centrality of China and the ability of Japan to remain outside of China's grasp. The traditional Chinese conception of geopolitics was based on the idea of a universal empire: All territories were supposed to be a part of the Chinese Empire, pay tribute to it and acknowledge its supremacy, or stand outside the system altogether. Until the 1800s, the Chinese government would not recognize any other as its diplomatic equal. When China could no longer maintain its power in the face of European aggression, the East Asian political system began to fall apart. As European power declined in the 1900s, China and Japan competed for regional leadership. After World War II, East Asia was split by larger **Cold War** rivalries (Figure 11.24).

The Evolution of China

The original core of Chinese civilization was the North China Plain and the Loess Plateau. For many centuries, periods of unification alternated with times of division into competing states. The most important episode of unification occurred in the

Ethnic tensions. The people of Xinjiang, like those of Tibet, have called for more autonomy. With the opening of China in the last several decades, the people of Xinjiang, mostly Muslim Uygurs, have become more vocal about their demands, resulting in numerous outbreaks of violence.

Border dispute. The 1991 Russia–China Border Agreement established the Amur and Ussuri rivers as the official border, but the two countries have squabbled over the possession of islands in the rivers.

Territorial claims. Japan claims the four southernmost Kuril Islands, which were annexed by Russia at the end of World War II.

India–China border. Aksai Chin, an uninhabited section of the Tibetan Plateau, is claimed by India but controlled by China.

The Korean DMZ. The 148-mile-long border between North and South Korea was established in 1953 following the Korean conflict. Though there was no formal agreement concluding the war and officially creating the boundary, it is well demarcated and is ringed by a 2.4-mile-wide demilitarized zone.

Taiwan–China tensions. The democratic Taiwanese government claims to be the legitimate government of China, while mainland China views Taiwan as a renegade province and refuses to recognize its government. The two nations are in a latent state of war and gunfire has been exchanged on Quemoy and Matsu, two islands off the Chinese mainland controlled by Taiwan.

U.S. military bases. The United States has maintained several large military bases on the island of Okinawa, causing much resentment among many islanders who wish to see most, if not all, of the bases closed.

Occupied Tibet. The region of Tibet was first occupied by China during the Manchu dynasty and was virtually independent from 1912 to 1950. China invaded and reoccupied the territory in 1959 and has refused Tibetans' calls for more autonomy since. Tibet's spiritual leader, the Dalai Lama, went into exile in India following the 1959 invasion.

Hong Kong. The former British colony of Hong Kong was returned to China on July 1, 1997, and China has promised no change with Hong Kong's capitalist economic system.

Island claims. The Paracel Islands are claimed by China, Taiwan, and Vietnam.

China–India border tensions. The McMahon line was proposed in 1913 at the watershed of the Himalayas and is the current boundary between China and India. China has never accepted this boundary and unsuccessfully invaded across this border into India in 1962 before being pushed back. Tensions are now easing in this area.

Back to China. Macau, a Portuguese colony, returned to China on December 20, 1999.

Spratly Islands. The Spratly Islands are claimed by China, Taiwan, Vietnam, Malaysia, and the Philippines. These islands, as well as the Paracel Islands, potentially hold petroleum reserves beneath the sea.

— Autonomous regions

FIGURE 11.24 • GEOPOLITICAL ISSUES IN EAST ASIA East Asia remains one of the world's geopolitical hot spots. Tensions are particularly severe between capitalist, democratic South Korea and the isolated communist regime of North Korea, and between China and Taiwan. China has had several border disputes, which include a number of small islands in the South China Sea. Japan and Russia have not been able to resolve their quarrel over the southern Kuril Islands.

third century B.C. Once political unity was achieved, the Chinese Empire began to expand vigorously to the south of the Yangtze Valley. Subsequently, the ideal of the unity of China triumphed, helping to link the Han Chinese into a single people.

Various Chinese dynasties attempted to conquer Korea, but the Koreans resisted. Eventually China and Korea worked out an arrangement whereby Korea paid token tribute and acknowledged the supremacy of the Chinese Empire and in return received trading privileges and retained independence. When foreign armies invaded Korea—as did those of Japan in the late 1500s—China sent troops to support its "vassal kingdom."

The Manchu Ch'ing Dynasty   The most significant conquest of China occurred in 1644, when the Manchus toppled the Ming Dynasty and replaced it with the Ch'ing (also spelled Qing) Dynasty. As earlier conquerors did, the Manchus retained the

Chinese bureaucracy and made few institutional changes. Their strategy was to adapt themselves to Chinese culture, yet at the same time to preserve their own identity as an elite military group. Their system functioned well until the mid-nineteenth century, when the empire began to crumble at the hands of European and, later, Japanese power.

The Modern Era   From its height in the 1700s, the Chinese Empire declined rapidly in the 1800s as it failed to keep pace with the technological progress of Europe. Threats to the empire had always come from the north, and Chinese and Manchu officials saw little danger from European merchants operating along their coastline. But the Europeans were distressed by the amount of silver needed to obtain Chinese silk, tea, and other products. In response to their lack of alternative goods, the British began to sell opium, which Chinese authorities viewed as a threat to their nation. When the imperial government tried to suppress the opium trade in the 1840s, the British attacked and quickly prevailed.

This first "opium war" introduced a century of political and economic chaos in China. The British demanded and received free trade in selected Chinese ports. As European businesses penetrated China and weakened local economic interests, anti-Manchu rebellions began to break out. At first, all such uprisings were crushed, but not before causing tremendous destruction. Meanwhile, European power continued to advance. In 1858 Russia annexed the northernmost reaches of Manchuria, and by 1900 China had been divided—as shown in Figure 11.25—into separate **"spheres of influence"** in which different European economic interests prevailed. (In a "sphere of influence," the colonial power had no formal political authority, but it did have informal influence and tremendous economic clout.)

A successful rebellion in 1911 finally toppled the Manchus and destroyed the empire, but subsequent efforts to establish a unified Chinese Republic were not successful. In many parts of the country, local military leaders ("warlords") grabbed power for themselves. By the 1920s it appeared that China might be completely torn apart. The Tibetans had gained autonomy; Xinjiang was under Russian influence; and in China proper Europeans and local warlords vied with the weak Chinese Republic for power. Japan was also increasing its demands and seeking to expand its territory.

The Rise of Japan

Japan did not emerge as a unified state until the seventh century, more than 2,000 years later than China. From its earliest days, Japan looked to China (and, at first, to Korea as well) for intellectual and political models. Its offshore location, however, insulated Japan from the threat of rule by the Chinese Empire. The Japanese viewed their islands as a separate empire, equal in certain respects to China. Between 1000 and 1580, however, Japan had no real unity, being divided into a number of small states continually at war with each other.

The Closing and Opening of Japan   By the early 1600s, Japan had been reunited by the armies of the Tokugawa **Shogunate** (a shogun is a military leader who in theory remains under the emperor but who actually holds power). At this time, Japan attempted to isolate itself from the rest of the world. Until the 1850s, Japan traded with China mostly through the Ryukyu islanders and with Russia only through Ainu go-betweens. The only Westerners allowed to trade in Japan were the Dutch, and their activities were strictly limited.

Japan remained largely closed to foreign commerce and influence until U.S. gunboats sailed into Tokyo Bay in 1853 to demand trade access. Aware that China was losing power, Japanese leaders set about modernizing their economic, administrative, and military systems. This effort accelerated when the Tokugawa Shogunate was toppled in 1868 by the Meiji Restoration. (It is called a *restoration* because it was carried out in the emperor's name, but it did not give the emperor any real power.) Unlike China, Japan successfully strengthened its government and economy.

Scale:
0   500   1,000 mi
0   500   1,000 km

19th-Century European Spheres of Influence

- German sphere of influence
- French sphere of influence
- French colonial possessions
- Russian sphere of influence
- Russian territory
- Russian Influence, eventually detached from China (Mongolia)
- British sphere of influence
- British colonial possessions
- ● Hong Kong, British colonial possession
- ★ Macau, Portuguese colonial possession
- ○ Initial European treaty ports

Chinese Territorial Losses in the 19th and 20th Centuries

- Direct territorial losses
- Former tributary states detached from Chinese sphere of influence

Expansion of the Japanese Empire

- Japanese Empire circa 1910
- Japanese Empire circa 1943
- Present-day Japan

FIGURE 11.25 • NINETEENTH-CENTURY EUROPEAN COLONIALISM The Chinese lost influence and territory in the nineteenth century as European power expanded. Although China regained its autonomy and most of its territory in the 1900s, Russia retained large areas that were formerly under Chinese control. The first half of the twentieth century saw the rapid expansion of the Japanese Empire, which ended with the defeat of Japan in World War II.

The Japanese Empire Japan's new rulers realized that their country remained threatened by European imperial power. They therefore decided that the only way to meet the challenge was to expand their own territory. Japan soon took control over Hokkaido and began to move farther north into the Kuril Islands and Sakhalin.

In 1895 the Japanese government tested its newly modernized army against China, winning a quick victory that gave it control of Taiwan. Tensions then mounted with Russia as the two countries competed for power in Manchuria and Korea. The Japanese defeated the Russians in 1905, giving them considerable influence in northern China. With no strong rival in the area, Japan annexed Korea in 1910. Alliance with Britain, France, and the United States during World War I brought further gains, as Japan was awarded Germany's island colonies in Micronesia at the end of the war.

The 1930s brought a global depression, greatly reducing world trade and putting a resource-dependent Japan in a difficult situation. The country's leaders sought a military solution, and in 1931 Japan conquered Manchuria. In 1937 Japanese armies moved south, occupying the North China Plain and the coastal cities of southern China. During this period, Japan's relations with the United States steadily deteriorated. When the United States cut off the export of scrap iron, Japan began to experience a resource crunch.

In 1941 Japan's leaders decided to destroy the American Pacific fleet in order to clear the way for the conquest of resource-rich Southeast Asia. Their grand strategy was to unite East and Southeast Asia into a "Greater East Asia Co-Prosperity Sphere." This "sphere" was to be ruled by Japan, however, and was designed to keep the Americans and Europeans out.

Postwar Geopolitics

With the defeat of Japan at the end of World War II, East Asia became dominated by rivalry between the United States and the Soviet Union. Initially, American interests prevailed in Japan, South Korea, and Taiwan, while Soviet interests advanced on the mainland. Soon, however, East Asia began to experience its own revival.

Japan's Revival   Japan lost its colonial empire when it lost the war. Its territory was reduced to the four main islands plus the Ryukyu Archipelago. In general, the Japanese government agreed to this loss of land. The only remaining territorial conflict concerns the four southernmost islands of the Kuril chain, which were taken by the Soviet Union in 1945. Although Japan still claims these islands, Russia refuses even to discuss giving up control, causing considerable strains in Russo-Japanese relations.

After losing its overseas possessions, Japan was forced to rely on trade to obtain the resources needed for its economy. Here it proved remarkably successful. Japan's military power, on the other hand, was limited by the constitution imposed on it by the United States. Because of these restrictions, Japan has relied in part on the U.S. military for its defense needs. The U.S. Navy patrols many of its vital sea-lanes, and U.S. armed forces maintain several bases within Japan. This U.S. military presence, however, is becoming increasingly controversial. Many Japanese citizens believe that their country ought to provide its own defense, and they resent the presence of U.S. troops. Slowly but steadily, meanwhile, Japan's own military has emerged as a strong regional force despite the constitutional limits imposed on it.

The Division of Korea   The end of World War II brought much greater changes to Korea than to Japan. As the end of the war approached, the Soviet Union and the United States agreed to divide the country; Soviet forces were to occupy the area north of the thirty-eighth parallel, whereas U.S. troops would occupy the south. This soon resulted in the establishment of two separate governments. In 1950 North Korea invaded South Korea, seeking to reunify the country. The United States, with support from the United Nations, supported the south, while China aided the north. The war ended in a stalemate, and Korea remained a divided country, its two governments locked in a bitter cold war.

Large numbers of U.S. troops remained in the south after the war. South Korea in the 1960s was a poor agrarian country that could not defend itself. Over the past 30 years, however, the south has emerged as a wealthy trading nation while the fortunes of the north have significantly declined. Many South Korean students resent the presence of U.S. forces and seek peace with the north. Their periodic and sometimes violent demonstrations have been a significant force in South Korean politics. The South Korean government long regarded North Korea as a dangerous state that could attack at any time (Figure 11.26), but more recently it has been trying to reach some sort of agreement. A crisis was reached in late 1990s when North Korea refused to allow international inspection of its nuclear power plants. Apparently, North Korea does have nuclear weapons capability, as well as sophisticated missile technology. Thus far, however, all efforts to negotiate an ending of the North Korean nuclear program have ended in failure.

FIGURE 11.26 • THE DEMILITARIZED ZONE IN KOREA   North and South Korea were divided along the thirty-eighth parallel after World War II. Today, even after the conflict of the early 1950s, the demilitarized zone, or DMZ (which runs near the parallel), separates these two states. U.S. armed forces are active in patrolling the demilitarized zone. *(Yonhap/AP/Wide World Photos)*

The Division of China World War II brought tremendous destruction and loss of life to China. Before the war began, China had already been engaged in a civil conflict between nationalists (who favored an authoritarian capitalist economy) and communists, both of whom hoped to unify the country. After the Japanese invaded China proper in 1937, the two camps cooperated, but as soon as Japan was defeated, China again found itself involved in civil war. In 1949 the communists proved victorious, forcing the nationalists to retreat to Taiwan.

A dormant state of war between China and Taiwan has persisted ever since. Although no battles have been fought, gunfire has periodically been exchanged over Kinmen and Matsu, two small Taiwanese islands just off the mainland. The Beijing government still claims Taiwan as an integral part of China and vows eventually to reclaim it. The Taiwanese nationalists, for their part, insist that they represent the true government of China. It was actually made a crime in Taiwan to support Taiwanese independence, because the fiction had to be maintained that Taiwan was merely one province of a temporarily divided China. Taiwan is thus equipped with a provincial government that governs exactly the same territory that its national government does.

The idea of Chinese unity continues to be influential both in China and abroad. In the 1950s and 1960s, the United States recognized Taiwan as the only legitimate government of China, but its policy changed after U.S. leaders decided that it would be more useful—and more realistic—to recognize mainland China. Soon China entered the United Nations, and Taiwan found itself diplomatically isolated, with little international recognition. In reality, however, Taiwan is a separate country, and increasing numbers of its citizens would like to proclaim it an independent republic. China, however, threatens to invade if Taiwan declares independence. Some observers think that China might try to reclaim Taiwan, pointing to China's overwhelming military lead; others view such a scenario as highly unlikely, pointing instead to close economic connections between the island and the mainland that would be destroyed by war.

The Chinese Territorial Domain Despite the fact that it has been unable to regain Taiwan, China has been successful in retaining the territories that the Manchus formerly controlled. In the case of Tibet, this has required considerable force; resistance by the Tibetans compelled China to launch a full-scale invasion in 1959. The Tibetans, however, have continued to struggle for real autonomy if not actual independence, as they fear that the Han Chinese now moving to Tibet will eventually outnumber them and weaken their culture (Figure 11.27).

The postwar Chinese government also retained control over Xinjiang in the northwest, as well as Inner Mongolia (or Nei Monggol), a vast territory stretching along the Mongolian border. The native peoples of Xinjiang prefer to call the region *Eastern Turkestan* to emphasize its Turkish heritage. The Han Chinese, however, reject this term because it challenges the unity of China by suggesting that the area is more Turkish than Chinese. Like Tibet, Nei Monggol and Xinjiang are classified as autonomous regions. The peoples of Xinjiang are increasingly asserting their religious and ethnic identities, and separatist attitudes are growing. Most Han Chinese, however, regard Nei Monggol and Xinjiang as integral parts of their country, and they regard any talk of independence as treasonous.

One territorial issue was finally resolved in 1997 when China reclaimed Hong Kong from Britain (Figure 11.28). In the isolationist 1950s, 1960s, and 1970s, Hong Kong acted as China's window on the outside world, and it grew wealthy as a capitalist enclave. As Chinese relations with the outer world opened in the 1980s, Britain decided to honor its treaty provisions and return Hong Kong to China. China in turn promised that Hong Kong would retain its fully capitalist economic system for at least 50 years. Civil liberties not enjoyed in China itself also remained protected in Hong Kong.

In 1999 Macao, the last colonial territory in East Asia, was returned to China. This small Portuguese enclave, located across the estuary from Hong Kong, has functioned largely as a gambling refuge. It remains to be seen whether Macao will continue to serve as the Las Vegas of East Asia.

FIGURE 11.27 • CHINESE SOLDIERS IN TIBET Following a full-scale invasion of Tibet in 1959, China continues to increase its presence through its military forces, the relocation of migrants into the area from other parts of China, and rebuilding programs that mask the traditional Tibetan landscape. Here Chinese soldiers observe a praying Tibetan while eating ice cream bars. *(Galen Rowell/Corbis)*

FIGURE 11.28 • CHINA RECLAIMS HONG KONG Fireworks celebrate the return of Hong Kong to China in 1997 after a long period under British colonial rule. Britain agreed to honor its treaty provisions and return Hong Kong to China under the promise that Hong Kong's capitalistic system would remain intact for at least 50 years. *(D. Groshong/Corbis/Sygma)*

The Global Dimension of East Asian Geopolitics

In the early 1950s East Asia was divided into two hostile Cold War camps: China and North Korea were allied with the Soviet Union, while Japan, Taiwan, and South Korea were linked to the United States. The Chinese–Soviet alliance soon deteriorated into mutual hostility, however, and in the 1970s China and the United States found that they could work with each other, sharing as they did an enemy in the Soviet Union.

The end of the Cold War, coupled with the rapid economic growth of China, again altered the balance of power in East Asia. The United States no longer needs China to offset the Soviet Union, and the U.S. military has become increasingly worried about the growing power of the Chinese army. China's neighbors have also become more concerned. China now has the largest army in the world, nuclear capability, and sophisticated missile technology.

China is thus coming of age as a major force in global politics. Whether it is a force to be feared by other countries is a matter of considerable debate. Chinese leaders insist that they have no plans to expand their territory and no intention of interfering in the internal affairs of other countries. They do, however, regard concerns expressed by the United States and other countries about their human rights record, as well as their activities in Tibet, as excessive meddling in their internal affairs.

Economic and Social Development: An Emerging Core of the Global Economy

East Asia contains all levels of economic and social development. Japan's urban belt contains one of the world's greatest concentrations of wealth, whereas many interior districts of China remain among the world's poorest places. Overall, however, East Asia experienced rapid economic growth in the 1970s, 1980s, and 1990s, and two of its economies, those of Taiwan and South Korea, jumped from the ranks of underdeveloped to developed. Increasingly, East Asia functions as a global economic core (Figure 11.29). But again, growth has not been evenly distributed. For example, during the same period, North Korea experienced a significant decline.

In regard to social development, the picture is brighter. Even in the poorer parts of China, most people are reasonably healthy and well educated. As China moves to a market economy, however, such "modern" problems as unemployment and homelessness are beginning to appear.

Japan's Economy and Society

Japan was the pacesetter of the world economy in the 1960s, 1970s, and 1980s. In the 1990s, however, the Japanese economy experienced a major setback, and growth has been slow ever since. Despite its current problems, Japan is still the world's second largest economic power.

Japan's Boom and Bust   Although Japan's heavy industrialization began in the late 1800s, most of its people remained poor. The 1950s, however, saw the beginnings of the Japanese "economic miracle." With its empire gone, Japan was forced to export manufactured materials. Beginning with inexpensive consumer goods such as clothing and toys, Japanese industry moved to more sophisticated products, including automobiles, cameras, electronics, machine tools, and computer equipment. By the 1980s it was the leader in many segments of the global high-tech economy, and its currency, the yen, emerged as one of the world's strongest.

The early 1990s saw the collapse of Japan's inflated real estate market, leading to a banking crisis. At the same time, many Japanese companies discovered that producing labor-intensive goods at home had become too expensive. They therefore began to relocate factories to Southeast Asia and China. Because of these and related difficulties, Japan's economy slumped through the 1990s and into the first years of the new millennium. The Japanese government made several attempts to revitalize the economy through massive state spending, resulting in large public deficits.

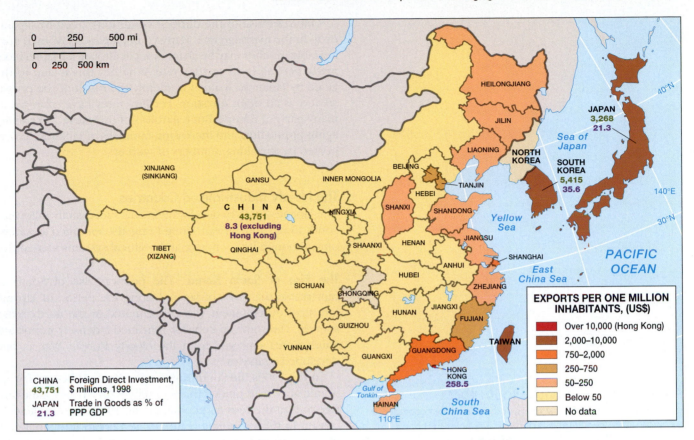

FIGURE 11.29 • EAST ASIA'S GLOBAL TIES   As this map shows, Hong Kong is one of the world's most important trading ports, while Taiwan, Japan, and South Korea are also highly integrated into global economic networks. In the rest of China, Guangdong and Shanghai are relatively well connected to the global economy through trade. Interior portions of China, on the other hand, are still largely isolated from the world economy.

Despite its downturn in the 1990s, Japan remains a core country of the global economic system. Its economic influence increasingly spans the globe, as Japanese multinational firms invest heavily in production facilities in North America and Europe, as well as in Third World countries. Japan remains a world leader in a range of high-tech fields, including robotics, optics, and machine tools for the semiconductor industry. It is also the world's largest lender of money to foreign firms and governments, owning a large percentage of U.S. government bonds.

Living Standards and Social Conditions in Japan

Despite its difficulties in the 1990s, Japan still has one of the world's most productive economies. Living standards are, however, somewhat lower in Japan than in the United States. Housing, food, transportation, and services are particularly expensive in Japan. Certain amenities that are almost standard in the United States, such as central heating, remain relatively rare.

Although the Japanese may live in cramped quarters and pay high prices for basic products, they also enjoy many benefits unknown in the United States. Unemployment remains lower than in the United States, health care is provided by the government, and crime rates are extremely low. By such social measures as literacy, infant mortality, and average longevity, Japan surpasses the United States by a comfortable margin. Japan also lacks the extreme poverty found in certain pockets of American society. The disparities of wealth between the haves and the have-nots, while still substantial, are not nearly as great as in the United States.

Women in Japanese Society

Critics often point out that Japanese women have not shared the benefits of their country's success. Advanced career opportunities remain limited for women, especially those who marry and have children. The expectation remains that mothers will devote themselves to their families and to their children's education. Japanese businessmen often work, or socialize with their coworkers, until late every evening, and thus contribute little to child care. Japan's recession of the 1990s seems to have resulted in further reductions in career opportunities for women.

One response to the difficult conditions faced by Japanese women has been a drop in the marriage rate. Many young Japanese women are delaying marriage, and a sizable number may be abandoning it altogether. Japan has seen an even more dramatic decline in its fertility rate. Whether this is due to the domestic difficulties faced by Japanese women or merely the result of the pressures of a postindustrial society is an open question. Fertility rates have, after all, dropped even lower in many parts of Europe. But regardless of the cause, a shrinking population means an aging population, and increasing numbers of Japanese retirees will have to be supported by shrinking numbers of workers.

The Newly Industrialized Countries

The Japanese path to development was successfully followed by its former colonies, South Korea and Taiwan. Hong Kong also emerged as a newly industrialized economy, although its economic and political systems were different.

The Rise of South Korea The postwar rise of South Korea was even more remarkable than that of Japan. During the period of Japanese occupation, Korean industrial development was concentrated in the north, which is rich in natural resources. The south, in contrast, remained a densely populated, poor agrarian region. South Korea emerged from the bloody Korean War as one of the world's least-developed countries.

In the 1960s the South Korean government began a program of export-led economic growth. It guided the economy with a heavy hand and denied basic political freedom to the Korean people. By the 1970s such policies had proved highly successful in the economic realm. Huge Korean industrial conglomerates, known as *chaebol*, moved from exporting inexpensive consumer goods to heavy industrial products and then to high-tech equipment.

At first South Korean firms remained dependent on the United States and Japan for basic technology. By the 1990s, however, this was no longer the case, as South Korea became one of the world's main producers of semiconductors. South Korean wages have also risen at a rapid rate. The country has invested heavily in education (by some measures it has the world's most demanding educational system), which has served it well in the global high-tech economy. Increasingly, South Korean companies are themselves becoming **multinational**, building new factories in the low-wage countries of Southeast Asia and Latin America, as well as in the United States and Europe (a multinational firm operates and manufactures in more than one country).

Contemporary South Korea The political and social development of South Korea has not been nearly as smooth as its economic progress. Throughout the 1960s and 1970s, student-led protests against the dictatorial government were repressed. As the South Korean middle class expanded and prospered, it began to pressure the government for democratization, and in the late 1980s the government gave in. But even though open elections are now held and basic freedoms allowed, political tension has not disappeared (Figure 11.30). South Korea's transition from an underdeveloped country ruled by a dictator to a prosperous democracy has, in other words, been both rapid and troubled.

Taiwan and Hong Kong Taiwan and Hong Kong have also experienced rapid economic growth since the 1960s. Both, in fact, have substantially higher per capita **gross domestic product (GDP)** levels than South Korea. The Taiwanese government, like that of South Korea and Japan, has guided the economic development of the country. Taiwan's economy, however, is organized not around large conglomerates and linked business firms, but rather around small to mid-sized family firms. This characteristically Chinese form of business organization is sometimes said to give Taiwan greater economic flexibility than its northern neighbors.

FIGURE 11.30 • PROTESTS IN SOUTH KOREA During the economic crisis of 1997, workers and students demonstrated against government financial and trade policies that had—in their minds, at least—led to the sudden downturn in the country's economic state. This protest rally is in the capital city, Seoul. *(Yun Jai-hyoung/AP/Wide World Photos)*

Hong Kong, unlike its neighbors, has been characterized by one of the most **laissez-faire** economic systems in the world (*laissez-faire* refers to market freedom, with little governmental control). State involvement has been minimal, which is one reason why the city's business elite was nervous about the transition to Chinese rule. Hong Kong traditionally functioned as a trading center, but in the 1960s and 1970s it became a major producer of textiles, toys, and other consumer goods. By the 1980s, however, such cheap products could no longer be made in such an expensive city. Hong Kong industrialists subsequently began to move their plants to southern China, while Hong Kong itself increasingly specialized in business services, banking, telecommunications, and entertainment. Fears that its economy would falter after the Chinese takeover in 1997 turned out to have been exaggerated.

Both Taiwan and Hong Kong have close overseas economic connections. Linkages are particularly tight with Chinese-owned firms located in Southeast Asia and North America. Taiwan's high-technology businesses are also intertwined with those of the United States; there is a constant back-and-forth flow of talent, technology, and money between Taipei and Silicon Valley. Hong Kong's economy is also closely bound with that of the United States (as well as those of Canada and Britain), but its closest connections, not surprisingly, are with the rest of China.

Chinese Development

China dwarfs all of the rest of East Asia in both physical size and population. Its economic takeoff is thus reshaping the economy of the entire region. Despite its recent growth, however, China's economy has a number of serious weaknesses. For example, the vast interior remains trapped in poverty, and many of its largest industries are not competitive. The future of the Chinese economy is thus one of the biggest uncertainties facing both East Asia and the world economy as a whole.

China Under Communism   More than a century of war, invasion, and near-chaos in China ended in 1949 when the communist forces led by Mao Zedong seized power. The new government, inheriting a weak economy, set about nationalizing private firms and building heavy industries. Their plans were most successful in Manchuria, where a large amount of heavy industrial equipment had been left by the Japanese.

In the late 1950s, however, China experienced an economic disaster ironically called the "Great Leap Forward." One of the main ideas behind this plan was that small-scale village workshops could produce the large quantities of iron needed for sustained industrial growth. Communist Party officials demanded that these inefficient workshops meet unreasonably high production quotas. In some cases the only way they could do so was to melt down peasants' agricultural tools for the iron that they contained. Peasants were also forced to contribute such a large percentage of their crops to the state that many went hungry. The result was a horrific famine that may have killed 20 million persons.

The early 1960s saw a return to more practical policies, but toward the end of the decade a new wave of radicalism swept through China. This "Cultural Revolution" aimed at mobilizing young people to stamp out the remaining traces of capitalism. Thousands of experienced industrial managers and college professors were expelled from their positions. Many were sent to villages to be "reeducated" through hard physical labor; others were simply killed. The economic consequences of such policies were devastating.

Toward a Postcommunist Economy   When Mao Zedong, who had been revered as an almost superhuman being, died in 1976, China faced a crucial turning point. Its economy was nearly stagnant and its people desperately poor. However, the economy of Taiwan, its rival, was booming. This led to a political struggle between pragmatists hoping for change and dedicated communists. The pragmatists emerged victorious, and by the late 1970s it was clear that China would embark on a different economic path. The new China would seek closer connections with the world economy and take a modified capitalist road to development (Figure 11.31).

FIGURE 11.31 • INDUSTRIAL EXPANSION IN COASTAL CHINA   One of the important economic reforms leading to China's recent development was the creation of Special Economic Zones (SEZs) along its eastern coast. Here workers in a coastal automobile plant assemble cars for the rapidly expanding domestic market. *(Serge Attal/Getty Images, Inc.—Liaison)*

China did not, however, transform itself into a fully capitalist country. The state continued to run most heavy industries, and the Communist Party kept a monopoly on political power. Instead of suddenly abandoning the communist model, as the former Soviet Union did, China allowed cracks to appear in which capitalist businesses could take root and thrive.

Industrial Reform An important early industrial reform involved opening **Special Economic Zones (SEZs)**, in which foreign investment was welcome and state interference minimal. The Shenzhen SEZ, adjacent to Hong Kong, proved particularly successful after Hong Kong manufacturers found it a convenient source of cheap land and labor. Additional SEZs were soon opened, mostly in the coastal region. The basic strategy was to attract foreign investment that could generate exports, the income from which could supply China with the capital that it needed to build its infrastructure (roads, electrical and water systems, telephone exchanges and the like) and thus achieve conditions for sustained economic growth.

Other market-oriented reforms followed. Former agricultural cooperatives were allowed to produce for the market. Many of these "township and village enterprises" proved highly successful. By the early 1990s, the Chinese economy was growing at some 8 to 15 percent a year, perhaps the fastest rate of expansion the world has ever seen. China emerged as a major trading nation, and by the mid-1990s it had huge trade surpluses, especially with the United States. Seeking to strengthen its connections with the global economic system, China joined the World Trade Organization (WTO), a body designed to facilitate free trade and provide ground rules for international economic exchange in November 2001.

By the early 2000s, China's annual economic growth rate slowed to roughly 6 to 7 percent, still high by global standards. Some evidence indicates, however, that China's banking system has serious problems, making the country vulnerable to recession. Critics contend that China must fully abandon centralized planning if its economic expansion is to continue. China's leadership, however, has made it clear that economic reform will be a gradual process.

Social and Regional Differentiation The Chinese economic surge brought about by the reforms of the late 1970s and 1980s resulted in growing **social and regional differentiation**. In other words, certain groups of people—and certain portions of the country—prospered, while others declined. Despite its socialist government, the Chinese state encouraged the formation of an economic elite, having concluded that only wealthy individuals can adequately transform the economy. The least-fortunate Chinese citizens were sometimes left without work, and many millions migrated from rural villages to seek employment in the booming coastal cities. The government attempted to control the transfer of population, but with only partial success. Shantytowns, as well as homeless populations, began to emerge around some of China's cities.

China's Booming Coastal Region Most of the benefits from China's economic transformation have flowed to the coastal region and to the capital city of Beijing. The first to benefit were the southern provinces of Guangdong and Fujian. These provinces have profited from their close connections with the overseas Chinese communities of Southeast Asia and North America. (The vast majority of overseas Chinese emigrants came from these provinces.) Their close location to Taiwan and especially Hong Kong also proved helpful. Huge amounts of capital have flowed to the south coastal region since the 1980s from overseas Chinese business networks. American, Japanese, and European firms have also invested heavily in the region.

By the 1990s the Yangtze Delta, centered on the city of Shanghai, reemerged as the most economically active region of China. The delta was the traditional economic (and intellectual) core of China, and before the communist takeover, Shanghai had been its leading industrial and financial center. The Chinese government, moreover, has encouraged the development of huge industrial, commercial, and residential complexes, hoping to take advantage of the region's vitality. The Suzhou Industrial Park is now becoming a hypermodern city of more than a half million people, thanks largely to a $20 billion investment, most of it from Singapore. Shanghai's Pudong industrial development zone has attracted $10 billion, much of it going to the construction of a new airport and subway system.

The Beijing–Tianjin region has also played a major role in China's economic boom. Its main advantage is its proximity to political power and its position as the gateway to northern China. The other coastal provinces of northern China have also done relatively well.

Interior and Northern China   Most of the other parts of China, in contrast, have seen relatively little economic expansion. Central and northern Manchuria remain relatively well-off as a result of fertile soils and early industrialization but have not participated much in the recent boom. Many of the state-owned heavy industries of the Manchurian **"rust belt,"** or zone of decaying factories, are relatively inefficient today.

Most of the interior provinces of China have likewise missed the wave of growth throughout the 1980s and 1990s. In many areas, rural populations continue to grow while the natural environment deteriorates because of social erosion, water shortages, and deforestation. One consequence is high levels of underemployment and out-migration. By most measures, poverty increases with distance from the coast (Figure 11.32). In 1997 per capita GDP in Guizhou in the south-central area stood at 2,215 yuan (China's currency), whereas in Shanghai it had reached 25,750 yuan. As a result of such differences, China is encouraging development in the interior, but thus far its efforts have been mostly limited to transportation improvement and natural resource extraction.

FIGURE 11.32 • ECONOMIC AND SOCIAL DIFFERENTIATION IN CHINA   Although China has seen rapid economic expansion in recent years, the benefits of growth have not been evenly distributed throughout the country. Economic wealth and social development are concentrated on the coast, especially in Shanghai, Guangdong, Beijing, and Tianjin. Most of the interior remains mired in poverty. The poorest part of China is the upland region of Guizhou in the south-central part of the country. *(Benewick and Donald, 1999, The State of China Atlas, p. 35, New York: Penguin Reference)*

FIGURE 11.33 • CHINA'S POPULATION POLICIES One aspect of China's population policy is the expansion of child-care facilities so that mothers can be near their children while at work. This enables women to resume participating in the workforce soon after giving birth. This photo shows a typical day-care center attached to an industrial plant in Guangdong province in coastal China. *(Xinhua/Getty Images, Inc.—Liaison)*

Social Conditions in China

Despite its pockets of persistent poverty, China has made significant progress in social development. Since coming to power in 1949, the communist government has made large investments in medical care and education, and today China has impressive health and longevity figures. The literacy rate remains fairly low, but because 97 percent of children supposedly attend elementary school, it will probably rise substantially in the coming years. However, human well-being in China varies from region to region. For example, the literacy rate remains relatively low in many of the poorer parts of China, including the uplands of Yunnan and Guizhou and the interior portions of the North China Plain. Literacy is much more widespread in Manchuria, the Yangtze Delta, and most major urban areas. Such regional disparities may increase as the gap between the wealthy and the poor grows.

China's Population Quandary Population policy also remains an unsettling issue for China. With more than 1.2 billion people highly concentrated in less than half of its territory, China is very densely populated. By the 1980s its government had become so concerned that it instituted the famous "one-child policy." Under this plan, couples in normal circumstances are expected to have only a single offspring and can suffer financial penalties if they do not comply (Figure 11.33). This strategy has been successful; the average fertility level is now only 1.8, and the population is growing at the relatively slow rate of 0.9 percent a year. Still, when one considers how large the population already is, even a 0.9 percent annual growth rate is worrisome. China will likely reach much more than 1.5 billion people before stabilization occurs.

Fertility levels in China, as might be expected, vary from province to province. Birthrates are relatively low in most large cities, the Yangtze Delta, parts of the Sichuan Basin, and parts of the North China Plain and are relatively high in many upland areas of south China, the Loess Plateau, and northwestern China. Higher levels of fertility in these poorer and more rural areas will probably lead to increased migration to the booming coastal cities.

While China's population policy has reduced its growth rate, it has also generated social tensions and human-rights abuses. Particularly troubling is the growing gender imbalance in the Chinese population. Baby boys in the country now far outnumber baby girls. This gap reflects the practice of honoring one's ancestors; because family lines are traced through male offspring, one must produce a male heir to maintain one's lineage. Many couples are therefore desperate to produce a son. Some decide to bear more than one child, regardless of the penalties they may face. Another option is gender-selective abortion; if ultrasound reveals a female fetus, the pregnancy is sometimes terminated. International women's organizations, as well as anti-abortion groups, are concerned about these effects of China's population policy. Many environmentalists, however, applaud China for lowering its birthrate.

The Position of Women Women have historically had a relatively low position in Chinese society, as is true in most other civilizations. One traditional expression of this was the practice of foot binding: The feet of elite girls were usually deformed by breaking and binding them in order to produce a dainty appearance. This crippling and painful practice was eliminated only in the twentieth century. In certain areas of southern China, it was also common in traditional times for girls to be married, and hence to leave their own families, when they were mere toddlers (such marriages, of course, would not be consummated for many years).

Not all women suffered such disabilities in pre-modern China. Some individuals achieved fame and fortune—a few even through military service. Both the nationalist and communist governments have, moreover, sought to begin equalizing the relations between the sexes. Many of their measures have been successful, and women now have a relatively high level of participation in the Chinese workforce. But it is still true that throughout East Asia—in Japan no less than in China—few women have achieved positions of power in either business or government. As China modernizes and its urban economy grows, the position of its women will probably improve.

CONCLUSION

East Asia is united by deep cultural and historical bonds. China has had a particularly large influence because, at one time or another, it covered nearly the entire region. Although Japan has never been under Chinese rule, even it has profound historical connections to Chinese civilization. Because of these deep regional ties, some observers predict the formation of a "Confucian bloc" that will challenge the West for global dominance. Others doubt that such a block could form, given the tensions and mutual hostility between countries in the region. In addition, the economies and cultures of East Asia, including China, are increasingly connected to the global system.

A few observers even question whether China itself will persist as a unified state. The Tibetans and other non-Han peoples of western China long for independence, although this probably remains an unrealistic goal considering the power of China and its determination to keep these areas.

Internal tensions also seem to be mounting within China proper. The relatively wealthy coastal provinces increasingly resent the control of Beijing and the flow of their tax receipts to the national government. The old split between China's more capitalistic south and its more bureaucratic north, which seemed to disappear under the communist system, may also be reemerging. Chinese provinces are starting to set their own economic policies, rather than waiting for orders from the center. Contributing to this centrifugal process (one leading to a spreading out or a breaking apart) is the explosive growth of the private economy coupled with the near-stagnation of the state-owned sector. Some China watchers believe that this growing imbalance may eventually sap the strength of the Chinese Communist Party, undermining the country's central authority.

It is, however, difficult to understand present conditions in China, let alone to predict the political future. Just how strong is popular sentiment for democratization in China? How intense are feelings of provincial, as opposed to national, loyalty? Such questions are not easy to address in a country in which freedom of expression is limited. China does, however, have a several-thousand-year legacy of unity under a single government. Much evidence suggests that, moreover, nationalism is now growing stronger in China. It is therefore reasonable to expect China to remain united. If it does, and if its economy continues to grow at current rates, China will clearly be one of the world's leading countries later in the twenty-first century.

The most important questions about East Asia's future center on China, but significant issues face the region's other countries. It remains to be seen, for example, whether South Korea can successfully manage the transition from being a low-wage exporter to becoming a high-wage technological powerhouse. South Korea also faces a major challenge to the north. As long as the present regime remains in power in Pyongyang, South Korea will feel threatened, despite the peace overtures that its government is now making.

Japan, for its part, now stands at a crossroads brought on by the decline of its long-lasting economic miracle. For all of its economic power and prestige, Japan has yet to find its place in the world. Will Japan's political influence ever begin to match its economic power? This question is hotly debated both in Japan and abroad. Equally important is whether Japan will be able to retain its unique economic and social institutions in the face of mounting global competition. There is a deep sense of uncertainty in the country today; young people no longer feel assured about their futures, and leaders debate whether significant institutional changes are required.

KEY TERMS

anthropogenic landscape (page 290)
autonomous region (page 298)
China proper (page 280)
Cold War (page 300)
Confucianism (page 294)
desertification (page 288)
diaspora (page 297)

geomancy (page 295)
gross domestic product (GDP) (page 308)
hiragana (page 294)
ideographic writing (page 293)
kanji (page 294)
laissez-faire (page 309)
loess (page 283)

Marxism (page 296)
multinational (page 308)
pollution exporting (page 284)
rust belt (page 311)
sediment load (page 282)
Shogun, Shogunate (page 302)
social and regional differentiation (page 310)

Special Economic Zones (SEZs) (page 310)
spheres of influence (page 302)
superconurbation (page 292)
tonal (page 298)
tribal peoples (page 298)
urban primacy (page 292)

South Asia

FIGURE 12.1 South Asia

This region is the second most populated in the world, primarily because of India's more than 1 billion residents. Bordering India on the west and east are Pakistan and Bangladesh, two large countries with predominantly Muslim populations. The two Himalayan countries of Nepal and Bhutan, along with the island nations of Sri Lanka and the Maldives, round out the region.

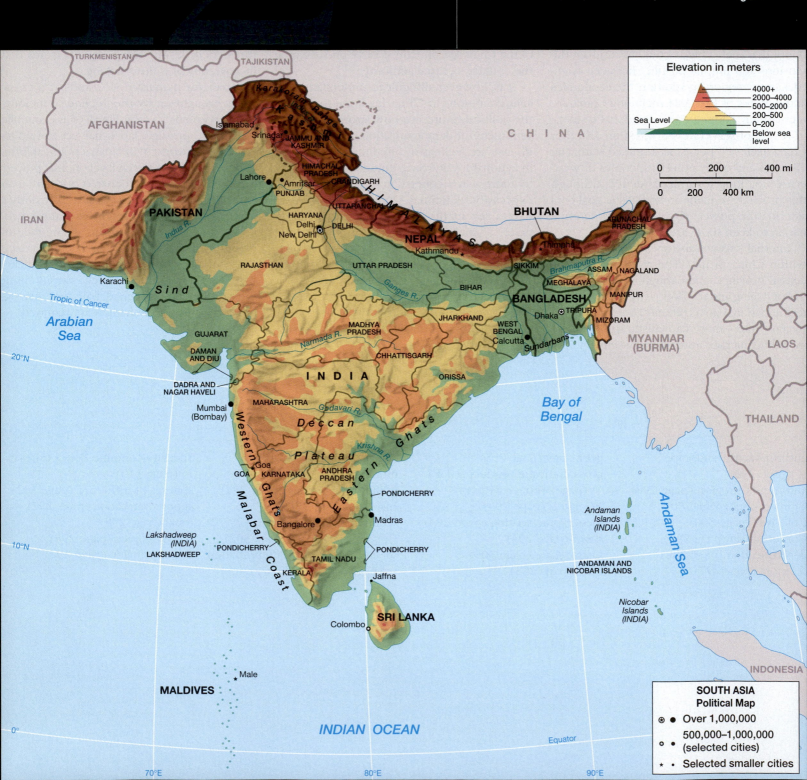

Elevation in meters

4000+	
2000–4000	
500–2000	
200–500	
0–200	
Below sea level	

Sea Level

SOUTH ASIA
Political Map

⊗ ● Over 1,000,000

○ ⊙ 500,000–1,000,000
(selected cities)

★ ● Selected smaller cities

SETTING THE BOUNDARIES

South Asia forms a distinct landmass separated from the rest of the Eurasian continent by a series of sweeping mountain ranges, including the Himalayas—the highest in the world. For this reason it is often called the Indian **subcontinent**, in reference to its largest country. To the south, South Asia includes a number of islands in the Indian Ocean, including the countries of Sri Lanka and the Maldives, as well as the Indian territories of the Lakshadweep and the Andaman and Nicobar Islands.

India is by far the largest South Asian country, both in size and in population. Covering more than 1 million square miles from the Himalayan crest to the southern tip of the peninsula at Cape Comorin, India is the world's seventh largest country in terms of area and, with more than 1 billion inhabitants, second only to China in population. Although mostly a Hindu country, India contains a great amount of religious, ethnic, linguistic, and political diversity.

Pakistan, the next largest country, is less than one-third the size of India. Stretching from the high northern mountains to the arid coastline on the Arabian Sea, its population of 145 million is only about 15 percent of India's. Despite this imbalance, these two countries have been locked in a tense power struggle, especially over the disputed territory of Kashmir. Until independence in 1947, Pakistan was one portion of a larger undivided British colonial realm simply called *India*. Because of its strong ties to Islam, however, some Pakistanis argue that their country is now more closely connected to its Muslim neighbors in Southwest Asia than it is to India and the rest of South Asia.

Bangladesh, on India's eastern shoulder, is also a largely Muslim country. Originally created as East Pakistan in the hurried division of India in 1947, it achieved independence after a brief civil war in 1971. Although a small country in area (54,000 square miles), Bangladesh is one of the most densely populated in the world—and also one of the poorest—with a population of 133.5 million in an area about the size of Wisconsin. Bangladesh has a short border with Burma (Myanmar), but it is otherwise bordered only by India.

Nepal and Bhutan are both located in the Himalayan Mountains, sandwiched between India and the Tibetan plateau of China. Nepal, with some 23.5 million people, is much larger in both area and population, and is far more open to and engaged with the contemporary world. Bhutan, on the other hand, has purposely disconnected itself from the global system, remaining a relatively isolated Buddhist kingdom of less than 1 million inhabitants.

The two island countries of Sri Lanka (formerly Ceylon) and the Maldives round out South Asia. Each of these countries has its own problems that cloud the future. Sri Lanka (population of 19.5 million) has since 1983 been mired in a civil war. The predicament facing the small island nation of the Maldives is quite different. If global warming continues and if sea levels rise as predicted, this island nation—where the highest point is only 6 feet above sea level—will be entirely flooded and its 300,000 inhabitants will have to seek higher ground somewhere else.

South Asia, a land of deep historical and cultural interconnections, has recently experienced intense political conflict. Since independence from Britain in 1947, the two largest countries, India and Pakistan, have fought several wars and remain locked in a bitter struggle. This political tension reached such heights that many experts consider South Asia the leading candidate for a nuclear war. Religious divisions lie beneath this geopolitical turmoil, for India is primarily a Hindu country (with a large Muslim minority), while neighboring Pakistan and Bangladesh are both predominantly Muslim (Figure 12.1; see "Setting the Boundaries").

South Asia also has its share of economic and demographic problems. Given its current rate of growth, South Asia could soon surpass East Asia as the world's most populous region. Although agricultural production has increased slightly faster than population, some experts think that farming improvements are approaching their limit. Compounding this serious situation is the widespread poverty of South Asia, which is one of the poorest parts of the world. Roughly half of India's people live on less than one dollar a day.

South Asia is less connected to the contemporary globalized world than are East or Southeast Asia. South Asia is, however, beginning to have a significant global impact, based on high levels of scientific and technical skills, the international links that the region has established through migration, and the enormous size of its local markets. Parts of India, for example, have recently become major players in the global computer industry.

Environmental Geography: Diverse Landscapes, from Tropical Islands to Mountain Rim

South Asia's diverse environmental geography ranges from the highest mountains in the world to densely populated delta islands barely above sea level; from some of the wettest places on Earth to dry, scorching deserts; and from tropical rain forests

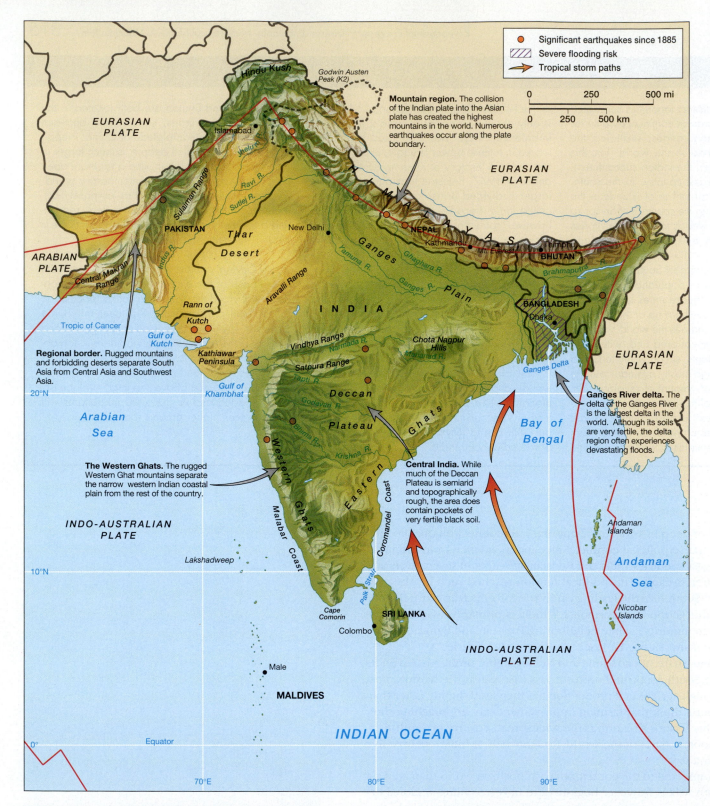

FIGURE 12.2 • PHYSICAL GEOGRAPHY OF SOUTH ASIA
This region is composed of four extensive physical subregions: the high Himalayan mountains in the north; the expansive Indus-Ganges lowland that reaches from Pakistan in the west to the delta lands of Bangladesh; peninsular India, dominated by the Deccan Plateau; and the island realm that includes Sri Lanka and the Maldives. The region is subject to a number of natural hazards, including earthquakes, flooding, and typhoons from the Bay of Bengal.

to eroded scrublands to coral reefs (Figure 12.2). All of these ecological zones have their own distinct and complex environmental problems. To illustrate the complexity of South Asian environmental issues, let us begin with the gripping story of Rajkumar and Veerappan—the film star and the poacher king.

The Film Star and the Poacher King

In southern India, where the states of Kerala, Tamil Nadu, and Karnataka converge, lies a large area of steep mountains and wild forests containing elephants, leopards, tigers, bears, and other rare species. Although several national parks have been established in this area, access is still difficult and dangerous, in part because these forests are also the haunts of one of India's most notorious outlaw gangs, led by the famed Koose Veerappan. As a result of Veerappan's poaching activities, ivory-bearing male elephants are extremely rare, leaving the remaining herds largely composed of tuskless females. By the late 1990s, as ivory grew scarce and the Indian government clamped down on exports, Veerappan turned increasingly to the profitable business of poaching sandalwood, a rare tree that yields a fragrant oil highly valued in much of Asia.

On July 30, 2000, Veerappan—who is viewed as a Robin Hood figure by some—shocked India by kidnapping one of the country's most noted actors, Rajkumar. Although nationally famous, Rajkumar is especially beloved in his home state of Karnataka, where he has been a supporter of the local language and culture. Such activities did not escape the attention of Veerappan, whose initial ransom demands were basically political. The bandit chief is an equally passionate defender of his native state of Tamil Nadu, and he holds a grudge against Karnataka. Not only did he demand that Karnataka release larger amounts of water from the Cauvery River to downstream areas in Tamil Nadu, but he also asked for Tamil to be made an official language in Karnataka.

Rajkumar was released four months later without any of these demands having been met. The evidence now suggests that Veerappan's political claims were at least in part a smokescreen. Rajkumar seems to have angered Veerappan because of his son's shady dealings in the high-quality granite business on lands controlled by the bandit group. The woman who negotiated the actor's release, moreover, had recently been accused of cheating an Italian granite importer out of a large amount of money. Although the details of the case will probably remain hidden, it is likely that Veerappan was mainly seeking a larger cut of this profitable internationally oriented business.

The story of Rajkumar and Veerappan tells a lot about contemporary India, with its diverse but threatened biological treasures, serious water shortages, and rapidly increasing global economic connections. It also reveals the intensity of local ethnic sentiments in a country where each state prizes its own language and distinctive cultural heritage. That, however, is a story for later in this chapter. For the moment, let us examine in detail a few key environmental issues faced by contemporary South Asia.

Environmental Issues in South Asia

As is true in other poor and densely settled regions of the world, a number of serious environmental issues are present in South Asia (Figure 12.3). The region suffers from natural hazards, especially flooding in the region's large river deltas; deforestation, both for agricultural expansion and commercial export purposes; and widespread water and air pollution, which often accompany early industrialization and manufacturing. South Asia has also suffered from some of the world's worst environmental disasters. The 1984 explosion of a fertilizer plant in Bhopal, India, for example, killed more than 2,500 persons and maimed many more. Compounding all of these problems are the immense numbers of new people added each year through natural population growth.

Natural Hazards in Bangladesh   The link between population pressure and environmental problems is nowhere clearer than in the delta area of Bangladesh, where the search for fertile land has driven people into hazardous areas, putting millions at risk from seasonal flooding as well as from the powerful *cyclones* that form over the Bay of Bengal. For thousands of years, drenching monsoon rains have eroded and transported huge quantities of sediment from the Himalayan slopes to the sea by the Ganges and Brahmaputra rivers, gradually building this low-lying delta environment. While farming in the fertile delta supports Bangladesh's large population, it increases the number of people affected by natural disasters in the delta area.

Although periodic flooding is a natural, even beneficial, phenomenon that enlarges deltas by depositing fertile river-borne sediment, flooding has become a serious problem for people living in these low-lying areas. In September 1998, for example, more

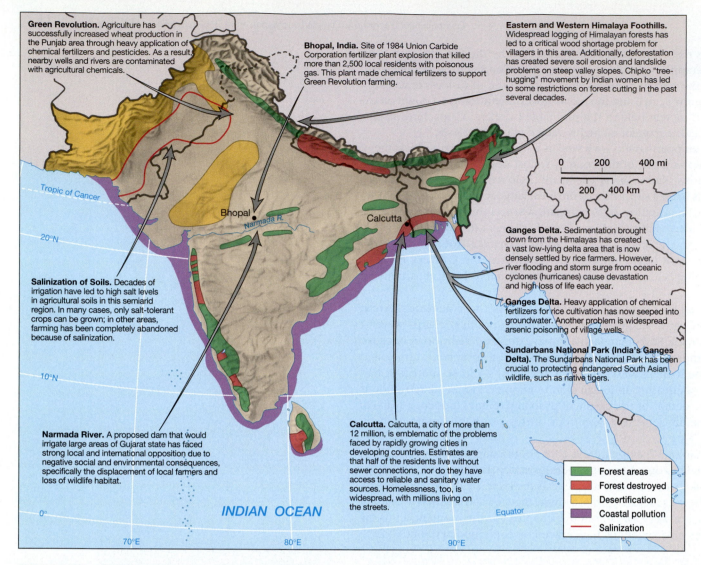

Green Revolution. Agriculture has successfully increased wheat production in the Punjab area through heavy application of chemical fertilizers and pesticides. As a result, nearby wells and rivers are contaminated with agricultural chemicals.

Bhopal, India. Site of 1984 Union Carbide Corporation fertilizer plant explosion that killed more than 2,500 local residents with poisonous gas. This plant made chemical fertilizers to support Green Revolution farming.

Eastern and Western Himalaya Foothills. Widespread logging of Himalayan forests has led to a critical wood shortage problem for villagers in this area. Additionally, deforestation has created severe soil erosion and landslide problems on steep valley slopes. Chipko "tree-hugging" movement by Indian women has led to some restrictions on forest cutting in the past several decades.

Salinization of Soils. Decades of irrigation have led to high salt levels in agricultural soils in this semiarid region. In many cases, only salt-tolerant crops can be grown; in other areas, farming has been completely abandoned because of salinization.

Ganges Delta. Sedimentation brought down from the Himalayas has created a vast low-lying delta area that is now densely settled by rice farmers. However, river flooding and storm surge from oceanic cyclones (hurricanes) cause devastation and high loss of life each year.

Ganges Delta. Heavy application of chemical fertilizers for rice cultivation has now seeped into groundwater. Another problem is widespread arsenic poisoning of village wells.

Sundarbans National Park (India's Ganges Delta). The Sundarbans National Park has been crucial to protecting endangered South Asian wildlife, such as native tigers.

Narmada River. A proposed dam that would irrigate large areas of Gujarat state has faced strong local and international opposition due to negative social and environmental consequences, specifically the displacement of local farmers and loss of wildlife habitat.

Calcutta. Calcutta, a city of more than 12 million, is emblematic of the problems faced by rapidly growing cities in developing countries. Estimates are that half of the residents live without sewer connections, nor do they have access to reliable and sanitary water sources. Homelessness, too, is widespread, with millions living on the streets.

Legend:
- Forest areas
- Forest destroyed
- Desertification
- Coastal pollution
- Salinization

FIGURE 12.3 • ENVIRONMENTAL ISSUES IN SOUTH ASIA
As might be expected in a highly diverse and densely populated region, there are a wide range of environmental problems. These range from salinization of irrigated lands in the dry lands of Pakistan and western India to groundwater pollution from Green Revolution fertilizers and pesticides. Additionally, deforestation and erosion are widespread in upland areas.

than 22 million Bangladeshis were made homeless when water covered two-thirds of the country (Figure 12.4). With the populations of both Bangladesh and northern India growing rapidly, there is a strong possibility that flooding will take even higher tolls in the next decade as desperate farmers relocate into the hazardous lower floodplains. Deforestation of the Ganges and Brahmaputra headwaters magnifies the problem. Since forest cover and ground vegetation intercept rainfall and slow runoff, deforestation in the river headwaters results in increased flooding.

Forests and Deforestation Tropical monsoon forests and savannah woodlands once covered most of the region, except for the desert areas in the northwest, but in most areas tree cover has vanished as a result of human activities. The Ganges Valley and coastal plains of India, for example, were largely deforested thousands of years ago to make room for agriculture. Elsewhere, forests were cleared more gradually for agricultural, urban, and industrial expansion. More recently, hill slopes in the Himalayas and elsewhere have been logged for commercial purposes. Extensive forests can still be found, however, in the far northern, the southwestern, and the east-central areas.

FIGURE 12.4 • FLOODING IN BANGLADESH Devastating floods are common in the low-lying delta lands of Bangladesh. Heavy rains come with the southwest monsoon, especially to the Himalayas, and powerful cyclones often develop over the Bay of Bengal. *(Baldev/Corbis/Sygma)*

As a result of deforestation, many villages of South Asia suffer from a shortage of fuelwood for household cooking, forcing people to burn dung cakes from cattle. While this low-grade fuel provides adequate heat, it also prevents manure from being used as fertilizer. Where wood is available, collecting it may involve many hours of female labor because the remaining sources of wood are often far from the villages. In many areas, extensive eucalyptus stands have been planted to supply fuelwood and timber. These nonnative Australian trees support little if any wildlife, thus adding to problems of declining biodiversity throughout the region.

South Asia's Monsoon Climates

The dominant climatic factor for most of South Asia is the **monsoon**, the distinct seasonal change of pressure and wind direction, which corresponds to wet and dry periods. This monsoon pattern is caused by large-scale climatic processes that affect much of Asia. During the Northern Hemisphere's winter, a large high-pressure system forms over the cold Asian landmass. Since winds flow from high pressure to low (just as water flows from high elevation to low), cold, dry winds flow outward from the continental interior, over the Himalayas and down across South Asia. This is the cool and dry season, extending from November until February. Only a few areas get rainfall during this otherwise dry time. As winter turns to spring, these winds diminish, resulting in the hot, dry season of March through May. Eventually the buildup of heat over South and Southwest Asia produces a large low-pressure cell. By early June the low-pressure is strong enough to cause a shift in wind direction so that warm, moist air from the Indian Ocean moves toward the continental interior. This signals the onset of the warm and rainy season of the southwest monsoon, which lasts from June through October and affects the entire region, from the tip of peninsular India northward to the Himalayas (Figure 12.5).

Orographic rainfall results from the uplifting and cooling of moist monsoon winds over the Western Ghats and the Himalayan foothills. As a result, some areas receive more than 200 inches (508 centimeters) of rain during the four-month wet season (Figure 12.6). Cherrapunji, in northestern India, is one of the world's wettest places, with an average rainfall of 451 inches (1,128 centimeters). On the Deccan Plateau, however, rainfall is dramatically reduced by a strong **rain-shadow effect**. A rain shadow is the area of low rainfall found on the leeward (or downwind) side of a mountain range. As winds move down slope, the air becomes warmer and dry conditions usually prevail.

Physical Subregions of South Asia

To better understand environmental conditions in this diverse region, South Asia can be broken down into four physical subregions, starting with the high mountain ranges of its northern edge and extending to the tropical islands of the far south. Lying south of the mountains are the extensive river lowlands that form the heartland of both India and Pakistan. Between river lowlands and the island countries is the vast area of peninsular India, extending more than 1,000 miles (1,600 kilometers) from north to south (Figure 12.7).

Mountains of the North   South Asia's northern rim of mountains is dominated by the great Himalayan Range, forming the northern borders of India, Nepal, and Bhutan. More than two dozen peaks exceed 25,000 feet (7,620 meters), including the world's highest mountain, Everest, on the Nepal–China (Tibet) border at 29,028 feet (8,848 meters). To the east are the lower Arakan Yoma Mountains, forming the border between India and Burma (Myanmar) and separating South Asia from Southeast Asia.

These mountain ranges are a result of the dramatic collision of northward-moving peninsular India with the Asian landmass. The entire region is still geologically active, putting all of northern South Asia in serious earthquake danger (see Figure 12.2). While most of South Asia's northern mountains are too rugged and high to support dense human settlement, there are major population clusters in the Katmandu Valley of Nepal, situated at 4,400 feet (1,341 meters), and the Valley, or Vale, of Kashmir in northern India, at 5,200 feet (1,585 meters).

FIGURE 12.5 • MONSOON RAIN   During the summer monsoon, some Indian cities such as Mumbai (Bombay) receive more than 70 inches of rain in just three months. These daily torrents cause floods, power outages, and daily inconvenience. However, these monsoon rains are crucial to India's agriculture. If the rains are late or abnormally weak, crop failure often results. *(Sharad J. Devare/Dinodia Picture Agency)*

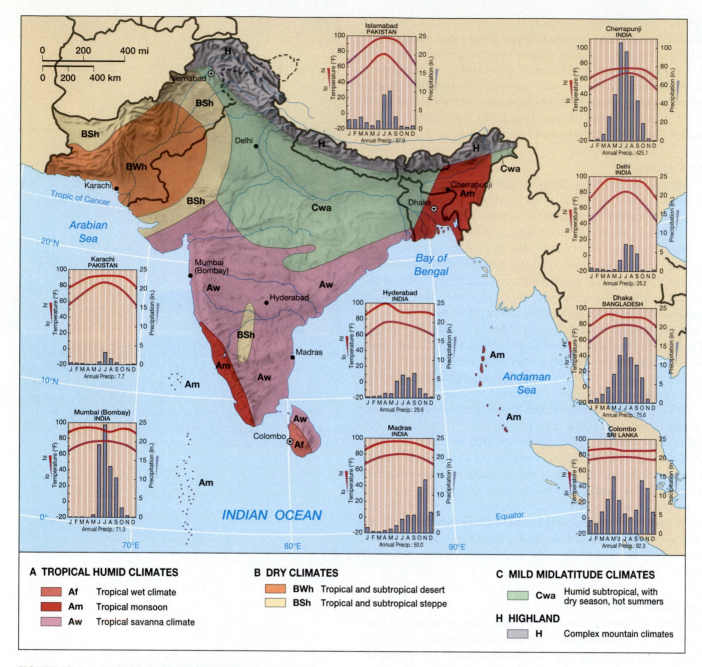

A TROPICAL HUMID CLIMATES

Af	Tropical wet climate
Am	Tropical monsoon
Aw	Tropical savanna climate

B DRY CLIMATES

BWh	Tropical and subtropical desert
BSh	Tropical and subtropical steppe

C MILD MIDLATITUDE CLIMATES

Cwa	Humid subtropical, with dry season, hot summers

H HIGHLAND

H	Complex mountain climates

FIGURE 12.6 • CLIMATES OF SOUTH ASIA Except for the extensive Himalayas, South Asia is dominated by tropical and subtropical climates. Many of these climates show a distinct summer rainfall season that is associated with the southwest monsoon. The climographs for Mumbai (Bombay) and Delhi are excellent illustrations.

Indus-Ganges-Brahmaputra Lowlands South of the northern mountains lie large lowlands created by three major river systems that have deposited sediments to build huge alluvial plains of fertile and easily farmed soils. These river lowlands are densely settled and constitute the population core areas of Pakistan, India, and Bangladesh.

Of these three rivers the Indus is the longest, covering more than 1,800 miles (2,880 kilometers) as it flows southward from the Himalayas through Pakistan to the Arabian Sea, providing much-needed irrigation waters to the desert areas in the southern portions of that country. More famous, however, is the Ganges, which, after flowing out of the Himalayas, travels southeasterly some 1,500 miles (2,400 kilometers) to empty into the Bay of Bengal. The Ganges has provided the fertile alluvial soil that has made northern India one of the world's most densely settled areas. Given the central role of this important river in South Asia's past and present, it is understandable why Hindus consider the Ganges sacred. Finally, the Brahmaputra River rises on the Tibetan Plateau and flows easterly, then southward and westerly over 1,700 miles (2,720 kilometers), joining the Ganges in central Bangladesh and spreading out over the largest delta in the world.

Peninsular India Extending southward is the familiar shape of peninsular India, made up primarily of the Deccan Plateau, which is bordered on each coast by narrow coastal plains backed by north–south mountain ranges. On the west are the higher Western Ghats, which are generally about 5,000 feet (1,524 meters); to the east, the Eastern Ghats are lower and less continuous. On both coastal plains, fertile soils and an adequate water supply support population densities comparable to the Ganges lowland to the north.

Soil quality ranges from fair to poor over much of the Deccan Plateau, but in Maharashtra state lava flows have produced particularly fertile black soils. A reliable water supply for agriculture, however, is a major problem in most areas. Much of the western plateau lies in the rain shadow of the Western Ghats, giving it a somewhat dry climate. Small reservoirs or tanks have been the traditional method for collecting monsoon rainfall for use during the dry season. More recently, the installation of deep wells and powerful pumps allow groundwater development to support irrigated crops and village water needs.

Partly because of the overuse of these groundwater resources, the Indian government plans a series of large dams to provide for irrigation. These plans, however, are controversial because the reservoirs will displace hundreds of thousands of rural residents. A case in point is the Sardar Sarovar Dam project on the Narmada River (see Figure 12.3) in the state of Madhya Pradesh, which alone will displace more than 100,000 people. Local residents and activists throughout India have joined forces in opposition, but farmers in neighboring Gujarat, who will benefit, strongly support the project.

The Southern Islands At the southern tip of peninsular India lies the island country of Sri Lanka. Sri Lanka is ringed by extensive coastal plains and low hills, but mountains reaching more than 8,000 feet (2,438 meters) occupy the southern interior, providing a cool, moist climate. Because the main monsoon winds arrive from the southwest, that portion of the island is much wetter than the rain shadow area of the north and east.

Forming a separate country are the Maldives, a chain of more than 1,200 islands stretching south to the equator some 400 miles (640 kilometers) off the southwestern tip of India. The combined land area of these islands is only about 116 square miles (290 square kilometers), and only a quarter of the islands are actually inhabited. Like many islands of the south Pacific, the Maldives are mainly flat, low coral atolls. With its highest elevation just over 6 feet (2 meters) above sea level, the country plays a prominent role in the international debate about global warming and the accompanying rise in sea level (see Chapter 2).

FIGURE 12.7 • SOUTH ASIA FROM SPACE The four physical subregions of South Asia are clearly seen in this satellite photograph, from the snow-clad Himalayan mountains in the north to the islands of the south. The Deccan Plateau is dark, fringed by white clouds as moist air is lifted over the uplands of the Western Ghats. *(Earth Satellite Corporation/Science Photo Library/Photo Researchers, Inc.)*

Population and Settlement: The Demographic Dilemma

South Asia could soon surpass East Asia as the world's most populated region (Figure 12.8). India alone is home to more than 1 billion people, second only to China in population, while Pakistan and Bangladesh, with 145 and 133.5 million residents, respectively, rank among the world's 10 most populous countries. Furthermore, South Asia is growing more than twice as fast as East Asia. India alone adds some 18 million people each year.

Although South Asia has made remarkable agricultural gains over the last several decades, there is still widespread concern over its ability to feed itself. The threat of crop failure, although much reduced, remains, in part because much South Asian farming is vulnerable to the unpredictable monsoon rains. Continuing population growth is also an issue. While all South Asian countries have family planning programs, the commitment to these policies—along with the results—varies widely from place to place.

Widespread concern over India's population growth began in the 1960s. To some extent, the measures to control population growth taken over the last 40 years have

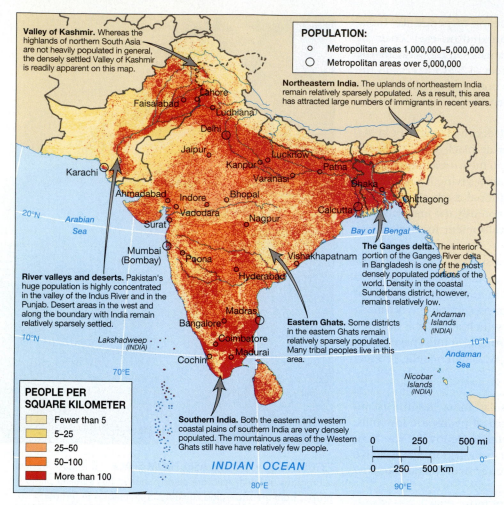

Valley of Kashmir. Whereas the highlands of northern South Asia are not heavily populated in general, the densely settled Valley of Kashmir is readily apparent on this map.

POPULATION:
○ Metropolitan areas 1,000,000–5,000,000
◯ Metropolitan areas over 5,000,000

Northeastern India. The uplands of northeastern India remain relatively sparsely populated. As a result, this area has attracted large numbers of immigrants in recent years.

Faisalabad
Lahore
Ludhiana
Delhi
Jaipur
Lucknow
Kanpur
Varanasi
Patna
Dhaka
Chittagong
Karachi
Ahmadabad
Indore
Bhopal
Calcutta
Vadodara
Nagpur
Surat
Mumbai (Bombay)
Poona
Vishakhapatnam
Hyderabad
Madras
Bangalore
Coimbatore
Madurai
Cochin

Arabian Sea
20°N
Lakshadweep (INDIA)
70°E
Bay of Bengal
Andaman Islands (INDIA)
10°N
Andaman Sea
Nicobar Islands (INDIA)
INDIAN OCEAN
0°
80°E
90°E

River valleys and deserts. Pakistan's huge population is highly concentrated in the valley of the Indus River and in the Punjab. Desert areas in the west and along the boundary with India remain relatively sparsely settled.

The Ganges delta. The interior portion of the Ganges River delta in Bangladesh is one of the most densely populated portions of the world. Density in the coastal Sunderbans district, however, remains relatively low.

Eastern Ghats. Some districts in the eastern Ghats remain relatively sparsely populated. Many tribal peoples live in this area.

PEOPLE PER SQUARE KILOMETER
Fewer than 5
5–25
25–50
50–100
More than 100

Southern India. Both the eastern and western coastal plains of southern India are very densely populated. The mountainous areas of the Western Ghats still have have relatively few people.

0   250   500 mi
0   250   500 km

FIGURE 12.8 • POPULATION MAP OF SOUTH ASIA Except for the desert areas of the west and the high mountains of the north, South Asia is a densely populated region. Particularly high densities of people are found on the fertile plains along the Indus and Ganges rivers and in India's coastal lowlands. In rural areas the population is typically clustered in villages, often located near water sources, such as streams, wells, canals, or small tanks that store water between monsoon rains.

been successful; the total fertility rate (TFR) dropped from 6 in the 1950s to the current rate of 3.2. Fertility rates vary widely within India, from lows of 1.9 and 2.0 in the states of Goa and Kerala, to a high of 4.8 in Uttar Pradesh. As is the case in China, a distinct cultural preference for male children is found in most of South Asia, a tradition that further complicates family planning. Where allowed, sex determination clinics provide couples with information about the sex of the fetus, resulting in a higher rate of abortion for female fetuses. In southern South Asia (particularly Sri Lanka and the Indian state of Kerala), where women have a much higher social position, sex ratios are balanced and birthrates are much lower.

Pakistan, on the other hand, lacks an effective, coordinated family planning program. As a result, the TFR remains very high at 5.6, and the current rate of natural increase of 2.8 percent adds about 4 million children each year. While some attribute Pakistan's high birthrate to its strong Muslim culture, the two are not necessarily linked. Although Bangladesh is predominantly Muslim, it has made significant strides in family planning (Figure 12.9). As recently as 1975 the TFR was 6.3, but it had dropped to 3.3 by the late 1990s. The success of family planning can be attributed to strong support from the Bangladesh government, advertised through radio and billboards.

Migration and the Settlement Landscape

South Asia is one of the least urbanized regions in the world, with only a quarter of its huge population living in cities. The majority of its people live in compact rural villages. Rapid migration from villages to large cities, however, is occurring. This often results as much from desperate conditions in the countryside as from the attraction of employment in the city. A primary cause of urbanization is changes in agriculture: increased mechanization with resulting unemployment; the higher costs of farming modern crops; expansion of large farms at the expense of subsistence farming; and environmental deterioration.

The most densely settled areas of South Asia still coincide with areas of fertile soils and dependable water supplies. The largest rural populations are found in the core area of the Ganges and Indus river valleys and on the coastal plains of India. Settlement is less dense on the Deccan Plateau and is relatively sparse in the highlands of the far north and the arid lands of the northwest.

As is true in most other parts of the world, South Asians have historically migrated from poor and densely populated areas to places that are either less densely populated or wealthier. Migrants are often attracted to large cities such as Mumbai (Bombay), but those from Bangladesh are settling in large numbers in rural portions of adjacent Indian states, creating ethnic and religious tensions. Sometimes migrants are forced out by war; a large number of Tamils from Sri Lanka and Hindus from Kashmir have recently sought security away from their battle-scarred homelands.

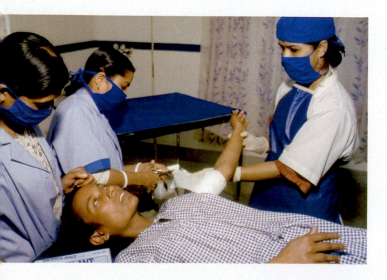

FIGURE 12.9 • FAMILY PLANNING IN BANGLADESH
Bangladesh has been one of the most successful nations in South Asia in reducing its fertility rate through family planning. Many women in Bangladesh use oral contraceptives. This photo shows a woman receiving a contraceptive implant. *(Peter Barker/Panos Pictures)*

Agricultural Regions and Activities

South Asian agriculture has historically been relatively unproductive, especially when compared with that of East Asia. Since the 1970s, however, agricultural production has grown faster than the population, primarily because of the **Green Revolution**, agricultural cultivation techniques based on new crop varieties and the heavy use of industrial fertilizers and chemical pesticides. Because the Green Revolution also carries significant social and environmental costs, it has also been highly controversial.

Crop Zones   South Asia can be divided into several distinct agricultural regions, all with different problems and potentials. These regions are based on the production of three subsistence crops—rice, wheat, and millet.

Rice is the main crop and foodstuff in the lower Ganges Valley, along the lowlands of India's eastern and western coasts, in the delta lands of Bangladesh, along Pakistan's lower Indus Valley, and in Sri Lanka (Figure 12.10). This distribution reflects the large volume of irrigation water needed to grow rice. The sheer amount of rice grown in South Asia is impressive: India ranks behind only China in world rice production, and Bangladesh is the fourth largest producer.

Wheat is the principal crop of the northern Indus Valley and in the western half of India's Ganges Valley. South Asia's "breadbasket" is the northwestern Indian state of Punjab and adjacent areas in Pakistan. Here the Green Revolution has been particularly successful in increasing grain yields. In the less fertile areas of central India, millet and sorghum are the main crops, along with root crops such as manioc. In general, wheat and rice are the preferred staples throughout South Asia, and it is generally poorer people who subsist on millet.

The Green Revolution   The main reason South Asian agriculture has been able to keep up with population growth is the Green Revolution, which originated during the 1960s in agricultural research stations established by international development agencies. By the 1970s it was clear that these efforts to breed high-yield varieties of rice and wheat had succeeded in reaching their initial goals. As a result, South Asia was transformed from a region of food deficiency to one of self-sufficiency. India more than doubled its annual grain production between 1970 and the mid-1990s (Figure 12.11).

While the Green Revolution was clearly an agricultural success, many argue that it has been an ecological and social disaster. Serious environmental problems result from the chemical dependency of the new crop strains. Not only do they typically need large quantities of industrial fertilizer, which is both expensive and polluting, but they also require frequent pesticide applications because they lack natural resistance to local plant diseases and insects.

Social problems have also followed the Green Revolution. In many areas only the wealthiest farmers are able to afford the new seed strains, irrigation equipment, farm machinery, fertilizers, and pesticides. As a result, poorer farmers have sometimes been forced from their lands. Many of these people either become wage laborers for their more successful neighbors or migrate to the region's already crowded cities.

Although the Green Revolution has been subjected to serious criticism, it also has been strongly defended. Some advocates argue that its environmental dangers have been reduced as farmers learn to grow new crop varieties using traditional methods of fertilization and pest control. Others contend that overall poverty and malnutrition have decreased markedly in areas where the new techniques have been adopted.

While the Green Revolution has fed South Asia's expanding population over the past several decades, it remains unclear whether it will be able to continue doing so. Another option for increasing agricultural yield is to expand water delivery systems (either through canals or wells), as many fields remain unirrigated. Irrigation, however, brings its own problems. In much of Pakistan and northwestern India, where irrigation has been practiced for generations, soil **salinization**, or the buildup of salt in agricultural fields, is already a major problem (see Figure 12.3). Additionally, groundwater is being depleted, especially in the Punjab, India's breadbasket.

FIGURE 12.10 • RICE CULTIVATION   A large amount of irrigation water is needed to grow rice, as is apparent from this photo from Sri Lanka. Rice is also the main crop in the lower Ganges Valley and delta, along the lower Indus River of Pakistan, and in India's coastal plains. *(Mahaux Photography/Getty Images, Inc.— Image Bank)*

FIGURE 12.11 • GREEN REVOLUTION FARMING   Because of "miracle" wheat strains that have increased yields in the Punjab area, this region has become the breadbasket of South Asia. India more than doubled its wheat production in the last 25 years and has moved from continual food shortages to self-sufficiency. Increased production, however, has led to both social and environmental problems. *(Earl Kowall/Corbis)*

FIGURE 12.12 • MUMBAI (BOMBAY) CENTRAL CITY The heart of this metropolitan area of 16 million is Bombay Peninsula. Because space is limited and building restrictions are severe, most recent growth has been in the east and to the north of the city center. *(Rob Crandall/Rob Crandall, Photographer)*

FIGURE 12.13 • MUMBAI HUTMENTS Hundreds of thousands of people in Mumbai live in crude hutments, with no sanitary facilities, built on formerly busy sidewalks. Hutment construction is forbidden in many areas, but wherever it is allowed, sidewalks quickly disappear. *(Rob Crandall/Rob Crandall, Photographer)*

Urban South Asia

Although South Asia is one of the least urbanized regions of the world, with about a quarter of its population living in cities, this does not mean that cities are unimportant. South Asia actually has some of the largest urban areas in the world. India alone lists more than 30 cities with populations greater than a million, most of which are growing rapidly.

Because of this rapid growth, South Asian cities have large problems with homelessness, poverty, congestion, water shortages, air pollution, and sewage disposal. In Calcutta, perhaps half a million people sleep on the streets each night. Throughout South Asia, sprawling squatter settlements, or **bustees**, exist in and around urban areas, providing temporary shelter for many urban migrants. A brief survey of the region's major cities reveals the problems and prospects of the region's urban areas.

Mumbai (Bombay) The largest city in South Asia, Mumbai (Bombay) is India's financial, industrial, and commercial center (Figure 12.12). The major port on the Arabian Sea, Mumbai is responsible for half the country's foreign trade. Long noted as the center of India's textile manufacturing, the city is also the center of its film industry, the largest in the world. Mumbai's economic vitality draws people from all over South Asia, resulting in ethnic tensions. Partly in reaction to this ethnic change, the nationalist party that governs the city asserted its local identity by officially changing the city's name from Bombay—a colonial name—to Mumbai, after the Hindu goddess Mumba.

Because of restricted space, most growth has taken place to the north and east of the historic city. Building restrictions in downtown Mumbai have also resulted in skyrocketing commercial and residential rents, which are now some of the highest in the world. Even members of the city's thriving middle class have a difficult time finding adequate housing. Hundreds of thousands of less-fortunate immigrants, eager for work in central Mumbai, live in "hutments," crude shelters built on formerly busy sidewalks (Figure 12.13). The least fortunate sleep on the street or in simple plastic tents, often placed along busy roadways.

Despite Mumbai's extraordinary contrasts of wealth and poverty, it remains in many ways an orderly and relatively crime-free city. It is less dangerous to walk the streets of central Mumbai than those of many major North American cities. Organized crime is a problem, especially in the massive film industry, but it has few effects on the lives of the average people.

Delhi Delhi, the sprawling capital of India, has more than 11 million people in its greater urban area. It consists of two contrasting landscapes expressing its past: Delhi (or old Delhi), a former Muslim capital, is a congested town of tight neighborhoods; New Delhi, in contrast, is a city of wide boulevards, monuments, parks, and expansive residential areas. New Delhi was born as a planned city when the British moved their colonial capital from Calcutta in 1911. Located here are the embassies, luxury hotels, government office buildings, and airline offices necessary for a vibrant political capital.

Calcutta To many, Calcutta symbolizes the problems faced by rapidly growing cities in developing countries. Not only is homelessness widespread, but this city of more than 12 million falls far short of supplying its residents with water, power, or sewage treatment. Electrical power is so inadequate that every hotel, restaurant, shop, and small business has to have some sort of standby power system. During the wet season, many streets are routinely flooded. No other Indian city has problems of this magnitude.

With rapid growth as migrants pour in from the countryside, a mixed Hindu-Muslim population that generates ethnic tension, a declining economic base, and an overloaded infrastructure, Calcutta faces a problematic future. Yet it remains a culturally vibrant city noted for its fine educational institutions, its theaters, and its successful publishing firms.

Karachi   Karachi, a rapidly growing port city of more than 7 million people, is Pakistan's largest urban area and its commercial core. It also served as the country's capital until 1963, when the new city of Islamabad was created in the northeast. Karachi, however, has suffered little from the departure of government functions; it is the most cosmopolitan city in Pakistan, its streets lined with high-rise buildings, hotels, banks, and travel agencies.

Karachi, however, suffers from serious political and ethnic tensions that have turned parts of the city into armed camps, if not battlegrounds. During the worst of the violence in 1995, more than 200 people were killed on the city's streets each month, and army bunkers were placed at the largest urban crossroads. The major problems are between the Sindis, who are the region's native inhabitants, and the Muhajirs, the Muslim refugees from India who settled in and around the city after independence and division from India in 1947. There are also clashes between Sunni and Shiite Muslims. As a result, travel both within and beyond the city is often difficult, particularly for international travelers, causing much damage to Karachi's economy.

Cultural Coherence and Diversity: A Common Heritage Undermined by Religious Rivalries

Historically, South Asia is a well-defined cultural region. A thousand years ago, virtually the entire area was united by the religion of Hinduism. The subsequent arrival of Islam added a new religious dimension but did not really undercut the region's cultural unity. British imperialism later added a number of cultural features to the entire region, from the widespread use of English to a common passion for cricket. Since the mid-twentieth century, however, religious strife has intensified, leading some to question whether South Asia can still be considered a culturally unified world region.

India has been a secular state since its creation, with the Congress Party, its guiding political organization until the 1980s, struggling to keep politics and religion separate. Since the 1980s, this secular political tradition has come under increasing pressure from the growth of Hindu "fundamentalism," which is better referred to as **Hindu nationalism**. These nationalists promote Hindu values as the essential basis of Indian society. Hindu nationalists have gained considerable political power through the Bharatiya Janata Party (BJP). Open campaigning against the country's Muslim minority increased in the mid-1990s. In several high-profile instances, Hindu mobs demolished Muslim mosques that had allegedly been built on the sites of ancient Hindu temples (Figure 12.14).

Fundamentalism, Islamic in this case, has also been a divisive issue in Pakistan. Powerful fundamentalist leaders want to make Pakistan a religious state under Islamic law, a plan rejected by the country's secular intellectuals and international businesspeople. The government has attempted to mediate between the two groups, but until recently it has often been viewed as favoring the fundamentalists.

Origins of South Asian Civilizations

Many scholars think that the roots of South Asian culture extend back to the Indus Valley civilization, which flourished more than 5,000 years ago in what is now Pakistan. This remarkable urban-oriented society vanished almost entirely around 1800 B.C. By 800 B.C., however, a new urban focus had emerged in the middle Ganges Valley.

Hindu Civilization   The religion that emerged out of this early Ganges Valley civilization was Hinduism, a complicated faith that lacks a single, uniformly accepted system of belief. Certain deities are recognized, however,

FIGURE 12.14 • DESTRUCTION OF THE AYODHYA MOSQUE
A group of Hindu nationalists are seen listening to speeches urging them to demolish the mosque at Ayodhya, allegedly built on the site of a former Hindu temple. The mosque was later destroyed by Hindu fundamentalists. *(Sunil Malhotra/Reuters/Getty Images, Inc.)*

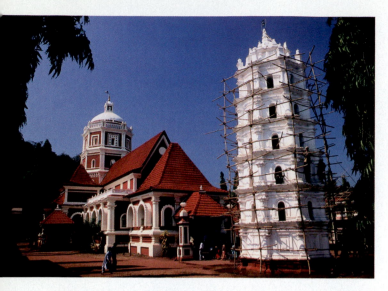

FIGURE 12.15 • HINDU TEMPLE Hindu temples dot India's landscape, taking many different forms, and are devoted to a wide range of deities. This is the Shantadurga temple in Ponda, Goa. *(Rob Crandall/Rob Crandall, Photographer)*

by all believers, as is the notion that these various gods are all expressions of a single divine being (Figure 12.15). All Hindus, moreover, share a common set of epic stories, usually written in **Sanskrit**, the sacred language of this religion. Hinduism is noted for its mystical tendencies, which have long inspired many men (and few women) to seek an ascetic lifestyle, renouncing property and sometimes all regular human relations. One of its hallmarks is a belief in the transmigration of souls from being to being through reincarnation. Hinduism also gave birth to India's **caste system**, the strict division of society into hereditary groups that are ranked as superior or inferior to each other.

Buddhism Ancient India's caste system was challenged from within by Buddhism. Siddhartha Gautama, the Buddha, was born in 563 B.C. in an elite caste. He rejected the life of wealth and power, however, and sought instead to attain enlightenment, or mystical union with the universe. He preached that the path to such enlightenment (or "nirvana") was open to all, regardless of social position. His followers eventually established Buddhism as a new religion. Buddhism spread through most of South Asia and later expanded through most of East, Southeast, and Central Asia. But for all of its successes abroad, Buddhism never replaced Hinduism in India. By A.D. 500 Buddhism was on the retreat throughout South Asia, and within another 500 years it had disappeared from most of the region.

Arrival of Islam The next major challenge to Hindu society—Islam—came from the outside. Around the year 1000, Turkish-speaking Muslims began to invade from Central Asia. By the 1300s, most of South Asia lay under Muslim power, although Hindu kingdoms remained in southern India and in the arid lands of Rajasthan. Later, during the sixteenth and seventeenth centuries, the **Mughal Empire**, the most powerful of the Muslim states, dominated much the region from its power center in the upper Indus-Ganges basin.

At first Muslims formed a small ruling elite, but over time increasing numbers of Hindus converted to the new religion. Conversions were most pronounced in the northwest and northeast, and eventually the areas now known as Pakistan and Bangladesh became predominantly Muslim.

The Caste System Caste is one of the historically unifying features of South Asia, as certain aspects of caste organization are even found among the Muslim populations of Pakistan and Bangladesh. Islam, however, has gradually reduced its significance, and even in India, caste is now being de-emphasized, especially among the more educated people. But caste remains significant throughout India, especially in the rural areas. Marriage across caste lines, for example, is still relatively rare.

Caste is actually a rather clumsy term referring to the complex social order of the Hindu world. It combines two distinct local concepts: *varna* and *jati. Varna* refers to the ancient fourfold social hierarchy of the Hindu world, which distinguishes the Brahmins (priests), Kshatriyas (warriors), Vaishyas (merchants), and Sundras (farmers and craftsmen) in declining order of supposed purity (members of higher castes are thus traditionally thought to suffer from "spiritual pollution" if they eat food that has been touched by members of lower castes). Standing outside of this traditional order are the so-called untouchables, now usually called **Dalits**, whose ancestors held supposedly impure jobs, such as those associated with leather-working. *Jati*, on the other hand, refers to the hundreds of local endogamous ("marrying within") groups that exist at each *varna* level (different *jati* groups are thus usually called *subcastes*).

Contemporary Geographies of Religion

South Asia thus has a mainly Hindu heritage overlain by a significant Muslim presence. Such a picture fails, however, to capture the enormous diversity of modern religious expression in contemporary South Asia. The following discussion looks specifically at the geographical patterns of the region's main faiths (Figure 12.16).

Hinduism Fewer than 1 percent of the people of Pakistan are Hindu, and in Bangladesh and Sri Lanka Hinduism is a minority religion. Almost everywhere in India, however—and in Nepal, as well—Hinduism is very much the faith of the majority. In east-central India, more than 95 percent of the population is Hindu. Hinduism is itself a geographically complicated religion, with different aspects of faith varying between different parts of India.

Islam Islam may be considered a "minority" religion for the region as a whole, but such a designation hides the tremendous importance of this religion in South Asia. With some 400 million members, the South Asian Muslim community is one of the largest in the world. Bangladesh and especially Pakistan are overwhelmingly Muslim. India's Muslim community, although constituting only some 15 percent of the country's population, is still roughly 150 million strong.

Muslims live in almost every part of India (see Figure 12.16). They are, however, concentrated in four main areas: in most of India's cities; in Kashmir, in the far north, particularly in the densely populated Vale of Kashmir (more than 80 percent of the population here follows Islam); in the central Ganges plain, where Muslims constitute 15 to 20 percent of the population; and in the southwestern state of Kerala, which is approximately 25 percent Muslim.

Interestingly, Kerala was one of the few parts of India that never experienced prolonged Muslim rule. Islam in Kerala is historically connected not to Central Asia, but rather to trade across the Arabian Sea. Kerala's Malabar Coast historically supplied spices and other luxury products to Southwest Asia, encouraging many Arabian traders to settle there. Gradually many of Kerala's native residents converted to the new religion as well. Sri Lanka is approximately 9 percent Muslim, and the Maldives is almost entirely Muslim.

Sikhism The tension between Hinduism and Islam in northern South Asia gave rise to a new religion called **Sikhism**. Sikhism originated in the late 1400s in the Punjab, near the modern boundary between India and Pakistan. The Punjab was the site of religious competition at the time; Islam was gaining converts and Hinduism was increasingly on the defensive. The new faith combined elements of both religions. Many orthodox Muslims at the time viewed Sikhism as dangerous because it incorporated elements of their own religion in a manner that was contrary to accepted beliefs. Periodic persecution by Muslim rulers led the Sikhs to adopt a militantly defensive stance. Even today, Sikh men are noted for their work as soldiers and bodyguards.

At present the Indian state of Punjab is approximately 60 percent Sikh. Small but often influential groups of Sikhs are scattered across the rest of India. Devout Sikh men are immediately visible, since they do not cut their hair or their beards. Instead, they wear their hair wrapped in a turban and often tie their beards close to their faces.

Buddhism and Jainism Although Buddhism virtually disappeared from India in medieval times, it persisted in Sri Lanka. Among the island's dominant Singhalese people, Theravada Buddhism developed into a something of a national religion. In the high valleys of the Himalayas, Buddhism also survived as the majority religion

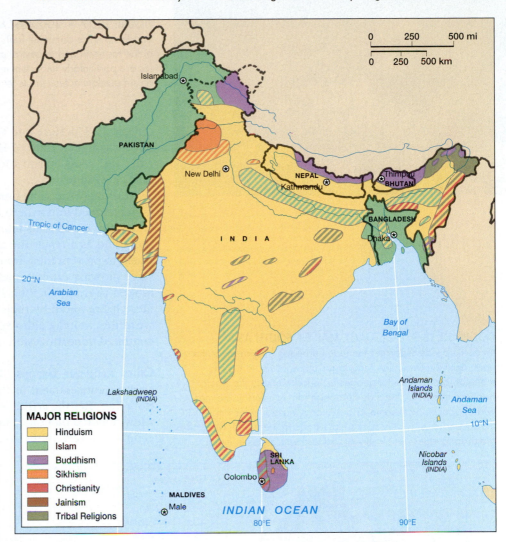

FIGURE 12.16 • RELIGIOUS GEOGRAPHY OF SOUTH ASIA
Hindu-dominated India is bordered by the two important Muslim countries of Pakistan and Bangladesh. Some 150 million Muslims, however, live within India, making up roughly 15 percent of the total population. Of particular note are the Muslims in northwest Kashmir and in the Ganges Valley. Sikhs form the majority population in India's state of Punjab. Also note the Buddhist populations in Sri Lanka, Bhutan, and northern Nepal, the areas of tribal religion in the east, and the centers of Christianity in the southwest.

FIGURE 12.17 • BUDDHIST MONASTIC LANDSCAPE
Buddhism is the dominant faith in most of the Himalayan districts of the far north and in central and southern Sri Lanka. Monasteries are particularly important in the north, where Tibetan Buddhism predominates. *(Linde Waidhofer/Getty Images, Inc.—Liaison)*

(Figure 12.17). Here one finds Tibetan Buddhism. The town of Dharmsala in the northern Indian state of Himachal Pradesh is the seat of Tibet's government-in-exile and of its spiritual leader, the Dalai Lama, who fled Tibet in 1959 after an unsuccessful revolt.

At roughly the same time as the birth of Buddhism (circa 500 B.C.), another religion emerged in northern India: **Jainism**. This religion also stressed nonviolence, taking this creed to its ultimate extreme. Jains are forbidden to kill any living creatures, and as a result the most devoted members of the community wear gauze masks to prevent them from inhaling small insects. Agriculture is forbidden to Jains, since plowing can kill small creatures. As a result, most members of the faith have looked to trade for their livelihoods. Many have prospered, aided by the frugal lifestyles required by their religion. The small Jain community is now one of the wealthiest in India, although it includes many poor individuals. Today Jains are concentrated in northwestern India, particularly Gujarat.

Other Religious Groups Even wealthier than the Jains are the Parsis, or Zoroastrians, concentrated in the Mumbai area. The Parsis arrived as refugees, fleeing from Iran after the arrival of Islam in the seventh century. Although numbering only a few hundred thousand, the Parsi community has had a major impact on the Indian economy. Several of the country's largest industrial firms are still controlled by Parsi families. Intermarriage and low fertility, however, now threaten the survival of the community.

Indian Christians are more numerous than either Parsis or Jains. Their religion arrived some 1,800 years ago as missionaries from Southwest Asia brought Christianity to India's southwestern coast. Today, roughly 20 percent of the population of the state of Kerala follows Christianity. Several Christian sects are represented, but the largest are affiliated with the Syrian Christian Church of Southwest Asia. Another stronghold of Christianity is the small Indian state of Goa, a former Portuguese colony. Here Roman Catholics and Hindus each make up roughly half of the population.

During the colonial period, British missionaries went to great efforts to convert South Asians to Christianity. They had very little success, however, in Hindu, Muslim, and Buddhist communities. The remote tribal districts of British India, on the other hand, proved to be more receptive to missionary activity. In the uplands of India's extreme northeast, entire communities abandoned their traditional animist faith in favor of Protestant Christianity.

Geographies of Language

South Asia's linguistic diversity matches its religious diversity. In fact, one of the world's most important linguistic boundaries runs directly across India (Figure 12.18). North of the line, languages belong to the Indo-European family, the world's largest. The languages of southern India, on the other hand, belong to the **Dravidian** family, a linguistic group that is unique to South Asia. Along the

FIGURE 12.18 • LANGUAGE MAP OF SOUTH ASIA
A major linguistic divide separates the Indo-European languages of the north from the Dravidian languages of the south. In the Himalayan areas, most languages instead belong to the Tibeto-Burmese family. Of the Indo-European family, Hindi is the most widely spoken, with some 480 million speakers, which makes it the second most widely spoken language in the world. Most other major languages are closely associated with states in India.

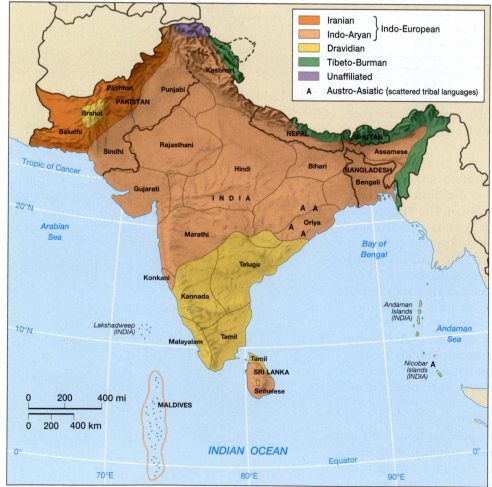

mountainous northern rim of the region a third linguistic family, Tibeto-Burman, is dominant, but this area is marginal to South Asia. Within these broad divisions there are many different languages, each associated with a distinct culture. In many parts of South Asia, several languages are spoken within the same region or even city, and the ability to speak several languages is common everywhere.

Any modern Indo-European language of India, such as Hindi or Bengali, is more closely related to English than it is to any Dravidian language of southern India, such as Tamil. But it must also be noted that South Asian languages on both sides of this linguistic divide do share a number of features. All of the major Dravidian languages, for example, have borrowed many words from Sanskrit, particularly those associated with religion and scholarship.

Each of the major languages of India is associated with an Indian state, since the country deliberately structured its political subdivisions along linguistic lines a decade after attaining independence. As a result, one finds Gujarati in Gujarat, Marathi in Maharashtra, Oriya in Orissa, and so on. Two of these languages, Punjabi and Bengali, extend into Pakistan and Bangladesh, respectively, since these borders were established on religious rather than linguistic lines. Nepali, the national language of Nepal, is largely limited to that country. Minor dialects and languages, most of which are neither written nor standardized, are common in many parts of the region.

The Indo-European North The most widely spoken language of South Asia is **Hindi**—not to be confused with the Hindu religion. With some 480 million speakers, Hindi is the second most widely spoken language in the world. It occupies a prominent role in present-day India, both because so many people speak it and because it is the main language of the Ganges Valley, the historical and demographic core of India. Many northerners would like Hindi to become the country's common language, but many other Indians, particularly those living in the south, resist the idea. Still, most Indian students learn some Hindi, often as their second or third language.

Bengali is the second most widely spoken language in South Asia. It is the national language of Bangladesh and the main language of the Indian state of West Bengal. Spoken by almost 200 million people, Bengali is the world's ninth most widely spoken language. Its significance extends beyond its official status in Bangladesh and its large number of speakers. Bengali is also the language of an extensive amount of South Asian literature; West Bengal, particularly its capital of Calcutta, has long been one of South Asia's main literary and intellectual centers. Calcutta may be noted for its poverty, but it also has one of the highest levels of cultural production in the world, as measured by the output of drama, poetry, fiction, and film.

The Punjabi-speaking zone in the west was similarly split at the time of independence, in this case between Pakistan and the Indian state of Punjab. While an estimated 100 million people speak Punjabi, it does not have the significance of Bengali. Punjabi did not become the national language of Pakistan, even though it is the day-to-day language of some two-thirds of the country's people. Instead, that position was given to Urdu.

Urdu, like Hindi, originated on the plains of northern India. The difference between the two was largely one of religion: Hindi was the language of the Hindu majority, Urdu that of the Muslim minority, including the former ruling class. Because of this distinction, Hindi and Urdu are written differently—the former in the Devanagari script (derived from Sanskrit) and the latter in the Arabic script. Although Urdu contains many words borrowed from Persian, its basic grammar and vocabulary are almost identical to those of Hindi. With independence in 1947, millions of Urdu-speaking Muslims from the Ganges Valley fled to the new state of Pakistan. Since Urdu had a higher status than Pakistan's native tongues, it was quickly established as the new country's official language.

Languages of the South The four main Dravidian languages are confined to southern India and northern Sri Lanka. As in the north, each language is closely associated with an Indian state: Kannada in Karnataka, Malayalam in Kerala, Telugu in Andhra Pradesh, and Tamil in Tamil Nadu. Tamil is usually considered the most

FIGURE 12.19 • MULTILINGUALISM This four-language sign, in Malayalam, Tamil, Kannada, and English, shows the multilingual nature of contemporary South Asia. In many ways, English, the colonial language of British rule, still serves to bridge the gap between the many different languages of the region. *(Rob Crandall/Rob Crandall, Photographer)*

important member of the family because it has the longest history and the largest literature. Tamil poetry dates back to the first century A.D., making it one of the world's oldest written languages.

Although Tamil is spoken in northern Sri Lanka, the country's majority population, the Singhalese, speak an Indo-European language. Apparently the Singhalese migrated from northern South Asia several thousand years ago. The migrants evidently settled on the island's fertile and moist southwestern coastal and central highland areas, which formed the core of a number of Singhalese kingdoms. These same people also migrated to the Maldives, where the national language, Divehi, is essentially a Singhalese dialect. The drier north and east of Sri Lanka, on the other hand, were settled many hundreds of years ago by Tamils from southern India.

Linguistic Dilemmas The multilingual countries of Sri Lanka, Pakistan, and India are all troubled by linguistic conflicts. Such problems are most complex in India, simply because India is so large and has so many different languages. India's linguistic environment is changing in many ways, pushed along by modern economic and political forces (Figure 12.19).

Indian nationalists have long dreamed of a national language, one that could help make their country into a more unified nation. But **linguistic nationalism**, or the linking of a specific language with political goals, faces the resistance of many people. The obvious choice for a national language would be Hindi, and Hindi was indeed declared as such in 1947. Raising Hindi to this position, however, angered many non-Hindi speakers, especially in the Dravidian south. As a result, in the 1950 Indian constitution Hindi was demoted to sharing the position of "official language" of India with 14 other languages.

Regardless of opposition, the role of Hindi is expanding, especially in the Indo-European north. Here local languages are fairly closely related to Hindi, which can therefore be learned without too much difficulty. Hindi is spreading through education, but even more significantly television and motion pictures. Films and television programs are made in several northern languages, but Hindi remains primary. In a poor but modernizing country such as India, where many people experience the wider world largely through moving images, the influence of a national film and television culture can be tremendous.

Even if Hindi is spreading, it still cannot be considered anything like a common national language, even in northern India. In the Dravidian south, more importantly, its role remains strictly secondary. National-level political, journalistic, and academic communication thus cannot be conducted in Hindi. Only English, an "associate official language" of contemporary India, serves this function.

Before independence, many educated South Asians learned English for its political and economic benefits under colonialism. Today many nationalists wish to deemphasize English. Others, however, and particularly southerners, advocate English as a neutral national language, since all parts of the country have an equal stake in it. Furthermore, English gives substantial international benefits. English-medium schools abound in all parts of the region, and many children of the elite learn this global language well before they begin their schooling.

South Asians in a Global Cultural Context

The widespread use of English in South Asia not only aids the spread of global culture into the region, but it has also helped South Asians' cultural production to reach a global audience. The global spread of South Asian literature, however, is nothing new. As early as the turn of the twentieth century, Rabindranath Tagore gained international acclaim for his poetry and fiction, earning the Nobel Prize for Literature in 1913.

The spread of South Asian culture abroad has been accompanied by the spread of South Asians themselves. Migration from South Asia during the time of the British Empire led to the establishment of large communities in such distant places as eastern Africa, Fiji, and the southern Caribbean (Figure 12.20). Subsequent migration has been aimed more at the developed world; there are now several million people of South Asian descent living in Britain and a similar number in North America. Many

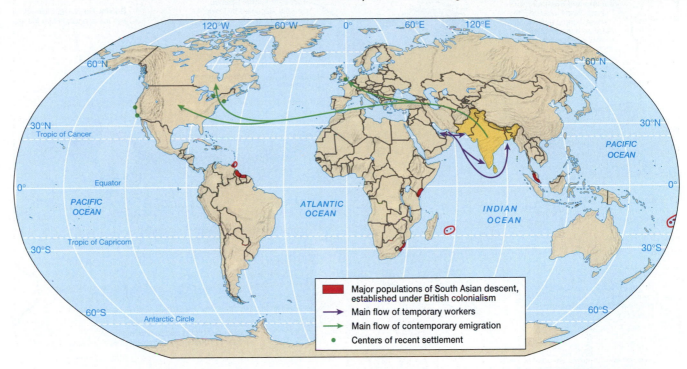

FIGURE 12.20 • THE SOUTH ASIAN GLOBAL DIASPORA
During the British imperial period, large numbers of South Asian workers settled in other colonies. Today, roughly 50 percent of the population of such places as Fiji and Mauritius are of South Asian descent. More recently, large numbers have settled, and are still settling, in Europe (particularly Britain) and North America. Large numbers of temporary workers, both laborers and professionals, are employed in the wealthy oil-producing countries of the Persian Gulf.

present-day migrants to the United States are doctors, software engineers, and other professionals, making Indian-Americans one of the country's wealthiest ethnic groups.

In South Asia itself, the globalization of culture has brought tensions as severe as those felt anywhere in the world. Traditional Hindu and Muslim religious norms frown on any overt display of sexuality—a staple feature of global popular culture. Religious leaders thus often criticize Western films and videos as immoral. Still, the pressures of internationalization are hard to resist. Many Indians were surprised in the mid-1990s when the fundamentalist Hindu government in Mumbai relented and allowed Michael Jackson to perform in public. In the tourism-oriented Indian state of Goa, such cultural tensions are on full display. There, German and British sun-worshipers often wear nothing but thong bikini-bottoms, whereas Indian women tourists go into the ocean fully clothed. Young Indian men, for their part, often simply walk the beach and gawk, good-naturedly, at the semi-naked foreigners (Figure 12.21).

FIGURE 12.21 • GOA BEACH SCENE The liberal Indian state Goa, formerly a Portuguese colony, is now a major destination for tourists, both from within India and from Europe and Israel. European tourists come in the winter for sunbathing and for "Goan Rave parties," where ecstasy and other drugs are widely available. Indian tourists typically find the scantily clad foreigners unusual if not bizarre. *(Rob Crandall/Rob Crandall, Photographer)*

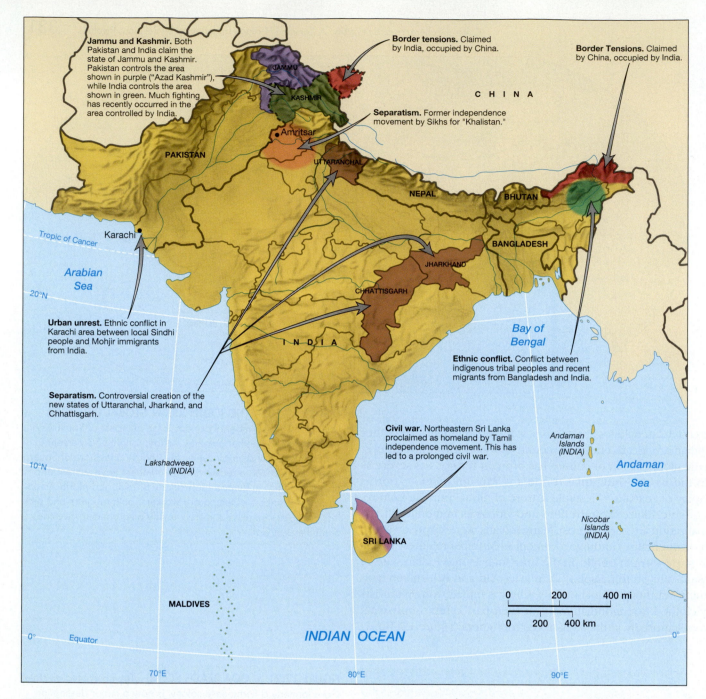

Jammu and Kashmir. Both Pakistan and India claim the state of Jammu and Kashmir. Pakistan controls the area shown in purple ("Azad Kashmir"), while India controls the area shown in green. Much fighting has recently occurred in the area controlled by India.

Border tensions. Claimed by India, occupied by China.

Border Tensions. Claimed by China, occupied by India.

Separatism. Former independence movement by Sikhs for "Khalistan."

Urban unrest. Ethnic conflict in Karachi area between local Sindhi people and Mohjir immigrants from India.

Separatism. Controversial creation of the new states of Uttaranchal, Jharkand, and Chhattisgarh.

Ethnic conflict. Conflict between indigenous tribal peoples and recent migrants from Bangladesh and India.

Civil war. Northeastern Sri Lanka proclaimed as homeland by Tamil independence movement. This has led to a prolonged civil war.

FIGURE 12.22 • GEOPOLITICAL ISSUES IN SOUTH ASIA
Given the cultural mosaic of South Asia, it is not surprising that ethnic tensions have created numerous geopolitical problems in the region. Particularly troubling are ethnic tensions in Sri Lanka, Kashmir, and the Punjab.

Geopolitical Framework: A Deeply Divided Region

Before the coming of British imperial rule, South Asia had never been politically united. While a few empires at various times ruled most of the subcontinent, none covered its entire extent. Whatever unity the region had was cultural, not political. The British, however, brought the entire region into a single political system by the middle of the nineteenth century. Independence in 1947 witnessed the separation of Pakistan from India; in 1971 Pakistan itself was divided with the independence of Bangladesh, formerly East Pakistan. Even today, serious geopolitical issues continue to exist in the region (Figure 12.22)

South Asia Before and After Independence in 1947

During the 1500s, when Europeans began to arrive on the coasts of South Asia, most of the northern subcontinent came under the power of the Islamic Mughal Empire (Figure 12.23). Southern India remained under the control of a powerful Hindu

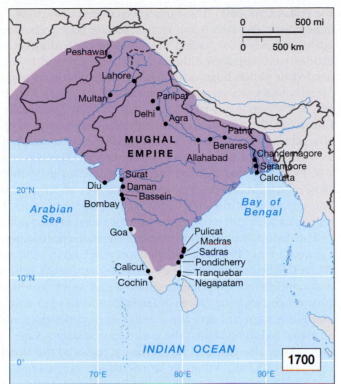

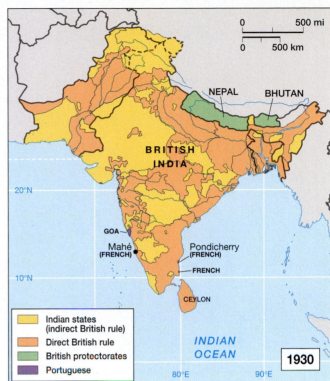

FIGURE 12.23 • GEOPOLITICAL CHANGE   At the onset of European colonialism before 1700, much of South Asia was dominated by the powerful Mughal Empire. Under Britain, the wealthiest parts of the region were ruled directly, but other lands remained under the partial authority of indigenous rulers. Independence for the region came after 1947, when the British abandoned their extensive colonial territory. Bangladesh, which was formerly East Pakistan, gained its independence in 1971 after a short struggle against centralized Pakistani rule from the west.

kingdom called *Vijayanagara*. European merchants, eager to obtain spices, textiles, and other Indian products, established a series of coastal trading posts. The Portuguese carved out an enclave in Goa on the west coast, while the Dutch gained control over much of Sri Lanka in the 1600s, but neither was a significant threat to the Mughals. In the early 1700s the Mughal Empire weakened rapidly. A number of competing states, some ruled by Muslims, others by Hindus, and a few by Sikhs, emerged in former Mughal territories. As a result, the 1700s was a century of political and military turmoil, often bordering on chaos.

The British Conquest   These unsettled conditions provided an opening for European imperialism. The British and French, having largely displaced the Dutch and Portuguese, competed for trading posts. Before the Industrial Revolution, Indian cotton textiles were considered the best in the world, and British and French merchants needed to obtain huge quantities for their global trading networks. With Britain's overwhelming victory over France in the Seven Years' War (1756–63), France retained only a few minor coastal cities. Britain, or more specifically the **British East India Company**, the private organization that acted as an arm of the British government, now monopolized trade in the area and was free to stake out a South Asian empire of its own. By the 1840s, British control of South Asia was essentially completed. Valuable local allies, such as the ruler of Hyderabad in southern India, however, were allowed to remain in power, provided that they no longer threatened British interests. The territories of these indigenous (or "princely") states, however, were gradually reduced, and British advisors increasingly dictated their policies.

The continuing reduction in size of the indigenous states, coupled with the growing arrogance of British officials, led to a rebellion in 1856 across much of South Asia. When this uprising (called the *Sepoy Mutiny* by the British) was finally crushed, a new political order was implemented. South Asia was now to be ruled by the British government, with the Queen of England as its head of state. After this, the British enjoyed direct control over the region's most productive and most densely populated areas, including almost the entire Indus-Ganges Valley and most of the coastal plains. They also ruled Sri Lanka, having replaced the Dutch in the 1700s.

British officials continually worried about threats to their immensely profitable Indian colony, particularly from the Russians advancing across Central Asia. In response, they attempted to expand as far to the north as possible. In some cases this merely required making alliances with local rulers. In such a manner Nepal and Bhutan retained their independence, although they would no longer be free of British interference. In the extreme northeast, a number of small states and tribal territories, most of which had never been part of the South Asian cultural sphere, were taken over by the British Empire, hence into India. A similar policy was conducted on the vulnerable northwestern frontier.

Independence and Partition   The framework of British India began to unravel in the early twentieth century as the people of South Asia increasingly demanded independence. The British, however, were determined to stay, and by the 1920s South Asia was caught up in massive political protests.

The leaders of the rising nationalist movement faced a dilemma in attempting to organize a potentially independent country. Many leaders, including Mohandas Gandhi—the father-figure of Indian independence—favored a unified state that would include all British territories in mainland South Asia. Most Muslim leaders, however, feared that a unified India would leave their people in a vulnerable position. They therefore argued for the division of British India into two new countries: a Hindu-majority India and a Muslim-majority Pakistan. In several parts of northern South Asia, however, Muslims and Hindus were settled in roughly equal numbers. A more significant problem was the fact that the areas of clear Muslim majority were located on opposite sides of the subcontinent, in present-day Pakistan and Bangladesh.

The British withdrew from South Asia in 1947. As this occurred, the region was indeed divided into two countries: India and Pakistan. Partition itself was a horrific event; not only were millions of people displaced, but hundreds of thousands were killed in the process. Millions of Hindus and Sikhs fled from Pakistan, to be replaced by millions of Muslims fleeing India (Figure 12.24).

The Pakistan that emerged from partition was for several decades a clumsy two-part country, its western section in the Indus Valley, its eastern portion in the Ganges Delta. The Bengalis, occupying the poorer eastern section, complained that they were treated as second-class citizens. In 1971 they launched a rebellion, and, with the help of India, quickly prevailed. Bangladesh then emerged as a new country. This second partition did not solve Pakistan's problems, however, as it remained politically unstable and prone to military rule. It is also important to note that Pakistan kept the British

FIGURE 12.24 • PARTITION, 1947   Following Britain's decision to leave South Asia, violence and bloodshed broke out between Hindus and Muslims in much of the region. With the creation of Pakistan (originally in two different sectors, west and east), millions of people relocated both to and from the new states. Many were killed in the process, damaging relations between India and Pakistan. *(Bettmann/Corbis)*

policy of allowing almost full autonomy to the Pashtun tribes living along the border with Afghanistan, a relatively lawless area marked by clan fighting and vengeance feuds. This area later lent much support to Afghanistan's Taliban regime and to Osama bin Laden's Al Qaeda organization.

Geopolitical Structure of India   The leaders of newly independent India were committed to democracy but faced a major challenge in organizing such a large and culturally diverse country. They decided to pursue a middle ground between centralization and local autonomy. India itself was thus organized as a **federal state**, with a significant amount of power being given to its individual states. The national government, however, retained full control over foreign affairs and a large degree of economic authority.

Following independence, India's constituent states were reorganized to follow linguistic geography. Yet only the largest groups received their own territories, which has led to recurring demands from smaller groups that have felt politically excluded. Over time, several smaller states have been added to the map. Goa, following the forced eviction of the Portuguese in 1961, became a separate state in 1987. In 2000 three new states were added: Jharkand, Uttaranchal, and Chhattisgarh.

Ethnic Conflicts in South Asia

The movement for new states in India is rooted in ethnic tensions. Unfortunately, more violent ethnic conflicts have emerged in other parts of South Asia, specifically in Kashmir, the Punjab, the far northeast, and Sri Lanka. Of these conflicts, the most complex—and perilous—is that in Kashmir.

Kashmir   Relations between India and Pakistan were hostile from the start, and the situation in Kashmir has kept the conflict burning (Figure 12.25). During the British period, Kashmir was a large princely state with a primarily Muslim core joined to a Hindu district in the south (Jammu) and a Tibetan Buddhist district in the far northeast (Ladakh). Kashmir was ruled by a Hindu **maharaja**, or a king subject to British advisors. During partition, Kashmir came under severe pressure from both India and Pakistan. After troops from Pakistan gained control of western and much of northern Kashmir, the maharaja decided to join India. India thus retained most of Kashmir. But neither Pakistan nor India would accept the other's control over any portion of Kashmir, and they have since fought several wars over the issue.

Although the Indo-Pakistani boundary has remained fixed, fighting in Kashmir intensified, reaching a peak in the 1990s. Many Muslim Kashmiris hope to join their homeland to Pakistan; others would rather see it became an independent country. Hindu militants and other Indian nationalists, on the other hand, are determined that Kashmir remain part of India, and the Indian government agrees. The result has been a low-level but periodically brutal war. Efforts have been made to reach a peaceful settlement, but they seem unlikely to succeed. The Vale of Kashmir, with its lush fields and orchards nestled among some of the world's most spectacular mountains, was once one of South Asia's premier tourist destinations; now, however, it is a battle-scarred war zone.

The Punjab   Religious conflict also lay at the root of violence in India's Punjab. The original Punjab, an area of Hindu, Muslim, and Sikh communities, was divided between India and Pakistan in 1947. During partition,

FIGURE 12.25 • CONFLICT IN KASHMIR   Unrest in Kashmir maintains hostility between the two nuclear powers of India and Pakistan. Under the British, this region of predominantly Muslim population was ruled by a Hindu maharaja, who managed to join the province to India upon partition. Today many Kashmiris wish to join Pakistan, while many others argue for an independent state.

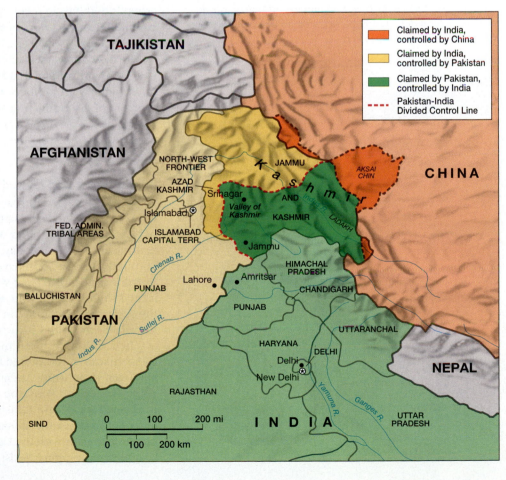

virtually all Hindus and Sikhs fled from Pakistan's allotted portion, just as Muslims left the Indian zone. Relations between Hindus and Sikhs had previously been friendly, but they too began to deteriorate. In the 1970s and 1980s, as the area prospered due to the Green Revolution, Sikh leaders began to strive for more self-government within India while Sikh radicals began to press for independence.

Since many Sikh men had maintained the military traditions of their ancestors, this secession movement soon became a strong fighting force. The Indian government reacted with military force, and tensions grew. Violence increased in 1984 when the Indian army raided the main Sikh temple at Amritsar, in which a group of militants had barricaded themselves (Figure 12.26). Shortly thereafter, the president of India, Indira Gandhi, was assassinated by her own Sikh bodyguards. As hostility escalated, the Indian government placed the Punjab under martial law. Such a policy eventually proved relatively successful in ending political violence, at least in the short run. Renewed conflict in the area, however, remains a possibility.

FIGURE 12.26 • SIKH TEMPLE AT AMRITSAR Separatism in Indian's Punjab region is strong because of hostility between the Sikh majority and the Indian government. These tensions were magnified in 1984 when the Indian army raided the Amritsar temple to dislodge Sikh militants. *(Daniel O'Leary/Panos Pictures)*

The Northeast Fringe A more complicated ethnic conflict emerged in the 1980s in the uplands of India's extreme northeast, particularly in the states of Arunachal Pradesh, Nagaland, Manipur, and portions of Assam. The underlying problem is rooted in population growth and cultural conflict. Much of this area is still relatively lightly populated and as a result has attracted hundreds of thousands of migrants from Bangladesh and nearby provinces of India. Many local people consider this movement a threat to their lands and their culture. On several occasions local guerillas have attacked newcomer villagers and, in turn, have suffered reprisals from the Indian military. This is a remote area, however, and relatively little information from it reaches the outside world.

Sri Lanka Ethnic violence in Sri Lanka has been especially severe. Here the conflict stems from both religious and linguistic differences. Northern Sri Lanka and parts of its eastern coast are dominated by Hindu Tamils, while the island's majority group is Buddhist in religion and Singhalese in language. Relations between the two communities have historically been fairly good, but tensions mounted soon after independence (Figure 12.27).

The basic problem is that Singhalese nationalists favor a centralized government, some of them arguing that Sri Lanka should be a Buddhist state. Most Tamils, on the other hand, want political and cultural autonomy, and they have accused the government of discriminating against them. In 1983 war erupted when the rebel force known as the *Tamil Tigers* attacked the Sri Lankan army. Northern Sri Lanka has been a war zone ever since, although occasional peace talks inspire some hope.

FIGURE 12.27 • CIVIL WAR IN SRI LANKA The majority of Sri Lankans are Singhalese Buddhist, many of whom maintain that their country should be a Buddhist state. A Tamil-speaking Hindu minority in the northeast strongly resists this idea. Tamil militants, who have waged war against the Sri Lankan government for several decades, hope to create an independent country in their northern homeland.

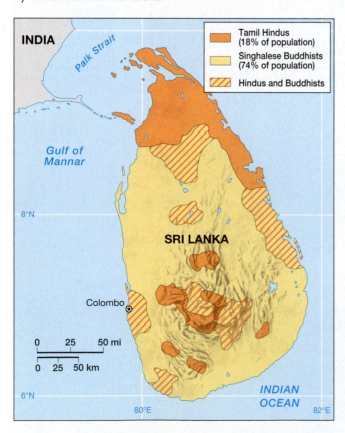

International and Global Geopolitics

South Asia's major international geopolitical problem is the continuing cold war between India and Pakistan (Figure 12.28). Since independence, these two countries have regarded each other as enemies, and both maintain large military forces to neutralize the other. Today, however, the stakes are considerably higher because both India and Pakistan have nuclear capabilities.

During the global Cold War, Pakistan allied itself with the United States and India leaned slightly toward the Soviet Union. Such alliances fell apart with the end of the superpower conflict in the early 1990s. Since then Pakistan has forged an informal alliance with China, which has long been in military competition with India.

The conflict between India and Pakistan became more complex after the terrorist attacks in the United States on September 11, 2001. Until that time, Pakistan had been supporting Afghanistan's Taliban regime. After the attack on the World Trade Center and Pentagon (orchestrated from Afghanistan by Osama bin Laden), the United States gave Pakistan a stark choice: Either it would assist the United States in its fight against the Taliban and receive financial aid in return, or it would completely lose favor with the United States. Pakistan's President Pervez Musharraf quickly agreed to help, and Pakistan offered valuable intelligence to the U.S. military.

Pakistan's decision to help the United States, however, came with large risks. Osama bin Laden had gained much popularity among Pakistan's more extreme Islamic fundamentalists. Both he and the Taliban, moreover, enjoy support among the Pashtun populace of Pakistan's wild North-West Frontier province. In 2003, several major but unsuccessful attempts were made to assassinate President Musharraf. More recently, Pakistan and India began serious peace negotiation, leading some observers to hope that an end to this conflict was at least possible.

The tension between India and Pakistan, and the relationships that this entails with China and the United States, overshadows all other international geopolitical issues in South Asia. Elsewhere in the region, the power of India is simply overwhelming, although often resented. In the late 1990s, a Marxist rebel movement emerged in Nepal, aimed in part at dissolving the country's close ties to India. Bangladesh, on the other hand, owes its very existence to Indian support, and the two countries have enjoyed relatively friendly relations.

FIGURE 12.28 • BORDER TENSIONS   Relationships between India and Pakistan have been extremely tense since independence in 1947. Moreover, with both countries now nuclear powers, the fear that border hostilities will escalate into wider warfare has become a nightmarish possibility. *(Deepak Sharma/ AP/Wide World Photos)*

Economic and Social Development: Burdened by Poverty

South Asia is, along with Sub-Saharan Africa, the poorest world region, yet it is also the site of great wealth. It has a sizable and growing middle class, yet large areas, and many social groups, remain virtually cut off from development. Many of South Asia's scientific and technological accomplishments are world-class, but it also has some of the world's highest illiteracy rates. While South Asia's high-tech businesses are closely integrated with the American economy, the South Asian economy as a whole has been, until recently, one of the world's most isolated.

South Asian Poverty

One of the clearest measures of human well-being is nutrition, and by this score South Asia ranks very low indeed. Probably nowhere else can one find so many undernourished people. More than 300 million Indian citizens live below their country's official poverty line, which is set at a very low level. Bangladesh is poorer still. This country is so impoverished, and has so many environmental problems, that some critics recently regarded it as almost hopeless. According to World Bank statistics, 67 percent of Bangladeshi children are underweight—the highest figure in the world. By measures such as infant mortality and average longevity, Nepal and Bhutan are in even worse condition (Figure 12.29).

Despite such deep and widespread poverty, South Asia should not be regarded as a zone of misery. India especially has a large and growing middle class, as well as a small but wealthy upper class. Roughly 100 million Indians are able to purchase such modern goods as televisions, motor scooters, and washing machines. This is a large market, and it has begun to interest corporate executives worldwide. India's economy has grown since the 1950s at a moderate but accelerating pace, and certain places and sectors are now booming. But if several Indian states have shown marked economic progress, others have seen little economic development. Similarly, some parts of South Asia have made impressive gains in social well-being, but others have made only modest improvements.

FIGURE 12.29 • FOOD AID   Partly because of the large population, food problems have long plagued South Asia. In this photo, military personnel dispense wheat mix and dried skimmed milk to people in Nepal as aid against a famine. *(Jeremy Hartley/Panos Pictures)*

Geographies of Economic Development

After independence, the governments of South Asia attempted to build new economic systems that would benefit their own people rather than foreign countries or corporations. Planners initially stressed heavy industry and economic self-sufficiency. While some major gains were realized, the overall pace of development remained slow. Since the 1990s, governments in the region, and especially that of India, have gradually opened their economies to the global economic system. In the process, core areas of development and social progress have emerged, surrounded by large peripheral areas that have lagged behind.

The Himalayan Countries   Both Nepal and Bhutan are disadvantaged by their rugged terrain and remote locations and by the fact that they have been relatively isolated from modern technology and infrastructure. But such measurements are somewhat misleading, especially for Bhutan, since many areas in the Himalayas are still largely subsistence-oriented and are not full participants in market economies.

Bhutan has purposely remained virtually disconnected from the modern world economy, and its small population lives in a relatively pristine natural environment. Indeed, Bhutan is so isolationist that it has only recently allowed tourists to enter—provided that they spend at least $165 a day. Nepal, on the other hand, is more heavily populated and suffers much more severe environmental degradation. It is also more closely integrated with the Indian economy. Nepal also relies heavily on international tourism, which has brought some prosperity to a few favored places, but often at the cost of ecological damage.

Bangladesh   By most measurements, Bangladesh is the poorest South Asian country. Environmental degradation and colonialism have contributed to Bangladesh's poverty, as did the partition of 1947. Most of pre-partition Bengal's businesses were located in the western area, which went to India. Bangladesh has also suffered because of its agricultural emphasis on jute, a plant that yields tough fibers useful for making ropes and burlap bags. Bangladesh failed to discover any major alternative export crops as synthetic materials undercut the global jute market.

Not all of the economic news coming from Bangladesh is negative. The country is internationally competitive in textile and clothing manufacture, in part because its wage rate is so low. Low-interest credit provided by the internationally acclaimed Grameen Bank has given hope to many poor women in Bangladesh, allowing the emergence of a number of healthy small-scale enterprises (Figure 12.30). By 2000, with the country's birthrate steadily falling, Bangladesh's economy was finally beginning to grow substantially faster than its population.

Pakistan   Pakistan also suffered the effects of partition in 1947. But by most economic measurements, Pakistan maintained for several decades a more productive economy than India. The country has a strong agricultural sector, as it shares the fertile Punjab with India. Pakistan also boasts a large textile industry, based in part on its huge cotton crop.

Pakistan's economy, however, is less dynamic than that of India, with a lower potential for growth. Part of the problem is that Pakistan is burdened by very high levels of defense spending. Additionally, a small but powerful landlord class that pays virtually no taxes to the central government controls much of its best agricultural lands. Unlike India, moreover, Pakistan has not been able to develop a successful high tech industry.

Sri Lanka and the Maldives   Sri Lanka's economy is one of the most highly developed in South Asia. Its exports are concentrated in textiles and agricultural products such as rubber and tea. By global standards, however, it is still a very poor country. Its progress, moreover, has been undercut by its ongoing civil war. If it did not suffer from this conflict, Sri Lanka would benefit more from the prime location

FIGURE 12.30 • GRAMEEN BANK   This innovative institution loans money to rural women so they can buy land, purchase homes, or start cottage industries. In this photo, taken in Bangladesh, women proudly repay their loans to a bank official as testimony to their success. *(John Van Hasselt/Corbis/Sygma)*

of the port of Colombo and from its high levels of education. The Maldives is the most prosperous South Asian country based on per capita economic output, but its total economy, like its population, is very small. Most of its revenues are gained from fishing and international tourism.

India's Less Developed Areas India's economy, like its population, dwarfs those of the other South Asian countries. While India's per capita GNI is roughly the same as that of Pakistan, its total economy is more than five times larger. As the region's largest country, India also has far more internal variation in economic development. The most basic economic division is that between India's more prosperous southern and western areas and its poorer districts in the north and east.

India's least developed, and most corrupt, state is Bihar, located in the lower Ganges Valley. Neighboring Uttar Pradesh, India's most populous state, is also extremely poor. Like Bihar, it is densely populated and has experienced little industrial development. While both Bihar and Uttar Pradesh have fertile soils, their agricultural systems have not profited as much from the Green Revolution as have those of the Punjab. Both states are also noted for their conservative outlooks and caste tensions.

India's Centers of Economic Growth The northwestern Indian states of Punjab and Haryana are showcases of the Green Revolution. Their economies rest largely on agriculture, but investments have been recently made in food processing and other industries. On Haryana's eastern border lies the capital district of New Delhi. India's political power and much of its wealth are concentrated here.

The west-central states of Gujarat and Maharashtra are noted for their industrial and financial power, as well as for their agricultural productivity. Gujarat was one of the first parts of South Asia to experience industrialization, and its textile mills are still among the most productive in the region. Gujaratis are well known as merchants and overseas traders, and they are heavily represented in the **Indian diaspora**, the migration of large numbers of Indians to foreign countries. As a result, cash remittances sent home from these emigrants help to strengthen the state's economy.

The state of Maharashtra is usually viewed as India's economic pacesetter. The huge city of Mumbai (Bombay) has historically been the financial center and media capital of India; Mumbai's port, moreover, carries some 25 percent of India's total foreign trade (Figure 12.31). Major industrial zones are located around Mumbai and in several other parts of Maharashtra. In recent years, Maharashtra's economy has grown more quickly than those of most other Indian states, reinforcing its primacy.

The center of India's fast-growing high-technology sector lies farther to the south, especially in Karnataka's capital of Bangalore. The Indian government selected the upland Bangalore area, which is noted for its pleasant climate, for technological investments in the 1950s. Other businesses soon followed. In the 1980s and 1990s, a quickly growing computer software and hardware industry emerged, earning Bangalore the label of "Silicon Plateau" (Figure 12.32). By 2003, large numbers of American software, accounting, and data-processing jobs were being transferred, or "outsourced," to Bangalore and other India cities, generating much criticism in the United States. At the same time, however, the U.S. government pledged to increase its technological ties with India.

India has proved especially competitive in software because software development does not require a sophisticated infrastructure. Computer code can be exported by way of wireless telecommunication systems without the use of modern roads or port facilities. What is necessary, of course, is technical talent, and this India

FIGURE 12.31 • MUMBAI (BOMBAY) STOCK EXCHANGE
Evidence of this city's central role in the globalizing South Asia economy is the stock exchange, upper right, located in the heart of the city. (Rob Crandall/Rob Crandall, Photographer)

FIGURE 12.32 • INDIA'S SILICON PLATEAU Since a significant proportion of its population is extremely well educated, India is suited for high-tech jobs both in computer assembly and manufacture and software development. Many of California's Silicon Valley firms draw upon facilities in India, which is 12 hours away in time zones, to run nonstop operations. (Chris Stowers/Panos Pictures)

has in great abundance. Many Indian social groups have long been highly committed to education, and India has been a major scientific power for decades. With the growth of the software industry, India's brain power has finally begun to create economic gains. Whether such developments can spread benefits beyond the rather small high-tech areas they presently occupy remains to be seen.

Globalization and India's Economic Future

South Asia is one of the world's least globalized regions. The volume of foreign trade is not large; foreign direct investment is still relatively small; and (with the exception of the Maldives) international tourists are few. But globalization is advancing rapidly, especially in India.

To understand the low level of globalization in the region, it is necessary to look at its recent economic history. After independence, India's economic policy was based on widespread private ownership combined with governmental control of planning, resource allocation, and certain heavy industrial sectors. India also established high trade barriers to protect its economy from global competition. This mixed socialist-capitalist system brought a fairly rapid development of heavy industry and allowed India to become nearly self-sufficient, even in the most technologically sophisticated goods.

By the 1980s, however, problems with this model were becoming apparent, and frustration with India's economic progress was growing among business and political leaders. Economic growth was only slightly faster than population growth. The percentage of Indians living below the poverty line, moreover, remained almost constant. At the same time, countries such as China and Thailand were experiencing much faster development after opening their economies to globalization. Many Indian businesspeople also disliked the governmental regulations that they believed undercut their ability to expand.

In response to these difficulties, the Indian government began to open up its economy in the early 1990s. Many regulations were modified and some were eliminated, and the economy was gradually opened to imports and international businesses. Other South Asian countries have followed a somewhat similar path. Pakistan, for example, began to privatize many of its state-owned industries in 1994.

This gradual internationalization and deregulation of the Indian economy has generated substantial opposition. Foreign competitors are now seriously challenging domestic firms. Cheap manufactured goods from China are seen as an especially serious threat. India's future economic policies thus remain uncertain. Between 2000 and 2003, the Indian economy was growing at roughly 6 percent a year, much faster than it had in the 1950s, 1960s, and 1970s, but still short of the 8 percent that is generally considered necessary to rapidly reduce poverty. And while globalization and market-oriented economies often generate faster economic growth, they also bring heightened insecurity and instability.

Social Development

South Asia has relatively low levels of health and education, which is hardly surprising considering its poverty. Levels of social well-being vary greatly across the region. As might be expected, people in the wealthier areas of western India are healthier, live longer, and are better educated, on average, than people in the poorer areas, such as the lower Ganges Valley. Bihar thus stands at the bottom of most social-development rankings, while Punjab, Gujarat, and Maharashtra stand near the top. Several key measurements of social welfare, moreover, are slightly higher in India than in Pakistan. Pakistan has done a particularly poor job of educating its people, which is one reason why extreme fundamentalist Islamic organizations have been able to recruit so effectively in much of the country. On the other hand, homelessness and malnutrition are not nearly as widespread in Pakistan as they are in India.

Several oddities stand out when one compares South Asia's map of economic development with its map of social well-being. Portions of India's extreme northeast, for example, show relatively high literacy rates despite their poverty; this is because of the educational efforts of Christian missionaries. In regard to health, longevity, and education, however, the south outpaces all other parts of the region. This is especially notable in the case of the Indian state of Kerala, which is economically troubled yet quite successful in terms of social development.

The Educated South   Southern South Asia's relatively high levels of social welfare are clearly visible when one examines Sri Lanka. Considering its poor economy and endless civil war, Sri Lanka must be considered a great success in terms of social development. Sri Lanka's average longevity of 72 years stands in favorable comparison with many of the world's industrialized countries, as does its literacy rate. Equally impressive is the fact that Sri Lanka's fertility rate has been reduced to the replacement level. The Sri Lankan government has achieved these results through universal primary education and inexpensive medical clinics.

On the mainland, Kerala in southwestern India has achieved even more impressive results. Kerala is not a prosperous state for it is extremely crowded and has long had some difficulty feeding its population. Its overall economic figures are roughly average for India. Kerala's level of social development, however, is the best in India and is comparable to that of Sri Lanka (Figure 12.33). Since Kerala is poorer than Sri Lanka, its social accomplishments must be considered all the more impressive.

Some observers attribute Kerala's social successes to its state policies. Kerala has long been led by a socialist party that has stressed education and community health care. While this has no doubt been an important factor, it does not seem to offer a complete explanation. Some researchers suggest that one of the key factors is the relatively high social position of women in Kerala.

The Status of Women   It is often said that South Asian women have a very low social position in both the Hindu and Muslim traditions. Throughout northern India, women traditionally leave their own families shortly after puberty to join those of their husbands. As outsiders, often in distant villages, young brides have few opportunities. Several indicators show that women in the Indus-Ganges basin suffer much discrimination. In Pakistan, Bangladesh, and such Indian states as Rajasthan, Bihar, and Uttar Pradesh, female levels of literacy are far lower than those of males. An even more disturbing statistic is that of gender ratios, the relative proportion of males and females in the population. All things being equal, there should be slightly more women than men in any population, since women generally have a longer life expectancy. Northern South Asia, however, contains many more men than women.

An imbalance of males over females often results from differences in care. In poor families of the region, boys typically receive better nutrition and medical care than do girls, which results in higher rates of survival. Economics play a major role. In rural households, boys are usually viewed as a blessing, since they typically remain with their families and work for its well-being. In the poorest groups, elderly people (especially widows) subsist largely on what their sons can provide. Girls, on the other hand, marry out of their families at an early age and must be provided with a dowry. They are thus seen as an economic liability. Considering the desperation of the South Asian poor, it is perhaps not surprising to see such unbalanced gender ratios.

Much evidence suggests that the social position of women is improving, especially in the more prosperous parts of western India where employment opportunities outside the family are emerging. But even in many of the region's middle-class households, women still experience discrimination. Indeed, dowry demands seem to be increasing in some areas, and there have been a number of well-publicized murders of young brides whose families failed to deliver an adequate dowry.

While the social bias against women across northern South Asia is striking, it is much less evident in southern India and Sri Lanka. In Kerala especially, women have relatively high status, regardless of whether they are Hindus, Muslims, or Christians. Here the gender ratio shows the normal pattern. Female literacy is very high in Kerala, which is one reason why the state's overall illiteracy rate is so low. Kerala's fertility rate is the lowest in India, which is another sign of women's social power.

FIGURE 12.33 • EDUCATION IN KERALA   India's southwestern state of Kerala, which has virtually eliminated illiteracy, is South Asia's most highly educated region. It also has the lowest fertility rate in South Asia. Because of this, many argue that women's education and empowerment is the best and most enduring form of contraception. *(Rob Crandall/Rob Crandall, Photographer)*

CONCLUSION

South Asia, a large and complex area of more than a billion people, has in many ways been overshadowed by neighboring world regions: by the economic growth and instability of Southeast Asia, the size and political weight of East Asia, and the geopolitical tensions of Southwest Asia. Much of that is changing, however, for South Asia now figures prominently in discussions of world problems and issues.

Geopolitical tensions in South Asia, both between countries and within them, remain worrisome. Because of South Asia's complex cultural geography, shaped by peoples speaking several dozen languages and following four major religions, cultural differences often result in geopolitical conflict. The long-standing feud between Pakistan and India escalated dangerously in the late 1990s with threats to use weapons of mass destruction.

Internal political tension is another cause for concern. Sri Lanka's ethnic conflict continues well into its second decade. Ethnic and religious fighting also periodically rages in the streets of Pakistan's largest city, Karachi. More than in most Muslim countries, in Pakistan an extremely fundamentalist Islamic movement challenges all aspects of the state and secular society. In India as well, internal politics sometimes result in violence. Religious strife between Muslims and Hindus continues; Hindu radicals attack Christian missionaries; Sikhs clash with Hindus in the Punjab. Hindu fundamentalist politicians—sometimes armed with funds raised by the Indian diaspora in the United States—work to impose their agenda on education and the arts.

Bangladesh, a country that international aid workers recently viewed with great pessimism because of its huge population, hazardous lowland environment, and limited resources, may actually have found some relief through economic globalization. As has been demonstrated many times, international industry will locate in countries with a large, low-paid labor force and a stable political environment. Currently, Bangladesh meets those needs. How long this will last and how beneficial it may actually prove remain to be seen.

Many argue that India is perfectly positioned to take advantage of economic globalization. Large segments of its huge labor force are well educated and speak excellent English, the major language of global commerce. But will these global connections help the vast numbers of India's poor or merely the small number of its economic elite? Can South Asia be saved from food shortages and ecological degradation as population growth threatens to outrun the resource base? As with all complicated world problems, the answers vary. Advocates of free markets and globalization tend to see a bright future, while environmentalists more often see growing problems. Perhaps it is fitting to conclude with the Hindu proverb: To make the gods laugh, attempt to predict the future.

KEY TERMS

British East India Company (page 334)
bustees (page 324)
caste system (page 326)
dalit (page 326)
Dravidian language (page 328)

federal state (page 335)
Green Revolution (page 323)
Hindi (page 329)
Hindu nationalism (page 325)
Indian diaspora (page 339)
Jainism (page 328)

linguistic nationalism (page 330)
maharaja (page 335)
monsoon (page 319)
Mughal Empire (also spelled *Mogul*) (page 326)
orographic rainfall (page 319)

rainshadow effect (page 319)
salinization (page 323)
Sanskrit (page 326)
Sikhism (page 327)
subcontinent (page 315)
Urdu (page 329)

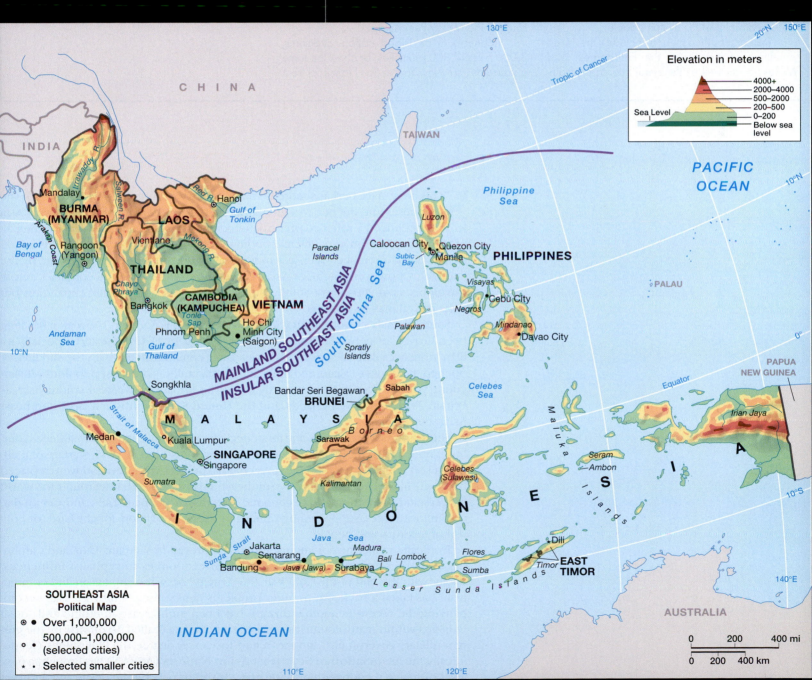

FIGURE 13.1   Southeast Asia

This region includes the large peninsula in the southeastern corner of Asia as well as a large number of islands scattered to the south and east. It is commonly divided into two subregions: mainland Southeast Asia, which includes Burma (Myanmar), Thailand, Laos, Cambodia, and Vietnam, and insular (or island) Southeast Asia, which includes Indonesia, the Philippines, Malaysia, Brunei, Singapore, and East Timor. Malaysia consists of the tip of the mainland peninsula and most of the northern part of the island of Borneo.

Southeast Asia

Elevation in meters

Sea Level

- 4000+
- 2000–4000
- 500–2000
- 200–500
- 0–200
- Below sea level

SOUTHEAST ASIA
Political Map

- ⊗ ● Over 1,000,000
- ○ ● 500,000–1,000,000 (selected cities)
- ✶ ● Selected smaller cities

0   200   400 mi

0   200   400 km

SETTING THE BOUNDARIES

The region of Southeast Asia traditionally consists of 10 countries that vary widely in spatial extent, population, cultural traits, and levels of economic and social development. With the recent emergence of East Timor as an independent state, the list must now be expanded to 11. Geographically, these countries are commonly divided into those on the Asian mainland and those on islands, or the insular realm. The mainland includes Burma (Myanmar), Thailand, Cambodia, Laos, and Vietnam. Although Burma is the largest in territory, Vietnam has the largest population of the mainland states, with almost 80 million people.

Insular Southeast Asia includes the large countries of Indonesia, the Philippines, and Malaysia, as well as the small countries of Singapore, Brunei, and East Timor. Although classified as part of the insular realm because of its cultural and historical background, Malaysia actually splits the difference between mainland and islands. Part of its national territory is on the mainland's Malay Peninsula and part is on the large island of Borneo, some 300 miles distant. Borneo also includes the Muslim sultanate of Brunei, a small but oil-rich island country of 300,000 people covering an area slightly larger than Rhode Island. Singapore is essentially a city-state, occupying a small island just to the south of the Malay Peninsula.

Indonesia is an island nation, stretching 3,000 miles (4,800 kilometers, or about the same distance as from New York to San Francisco) from the large island of Sumatra in the west to New Guinea in the east, and containing more than 13,000 separate islands. Not only does it dwarf all other Southeast Asian states in size, but it is by far the largest in population. With more than 200 million people, it is ranked as the world's fourth most populated country. Additionally, because most of its inhabitants are Muslim, Indonesia is also counted as the world's largest Islamic nation. Lying north of the equator is the Philippines, a country of 77 million people spread over some 7,000 islands, both large and small.

Southeast Asia, perhaps more than any other world region, illustrates both the promises and the perils of globalization. In the 1990s, much of Southeast Asia experienced a roller-coaster ride of economic boom and bust, descending almost overnight from a spectacular boom to a severe recession. In the 1980s Southeast Asia had often been seen as a model for world economic development. By the late 1990s, however, Southeast Asia's economic troubles affected countries across the globe. As of 2004, most Southeast Asian economies were again expanding, but their revival may be unstable.

Southeast Asia's involvement with the larger world is not new. The region has long been heavily influenced by external forces (Figure 13.1). Chinese and especially Indian influences date back many centuries. Later, commercial ties with the Middle East opened the doors to Islam, and today Indonesia ranks as the most populous Muslim country in the world. More recently came the impact of Western colonialism, as Britain, France, the Netherlands, and the United States controlled large Southeast Asian colonies.

Southeast Asia's resources and its strategic location made it a major battlefield during World War II. Yet long after world peace was restored in 1945, warfare of a different sort continued in this region. As colonial powers withdrew and were replaced by newly independent countries, Southeast Asia became a battleground for world powers and their competing economic systems. In Vietnam, Laos, and Cambodia, communist forces, supported by China and the Soviet Union, fought hard for control of territory and people.

Ironically, while the communist forces did prevail in Vietnam, Laos, and Cambodia, all of these countries later opened their economies to capitalist influences. Today, with the end of the Cold War, the struggle between capitalism and communism has taken a back seat to other problems confronting Southeast Asia. Accompanying the recent economic turmoil has been an increase in ethnic and social tensions within many countries, most notably Burma (called *Myanmar* by the current military government), Indonesia, and, to a lesser degree, the Philippines. Geopolitically, the **Association of Southeast Asian Nations (ASEAN)** has brought a new level of regional cooperation to the area, driven in part by the desire of Southeast Asia's countries to control—rather than be controlled by—external global forces (see "Setting the Boundaries").

Environmental Geography: A Once-Forested Region

The mountainous area along the border between northern Thailand and Burma is a rugged place. Slopes are steep and thickly wooded in most places, rainfall is often heavy, leeches are plentiful, and several strains of medicine-resistant malaria are common. These same uplands are also home to the Karen, a distinctive ethnic group numbering roughly 7 million. Unfortunately, the story of the Karen and their homeland is not a happy one. Much of their territory has been overrun by the Burmese army, and many of the Karen have been forced into neglected refugee camps in Thailand. Furthermore, the Karen's loss of land has been accompanied by the loss of the magnificent and highly valuable teak forests of upland Burma. The struggle of these people illustrates the connections among cultural, political, economic, and environmental forces.

The Tragedy of the Karen

The Karen were never fully included in the Burmese kingdom, which long ruled the lowlands of the country. With the beginning of British colonial rule in the 1800s, however, the Karen territory was joined to the Burmese lowlands. British and American missionaries educated many Karen and converted roughly 30 percent of them to Protestant Christianity. A number of Karen Christians obtained positions in Burma's colonial government. The Burmans of the lowlands, a strongly Buddhist people, resented this deeply, for they had long viewed the Karen as culturally inferior (according to conventional but confusing terminology, the term *Burmese* refers to all of the inhabitants of Burma, whereas *Burmans* refers only to the country's dominant, Burmese-speaking ethnic group). After independence, the Karen lost their favored position and soon grew to resent what they saw as Burman cultural and economic control.

By the 1970s the Karen were in open rebellion and soon managed to establish an semi-independent state of their own (Figure 13.2). They supported their rebellion by smuggling goods between Thailand and Burma, including rubies and sapphires. The Burmese army, however, began to make headway against the rebels in the early 1990s and by the end of the decade had overrun most of the Karen territory. Although some Karen continue to fight, others have made peace with the Burmese government. Crucial to Burma's success was an agreement made with Thailand to prevent Karen soldiers from finding refuge on the Thai side of the border. This agreement was reached in part so that Thai timber interests could have access to Burma's valuable teak forests.

The Deforestation of Southeast Asia

The story of Burma's teak forests is unique, but unfortunately deforestation and related environmental problems are major issues throughout most of Southeast Asia (Figure 13.3). Globalization has had a particularly important effect on the Southeast Asian environment. Although countries such as Indonesia look to their forest lands for increasing local food supplies through expanded agriculture and for relieving population pressure in their more densely settled areas, agriculture and population growth are usually not the main cause of deforestation. Most forests are cut so that the wood products can be exported to other parts of the world (Figure 13.4). Export-oriented logging companies have reached deep into the region's forests, cutting trees, damaging watersheds, destroying wildlife habitat, and dispersing vast quantities of atmospheric pollutants as cutover lands are burned. This forest burning aggravates often intolerable air pollution within the rapidly growing major cities of Southeast Asia. Pollution levels in Jakarta, for example, are almost three times higher than World Health Organization guidelines specify.

Although colonial powers cut Southeast Asian forests for tropical hardwoods and naval supplies, and local peoples have historically cleared small areas of forest for

FIGURE 13.2 • KAREN REBELS The Karen people have been in rebellion against Burma (Myanmar) since the 1970s. For several years they maintained a semi-independent state, with its own capital city and regular army. The Burmese military advanced in the 1990s, however, forcing the Karen back to a guerrilla-style war. *(Dean Chapman/Panos Pictures)*

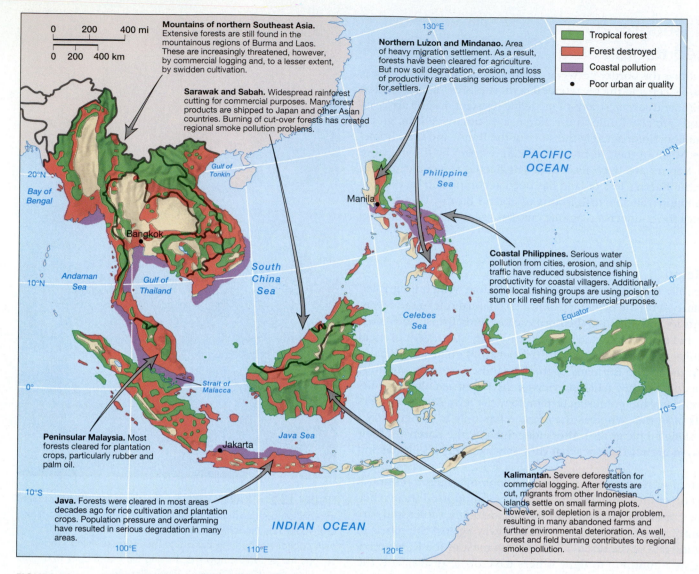

FIGURE 13.3 • ENVIRONMENTAL ISSUES IN SOUTHEAST ASIA Southeast Asia was once one of the most heavily forested regions of the world. Most of the tropical forests of Thailand, the Philippines, peninsular Malaysia, Sumatra, and Java, however, have been destroyed by a combination of commercial logging and agricultural settlement. The forests of Kalimantan (Borneo), Burma (Myanmar), Laos, and Vietnam, moreover, are now being rapidly cleared. Water and urban air pollution, as well as soil erosion, are also widespread in Southeast Asia.

agricultural use, widespread deforestation has come only in the last several decades with large-scale international commercial logging. This activity is largely driven by Asia's large appetite for wood products such as plywood and paper pulp. The demand for paper pulp alone has increased at about 6 percent per year over the last decade throughout Asia. The costs associated with commercial logging are high for rural peoples who rely on forest resources for their traditional way of life. In addition, the global effects of tropical forest clearing are also becoming increasingly problematic as more evidence links forest cutting to atmospheric warming and air pollution.

Malaysia has long been one of the leading exporters of tropical hardwoods from Southeast Asia. In recent decades, 60 percent of these log exports went to Japan. The cost to the environment has been high. Peninsular Malaysia was largely deforested by 1985, when a cutting ban was imposed. Since then, forest cutting has been concentrated in the states of Sarawak and Sabah on the island of Borneo, where the granting of logging concessions to Malaysian and foreign firms has caused considerable problems with local tribal people by disrupting their traditional resource base.

Thailand cut more than 50 percent of its forests between 1960 and 1980. This loss was followed by a series of logging bans so that by 1995 forest cutting was no longer legal. Damage to the landscape, however, was severe; flooding increased in lowland areas, and severe erosion on hill slopes led to problems such as the accumulation of silt in irrigation works and hydroelectric facilities. Increasingly, these cutover lands are being reforested with fast-growing Australian eucalyptus trees. Eucalyptus forests, however, are not native to the area and thus cannot support the local wildlife or maintain the area's biological diversity.

Indonesia, the largest country in Southeast Asia, has fully two-thirds of the region's forest area, including about 10 percent of the world's true tropical rain forests. Most of Sumatra's forests have been cut, however, and those of Kalimantan (Borneo) will not last long at present rates of cutting. Indonesia's last forestry frontier is on the island of New Guinea, where forests are still extensive.

Smoke and Air Pollution

Until recently, most of Southeast Asia's residents seemed unconcerned about the widespread air pollution created by a combination of urban smog and the smoke from tropical forest clearing. Then, late in the 1990s, the region suffered from two consecutive years of disastrous air pollution that served as a wake-up call (Figure 13.5). Because of global publicity, billions of tourism dollars were lost as potential visitors decided to travel to other parts of the world.

Several factors, both natural and economic, combined to produce the region's terrible air pollution problems. First, large portions of insular (or island) Southeast Asia suffered a severe drought caused by El Niño (discussed in Chapter 4), turning the normally wet tropical forests into tinderboxes. This drought also dried out the widespread peat bogs of coastal Kalimantan, which continued to burn for months after the fires began. Second, commercial forest cutting was—and is—responsible for many fires, as the leftover slash (branches, small trees, etc.) is often burned in order to clear out the land.

The third factor adding to the air pollution problem is Southeast Asia's rapidly growing cities, where cars, trucks, and factories emit huge quantities of pollutants. Regardless of forest burning, many cities, such as Bangkok, Jakarta, and Manila, already have unhealthy levels of pollution. Given the increase in auto traffic and other sources of urban pollution coupled with the increasing presence of smoke from forest fires, this unhealthy situation is likely to grow worse before any decisive action is taken.

Patterns of Physical Geography

Primarily two types of forest are of interest for commercial logging in Southeast Asia. The southern portion of the region—known as insular Southeast Asia—is one of the world's three main zones of tropical rain forest (the others are found in Latin America and Sub-Saharan Africa). The northern part of the region, known as mainland Southeast Asia, is located in the tropical wet-and-dry zone that is noted for particularly valuable timber species, such as teak. The distribution of these forest types is closely linked to the landforms and climates of these two distinct areas of the region.

Mainland Environments   Mainland Southeast Asia is an area of rugged uplands mixed together with broad lowlands associated with large rivers (Figure 13.6). The region's northern boundary lies in a cluster of mountains connected to the highlands of western Tibet and south-central China. In the far north of Burma (Myanmar), peaks reach 18,000 feet (5,500 meters). From this point, a series of distinct mountain ranges spreads out, extending through western Burma, along the Burma–Thailand border, and through Laos into southern Vietnam.

Several large rivers flow southward out of Tibet and its adjacent highlands into mainland Southeast Asia. The valleys and deltas of these rivers are the centers of both population and agriculture in mainland Southeast Asia. The longest river is the Mekong, which flows through Laos and Thailand, then across Cambodia before entering the South China Sea through a large delta in southern Vietnam. Second longest is the Irrawaddy at about 1,300 miles (2,100 kilometers), which flows through Burma's central plain before reaching the Bay of Bengal. This river also has a large delta. Two smaller rivers are equally significant: the Red River, which

FIGURE 13.4 • COMMERCIAL LOGGING   Southeast Asia has historically been the world's most important supplier of tropical hardwoods. Unfortunately, most of the tropical forests of the Philippines and Thailand, as well as the Indonesian islands of Java and Sumatra, have been destroyed by the logging process. *(Jean-Leo Dugast/Panos Pictures)*

FIGURE 13.5 • URBAN AIR POLLUTION   Air pollution has reached a crisis stage in the rapidly industrializing cities of Southeast Asia, particularly in Bangkok, Manila, and Jakarta. People sometimes resort to using face masks to filter out soot and other forms of particulate matter. Forest fires, which often follow logging, increase the problem. *(V. Miladinovic/Corbis/Sygma)*

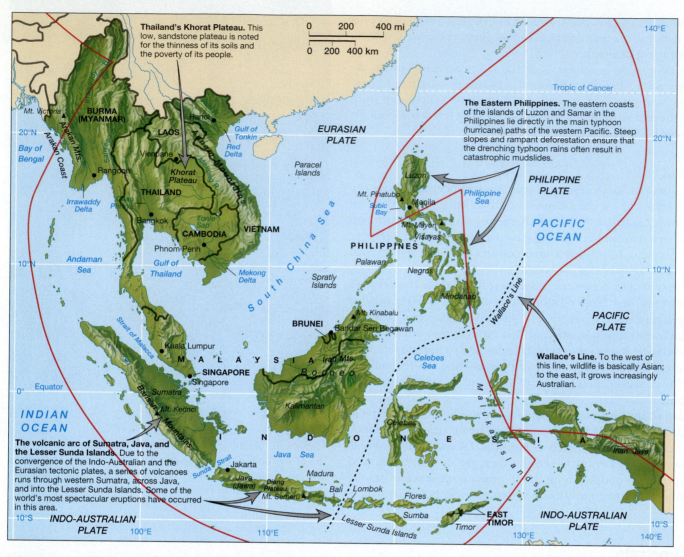

Thailand's Khorat Plateau. This low, sandstone plateau is noted for the thinness of its soils and the poverty of its people.

The Eastern Philippines. The eastern coasts of the islands of Luzon and Samar in the Philippines lie directly in the main typhoon (hurricane) paths of the western Pacific. Steep slopes and rampant deforestation ensure that the drenching typhoon rains often result in catastrophic mudslides.

Wallace's Line. To the west of this line, wildlife is basically Asian; to the east, it grows increasingly Australian.

The volcanic arc of Sumatra, Java, and the Lesser Sunda Islands. Due to the convergence of the Indo-Australian and the Eurasian tectonic plates, a series of volcanoes runs through western Sumatra, across Java, and into the Lesser Sunda Islands. Some of the world's most spectacular eruptions have occurred in this area.

FIGURE 13.6 • PHYSICAL GEOGRAPHY OF SOUTHEAST ASIA Southeast Asia is one of the world's most geologically active regions and includes a number of active volcanoes. Several mountain ranges radiate out from the rugged uplands of northern Southeast Asia, dividing the area's broad river valleys and deltas into several distinct physical regions.

forms a large and heavily settled delta in northern Vietnam, and the Chao Phraya, which has created the fertile alluvial plain of central Thailand (Figure 13.7).

The centermost area of mainland Southeast Asia is Thailand's Khorat Plateau, which is neither a rugged upland nor a fertile river valley. This low sandstone plateau averages about 500 feet (175 meters) in height and is noted for its thin, poor soils. Water shortages and droughts make settlement and agriculture difficult in this extensive area.

The Influence of the Monsoon Almost all of mainland Southeast Asia is affected by a pattern of seasonally shifting winds known as the *monsoon*. The climate of this area is characterized by a distinct hot and rainy season from May to October. This is followed by dry but still generally hot conditions from November to April (Figure 13.8). Only the central highlands of Vietnam and a few coastal areas receive significant rainfall during this period. In the far north, the winter months bring mild and sometimes rather cool weather.

FIGURE 13.7 • DELTA LANDSCAPE Southeast Asia has some of the world's largest delta landscapes. These environments are used for intensive irrigated rice cultivation, and support very high rural population densities. Delta wetlands are also used for aquaculture (fish farming) and other forms of intensive food production. Most of mainland Southeast Asia's large cities are located in delta areas, resulting in periodic flooding and other environmental problems. (John Elk III/Stock Boston)

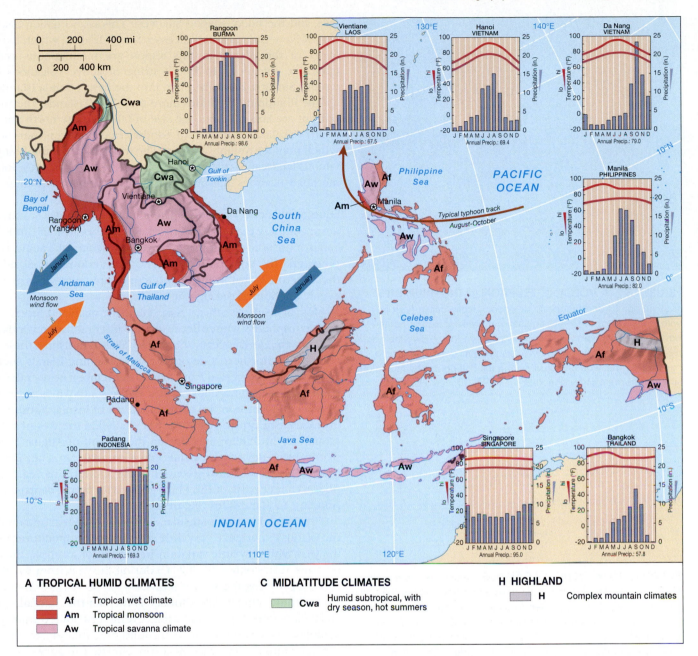

FIGURE 13.8 • CLIMATE MAP OF SOUTHEAST ASIA   Most of insular Southeast Asia is characterized by the constantly hot and humid climates of the equatorial zone. Mainland Southeast Asia, on the other hand, has the seasonally wet and dry climates of the tropical monsoon and tropical savanna types. Only in the far north are subtropical climates, with relatively cool winters, encountered. The northern half of the region is strongly influenced by the seasonally shifting monsoon winds. Northeastern Southeast Asia—and especially the Philippines—often experiences typhoons from August to October.

Two types of tropical climates are dominant in mainland Southeast Asia. Although both are affected by the monsoon, these two climates differ in the total amount of rainfall received during the year. Along the coasts and in the highlands, the tropical monsoon climate (Am) dominates. Rainfall totals for this climate are large, usually more than 100 inches (254 centimeters) each year. The greater portion of the mainland region falls into the tropical savanna (Aw) climate type, where annual rainfall totals are about half those of the Am rainforest region.

The original vegetation of most of the mainland was forest. Because of the dry winters caused by the monsoon, this forest was not as dense or as diverse as the equatorial rain forests in insular Southeast Asia. But until the late nineteenth century, this was one of the most heavily wooded portions of the world. Mainland forests include several valuable tree species, of which teak is the most important.

Insular Environments   The main feature of insular Southeast Asia is its island environment. Indeed, this is a region of countless islands of all sizes and shapes. Indonesia alone is said to contain more than 13,000 islands, dominated by the four great landmasses of Sumatra, Borneo (the Indonesian portion is called Kalimantan),

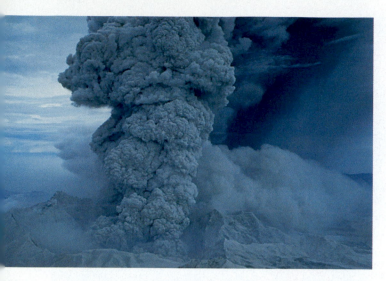

FIGURE 13.9 • VOLCANIC ERUPTION IN THE PHILIPPINES
The Philippines and Indonesia are geologically active regions with many volcanoes. Although several areas of insular Southeast Asia—most notably central and eastern Java—owe their fertile soils to past eruptions, volcanism remains a major threat to human life. This was demonstrated in the Philippines by the catastrophic eruption of Mt. Pinatubo in 1991. *(Durieux/SIPA Press)*

FIGURE 13.10 • TYPHOON DAMAGE Typhoons are a major environmental threat in the Philippines and parts of mainland Southeast Asia. Typhoons often result in wind damage, but flooding and mudslides cause the most destruction. Deforestation and associated soil erosion over the past 30 years have greatly increased the magnitude of floods and mudslides. *(Vogel/Getty Images, Inc.—Liaison)*

Java, and the oddly shaped Sulawesi. This island nation also includes the western half of New Guinea and the Lesser Sunda Islands, which extend to the east of Java. Many of the islands in this region are mountainous and include large, often explosive volcanoes. In fact, a string of active volcanoes extends the length of eastern Sumatra across Java and into the Lesser Sunda Islands.

The Philippines includes more than 7,000 islands. The two largest and most important are Luzon (about the size of Ohio) in the north and Mindanao (the size of South Carolina) in the south. Sandwiched between them are the Visayan Islands, which number roughly a dozen. The topography of the Philippines includes mountainous landscapes that reach elevations of 10,000 feet (3,500 meters), as well as numerous volcanoes that present hazards to human settlement (Figure 13.9).

Equatorial Island Climates The climates of insular Southeast Asia are somewhat more complex than those of the mainland because of three factors: a more complicated monsoon effect, the stronger influence of Pacific typhoons in the Philippines, and the equatorial location of the Indonesian islands. The overall effect of these factors is that in this island region precipitation is generally higher and more evenly distributed throughout the year.

Most of insular Southeast Asia, unlike the mainland, receives rain during the Northern Hemisphere's winter. Because the winter monsoon winds blow across large areas of warm equatorial ocean, where they absorb moisture, these winds can bring heavy rains to the northern side of east–west running islands such as Sumatra and Java. Then, during the period from May to September, heavy rains occur on the southern flanks of these same islands because of the south winds associated with the monsoon.

A second factor in the climate pattern of insular Southeast Asia are the tropical hurricanes, or **typhoons** as they are called in the western Pacific, that bring heavy rainfall to the Philippines during the period of August to October. These strong storms bring devastating winds and torrential rain. They develop east of the Philippines and then move westward into the South China Sea. Each year a number of typhoons affect parts of the Philippines with heavy damage and loss of life through flooding and landslides (Figure 13.10).

Finally, the location of insular Southeast Asia near and on the equator means that the islands experience very little seasonality. Temperatures remain high throughout the year, with very little variation. In Jakarta, for example, the average daily high temperature varies only 4 °F (2.2 °C) during the year, from 84 °F (29 °C) in January and February (during the winter monsoon) to 88 °F (31 °C) in September. Also associated with the equatorial influence is the fact that rainfall in most of Indonesia and Malaysia is both higher and more evenly distributed during the year than on the mainland. As a result of this year-round precipitation, most of island Southeast Asia is placed into the Af, or tropical wet, climate category.

Population and Settlement: Subsistence, Migration, Cities

The scale of Southeast Asia's population issue is quite different from those of its giant neighbors, China and India. With just over 500 million people, Southeast Asia is still *relatively* sparsely settled. Part of the reason can be attributed to extensive tracts of infertile soil and rugged mountains. Such areas generally remain thinly inhabited. In contrast, relatively dense populations are found in the region's deltas, coastal areas, and zones of fertile volcanic soil (Figure 13.11).

Many of the favored lowlands of Southeast Asia have experienced huge population growth over the past several decades. Demographic growth and family planning have thus become increasingly important issues through much of the region. Different Southeast Asian countries have responded to their demographic situations in very different ways. While Indonesia has promoted migration to its outer islands as a way to relieve population pressure in the core areas, Thailand has placed considerable emphasis on family planning.

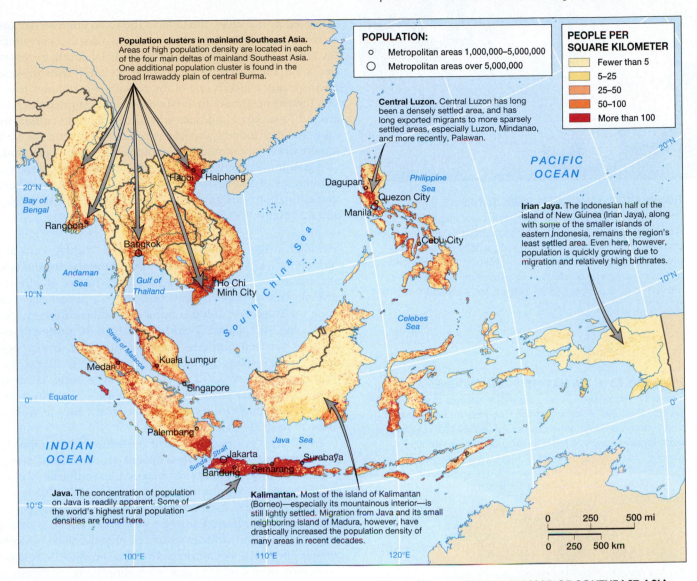

Population clusters in mainland Southeast Asia. Areas of high population density are located in each of the four main deltas of mainland Southeast Asia. One additional population cluster is found in the broad Irrawaddy plain of central Burma.

Central Luzon. Central Luzon has long been a densely settled area, and has long exported migrants to more sparsely settled areas, especially Luzon, Mindanao, and more recently, Palawan.

Irian Jaya. The Indonesian half of the island of New Guinea (Irian Jaya), along with some of the smaller islands of eastern Indonesia, remains the region's least settled area. Even here, however, population is quickly growing due to migration and relatively high birthrates.

POPULATION:
○ Metropolitan areas 1,000,000–5,000,000
○ Metropolitan areas over 5,000,000

PEOPLE PER SQUARE KILOMETER
Fewer than 5
5–25
25–50
50–100
More than 100

Java. The concentration of population on Java is readily apparent. Some of the world's highest rural population densities are found here.

Kalimantan. Most of the island of Kalimantan (Borneo)—especially its mountainous interior—is still lightly settled. Migration from Java and its small neighboring island of Madura, however, have drastically increased the population density of many areas in recent decades.

Settlement and Agriculture

Part of the reason for the historical lack of population in much of Southeast Asia is the infertile soil, which is unable to support intensive agriculture and high population densities. The island rain forests, though lush and biologically rich, grow on poor soils. Plant nutrients are locked up in the vegetation itself, rather than being stored in the soil where they would easily benefit agriculture. Furthermore, the constant rain of the equatorial zone tends to wash nutrients out of the soil. Agriculture must be carefully adapted to this limited soil fertility by constant field rotation or the application of heavy amounts of fertilizer.

There are, however, some notable exceptions to this generalization about soil fertility and settlement density in equatorial Southeast Asia. Unusually rich soils connected to volcanic activity are scattered through much of the region but are particularly widespread on the island of Java. With more than 50 volcanoes, Java is blessed with rich soils that support a range of tropical crops and a very high population density. The island is home to well over 100 million people—more than half the total population of Indonesia—in an area smaller than the state of Iowa. Dense populations are also found in pockets of fertile alluvial soils along the coasts of insular Southeast Asia, where people supplement land-based farming with fishing and trade activities.

The demographic patterns in mainland Southeast Asia are less complicated than those of the island realm. In all the mainland countries, population is concentrated in the agriculturally intensive valleys and deltas of the large rivers, whereas the uplands remain relatively lightly settled. The population core of Thailand, for example,

FIGURE 13.11 • POPULATION MAP OF SOUTHEAST ASIA
In mainland Southeast Asia, population is concentrated in the valleys and deltas of the region's large rivers. In the uplands, population density remains relatively low. In Indonesia, density is extremely high on Java, an island noted for its fertile soil and large cities. Some of Indonesia's outer islands, especially those of the east, remain lightly settled. Overall, population density is high in the Philippines, especially in central Luzon.

FIGURE 13.12 • SWIDDEN AGRICULTURE In the uplands of Southeast Asia, swidden (or "slash-and-burn") agriculture is widely practiced. When done by tribal peoples with low population densities, swidden is not environmentally harmful. When practiced by large numbers of immigrants from the lowlands, however, swidden can result in deforestation and extensive soil erosion. *(Paula Bronstein/Getty Images, Inc.—Liaison)*

FIGURE 13.13 • TEA HARVESTING IN INDONESIA Plantation crops, such as tea, are major sources of exports for several Southeast Asian countries. Coconut, rubber, palm oil, and coffee are other major cash crops. Many of these crops require large amounts of labor, particularly at harvest time. *(Dermot Tatlow/Panos Pictures)*

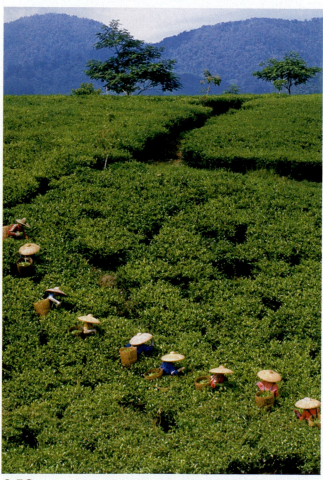

is formed by the valley and delta of the Chao Phraya River, just as Burma's is focused on the Irrawaddy. In Vietnam there are two distinct core areas: the Red River delta in the far north and the Mekong delta in the far south. In contrast to these densely settled areas, the middle reaches of the Mekong River provide only limited lowland areas in Laos, which is one of the reasons that country has a much smaller population than its neighbors. In Cambodia the largest population historically has clustered around Tonle Sap, a large lake with a very unusual seasonal flow reversal. During the rainy summer months, the lake receives water from the Mekong drainage, but during the drier winter months it contributes to the river's flow.

Agricultural practices and settlement forms vary widely across the complex environments of Southeast Asia. Generally speaking, however, three farming and settlement patterns are apparent: swidden in the upland areas, plantation agriculture, and rice cultivation in the lowlands.

Swidden in the Uplands Also known as shifting cultivation or "slash-and-burn" agriculture, swidden is practiced throughout the rugged uplands of both mainland and island Southeast Asia (Figure 13.12). In the **swidden** system, small plots of several acres of tropical forest or brush are periodically cut or "slashed" by hand. Then the fallen vegetation is burned to transfer nutrients to the soil before subsistence crops are planted. Yields remain high for several years, then drop off dramatically as the soil nutrients are exhausted and insect pests and plant diseases multiply. These plots are abandoned completely after a few years and allowed to return to woody vegetation. The cycle of cutting, burning, and planting is then moved to another small plot not far away—thus the term *shifting cultivation*. Families and small villages generally control a large amount of territory so they can rotate their fields on a regular basis.

Swidden is a sustainable form of agriculture when population densities remain relatively low and stable and when upland people control enough territory. Today, however, the swidden system is threatened in Southeast Asia for two reasons. First, it cannot easily support the increasing population that has resulted from relatively high human fertility and, in some cases, migration. With a higher population density, the swidden rotation period must be shortened, which is damaging to soil resources. Second, the upland swidden system is often damaged by commercial logging, which removes potential soil nutrients from the ecosystem as logs are exported. In some cases, moreover, swidden farmers are simply expelled from logging areas.

When swidden can no longer support the population, upland people often adapt by switching to a cash crop that will allow them to participate in the commercial economy. In the mountains of northern Southeast Asia, one of the main cash crops is opium, grown by local farmers for the global drug trade. This mountainous area is often called the **"Golden Triangle."** Burma is probably the world's second largest opium producer, after Afghanistan.

Plantation Agriculture With European colonization, Southeast Asia became a focus for commercial plantation agriculture, growing high-value specialty crops ranging from rice to rubber. Even in the nineteenth century, Southeast Asia was linked to a globalized economy through the plantation system. Forests were cleared and swamps drained to make room for these plantations; labor was supplied (often unwillingly) by native people or by laborers brought in from India or China. Over the years, with fluctuating world demand, competition from other world regions, and changing political conditions within the region, the assortment of plantation crops has changed (Figure 13.13).

Plantations are still an important part of Southeast Asia's geography and continue to play a major role in today's global economy. Most of the world's natural rubber, for example, is produced in Malaysia, Indonesia, and Thailand. Cane sugar has long been a plantation crop of the Philippines and parts of Indonesia. Further, Indonesia is the region's leading producer of tea, and Malaysia dominates the production of palm oil. Coconut oil and **copra** (dried coconut meat) are widely produced in the Philippines, Indonesia, and elsewhere.

Rice in the Lowlands   The lowland basins of mainland Southeast Asia are largely devoted to intensive rice cultivation. Throughout almost all of Southeast Asia, rice is the preferred staple food. Traditionally, rice was mainly cultivated on a subsistence basis by rural farmers. But as the number of wage laborers in Southeast Asia has grown due to economic development, so has the demand for commercial rice cultivation. Rice harvests are increasingly traded to meet the food needs of the region's expanding urban markets, and a large amount of rice is exported to the global system from Thailand. Three delta areas have been the focus for commercial rice cultivation: the Irrawaddy in Burma, the Chao Praya in Thailand, and the Mekong in Vietnam and Cambodia. The use of agricultural chemicals and high-yield crop varieties, along with improved water control and the expansion of irrigation facilities to permit dry-season cropping, have allowed production to keep pace with population growth. Such techniques have, however, also resulted in serious water pollution due to fertilizer runoff and pesticide contamination.

In those areas without irrigation and water control, yields remain relatively low. Rice growing on the Khorat Plateau, for example, depends largely on the uncertain rainfall, without the benefits of the more sophisticated water control methods available elsewhere in Thailand. In some lowland districts lacking irrigation, dry-field crops, especially sweet potatoes and manioc, form the staple foods of people too poor to buy market rice on a regular basis.

Recent Demographic Change

Because Southeast Asia is not facing the same kind of population pressure as East or South Asia, a wide range of government population policies exist. While several countries show concern about rapid growth and thus have strong family planning programs, others believe that their populations are too small. In those countries with rapid demographic expansion, internal relocation away from densely populated areas to outlying districts is a common policy.

Population Contrasts   The Philippines, the third most populous country in Southeast Asia, has a relatively high growth rate. Effective family planning here has been difficult to establish. When a popular democratic government replaced a dictatorship in the 1980s, the Philippine Roman Catholic Church, which played an active role in the peaceful revolution, pressured the new government to cut funding for family planning programs. As a result, many clinics and centers that had given out family planning information were closed. Although high birthrates are not always associated with Catholicism, the Church's outspoken stand against birth control gets in the way of the distribution of family planning information.

The highest total fertility rate (TFR) in Southeast Asia (5.4 children) is found in Laos, a country of Buddhist religious tradition. Here the high birthrate is best explained by the country's low level of economic and social development. Thailand, which shares cultural traditions with Laos yet is considerably more developed, demonstrates the other end of that range. Here the TFR has dropped dramatically within the last 30 years from 5.4 (in 1970) to 1.8, a figure that will soon bring population stability.

Indonesia, with the region's largest population at more than 206 million, has also seen a dramatic decline in fertility in recent decades, although its fertility rate remains above the replacement level. If the present trend continues, however, Indonesia will reach population stability well before most other large developing countries. As with Thailand, this drop in fertility seems to have resulted from a strong government family planning effort, coupled with improvements in education.

The city-state of Singapore stands out on the demographic charts with a fertility rate well below replacement levels. Unless the deficit is offset by immigration or a dramatic turnabout in the birthrate occurs, population will soon begin to decline. The government is concerned about this situation and is actively promoting marriage and childbearing, particularly among the most highly educated segment of its population. Ironically, this is the very segment that has undergone the most pronounced fertility decline.

FIGURE 13.14 • MIGRANT SETTLEMENT IN INDONESIA
Migration from densely settled to sparsely settled areas of Southeast Asia has resulted in the creation of thousands of new communities. Many of these communities have minimal transportation and communication facilities and receive few governmental services. Some of them struggle to survive. *(Charly Flyn/Panos Pictures)*

FIGURE 13.15 • BANGKOK Bangkok saw the development of an impressive skyline during its boom years from the late 1970s through the late 1990s. Unfortunately, transportation development did not keep pace with population and commercial growth, resulting in one of the most congested and polluted urban landscapes in the world. *(Robert Holmes/Robert Holmes Photography)*

Growth and Migration Indonesia has the most explicit policy of **transmigration**, or relocation of its population from one region to another within its national territory. Primarily because of migration from densely populated Java, the population of the outer islands of Indonesia has grown rapidly since the 1970s. The province of East Kalimantan, for example, experienced a huge growth rate of 30 percent per year during the last two decades of the 1900s (Figure 13.14). As a result of this shift in population, many parts of Indonesia outside of Java now have moderately high population densities, although many of the more remote districts remain lightly occupied.

As is the case worldwide, high social and environmental costs often accompany these relocation programs. Javanese peasants, accustomed to working the highly fertile soils of their home island, often fail in their attempts to grow rice in the former rain forest of Kalimantan. The rate of field abandonment is high after repeated crop failures. In some areas farmers have little choice but to adopt a semi-swidden form of cultivation, moving to new sites once the old ones have been exhausted, a process associated with further deforestation and soil degradation.

Urban Settlement

Despite the relatively high level of economic development found in Southeast Asia, the region is not heavily urbanized; only about 30 percent of its population lives in cities. Even Thailand maintains a rural flavor, which is somewhat unusual for a country that has experienced so much recent industrialization. Cities, however, are growing rapidly throughout the region, so the rate of urbanization will certainly increase significantly over the next decade.

Many Southeast Asian countries have **primate cities**, single, large urban settlements that overshadow all others. Thailand's urban system, for example, is dominated by Bangkok, just as Manila far surpasses all other cities in the Philippines. Both have grown recently into mega-cities with more than 10 million residents. More than half of all city-dwellers in Thailand live in the Bangkok metropolitan area (Figure 13.15). In both Manila and Bangkok, as is the case in most developing countries, explosive urban growth has led to housing problems, congestion, and pollution. With its rapidly growing number of private automobiles, Bangkok may now suffer from the worst traffic congestion in the world. In Manila, it is estimated that more than half of the city's population lives in squatter settlements, usually without basic water and electricity service.

Urban primacy is much less pronounced in other Southeast Asian countries. Vietnam, for example, has two main cities, Ho Chi Minh City (formerly Saigon) in the south, with more than 3 million people, and the capital city of Hanoi in the north, with slightly more than 1 million residents. Jakarta is the largest urban area in Indonesia, but the country has a number of other large and growing cities, including Bandung and Surabaya. Yangon (formerly Rangoon) is the capital and primate city of Burma. This city has doubled its population in the last two decades to more than 4 million residents. In Cambodia, the capital city of Phnom Penh has less than 1 million; the same is true of Vientiane, the capital and primate city of Laos.

Kuala Lumpur, the largest city in Malaysia, has received heavy investments from both the national government and the global business community in recent decades. This has produced a modern, forward-looking city that is free of most of the traffic, water, and slum problems that plague other Southeast Asian cities. As a symbol of Kuala Lumpur's modern outlook, the Petronas Towers, owned by the country's national oil company, were the world's tallest buildings at almost 1,500 feet (450 meters) when completed in 1996.

The independent republic of Singapore is essentially a city-state of 3 million people on an island of 240 square miles (600 square kilometers), about three times the size of Washington, D.C. While space is at a premium, Singapore has been very successful at developing high-tech industries that have brought it great wealth. Unlike other Southeast Asian cities, Singapore has no squatter settlements or slums. Only in the fast-disappearing Chinatown and historic colonial district does one find older buildings. Otherwise, Singapore is an extremely clean city of modern high-rise skyscrapers and space-intensive industry (Figure 13.16).

FIGURE 13.16 • SINGAPORE Singapore remains the economic and technological hub of Southeast Asia. It is famous for its clean, efficiently run, and very modern urban environment. Some residents complain, however, that Singapore has lost much of its charm as it has developed. *(Adina Tovy/Lonely Planet Images/Photo 20-20)*

Cultural Coherence and Diversity: A Meeting Ground of World Cultures

Unlike many other world regions, Southeast Asia lacks the historical dominance of a single civilization. Instead, the region has been a meeting ground for cultural influences from South Asia, China, the Middle East, Europe, and even North America. Abundant natural resources, along with the region's strategic location on oceanic trading routes connecting major continents, have long made Southeast Asia attractive to outsiders. As a result, the present-day cultural geography of this diverse region owes much to external influences.

The Introduction and Spread of Major Cultural Traditions

In Southeast Asia contemporary cultural diversity is related to the historical influence of the major religions of the region: Hinduism, Buddhism, Islam, and Christianity (Figure 13.17).

FIGURE 13.17 • RELIGION IN SOUTHEAST ASIA Southeast Asia is one of the world's most religiously diverse regions. Most of the mainland is Buddhist, with Theravada Buddhism dominant in Burma (Myanmar), Thailand, Laos, and Cambodia, and Mahayana Buddhism (combined with other elements of the so-called Chinese religious complex) prevailing in Vietnam. The Philippines is primarily Christian (Roman Catholic), but the rest of insular Southeast Asia is primarily Muslim. Substantial Muslim minorities are found in the Philippines, Thailand, and Burma. Animist and Christian minorities can be found in remote areas throughout Southeast Asia, especially in Indonesia's province of Irian Jaya.

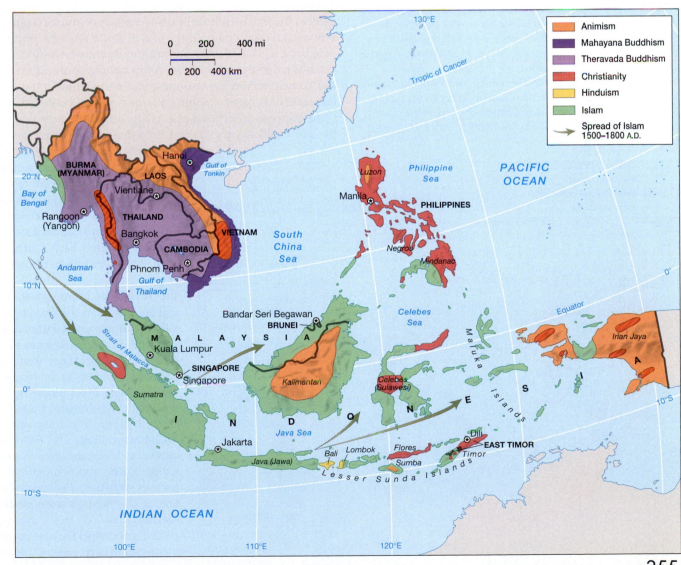

South Asian Influences

The first major external influence arrived from South Asia some 2,000 years ago when migrants from what is now India established Hindu kingdoms in coastal locations in Burma, Thailand, Cambodia, central and southern Vietnam, Malaysia, and western Indonesia. Although Hinduism faded away in most locations, this tradition is still found on the Indonesian island of Bali. However, the ancient Indian script formed the basis for many Southeast Asian writing systems, and in Muslim Java (Indonesia) the Hindu epic called the Ramayana remains a central cultural feature today.

A second wave of South Asian religious influence reached mainland Southeast Asia in the thirteenth century in the form of Theravada Buddhism, which is closely associated with Sri Lanka. Almost all of the people in lowland Burma, Thailand, Laos, and Cambodia converted to Buddhism at that time, and today it still forms the foundation for their social institutions. Saffron-robed monks, for example, are a common sight in Thailand and Burma, where Buddhist temples are common.

Chinese Influences

Unlike most other mainland peoples, the Vietnamese were not heavily influenced by South Asian civilization. Instead, their early connections were to East Asia. Vietnam was actually a province of China until about A.D. 1000, when the Vietnamese established a kingdom of their own. But while the Vietnamese rejected China's political rule, they kept many attributes of Chinese culture. The traditional religious and philosophical beliefs of Vietnam are, for example, centered on Mahayana Buddhism and Confucianism.

East Asian cultural influences in many other parts of Southeast Asia are directly linked to more recent immigration of southern Chinese. This migration reached a peak in the nineteenth and early twentieth centuries (Figure 13.18). China was then a poor and crowded country, which made sparsely populated Southeast Asia appear to be a place of great opportunity. Eventually, distinct Chinese settlements were established in every Southeast Asian country, especially in urban areas. In Malaysia the Chinese minority constitutes roughly one-third of the population, whereas in the city-state of Singapore some three-quarters of the people are of Chinese ancestry.

In many places in Southeast Asia, relationships between the Chinese minority and the native majority are strained. Even though their ancestors arrived many generations ago, many Chinese are still considered resident aliens because they maintain their Chinese citizenship. Probably a more significant source of tension is the fact that most Chinese communities in Southeast Asia are relatively wealthy. Many Chinese emigrants prospered as merchants, an occupation avoided by most local people. As a result, they often have a large economic influence on local affairs—influence that is often resented by others.

The Arrival of Islam

Muslim merchants from South and Southwest Asia arrived in Southeast Asia more than 1,000 years ago, and by the 1200s their religion began to spread. From an initial focus in northern Sumatra, Islam spread into the Malay Peninsula, through the main population centers in the Indonesian islands, and east to the southern Philippines. By 1650 Islam had largely replaced Hinduism and Buddhism throughout Malaysia and Indonesia. The only significant holdout was the small but fertile island of Bali, where thousands of Hindu musicians and artists fled from Java, giving the island a strong artistic tradition that has been maintained to the present day. Partly because of this legacy, Bali is today one of the region's main destinations of international tourism.

Today the world's most populous Muslim country is Indonesia, where some 87 percent of the nation's 206

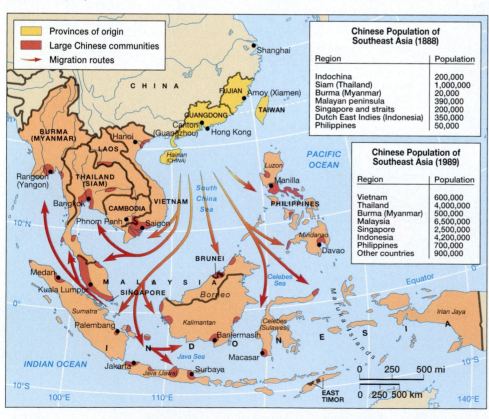

FIGURE 13.18 • CHINESE IN SOUTHEAST ASIA People from the southern coastal region of China have been migrating to Southeast Asia for hundreds of years, a process that reached a peak in the late 1800s and early 1900s. Most Chinese migrants settled in the major urban areas, but in peninsular Malaysia large numbers were drawn to the countryside to work in the mining industry and in plantation agriculture. Today Malaysia has the largest number of people of Chinese ancestry in the region. Singapore, however, is the only Southeast Asian country with a Chinese majority.

Provinces of origin
Large Chinese communities
Migration routes

Chinese Population of Southeast Asia (1888)

Region	Population
Indochina	200,000
Siam (Thailand)	1,000,000
Burma (Myanmar)	20,000
Malayan peninsula	390,000
Singapore and straits	200,000
Dutch East Indies (Indonesia)	350,000
Philippines	50,000

Chinese Population of Southeast Asia (1989)

Region	Population
Vietnam	600,000
Thailand	4,000,000
Burma (Myanmar)	500,000
Malaysia	6,500,000
Singapore	2,500,000
Indonesia	4,200,000
Philippines	700,000
Other countries	900,000

million inhabitants follow Islam (Figure 13.19). This figure, however, hides a significant amount of internal religious diversity. In some parts of Indonesia, such as in northern Sumatra (Aceh), orthodox forms of Islam took root. In others, such as central and eastern Java, a more lax form of worship emerged which included certain Hindu and even animistic beliefs. Islamic reformers, however, have long been striving to instill more orthodox forms of worship among the Javanese, and recently they have found some success.

Islam was still spreading eastward through insular Southeast Asia when the Europeans arrived in the sixteenth century. When the Spanish claimed the Philippine Islands in the 1570s, they found the southwestern portion of the archipelago to be thoroughly Islamic. The Muslims there resisted the Christian religion introduced by the Spanish. To this day, most of the far southwest Philippines is still Muslim. In the northern and central Philippines, Islam had arrived only a few decades before the Europeans and was wiped out rather quickly by Spanish priests and soldiers. Today the Philippines is roughly 85 percent Roman Catholic.

Christianity and Tribal Cultures   Christian missions spread through other parts of Southeast Asia in the late nineteenth and early twentieth century when European colonial powers controlled the region. While French priests converted many people in southern Vietnam to Catholicism, they had little influence elsewhere. Beyond Vietnam, missions failed to make headway in areas of Hindu, Buddhist, or Islamic heritage. Missionaries were, however, far more successful in Southeast Asia's highland areas, where they found many hill tribes who had never accepted the major religions. Instead, these people still held their native belief systems, which generally focus worship on nature spirits and ancestors. The general name for such religions is **animism**. While some modern hill tribes remain animist today, others were converted to Christianity. As a result, significant Christian concentrations are found in the Lake Batak area of north-central Sumatra, the mountainous borderlands between southern Burma and Thailand, the northern peninsula of Sulawesi, and the highlands of southern Vietnam.

Religion and Communism   By 1975 communism had triumphed in Vietnam, Cambodia, and Laos. In all three countries, religious practices were then strongly discouraged. At present, Vietnam's communist government is struggling against a revival of faith among the country's Buddhist majority and its 8 million Christians. Buddhist monks are sometimes harassed, and the government reserves for itself the right to appoint all religious leaders (Figure 13.20).

Geography of Language and Ethnicity
The linguistic geography of Southeast Asia is extremely complicated (Figure 13.21). The several hundred distinct languages of the region can all be placed into five major linguistic families, discussed below.

The Austronesian Languages   One of the world's most widespread language families is Austronesian, extending from Madagascar, off the coast of Africa, to Easter Island in the eastern Pacific. Today almost all insular Southeast Asian languages are within the Austronesian family. But despite this common linguistic family, more than 50 distinct languages are spoken in Indonesia alone. And in far eastern Indonesia, a variety of languages fall into the completely separate family of Papuan, closely associated with New Guinea.

The Malay language, however, overshadows all others in insular Southeast Asia. Malay is native to the Malay Peninsula, eastern Sumatra, and coastal Borneo, yet was spread historically throughout the region by merchants and seafarers. As a result, it became a common trade language, or **lingua franca**, understood and used by people of different languages throughout much of the insular realm. When Indonesia became an independent country in 1949, its leaders elected to use the lingua franca version of Malay as the basis for a new national language called "Bahasa Indonesia" (or more simply, "Indonesian"). Although Indonesian is slightly different from the

FIGURE 13.19 • INDONESIA'S LARGEST MOSQUE
Indonesia is often said to be the world's largest Muslim nation, because more Muslims reside here than in any other country. Islamic architecture in Indonesia is often modern in style, especially when contrasted with the more traditional styles found in Southwest Asia and North Africa. *(Dana Downie/Dana Downie Marketing Communications)*

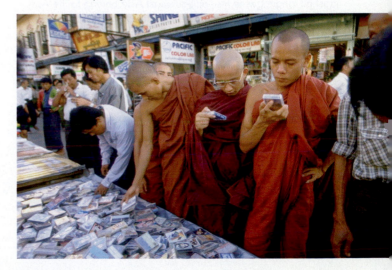

FIGURE 13.20 • MONKS IN A BURMESE MARKET
Buddhism remains important in mainland Southeast Asia, where many men become Buddhist monks. Religious traditionalism does not rule out engagement with modern global culture, as is evident in this photograph of monks in a Burmese market. *(Andres Hernandez/Getty Images, Inc.—Liaison)*

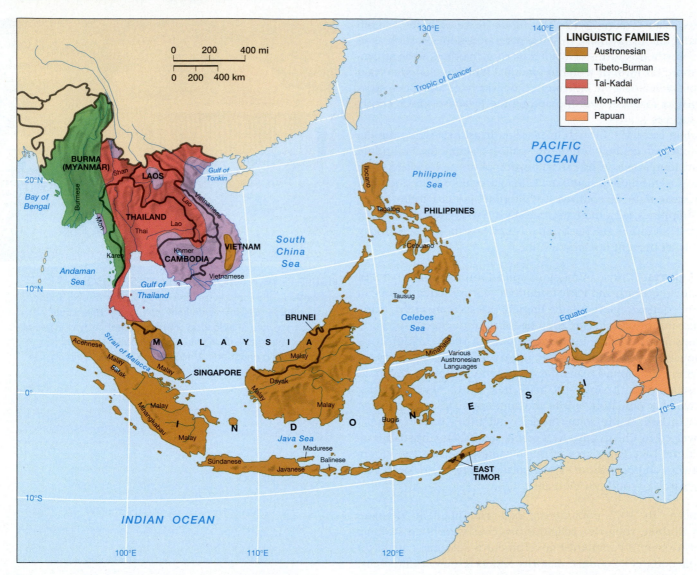

FIGURE 13.21 • LANGUAGE MAP OF SOUTHEAST ASIA A huge number of languages are found in Southeast Asia, but most are tribal tongues spoken by only a few thousand people. In mainland Southeast Asia—the site of three major language families—the central lowlands of each country are dominated by people speaking the national languages: Burmese in Burma, Thai in Thailand, Lao in Laos, and Vietnamese in Vietnam. Almost all languages in insular Southeast Asia belong to the Austronesian linguistic family. There were no dominant languages here before the creation of such national tongues as Filipino and Bahasa Indonesia in the mid-twentieth century.

Malaysian spoken in Malaysia, they do form a single, mutually understandable language. Both are now written in the Roman script.

The goal of the new Indonesian government was to offer a common language that could overcome ethnic differences throughout the huge state. In general, this policy has been successful, and more than 80 percent of all Indonesians (more than 160 million persons) now understand the language. Indonesian is also widely used in government, education, and entertainment. However, regionally based languages, such as Javanese, Balinese, and Sundanese, continue to be primary languages in most homes. More than 75 million people speak Javanese, which makes it one of the world's major tongues.

The Philippines is not as linguistically unified as either Malaysia or Indonesia, even though the eight major languages spoken in the archipelago are all Austronesian. Despite more than 300 years of Spanish colonialism, the Spanish language never became a unifying force for the islands. During the American period (1898–1946), English served as the language of government and education. After independence following World War II, Philippine nationalists searched for a language that could replace English and help unify the new country. They selected Tagalog, a language with a fairly large written literature that was dominant in the Manila area. The first task was to standardize and modernize Tagalog, which has many distinct dialects. After this was accomplished, it was renamed "Filipino" (or "Pilipino") and today, mainly because of its use in education, television, and films, it is gradually becoming a unifying national language.

Tibeto-Burman Languages   Each country of mainland Southeast Asia is closely identified with the national language spoken in its core territory. This does not mean, however, that all the residents of these countries speak these official languages on a daily basis. In the mountains and other remote districts, other languages are commonly spoken. This linguistic diversity strengthens ethnic diversity even in the face of educational programs designed to build national unity.

A good example comes from Burma. There the national language is Burmese, a language that is closely related to Tibetan. Some 22 million people speak Burmese (Figure 13.22). Although the nationalistic military government of Burma has sought to unify the population with one language, a major split has developed with several non-Burman "hill tribes" that live in the rough uplands on both sides of the Burmese-speaking Irrawaddy Valley. Although most of these tribal groups speak languages in the Tibeto-Burman family, they are quite distinctive from Burmese.

Tai-Kadai Languages   The Tai-Kadai linguistic family probably originated in southern China and then spread into Southeast Asia starting around 1200. Today closely related languages within the Tai subfamily are found through most of Thailand and Laos, in the uplands of northern Vietnam, in Burma's Shan Plateau, and in parts of southern China. Most Tai languages are quite localized, and a number are spoken by small tribal groups. But two of them, Thai and Lao, are important national languages of Thailand and Laos, respectively.

Historically, the main language of the Kingdom of Thailand, called Siamese (just as the kingdom was called Siam), was restricted to the lower Chao Phraya valley, the core region of the country. In the 1930s, however, the country changed its name to Thailand to emphasize the unity of all the peoples speaking the closely related Tai languages within its territory. Siamese was similarly renamed Thai, and it has gradually become the unifying language for the country. There are still a large number of variations in dialect, however, with those of the north often still considered to be separate languages. Even more distinctive is Lao, the Tai language that became the national tongue of Laos. More Lao speakers reside in Thailand than in Laos, however, where they form the majority population of the impoverished Khorat Plateau.

Mon-Khmer Languages   This language family probably once covered virtually all of mainland Southeast Asia. It contains two major languages, Vietnamese (the national tongue of Vietnam) and Khmer (the national language of Cambodia), as well as a host of minor languages spoken by hill peoples and a few lowland groups scattered throughout the area but concentrated in the uplands of Vietnam and Cambodia. Because of the historic Chinese influence in Vietnam, the Vietnamese language was written with Chinese characters until the French colonial government imposed the Roman alphabet, which remains in use. Khmer, like the other national languages of mainland Southeast Asia, is written in its own script, which originated in India.

The most important aspect of linguistic geography in mainland Southeast Asia is probably the fact that in each country the national language is spoken only in the core area of densely populated lowlands, whereas the peripheral uplands are populated by tribal peoples speaking separate languages. In Vietnam, for example, Vietnamese speakers occupy less than half of the national territory, even though they constitute a sizable majority of the country's population. This linguistic contrast between the lowlands and the uplands poses an obvious problem for national integration.

Southeast Asian Culture in Global Context

European colonial rule brought a new era of globalization to Southeast Asia, including European languages, Christianity, and new governmental, economic, and educational systems. During this period, Southeast Asian leaders lost their ability to make their own decisions. As a result, with political independence after World War II, many countries attempted to isolate themselves from the cultural and economic influences of the emerging global system. Burma, for example, retreated into its own form of Buddhist socialism, placing strict limits on foreign tourism. Although the

FIGURE 13.22 • BURMESE ROAD SIGNS   The Burmese language is written in a unique script, ultimately derived from South Asia. Roman letters and the English language are still used for some purposes. Burma has, however, discouraged the use of English, viewing it as the language of colonial oppression. *(Alain Evrard/Getty Images, Inc.—Liaison)*

country has become a bit more open to outsiders since the 1980s, the government of Burma remains wary of foreign influences.

Other Southeast Asian countries, however, have been receptive to foreign cultural influences. This is particularly true in the case of the Philippines, where U.S. colonialism may have led the country to many of the more popular forms of Western culture. Thailand, which was never subjected to colonial rule, appears to be the most receptive mainland Southeast Asian country to global culture, with its open policy toward tourism, mass media, and economic interdependence. But cultural globalization has been challenged in some Southeast Asian countries, especially in Malaysia. There one finds outspoken criticism, often by government officials, of American films and satellite television.

The use of English as the global language also causes controversy in many countries. On one hand, it is the language of popular culture that is opposed by many conservatives, yet on the other, it is a language that must be mastered if citizens are to participate in global business and politics. In Malaysia the widespread use of English grew increasingly controversial in the 1980s as nationalists stressed the importance of their native tongue. This worried the business community, which considers English vital to Malaysia's competitive position. It also troubled the influential Chinese communities, for which Malaysian is not a native language.

In Singapore, the situation is more complex. Mandarin Chinese, English, Malay, and Tamil (from southeastern India) are all official languages of this country. Furthermore, the languages of southern China are common in home environments, since 75 percent of Singapore's population is of southern Chinese ancestry. In recent years the Singapore government has encouraged the use of Mandarin Chinese, in part because it wishes to encourage the traditional Confucian cultural values associated with this language, and has discouraged the use of southern Chinese dialects.

In the Philippines, nationalists complain about the common use of English, even though widespread fluency has proved beneficial to the millions of Filipinos who have emigrated for better economic conditions. The Philippine government is now gradually replacing English with Filipino. At the same time, as is the case with many other languages, this national language is increasingly incorporating words and phrases from English, giving rise to a hybrid dialect known as "Taglish."

Geopolitical Framework: War, Ethnic Strife, and Regional Cooperation

Southeast Asia is sometimes defined as a geopolitical grouping of 10 different countries that have joined together under the umbrella organization of the Association of Southeast Asian Nations, or ASEAN. Today ASEAN, more than anything else, gives Southeast Asia regional coherence (Figure 13.23). Within this framework, however, many states are still struggling with serious ethnic and regional tension. In fact, in 1999 Indonesia surprised much of the world by giving up its control of the eastern portion of the island of Timor, which subsequently emerged as a new independent country.

Before European Colonialism

The modern countries of mainland Southeast Asia all existed in one form or another as kingdoms before European colonialism. Cambodia was the first to become a kingdom, and by the 1300s, independent kingdoms had been established by the Burmese, Siamese, Lao, and Vietnamese people. All of these realms were centered on major river valleys and deltas.

The situation in insular Southeast Asia was different from that of the mainland, with the premodern map bearing no resemblance to that of the modern nation-states. Many kingdoms existed on the Malay Peninsula and on the large islands of Sumatra and Java, but few were territorially stable. The countries of Indonesia, the Philippines, and Malaysia thus owe their territorial shape almost completely to the European colonial powers (Figure 13.24).

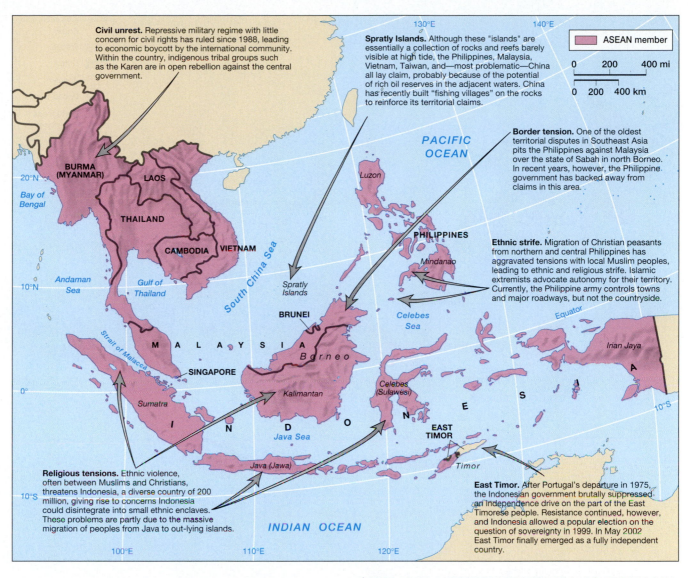

Civil unrest. Repressive military regime with little concern for civil rights has ruled since 1988, leading to economic boycott by the international community. Within the country, indigenous tribal groups such as the Karen are in open rebellion against the central government.

Spratly Islands. Although these "islands" are essentially a collection of rocks and reefs barely visible at high tide, the Philippines, Malaysia, Vietnam, Taiwan, and—most problematic—China all lay claim, probably because of the potential of rich oil reserves in the adjacent waters. China has recently built "fishing villages" on the rocks to reinforce its territorial claims.

Border tension. One of the oldest territorial disputes in Southeast Asia pits the Philippines against Malaysia over the state of Sabah in north Borneo. In recent years, however, the Philippine government has backed away from claims in this area.

Ethnic strife. Migration of Christian peasants from northern and central Philippines has aggravated tensions with local Muslim peoples, leading to ethnic and religious strife. Islamic extremists advocate autonomy for their territory. Currently, the Philippine army controls towns and major roadways, but not the countryside.

Religious tensions. Ethnic violence, often between Muslims and Christians, threatens Indonesia, a diverse country of 200 million, giving rise to concerns Indonesia could disintegrate into small ethnic enclaves. These problems are partly due to the massive migration of peoples from Java to out-lying islands.

East Timor. After Portugal's departure in 1975, the Indonesian government brutally suppressed an independence drive on the part of the East Timorese people. Resistance continued, however, and Indonesia allowed a popular election on the question of sovereignty in 1999. In May 2002 East Timor finally emerged as a fully independent country.

FIGURE 13.23 • GEOPOLITICAL ISSUES IN SOUTHEAST ASIA
The countries of Southeast Asia have managed to solve most of their border disputes and other sources of potential conflicts through ASEAN (the Association of Southeast Asia Nations). Internal disputes, however, mostly focused on issues of religious and ethnic diversity, continue to trouble several of the region's states, particularly Indonesia and Burma (Myanmar). ASEAN also experiences tension with China over the Spratly Islands of the South China Sea.

The Colonial Era

The Portuguese were the first Europeans to arrive (around 1500), lured by the cloves and nutmeg of the Maluku Islands (formerly the Spice Islands) in what is now eastern Indonesia. In the late 1500s, the Spanish conquered most of the Philippines, which they used as a base for their silver trade between China and the Americas. By the 1600s, the Dutch had started taking over territory, followed by the British. With superior naval weapons, the Europeans were able to conquer key ports and control strategic waterways. Yet for the first 200 years of colonialism, except in the Philippines, the Europeans made no major geopolitical changes.

By the 1700s, the Dutch had become the most powerful force in the region. As a result, a Dutch Empire in the "East Indies" (or Indonesia) began appearing on world maps. This empire continued to grow into the early twentieth century, when it defeated its last main enemy, the powerful Islamic state of Aceh in northern Sumatra. Later, the Dutch divided the island of New Guinea with Germany and Britain, rounding out their Indonesian colony.

The British, preoccupied with their empire in India, concentrated their attention on the sea-lanes linking South Asia to China. As a result, they established several fortified trading outposts along the strategically important Strait of Malacca, the most notable being on the island of Singapore. To avoid conflict, the British and Dutch agreed that the British would limit their attention to the Malay Peninsula and the northern portion of Borneo. The British allowed Muslim sultans to retain limited powers, much as they had done in parts of India.

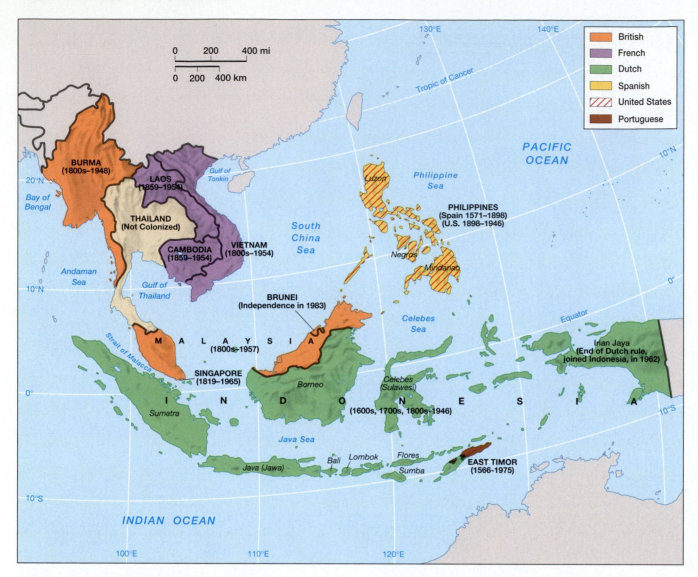

FIGURE 13.24 • COLONIAL SOUTHEAST ASIA With the exception of Thailand, all of Southeast Asia was under Western colonial rule by the early 1900s. The Netherlands had the largest empire in the region, covering the territory that was later to become Indonesia. France maintained a large colonial realm in Vietnam, Laos, and Cambodia, as did Britain in Burma and Malaysia (including Singapore and Brunei). The Philippines was initially colonized by Spain but passed to the control of the United States in 1898.

In the 1800s European colonial power spread through most of mainland Southeast Asia. The British conquered the kingdom of Burma, taking over a large area of upland territory that had never been under Burman rule. During the same period, the French moved into Vietnam's Mekong Delta, gradually expanding their territorial control to the west into Cambodia and north to China's border. This colonial empire was formally consolidated as the Union of French Indo-China in 1887. Thailand was the only country to avoid colonial rule, although it did lose substantial territories to the British in Malaysia and the French in Laos.

The final colonial power to enter the area was the United States, which took the Philippines first from Spain and then from Filipino nationalists between 1898 and 1900. The U.S. army later conquered the Muslim areas of the southwest, which had never been fully under Spanish authority.

Organized resistance to European rule began in the 1920s in mainland countries, but it took the Japanese occupation of World War II to show that European colonial power was vulnerable. After Japan's surrender in 1945, pressure for independence was renewed throughout Southeast Asia. As Britain realized that it could no longer control its South Asian empire, it also withdrew from adjacent Burma, which achieved independence in 1948. The British did not leave the Malay Peninsula until 1963, when the country of Malaysia was established. In northern Borneo, however, the Sultanate of Brunei became an independent state in its own right, backed by its substantial oil reserves. Singapore briefly joined Malaysia but then withdrew and became independent in 1965. In the Philippines, the United States granted long-promised independence on July 4, 1946, although it retained military bases, as well as considerable economic influence, for several decades. Although the Dutch attempted to re-establish their colonial rule after World War II, they were forced to acknowledge Indonesia's independence in 1949.

The Vietnam War and Its Aftermath

After World War II, France was determined to regain control of its Southeast Asian colonies. Resistance to French rule was organized primarily by communist groups based mainly in northern Vietnam. Open warfare between French soldiers and the communist forces went on for almost a decade until, in 1954, France agreed to withdraw. An international peace council then determined that Vietnam would be divided into two countries. As a result, the leaders of the communist rebellion came to power in North Vietnam, allying themselves with the Soviet Union and China. South Vietnam became an independent capitalist-oriented country with close political ties to the United States.

The Geneva peace accord did not, however, end the fighting. Communist guerrillas in South Vietnam fought to overthrow the new government and unite it with the north. North Vietnam sent troops and war materials across the border to aid the rebels. Most of these supplies reached the south over the Ho Chi Minh Trail, a confusing network of forest passages through Laos and Cambodia, thus steadily drawing these two countries into the conflict. In Laos the communist Pathet Lao forces challenged the government, while in Cambodia the **Khmer Rouge** guerrillas gained considerable power.

In Washington, D.C., the **domino theory** became accepted foreign policy. According to this notion, if Vietnam fell to the communists, then so would Laos and Cambodia; once those countries were lost, Burma and Thailand, and perhaps even Malaysia and Indonesia, would become members of the Soviet-dominated communist bloc. Fearing such an outcome, the United States was drawn ever deeper into the war. By 1965 thousands of U.S. troops had begun a brutal war against the communist guerrillas (Figure 13.25). However, despite superiority in arms and troops—and total domination of the air—U.S. forces gradually lost control over much of the countryside. As casualties mounted and the antiwar movement back home strengthened, the United States began secret talks in search of a negotiated settlement. U.S. troop withdrawals began in earnest by the early 1970s.

With the withdrawal of U.S. forces, the noncommunist governments began to collapse. Saigon fell in 1975, and in the following year Vietnam was officially reunited under the government of the north. Reunification was a traumatic event in southern Vietnam. Hundreds of thousands of people fled from the new regime to other countries, especially to the United States. The first wave of refugees consisted primarily of wealthy professionals and businesspeople, but later migrants included many relatively poor ethnic Chinese.

Vietnam proved fortunate compared to Cambodia. There the Khmer Rouge installed one of the most brutal regimes the world has ever seen. City-dwellers were forced into the countryside to become peasants, and most wealthy and highly educated people were executed. The Khmer Rouge's goal was to create a completely new agricultural society by returning to what they called "year zero." After several years of this bloodshed, neighboring Vietnam invaded Cambodia and installed a far less brutal, but still repressive, regime. Fighting between different groups continued for more than a decade. Most Vietnamese troops had left Cambodia by 1989 as the United Nations worked to end the fighting. Since that time, several coalition governments have brought a fragile peace to the shattered country.

Geopolitical Tensions in Contemporary Southeast Asia

A number of present-day conflicts in Southeast Asia are rooted in the region's colonial past. In several instances, local ethnic groups are struggling against national governments that inherited their territory from former colonial powers. Tension also results when tribal groups attempt to preserve their homelands from logging, mining, or interregional migration.

FIGURE 13.25 • U.S. SOLDIER AND VIETCONG PRISONERS
The United States maintained a substantial military presence in Vietnam in the 1960s and early 1970s. Although U.S. forces claimed many victories, they were ultimately forced to withdraw, leading to the victory of North Vietnam and the reunification of the country. (Getty Images, Inc.—Hulton Archive Photos)

FIGURE 13.26 • EAST TIMORESE DEMONSTRATION IN JAKARTA Indonesia invaded and claimed East Timor in 1975 when the Portuguese left their last colonial outpost in Southeast Asia. The East Timorese resisted the Indonesian takeover, and violence reached a climax in 1999 when the East Timorese voted for independence. Pro-Indonesian militias then attacked the Timorese leaders and large segments of Timorese society. *(V. Miladinovic/Corbis/Sygma)*

Conflicts in Indonesia When Indonesia gained independence in 1949, it included all of the former Dutch possessions in the region except western New Guinea. In 1962 the Dutch organized an election to see whether the people of this area wished to join Indonesia or form an independent country. The vote went for union, but many observers believe that the election was rigged by the Indonesian government. Tensions increased in the following decades as Javanese immigrants, along with mining and lumber firms, arrived in western New Guinea. Faced with the loss of their land and the degradation of their environment, the local people began to rebel. Rebel leaders demanded independence, or at least self-government, but they faced a far stronger force in the Indonesian army. The war in western New Guinea is small in scale but is occasionally quite bloody.

An even more brutal war erupted in 1975 on the island of Timor, in southeastern Indonesia (Figure 13.26). The eastern half of this poor and rather dry island had been a Portuguese colony (the only survivor of Portugal's sixteenth-century empire in the region) and had therefore evolved into a largely Christian society. The East Timorese expected independence when the Portuguese finally withdrew. Indonesia, however, viewed the area as rightfully its own, largely because of its geographical position, and immediately invaded. A brutal war followed, which the Indonesian army won in part by preventing food from reaching the province, thus starving the people of East Timor into submission.

After the economic crisis of 1997, Indonesia's power in the region slipped. A new Indonesian government promised an election in 1999 to see whether the East Timorese still wanted independence. At the same time, however, the Indonesian army began to organize militias in an attempt to intimidate the people of East Timor into voting to remain within the country. When it was clear that the vote would be strongly for independence, the militias began rioting, looting, and killing civilians. Under international pressure, Indonesia finally withdrew its armed forces, and the East Timorese began to build a new country. Considering the devastation caused by the militias, the process has not been easy.

Struggles for independence have occurred elsewhere in Indonesia as well. An especially violent struggle erupted in the late 1990s in the Aceh region of northern Sumatra. Many of the Acehnese, the most orthodox Muslim people of Indonesia, are demanding the creation of an independent Islamic state. While the Indonesian government has promised Aceh "special autonomy," it is determined to do whatever is necessary to prevent actual independence. More recently, the southern Maluku Islands, especially Ambon and Seram, have seen fighting between Muslims and Christians. Even the area's cities are divided into Muslim and Christian sectors, and crossing to the wrong side can be quite dangerous.

Indonesia has obviously had difficulties creating a unified nation throughout its extensive area and across its many distinct cultural and religious communities. Since Indonesia was basically created by Dutch colonialism, its sense of national unity has been relatively weak. Although some critics have viewed it as something of a Javanese empire, many non-Javanese individuals have risen to high governmental positions. It is undeniable, however, that the Indonesian state has dealt harshly with its opponents. While it is now basically democratic, the military remains politically powerful. Most Indonesian voters, moreover, live in Java and other central portions of the country, and most have little sympathy for independence movements in New Guinea, Aceh, and other peripheral areas.

Regional Tensions in the Philippines The Philippines has also suffered from regional independence movements, although not to the same extent as Indonesia. Its most persistent problem area is the Islamic southwest. The Philippine army generally maintains control over the area's main cities and roadways, but it has little power in the more remote villages. Since several extremist Muslim groups reportedly have close ties to Osama bin Laden and his Al Qaeda network, the Philippines quickly became a key site in the U.S.-led struggle against global terrorism. Many Filipino nationalists, however, are concerned that their country is again falling under U.S. domination.

Burma's Many Problems   Burma (Myanmar) has also suffered from ethnic conflict. Burma's simultaneous wars have pitted the central government, dominated by the Burmese-speaking ethnic group (the Burmans), against the country's varied non-Burman societies. Fighting intensified gradually after independence in 1948, and by the 1980s almost half of the country's territory had become a combat zone. Burma's troubles, moreover, are by no means limited to the country's ethnic minorities. Since 1988 Burma has been ruled by a very repressive military regime, one that has little popular support even among the majority Burmans.

The rebelling peoples of Burma want to maintain their cultural traditions, lands, and resources, and they generally see the national government as something of a Burman empire seeking to force its culture upon them. Most of the rebellious ethnic groups live in rugged and remote terrain, and many of them are animists or Christians. But even some lowland groups have rebelled. The Muslim peoples of the Arakan coast in far western Burma, moreover, have faced particular discrimination because of their religious convictions.

Several of Burma's ethnic rebellions have been financed largely by opium growing and heroin manufacture. This is especially true in the case of the Shan rebellion. The Shan are a Tai-speaking people inhabiting a plateau area within the "Golden Triangle" of drug production. In the late 1980s and early 1990s, a self-declared Shan state financed itself through the narcotics trade. For a number of years, this economic strategy proved to be successful, but by the mid-1990s it began to falter. Burmese agreements with Thailand reduced the Shan heroin trade, while military operations cut off the supply of raw opium reaching the Shan factories. Other ethnic groups, however, have now taken over much of the drug market.

As long as Burma retains its repressive Burman-dominated state structure, social unrest and ethnic turmoil are not likely to diminish. Protests against the government continue even in the country's core. Additionally, the international community is putting some pressure for democratic change on Burma's government with trade boycotts. The combination of internal and external pressures makes the future of this unstable state especially difficult to predict.

International Dimensions of Southeast Asian Geopolitics

Geopolitical conflicts in Southeast Asia have occurred not only within countries, but also between countries. Tensions have typically arisen when two countries claimed the same territory. The Philippines and Malaysia, for example, have quarrelled over their border, as have Malaysia and Thailand.

In recent years, however, the various countries of Southeast Asia have largely agreed to drop these and other border disputes. With the rise of the Association of Southeast Asian Nations (ASEAN) and the growth of trade relations, national leaders have concluded that friendly relations with one's neighbors are more important than the possible gain of additional territory. Militant Malays in southern Thailand were thoroughly demoralized in 1998, for example, when Malaysia offered to help the Thai government arrest them.

More difficult has been the dispute over the Spratly and Paracel islands in the South China Sea, two groups of rocks and reefs that are virtually submerged at high tide (Figure 13.27). The Philippines, Malaysia, and Vietnam have all claimed territory there, as have China and Taiwan. In the mid-1990s China began to strengthen its claims by building structures that it calls fishing shelters but which its neighbors refer to as military posts. Increasingly, the territorial conflict in the South China Sea pits Southeast Asia as a whole against China. One reason for the strong interest in these islands is that there is some evidence of large undersea oil reserves in their vicinity.

FIGURE 13.27 • THE SPRATLY ISLANDS   The Spratly Islands are small and barely above water at high tide, but they are geopolitically important. Oil may exist in large quantities in the surrounding areas, heightening the competition over the islands. Southeast Asian countries are especially concerned about China's military activities in the Spratlys. *(Nouvelle Chine/Getty Images, Inc.—Liaison)*

It is partly because of mutual concern over growing Chinese power that the countries of Southeast Asia are banding together. This is especially visible in the development and enlargement of ASEAN. While ASEAN attempts to address economic issues, its main role has been to encourage political cooperation and to present something of a united front against the rest of the world. Increasingly, the foreign power most at issue in Southeast Asia is China. While ASEAN is on friendly terms with the United States, another purpose of the organization is to prevent the United States—or any other country—from gaining too much influence in the region. For that reason, ASEAN leaders are eager to include all Southeast Asian countries within the association. In 1997 Burma gained membership, a move that disappointed the United States and other Western countries concerned about Burma's human-rights record. ASEAN's international policy, however, is negotiation rather than confrontation. For this reason, the ASEAN Regional Forum (ARF) was established in 1994 as an annual conference in which ASEAN leaders could meet with the leaders of both the East Asian and Western powers to attempt to ease tensions within the region.

Economic and Social Development: The Roller-Coaster Ride of Tiger Economies

Until the downturn of the late 1990s, economic development in Southeast Asia was often held up to the world as a model for a new globalized economy. With investment capital flowing from Japan, the United States, and other wealthy countries, Thailand, Malaysia, and Indonesia moved quickly into the ranks of booming "tiger" economies characterized by high annual growth rates. Because of the successful economic development of these countries, the notion of an "Asian model" evolved as an economic strategy that might also work in other world regions.

More recently, however, Southeast Asian countries have suffered from the roller-coaster ride of globalized economic development. Within a decade most economies have experienced unexpected highs and lows (Figure 13.28). Since the summer of 1997, regional economies have been in a state of periodic crisis. Several economies of the region later rebounded, but they remain unstable as the optimism of the boom years fades. However, some economies are still characterized by high unemployment, social unrest, and declining wealth.

Uneven Economic Development

Southeast Asia today is a region of very uneven economic and social development. While some countries, such as Indonesia, have experienced both boom and bust, others, such as Burma and Cambodia, resisted globalization and thus experienced relatively little change. Singapore, on the other hand, managed to avoid the economic crisis of the late 1990s and ranks as one of the world's most prosperous countries. As a result, the regional economic geography remains highly uneven.

The Philippine Decline Forty years ago, the Philippines was the most highly developed Southeast Asian country. It had the best-educated population in the region, and it seemed to be on the verge of rapid industrialization and economic development. By the late 1960s, however, Philippine development had been derailed. Through the 1980s and early 1990s, the country's economy failed to outpace its population growth, resulting in declining living standards for both the poor and the middle class. The Philippine people are still well educated and reasonably healthy by world standards, but even the country's educational and health systems declined during this period.

Why did the Philippines fail despite its earlier promise? While there are no simple answers, it is clear that dictator Ferdinand Marcos (who ruled from 1968 to 1986) wasted—and perhaps even stole—hundreds of millions of dollars while failing to create programs that would lead to genuine development. The Marcos regime

FIGURE 13.28 • BANK CLOSINGS The economic crisis of 1997–98 hit Southeast Asia's banking sector particularly hard. Many banks have been closed, and many others may soon be closed, as they are burdened with bad loans. As the availability of credit declined, the poor and lower middle class experienced particular hardships. *(Marcus Rose/Panos Pictures)*

instituted a kind of **crony capitalism** in which the president's many friends were given huge economic favors, while those believed to be enemies had their properties taken. After Marcos declared martial law in 1972 and suspended Philippine democracy, revolutionary activity intensified and the country began to fall into a downward spiral.

An elected democratic government finally replaced the Marcos dictatorship in 1986, but the Philippine economy continued to suffer through the 1990s. Many Filipinos have responded to the economic crisis by working overseas, either in the oil-rich countries of Southwest Asia, the wealthy cities of North America and Europe, or the newly industrialized nations of East and Southeast Asia. Men primarily work in the construction industry, and women work as nurses or domestic servants. Although foreign remittances have kept the Philippine economy afloat, this loss of labor represents in many respects a tragedy for the country as a whole. Tens of thousands of Filipina teachers, no longer able to support themselves on such small salaries, now work as maids and nannies in Singapore, Hong Kong, and Kuwait (Figure 13.29).

FIGURE 13.29 • FILIPINA MIGRANT WORKERS IN KUWAIT
The long period of economic stagnation in the Philippines has resulted in an outflow of workers from the country. Women from the Philippines often work as domestic servants in the Persian Gulf region and in Singapore and Hong Kong. *(Penny Tweedie/ Panos Pictures)*

The Regional Hub: Singapore   If the Philippines has been the biggest economic disappointment in Southeast Asia, Singapore and Malaysia have surely been the region's greatest developmental successes. Singapore has transformed itself from an **entrepôt** port city, a place where goods are imported, stored, and then transshipped, to one of the world's wealthiest and most modern states. Singapore is now the communications and financial hub of Southeast Asia, as well as a thriving high-tech manufacturing center. The Singaporean government has played an active role in the development process but has also allowed market forces freedom to operate. Singapore has encouraged investment by multinational companies (especially those involved in technology), and has itself invested heavily in housing, education, and some social services. The Singaporean government, however, remains only partly democratic, as the ruling party maintains a firm grip on all political processes.

The Malaysian Boom   Although not nearly as well-off as Singapore, Malaysia has also experienced very rapid economic growth. Development was initially concentrated in agriculture and natural resource extraction, focused on tropical hardwoods, plantation products, and tin. More recently, manufacturing, especially in labor-intensive high-tech sectors, has become the main engine of growth. As Singapore prospers, moreover, many of its companies are investing in neighboring Malaysia. Increasingly, Malaysia's economy is multinational; many Western high-tech firms operate in the country, while several Malaysian companies are themselves establishing branches in foreign lands.

The modern economy of Malaysia is not uniformly distributed across the country. One difference is geographical: most industrial development has occurred on the west side of peninsular Malaysia, with most of the rest of the country remaining largely agricultural. More important, however, are differences based on ethnicity. The industrial wealth generated in Malaysia has been concentrated in the Chinese community. Ethnic Malays remain less prosperous than Chinese-Malaysians, and those of South Asian descent are poorer still.

The unbalanced wealth of the local Chinese community is a feature of most Southeast Asian countries. The problem is particularly acute in Malaysia, however, because its Chinese minority is so large (some 30 percent of the country's total population). The government's response has been one of aggressive "affirmative action," by which economic power is transferred to the dominant Malay, or **Bumiputra** ("sons of the soil"), community. This policy has been reasonably successful. Since the economy as a whole grew from 1980 to 1997 at 8 percent a year, the Chinese community thrived even as its relative share of the country's wealth declined. Considerable resentment, however, is still felt by the Chinese. Especially annoying is the fact that most seats in the country's universities are reserved for students of Malay background.

Thailand: An Emerging Tiger?   Thailand, like Malaysia, climbed rapidly during the 1980s and 1990s into the ranks of the world's newly industrialized countries. Yet it also experienced a major downturn in the late 1990s that undercut much of this development through a devalued currency and the loss of investment capital. The Asian crisis of 1997 actually began in the overheated Thai economy and real estate market. Its subsequent recovery has been relatively slow.

Japanese companies were leading players in the earlier Thai boom. As Japan itself became too expensive for many manufacturing processes, Japanese firms began to relocate factories to such places as Thailand. They were particularly attracted by the country's low-wage, yet reasonably well-educated, workforce. Thailand is politically stable, lacking the severe ethnic tensions found in many other parts of Asia. Although Thailand's Chinese population is large and economically powerful, relations between the Thai and the Chinese have generally been good. Thailand also has a democratic government and a thriving free press, although it does have a legacy of military takeovers followed by periods of authoritarian rule. In recent years, however, Thailand's well-respected constitutional monarchy has acted to protect democracy.

To repeat a familiar story, however, Thailand's economic boom has by no means benefited the entire country to an equal extent. Most industrial development has occurred in the historical core, especially in the city of Bangkok itself. Yet even in Bangkok the blessings of progress have been mixed. As the city begins to choke on its own growth, industrial growth has begun to spread outward. The entire Chao Phraya lowland area shares to some extent in the general prosperity because of both its proximity to Bangkok and its rich agricultural resources.

Thailand's Lao-speaking northeast (the Khorat Plateau) is the country's poorest region. Because of the poverty of their homeland, northeasterners are often forced to seek employment in Bangkok. As Lao speakers, they sometimes experience ethnic discrimination. Men typically find work in the construction industry; northeastern women not uncommonly make their living as prostitutes.

Thailand has earned the doubtful distinction of being one of the world's prostitution centers. Thai prostitutes largely serve a domestic clientele, but they also attract many foreign customers. Most of these "sex tourists" come from wealthy countries such as Japan and Germany, but a significant number are Malaysian. Prostitutes, especially those who are underage, are frequently forced into their jobs. The "sex industry" of Thailand presents an ironic and tragic situation; in general, the social position of Thai women is relatively high, yet women in Thai brothels often experience severe exploitation.

Recent Economic Expansion in Indonesia   At the time of independence (1949), Indonesia was one of the poorest countries in the world. The Dutch had used their colony largely for its tropical crops and other resources and had invested little in transportation, health, or education. The population of Java expanded rapidly in the nineteenth and early twentieth centuries, and serious land shortages caused problems in peasant communities.

The Indonesian economy finally began to expand in the 1970s. Oil exports fueled the early growth, as did the logging of tropical forests. But unlike most other oil exporters, Indonesia continued to grow even after oil prices plummeted in the 1980s. Like Thailand and Malaysia, Indonesia proved attractive to multinational companies seeking to pay low wages. Large Indonesian firms, some three-quarters of them owned by local Chinese families, have also capitalized on the country's low wages and abundant resources. The national government has attempted to build technologically oriented businesses, but their success remains uncertain.

But despite rapid growth in the 1980s and 1990s, Indonesia remains a poor country. Its pace of economic expansion never matched that of Thailand, Singapore, and Malaysia, and it has remained much more dependent on the unsustainable exploitation of natural resources. The financial crisis of the late 1990s, moreover, hurt Indonesia more severely than any other country. Millions of Indonesians suddenly found themselves so poor that they could no longer afford rice, the country's basic food staple. Political instability is also a continuing concern. Economic recovery began by 2000 but has thus far been unsteady.

Persistent Poverty in Vietnam, Laos, and Cambodia   The three countries of former French Indochina—Vietnam, Cambodia, and Laos—experienced only modest economic expansion during the boom years of the 1980s and 1990s. This area endured almost continual warfare between 1941 and 1975, and fighting persisted until the late 1990s in Cambodia. Due to such conditions, few investments were made during this period.

Of these three countries, Vietnam is the wealthiest, but it is still quite poor. Postwar reunification in 1975 did not bring the anticipated growth. Conditions grew worse in the early 1990s after the fall of the Soviet Union, Vietnam's main supporter and trading partner. Until the mid-1990s, Vietnam remained under embargo by the United States and thus somewhat isolated from the world economy. Frustrated with their country's economic performance, Vietnam's leaders began to embrace market economics while retaining the political forms of a communist state. They have, in other words, followed the Chinese model (Figure 13.30). Vietnam now welcomes multinational corporations, which are attracted by its extremely low wages and relatively well-educated workforce. Merchants still complain of harassment by state officials, however, and one recent survey listed Vietnam as one of the most corrupt countries in Asia.

Laos and Cambodia face more serious problems than Vietnam. Laos faces special difficulties owing to its rough terrain and relative isolation, while in Cambodia, the ravages of war worsened an already difficult economic situation. Both countries also lack basic infrastructure; outside the few cities, paved roads and reliable electricity are rare. As a result, the economies of both Laos and Cambodia remain largely agricultural. The Laotian government is pinning its economic hopes on hydropower development. The country is mountainous with many large rivers and could therefore generate large quantities of electricity, which is in high demand in neighboring Thailand. The Laotian government, however, remains repressive, which discourages economic development. The Cambodian economy, on the other hand, is being increasingly integrated with that of Thailand. Several Cambodian border towns recently set themselves up as gambling centers, attracting major investments from Thai prostitution ringleaders frustrated with the fact that gambling is not allowed in Thailand.

Burma's Troubled Economy   Burma (Myanmar) stands with Cambodia and Laos at the bottom of the scale of Southeast Asian economic development. For all of its many problems, however, Burma is a land of great potential. It has abundant natural resources (including oil and other minerals, water, and timber), as well as a large expanse of fertile farmland. Its population density is moderate, and its people are reasonably well educated. Burma's economy, however, has remained relatively stagnant since independence in 1948.

Burma's economic woes can be traced in part to the continual warfare the country has experienced. Most observers also blame economic policy. Beginning in 1962, Burma attempted to isolate its economic system from global forces in order to achieve self-sufficiency under a system of Buddhist socialism. While its intentions may have been admirable, the experiment was not successful; instead of creating a self-contained economy, Burma found itself burdened by smuggling and black-market activities. Certainly its very lack of economic development has made Burma's capital of Yangon (Rangoon) a much more pleasant place than

FIGURE 13.30 • STOREFRONT CAPITALISM IN VIETNAM
Although Vietnam is a communist state, it has—like China—embraced many forms of capitalism. Private shops are now allowed, and foreign investment is welcome. In general, the market economy is much more highly developed in the south than in the north. *(Liba Taylor/Panos Pictures)*

its ancient rival Bangkok, as it is not nearly as congested and polluted. Some critics of industrial development, moreover, argue that the Burmese people have doubts about economic progress because of their religious beliefs. Other scholars counter that the Burmese people are thoroughly dissatisfied with their limited economic opportunities and that the government has been able to maintain its power only through repression.

Globalization and the Southeast Asian Economy

As the discussion above shows, Southeast Asia as a whole has undergone rapid integration into the global economy. Singapore has thoroughly staked its future to the success of multinational capitalism, as have several neighboring countries. Even Marxist Vietnam and once-isolationist Burma are opening their doors to the global system, although, in the case of Burma, with much hesitation.

Regardless of what happens in coming years, global economic integration has already brought about significant development in Singapore, Malaysia, Thailand, and even Indonesia. Over Southeast Asia as a whole, however, economic development has also resulted in massive environmental degradation and growing social inequality. Outside of Singapore and Malaysia, moreover, successful development has generally been based on labor-intensive manufacturing in which workers are paid miserably low wages and subjected to harsh discipline. Consumers in the world's wealthy countries are increasingly aware that many of their basic purchases, such as sneakers and clothing, are produced under exploitative conditions. Movements have thus begun in Europe, the United States, and elsewhere to pressure both multinational corporations and Southeast Asian governments to improve the working conditions of laborers in the export industries. Some Southeast Asian leaders, however, object, accusing Western activists of wanting to prevent Southeast Asian development under the excuse of concern over worker rights.

Issues of Social Development

As might be expected, several key indicators of social development in Southeast Asia are linked to levels of economic development. Singapore thus ranks among the world leaders in regard to health and education, as does the small, yet oil-rich country of Brunei. Laos and Cambodia, not surprisingly, come out near the bottom of the chart. The people of Vietnam, however, are healthier and better educated than might be expected on the basis of their country's overall economic performance. Although Vietnam's government has not yet been able to generate sustained economic growth, it has provided reasonably good social services.

With the exception of Laos, Cambodia, and Burma, Southeast Asia has achieved relatively high levels of social welfare. In Laos and Cambodia, however, life expectancy at birth hovers around 55 years (as compared to Thailand's 72 years), and literacy rates remain below 50 percent. But even the poorest countries of the region have made some improvements. War-torn Cambodia, however, has made relatively small gains, mainly because most of its budget is devoted to maintaining armed forces while little goes to social programs.

Most of the governments of Southeast Asia have placed a high priority on basic education. Literacy rates are relatively high in most countries of the region. Much less success, however, has been realized in university and technical education. As Southeast Asian economies continue to grow, this educational gap is beginning to have negative consequences, forcing many students to study abroad. In 1997, for example, more than 50,000 Malaysians were attending colleges in other countries. If Southeast Asian countries other than Singapore are to become fully developed, they will probably have to invest more money in their own human resources.

CONCLUSION

In many ways, Southeast Asia is the world region that best fits this book's focus on globalization and diversity. Diversity within the region is certainly pronounced, from the mix of different cultural traditions to continuing differences in economic and social development. Given this wide diversity, one might argue that modern globalization, for better or worse, has helped Southeast Asia find a new sense of regional identity through common struggles with economic development and geopolitics. Southeast Asia has gained an unusual degree of unity in recent years through ASEAN, while shared concerns about economic restructuring and global trading have also helped build regional identity. As a result, the historical and cultural unity that Southeast Asia has lacked may be forced upon it by twenty-first-century globalization.

The role of ASEAN and the economic unity of Southeast Asia, however, should not be exaggerated. Most of the region's trade is still directed outward toward the traditional centers of the global economy—North America, Europe, and especially Japan. This orientation is not surprising, considering the export-focused economic policies of most Southeast Asian countries. A significant question for Southeast Asia's future is whether the region will develop an integrated regional economy, or whether linkages with Japan and the United States will remain more important.

An equally important question concerns the future of the region's environment. Given the emphasis placed by global trade on exporting resources such as wood products, it is perhaps understandable that Southeast Asia has sacrificed so many of its forests to support other kinds of economic development. But in most of the region, forests are now seriously depleted. One of the questions Southeast Asia must address is whether to cooperate on forest conservation or, instead, to let special interests continue forest destruction. To further complicate the situation, as the larger Asian economy rebounds, market demand for Southeast Asian forest products is increasing. Southeast Asia must also meet the needs of its rapidly growing cities by improving traffic flow, curtailing water pollution, and reducing air pollution.

Finally, some degree of geopolitical change is needed to bring stability to the region. Specifically, the national governments of Indonesia, Burma, and Cambodia may need to reinvent themselves, seeking more cooperation between their people and their leaders. Indonesia has yet to adequately address internal ethnic and cultural tensions. Cambodia must begin genuine efforts at rebuilding and reconciliation. Last, Burma must achieve some sort of stability, coming to terms with both its separatist tribal peoples and the urban population that opposes the current military regime.

KEY TERMS

animism (page 357)
Association of Southeast Asia
    Nations (ASEAN) (page 344)
Bumiputra (page 367)

copra (page 352)
crony capitalism (page 367)
domino theory (page 363)
entrepôt (page 367)

Golden Triangle (page 352)
Khmer Rouge (page 363)
lingua franca (page 357)
primate cities (page 354)

swidden (page 352)
typhoons (page 350)
transmigration (page 354)

Australia and Oceania

FIGURE 14.1 Australia and Oceania
More water than land, the Australia and Oceania region covers the vast reaches of the western Pacific Ocean. Australia dominates the region, both in its physical size and in its economic and political influence. New Zealand and Australia both represent largely European settlement in the South Pacific. Elsewhere, however, the island subregions of Melanesia, Micronesia, and Polynesia contain important native populations that have mixed in varied ways with later European and American arrivals.

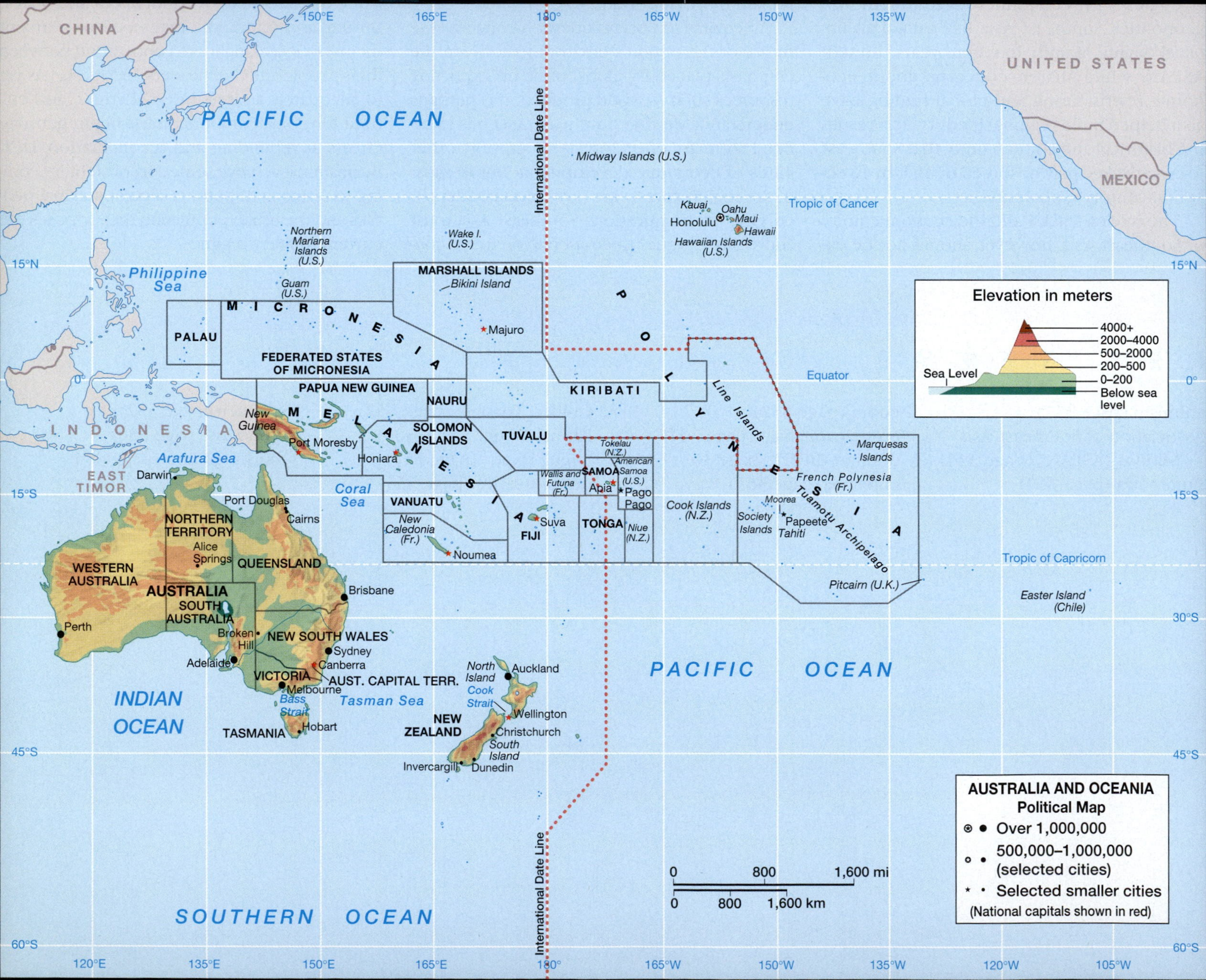

CHINA

UNITED STATES

MEXICO

PACIFIC OCEAN

Midway Islands (U.S.)

Tropic of Cancer

Kauai Oahu Maui
Honolulu Hawaii
Hawaiian Islands (U.S.)

Northern Mariana Islands (U.S.)

Wake I. (U.S.)

Philippine Sea

MARSHALL ISLANDS
Bikini Island

MICRONESIA

Guam (U.S.)

PALAU

Majuro

FEDERATED STATES OF MICRONESIA

Equator

PAPUA NEW GUINEA

NAURU

KIRIBATI

Line Islands

INDONESIA

New Guinea

MELANESIA

Port Moresby

SOLOMON ISLANDS

TUVALU

Honiara

Arafura Sea

EAST TIMOR

Darwin

Wallis and Futuna (Fr.)

Tokelau (N.Z.)

SAMOA
Apia

American Samoa (U.S.)
Pago Pago

Marquesas Islands

French Polynesia (Fr.)

Coral Sea

VANUATU

New Caledonia (Fr.)

Suva

Cook Islands (N.Z.)

Moorea
Society Islands
Papeete
Tahiti

Tuamotu Archipelago

POLYNESIA

Port Douglas

NORTHERN TERRITORY

Cairns

FIJI

TONGA

Niue (N.Z.)

Noumea

WESTERN AUSTRALIA

Alice Springs

QUEENSLAND

AUSTRALIA
SOUTH AUSTRALIA

Brisbane

Tropic of Capricorn

Pitcairn (U.K.)

Easter Island (Chile)

Perth

Broken Hill

NEW SOUTH WALES

Sydney

Adelaide

Canberra

AUST. CAPITAL TERR.

VICTORIA

Melbourne

North Island
Cook Strait

Auckland

PACIFIC OCEAN

INDIAN OCEAN

Bass Strait

Tasman Sea

NEW ZEALAND

Wellington

Christchurch
South Island

TASMANIA

Hobart

Invercargill Dunedin

International Date Line

SOUTHERN OCEAN

Elevation in meters
4000+
2000–4000
500–2000
200–500
0–200
Below sea level
Sea Level

0   800   1,600 mi
0   800   1,600 km

AUSTRALIA AND OCEANIA
Political Map
⊗ ● Over 1,000,000
○ 500,000–1,000,000 (selected cities)
✶ ● Selected smaller cities
(National capitals shown in red)

SETTING THE BOUNDARIES

The vast, watery distances of the Pacific, as well as shared elements of indigenous and colonial history that stretch from the South Pacific to Hawaii, help to define the boundaries of the Australia and Oceania region. Still, some of its regional boundaries are born from convenience, while others remain elusive and ill-defined. Australia (or "southern land") forms a coherent political unit and subregion and clearly dominates the region in both area and population. Across the Tasman Sea, New Zealand, while usually considered part of Oceania, is easily linked politically and economically to Australia, particularly in their parallel histories of European—mostly British—colonization and development. Both countries also have significant native populations, although Australia's Aborigines are culturally and ethnically distinct from New Zealand's Maori peoples.

Less clear is how those larger landmasses (often termed *Meganesia*) relate to the island worlds beyond, as well as to nearby portions of Southeast Asia. Increasing economic ties and their Pacific Ocean location link Australia and New Zealand to the smaller islands of Oceania. Still, the area's environmental and cultural diversity does not offer any easy defining characteristics. More broadly, however, traditional island cultures often intermingled, and they shared relationships with the sea and the varied resources of the tropical world. Later, Europeans and Americans came upon the scene. As they rushed to carve up the realm into colonial possessions, their strategies often had lasting consequences. Originally Polynesian, Hawaii, once it was securely in the U.S. sphere, evolved into a multicultural political outlier that expresses the complex globalization of the Pacific region. To the south, the French produced their own distinctive island subregion, mostly scattered across the eastern reaches of Polynesia. On the island of New Guinea, however, complex colonial-era boundaries produced a problematic regional border. Today, an arbitrary boundary line bisects the island, and while Papua New Guinea (eastern half) is usually considered a part of Oceania, neighboring Irian Jaya (western half) is a part of Indonesia and thus typically grouped with Southeast Asia.

Tﬁhis vast world region, dominated mostly by water, includes the island continent of Australia as well as **Oceania**, a collection of islands that reach from New Guinea and New Zealand to the U.S. state of Hawaii in the mid-Pacific (Figure 14.1; see "Setting the Boundaries"). Although native peoples settled the area long ago, more recent European and North American colonization began the process of globalization that now characterizes the region. Today, the region is caught up in global processes that are transforming rural and urban landscapes and producing new and sometimes unsettled cultural and political geographies.

Ongoing political and ethnic unrest in Fiji illustrates how the heat of twenty-first-century globalization has fired the cauldron of Pacific globalization (Figure 14.2). Currently, the country struggles to deal with new political and economic relationships between the two dominant ethnic groups, indigenous Fijians and the descendants of South Asian sugarcane workers (called Indo-Fijians) who were brought to the islands in the nineteenth century as a solution to labor shortages in the cane fields. Strong cultural and religious barriers exist between the two groups, resulting in extremely low rates of intermarriage. Generally speaking, the Indo-Fijians dominate the country's commercial life, causing resentment in the largely rural and village-based native Fijian community.

While tensions have long caused conflict between these two groups, this unrest has been heightened by contemporary globalization as foreign investors push for economic restructuring and market development. Achieving these economic goals requires a radical change away from the traditional land tenure system, which is primarily controlled by Fijian groups. As a result, island governance has been unsettled by a series of coups and counter-coups over the last two decades. In May 2000, armed Fijians took hostage the nation's first prime minister of Indian descent. Large numbers of Indo-Fijians fled the islands, leaving the commercial and economic life of the country in shambles. Despite calls from the international community to restore the democratically elected Fijian government, leaders of the military coup continue to hold power. By all accounts, these current tensions are more than a continuation of historic ethnic unrest between the two groups. Instead, observers attribute much of the problem to unrest within the traditional native Fijian community as its members became empowered by the riches of economic globalization. In many ways, the same tensions found in Fiji are also found in many other island countries of Oceania and, to a lesser degree, within Australia and New Zealand. This serves to remind us of the way environmental, settlement, cultural, geopolitical, and economic activities are inseparably linked in the contemporary world.

FIGURE 14.2 • UNREST IN FIJI Ethnic tensions between economically dominant Indo-Fijians, the descendants of South Asian sugarcane workers brought to the islands in the nineteenth century, and native Fijians often take a violent turn. Here, an Indo-Fijian family examines the ruins of their house after it was torched by Fijian rebels during the military takeover of August 2000. *(AFP/Corbis)*

FIGURE 14.3 • SYDNEY, AUSTRALIA Most Australians live in cities, and the country's urban landscapes often resemble their North American counterparts. This view of Sydney features its world-famous harbor and displays the dramatic interplay of land and water. *(Rob Crandall/Rob Crandall, Photographer)*

Australia and New Zealand share many geographical characteristics and dominate the regional setting. Major population clusters in both countries are located in the middle latitudes rather than the tropics. Australia's 19.4 million residents occupy a vast land area of 2.97 million square miles (7.69 million square kilometers), while New Zealand's combined North and South Islands (104,000 square miles, or 269,000 square kilometers) are home to 3.9 million people. Most residents of both countries live in urban settlements near the coasts (Figure 14.3). Australia's huge and dry interior, often termed the **outback**, is as thinly settled as North Africa's Sahara Desert, and much of the New Zealand countryside is a visually spectacular but sparsely occupied collection of volcanic peaks and rugged, glaciated mountain ranges.

Taken together, the land areas of these two South Pacific nations almost equal that of the United States, but their populations total less than 10 percent of their distant North Pacific neighbor. All three countries, however, share a European cultural heritage, the product of common global-scale processes that sent Europeans far from their homelands over the past several centuries. The highly Europeanized populations of both Australia and New Zealand also retain particularly close cultural links to Britain, and they maintain relatively high levels of income and economic development.

Punctuated with isolated chains of sand-fringed and sometimes mountainous islands, the blue waters of the tropical Pacific dominate much of the rest of the region. Three major subregions of Oceania each contain a surprising variety of human settlements and political units, although overall land areas and populations are small compared to Australia. Farthest west, **Melanesia** (meaning "dark islands") contains the culturally complex, generally darker-skinned peoples of New Guinea, the Solomon Islands, Vanuatu, and Fiji. The largest of these countries, Papua New Guinea (179,000 square miles, or 463,000 square kilometers), includes the eastern half of the island of New Guinea (the western half is part of Indonesia), as well as nearby portions of the northern Solomon Islands. Its population of 5 million people is slightly higher than that of New Zealand.

To the east, the small island groups, or **archipelagos**, of the central South Pacific are called **Polynesia** (meaning "many islands"), and this linguistically unified subregion includes French-controlled Tahiti in the Society Islands, the Hawaiian Islands, and smaller political states such as Tonga, Tuvalu, and Samoa. New Zealand is also often considered a part of Polynesia since its native peoples, known collectively as the **Maori**, share many cultural and physical characteristics with the somewhat lighter-skinned peoples of the mid-Pacific region. Finally, the more culturally diverse region

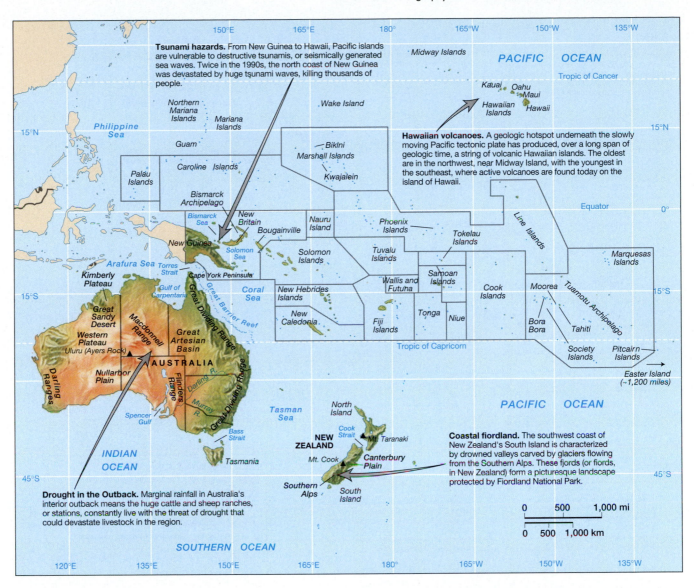

Tsunami hazards. From New Guinea to Hawaii, Pacific islands are vulnerable to destructive tsunamis, or seismically generated sea waves. Twice in the 1990s, the north coast of New Guinea was devastated by huge tsunami waves, killing thousands of people.

Hawaiian volcanoes. A geologic hotspot underneath the slowly moving Pacific tectonic plate has produced, over a long span of geologic time, a string of volcanic Hawaiian islands. The oldest are in the northwest, near Midway Island, with the youngest in the southeast, where active volcanoes are found today on the island of Hawaii.

Coastal fiordland. The southwest coast of New Zealand's South Island is characterized by drowned valleys carved by glaciers flowing from the Southern Alps. These fjords (or fiords, in New Zealand) form a picturesque landscape protected by Fiordland National Park.

Drought in the Outback. Marginal rainfall in Australia's interior outback means the huge cattle and sheep ranches, or stations, constantly live with the threat of drought that could devastate livestock in the region.

FIGURE 14.4 • PHYSICAL GEOGRAPHY OF AUSTRALIA AND OCEANIA A variety of physical processes have shaped the Australia and Oceania region. Active volcanoes from Hawaii to Papua New Guinea have produced some of Earth's newest landscapes. New Zealand is home to some of the region's most complex and varied physical settings. Dominant almost everywhere in the region, the waters of the blue Pacific shape land and life in fundamental ways.

of **Micronesia** (meaning "small islands") is north of Melanesia and west of Polynesia and includes microstates such as Nauru and the Marshall Islands, as well as the U.S. territory of Guam.

Environmental Geography: A Varied Natural and Human Habitat

The region's physical setting speaks to the power of space: The geology and climate of the seemingly limitless Pacific Ocean define much of the physical geography of Oceania, and the expansive interior of Australia shapes the basic physical setting of that island continent (Figure 14.4).

Environments at Risk

Despite their relatively small populations, many settings in Australia and Oceania face significant human-induced environmental problems. Some environmental challenges are caused by natural events that increasingly impact larger and more widely distributed human populations. For instance, Pacific Rim-earthquakes, periodic Australian droughts, and tropical cyclones now pose greater threats than they once did as new settlements have made increasing populations vulnerable to these problems. Other environmental issues, however, are even more directly related to human

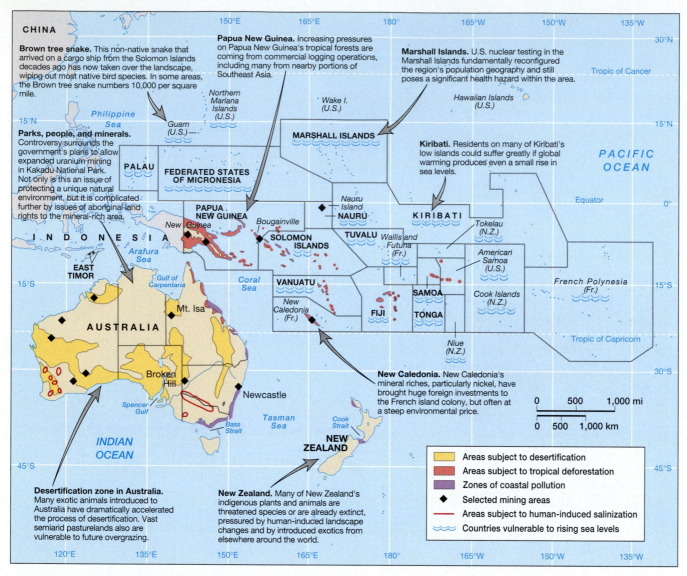

FIGURE 14.5 • ENVIRONMENTAL ISSUES IN AUSTRALIA AND OCEANIA Tropical deforestation and extensive mining have brought varied challenges to the region. Human settlements have also extensively modified the pattern of natural vegetation. Future environmental threats exist on low-lying Pacific Islands if global sea levels rise.

causes (Figure 14.5). Specifically, European colonization introduced many environmental threats, and recent economic globalization has further pressured the region's natural resource base.

Global Resource Pressures Globalization has exacted an environmental toll upon Australia and Oceania. Specifically, the region's considerable base of natural resources has been opened to development, much of it by outside interests. While gaining from the benefits of global investment, the region has also paid a considerable price for encouraging development, and the result is an increasingly threatened environment. Major mining operations have greatly impacted Australia, Papua New Guinea, New Caledonia, and Nauru. Some of Australia's largest gold, silver, copper, and lead mines are located in sparsely settled portions of Queensland and New South Wales, putting watersheds in these semiarid regions at risk to metals pollution. In Western Australia, huge open-pit iron mines dot the landscape, unearthing ore that is often bound for global markets, particularly Japan. To the north, Papua New Guinea's Bougainville copper mine has transformed the Solomon Islands, while even larger gold mining ventures have raised increasing environmental concerns on the island of New Guinea (Figure 14.6). Elsewhere, Micronesia's tiny Nauru has been virtually turned inside out as much of the island's jungle cover was removed to get at some of the world's richest phosphate deposits. Former Australian and New Zealand mine owners have already paid millions of dollars to settle environmental damage claims.

Deforestation is another major environmental threat across the region. Vast stretches of Australia's eucalyptus woodlands, for example, have been destroyed to produce better pastures. In addition, coastal rain forests in Queensland are only a fraction of their original area, although a growing environmental movement in the region is fighting to save the remaining forest tracts. Tasmania has also been an environmental battleground, particularly given the biodiversity of its midlatitude forest landscapes. While the island's earlier European and Australian development featured many logging and pulp mill operations, more than 20 percent of the island is now protected by national parks.

Deforestation is also a major problem in many parts of Oceania. With limited land areas, islands are subject to rapid tree loss, which in turn often leads to soil erosion. Sizable areas of rain forest are still found on most of the larger islands of Melanesia. Although rain forests still cover 70 percent of Papua New Guinea, more than 37 million acres (15 million hectares) have been identified as suitable for logging (Figure 14.7). Some of the world's most biologically diverse environments are being threatened in these operations, but landowners see the quick cash sales to loggers as attractive even though this nonsustainable practice is contrary to their traditional lifestyles.

Global Warming and Rising Sea Levels   Oceania's greatest future environmental threat may be global warming. If Earth grows significantly warmer in the near future, as many climate models predict, the impact on many low-lying atolls will be devastating. Higher global temperatures will partially melt polar ice caps, and resulting rises in sea levels could literally drown the region's low islands. Countries such as Kiribati and the Marshall Islands could simply disappear, forcing residents to flee elsewhere, probably to other overcrowded islands. Not surprisingly, the Pacific states have been major supporters of global treaties to limit the production of greenhouse gases. Unfortunately, these tiny island nations have little international power compared to developed countries such as the United States, which contribute most global warming pollution.

Exotic Plants and Animals   The introduction of exotic (nonnative) plants and animals has caused problems for endemic (native) species throughout the Pacific region. In Australia, some native fauna could not compete with nonnative animals. For example, nonnative rabbits successfully multiplied in an environment that lacked the diseases and predators that kept their numbers in check in Europe. Before long, rabbit populations had reached plague-like proportions, and large sections of land were almost completely stripped of vegetation. The animals were brought under control only through the purposeful introduction of the rabbit disease myxomatosis. Introduced sheep and cattle populations have also stressed the region's environment by increasing soil erosion and contributing to desertification.

The introduction of exotic plants and animals to island environments has also had similar effects. For example, many small islands possessed no native land mammals, and their native bird and plant species proved vulnerable to the ravages of introduced rats, pigs, and other animals. The larger islands of the region, such as those of New Zealand, originally supported several species of large, flightless birds that filled some of the ecological niches held by mammals on the continents. The largest of these, the moas, were substantially larger than ostriches. During the first wave of human settlement in New Zealand some 1,500 years ago, moa numbers fell rapidly as they were hunted, their habitat burned, and their eggs consumed by invading rats. By 1800, the moas had been completely exterminated.

FIGURE 14.6 • MINING IN PAPUA NEW GUINEA   Open-pit mining for gold, silver, copper, and lead mark the landscapes of Papua New Guinea, New Caledonia, and Nauru. Although bringing some economic benefit to local peoples, these activities also cause immense environmental damage to the region. In New Guinea, for example, sediments from upland mines have severely damaged the Fly River ecosystem. *(Michael Freeman/Corbis)*

FIGURE 14.7 • LOGGING IN OCEANIA   Foreign logging companies have made large investments in tropical Pacific settings such as Papua New Guinea. While bringing new jobs, these activities dramatically alter local environments as hardwood forests are harvested for export. *(David Austen/Woodfin Camp & Associates)*

FIGURE 14.8 • ISLAND PEST The brown tree snake, which arrived in Guam accidentally in the 1950s, has now taken over large parts of the island's forest lands and killed off most native bird species. These 10-foot-long snakes often climb along electrical wires, frequently causing power outages throughout Guam. *(John Mitchell/Photo Researchers, Inc.)*

The spread of nonnative species continues today, perhaps even at a greater pace. In Guam, for example, the brown tree snake, which arrived accidentally by cargo ship from the Solomon Islands in the 1950s, has taken over the landscape (Figure 14.8). In some forest areas, with more than 10,000 snakes per square mile, they have wiped out nearly all the native bird species. Additionally, the snakes cause frequent power outages as they crawl along electrical wires. While the brown tree snake has already done its damage to Guam, it threatens other islands as well, since it readily hides in cargo containers awaiting air or water shipment to island destinations.

Australian and New Zealand Environments

Curiously, Australia is one of the world's most urbanized societies, yet most people associate the country with its vast and arid outback, a sparsely settled land of sweeping distances, scrubby vegetation, and unusual animals (Figure 14.9). In contrast, the two small islands that make up New Zealand are known for their varied landscapes of rolling foothills and rugged mountains.

Regional Landforms Three major landform regions dominate Australia's physical geography (see Figure 14.4). The Western Plateau occupies more than half of the continent. Most of the region is a vast, irregular plateau that averages only 1,000 to 1,800 feet in height (305 to 550 meters). Further east, the Interior Lowland Basins stretch north to south for more than 1,000 miles from the swampy coastlands of the Gulf of Carpentaria to the Murray and Darling valleys, Australia's largest river system. Finally, more forested and mountainous country exists near Australia's Pacific coast. The Great Dividing Range extends from the Cape York Peninsula in northern Queensland to southern Victoria. Nearby, off the eastern coast of Queensland, the Great Barrier Reef offers a final dramatic subsurface feature: Over the past 10,000 years, one of the world's most spectacular examples of coral reef-building has produced a living legacy now protected by the Great Barrier Reef Marine Park (Figure 14.10).

Part of the Pacific Rim of Fire, New Zealand owes its geological origins to volcanic mountain-building that produced two rugged and spectacular islands in the South Pacific. The North Island's active volcanic peaks, reaching heights of more than 9,100 feet (2,775 meters), and geothermal features reveal the country's fiery origins (Figure 14.11). Even higher and more rugged mountains run down the western spine of the South Island. Often mantled by high mountain glaciers and surrounded by steeply sloping valleys, the Southern Alps are one of the world's most visually spectacular mountain ranges, complete with narrow, fjordlike valleys that indent much of the South Island's isolated western coast.

FIGURE 14.9 • THE AUSTRALIAN OUTBACK Arid and generally treeless, the vast lands of the Australian outback resemble some of the dry landscapes of the western United States. In this photo, wildflowers blossom along a dirt road near Tom Price, in the Pilbara region of Western Australia. *(Rob Crandall/Rob Crandall, Photographer)*

Climate Generally, zones of somewhat higher precipitation encircle Australia's arid center (Figure 14.12). In the tropical low-latitude north, seasonal changes are dramatic and unpredictable. For example, Darwin can experience drenching monsoonal rains in the summer (December to March), followed by bone-dry winters (June to September). Indeed, life across the region is shaped by this annual rhythm of "wet" and "dry." Along the east coast of Queensland, precipitation remains high (60 to 100 inches, or 153 to 254 centimeters), but it diminishes rapidly as one moves into the interior. Rainfall at interior locations such as the Northern Territory's Alice Springs averages less than 10 inches (25 centimeters) annually. South of Brisbane, more midlatitude influences dominate eastern Australia's climate. Coastal New South Wales, southeastern Victoria, and Tasmania experience the country's most dependable year-round rainfall, which averages 40 to 60 inches (102 to 152 centimeters) of precipitation per year. Nearby mountains see frequent winter snows. Farther west,

FIGURE 14.10 • THE GREAT BARRIER REEF   Stretching along the eastern Queensland coast, the famed Great Barrier Reef is one of the world's most spectacular examples of coral reef-building. Threatened by varied forms of coastal pollution, much of the reef is now protected in a national marine park. *(Hilarie Kavanagh/Getty Images, Inc.—Stone Allstock)*

summers are hot and dry in much of South Australia and in the southwest corner of Western Australia, producing a distinctively Mediterranean climate.

Climates in New Zealand are influenced by latitude, the moderating effects of the Pacific Ocean, and proximity to local mountains or mountain ranges. Most of the North Island is distinctly subtropical. The coastal lowlands near Auckland are mild and wet year-round. Still, local variations can be striking, as the area's volcanic peaks create their own microclimates. On the South Island, conditions become distinctly cooler as one moves poleward. Indeed, the island's southern edge feels the seasonal breath of Antarctic chill, as it lies more than 46° south of the equator. Mountain ranges on New Zealand's South Island also display incredible local variations in precipitation: West-facing slopes are drenched with more than 100 inches (254 centimeters) of precipitation annually, while lowlands to the east average only 25 inches (64 centimeters) per year. The Otago region, inland from Dunedin, sits partially in the rain shadow of the Southern Alps, and its rolling, open landscapes resemble the semiarid expanses of North America's Intermountain West (Figure 14.13).

FIGURE 14.11 • MT. TARANAKI   New Zealand's North Island contains several volcanic peaks, including Mt. Taranaki. The 8,000-foot (2,440-meter) peak offers everything from subtropical forests to challenging ski slopes and attracts both local and international tourists. *(Ken Graham/Ken Graham Agency)*

An Unusual Zoogeography   Isolation and genetics have combined to produce Australia's animal kingdom, surely one of the most unusual on the planet. Long ago separated from the Asia continent, animal species on the Australian landmass have evolved in unique ways. Marsupials, or animals that carry their young in pouches, never faced much competition from other species entering the region and adapted well to the continent's varied natural settings. More than 120 species of marsupials are present in the country. Europeans such as Captain Cook expressed amazement as they gazed upon leaping kangaroos, duck-billed platypuses, hairy-nosed wombats, and snarling Tasmanian devils during their early exploratory visits. Many of these animals, such as the red and gray kangaroo, thrive on the drought-resistant vegetation of the Australian outback.

As in Australia, New Zealand's isolation offered opportunities for the development of unique plant and animal species. Eighty-five percent of the country's native trees

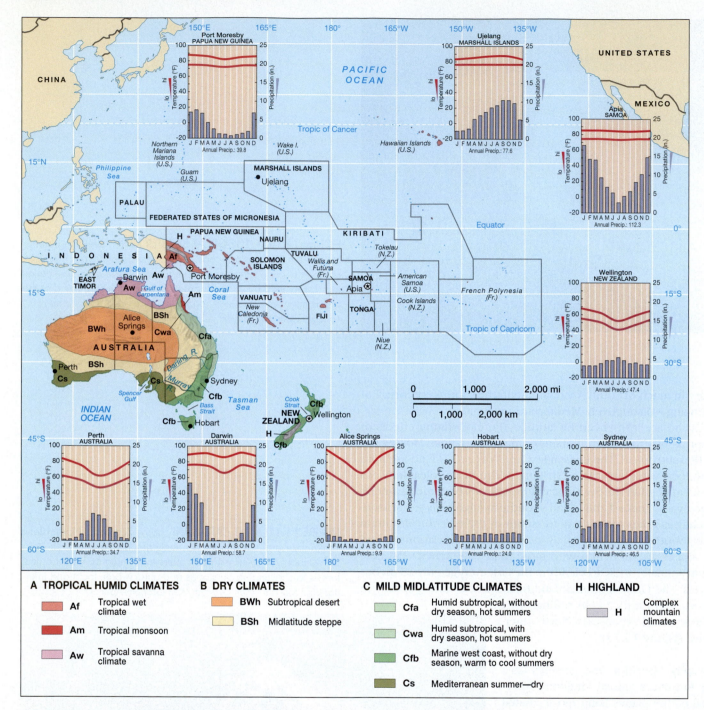

A TROPICAL HUMID CLIMATES

	Af	Tropical wet climate
	Am	Tropical monsoon
	Aw	Tropical savanna climate

B DRY CLIMATES

	BWh	Subtropical desert
	BSh	Midlatitude steppe

C MILD MIDLATITUDE CLIMATES

	Cfa	Humid subtropical, without dry season, hot summers
	Cwa	Humid subtropical, with dry season, hot summers
	Cfb	Marine west coast, without dry season, warm to cool summers
	Cs	Mediterranean summer—dry

H HIGHLAND

	H	Complex mountain climates

FIGURE 14.12 • CLIMATE MAP OF AUSTRALIA AND OCEANIA
Latitude and altitude shape the climatic patterns of the region. Equatorial portions of the Pacific bask in all-year warmth and humidity, while the Australian interior is desert because of the dominance of subtropical high pressure. Cool and moisture-bearing storms of the southern Pacific Ocean bring more moist conditions to New Zealand and certain portions of Australia.

and seed plants are found nowhere else on Earth. Bats are the region's only native mammals, while ancient tuatara reptiles, kiwi birds, and the flightless and now-extinct moas illustrate the country's unique biological legacy.

The Oceanic Realm

The vast expanse of Pacific waters reveals another set of environmental settings that are as rich and complex as they are fragile. Oceanic currents and wind patterns have historically served as natural highways of movement between these island worlds. Those same forces define the rhythm of weather patterns across the region in a broad band that extends more than 20° north (Hawaiian Islands) and south (New Caledonia) of the equator.

Creating Island Landforms   Much of Melanesia and Polynesia is part of the seismically active Pacific Basin. As a result, volcanic eruptions, major earthquakes, and **tsunamis**, or earthquake-induced sea waves, are not uncommon across the region, and they impose major environmental hazards upon the population. For example, volcanic eruptions and earthquakes on the island of New Britain (Papua New Guinea) forced more than 100,000 people from their homes in 1994. Only four years later, a massive tsunami triggered by an offshore earthquake swept across the north coast of New Guinea, killing 3,000 residents and destroying numerous villages. Such events are unfortunately a part of life in this geologically active part of the world.

Most of the islands of Polynesia and Micronesia, however, are truly oceanic, having originated from volcanic activity on the ocean floor. The larger active and recently active volcanoes form **high islands**, which often rise to a considerable elevation and cover a large area. The island of Hawaii, the largest and youngest of the Pacific's high islands, is more than 80 miles (128 kilometers) across and rises to a height of more than 13,000 feet (3,980 meters). Indeed, the entire Hawaiian archipelago exemplifies a geological **hot spot** where moving oceanic crust passes over a supply of magma from Earth's interior, thus creating a chain of volcanic islands. Many of the islands of French Polynesia, including Bora Bora, are smaller examples of high islands (Figure 14.14). Indeed, high islands are widely scattered throughout Micronesia and Polynesia. In tropical latitudes, most high islands are ringed by coral reefs, which quickly grow in the shallow waters near the shore.

The combination of narrow sandy islands, barrier coral reefs, and shallow central lagoons is also known as an **atoll**. The islands and reefs of the atoll characteristically form a circular or oval shape, although some are quite irregular. The world's largest atoll, Kwajalein in Micronesia's Marshall Islands, is 75 miles (120 kilometers) long and 15 miles (24 kilometers) wide. Polynesia and Micronesia are dotted with extensive atoll systems, and a number are found in Melanesia as well.

Patterns of Climate   Many Pacific islands receive abundant precipitation, and high islands in particular are often noted for their heavy rainfall and dense tropical forests (see Figure 14.12). Much of the zone is located in the rainy tropics or in a tropical wet-dry climate region where abundant summer rains and even tropical cyclones can bring heavy seasonal precipitation. Low-lying atolls usually receive less precipitation than high islands and very often experience water shortages. During dry periods, the limited stores of water on these islands are quickly depleted.

Population and Settlement: A Diverse Cultural Landscape

Modern population patterns across the region reflect the combined influences of indigenous and European settlement. In settings such as Australia, New Zealand, and the Hawaiian Islands, European migrations have structured the distribution and concentration of contemporary populations. In contrast, on smaller islands elsewhere in Oceania, population geographies are determined by the needs of native peoples.

Contemporary Population Patterns

Despite the popular stereotypes of life in the outback, modern Australia has one of the most highly urbanized populations in the world. Indeed, 40 percent of the country's residents live within either the Sydney or Melbourne metropolitan areas (Figure 14.15). Australia's eastern and southern areas are home to the majority of its almost 20 million people. Inland, population densities decline as rapidly as the rainfall: Semi-arid hills west of the Great Dividing Range still contain significant rural settlement, but the state's southwestern periphery remains sparsely peopled. New South Wales is the country's most heavily populated state, and its sprawling capital city of Sydney (4 million), focused around one of the world's most magnificent natural harbors, is

FIGURE 14.13 • CENTRAL OTAGO, SOUTH ISLAND   On New Zealand's South Island, the Southern Alps capture rainfall on the west coast but leave areas to the east in a drier rain shadow. As a result, the Central Otago region has a semiarid landscape resembling portions of the western United States. *(John Lamb/Getty Images, Inc. — Stone Allstock)*

FIGURE 14.14 • BORA BORA   Jewel of French Polynesia, Bora Bora displays many of the classic features of Pacific high islands. As the island's central volcanic core erodes away, surrounding coral reefs produce a mix of wave-washed sandy shores and shallow lagoons. *(Paul Chesley/Getty Images, Inc. — Stone Allstock)*

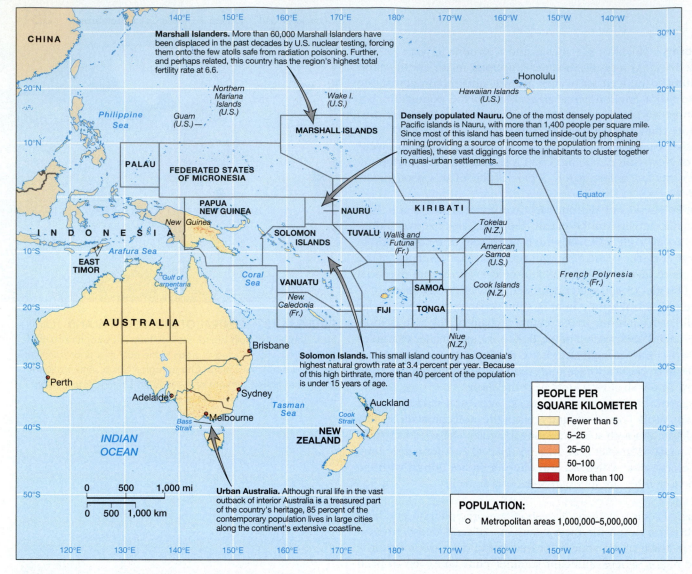

Marshall Islanders. More than 60,000 Marshall Islanders have been displaced in the past decades by U.S. nuclear testing, forcing them onto the few atolls safe from radiation poisoning. Further, and perhaps related, this country has the region's highest total fertility rate at 6.6.

Densely populated Nauru. One of the most densely populated Pacific islands is Nauru, with more than 1,400 people per square mile. Since most of this island has been turned inside-out by phosphate mining (providing a source of income to the population from mining royalties), these vast diggings force the inhabitants to cluster together in quasi-urban settlements.

Solomon Islands. This small island country has Oceania's highest natural growth rate at 3.4 percent per year. Because of this high birthrate, more than 40 percent of the population is under 15 years of age.

Urban Australia. Although rural life in the vast outback of interior Australia is a treasured part of the country's heritage, 85 percent of the contemporary population lives in large cities along the continent's extensive coastline.

PEOPLE PER SQUARE KILOMETER
- Fewer than 5
- 5–25
- 25–50
- 50–100
- More than 100

POPULATION:
○ Metropolitan areas 1,000,000–5,000,000

FIGURE 14.15 • POPULATION MAP OF AUSTRALIA AND OCEANIA Fewer than 30 million people occupy this world region. While Papua New Guinea and many Pacific islands feature mostly rural settlements, most regional residents live in the large urban areas of Australia and New Zealand. Sydney and Melbourne account for almost half of Australia's population, and most New Zealand residents live on the North Island, home to the cities of Auckland and Wellington.

the largest metropolitan area in the entire South Pacific. In the nearby state of Victoria, Melbourne's 3.3 million residents have long competed with Sydney for status as Australia's premiere city, claiming cultural and architectural supremacy over their slightly larger neighbor. In between these two metropolitan giants, the location of the much smaller federal capital of Canberra (population 346,000) represents a classic geopolitical compromise in the same spirit that created Washington, D.C., midway between the populous southern and northern portions of the United States.

Smaller clusters of population are found inland from Australia's eastern coast. More fertile farmlands in the interior of New South Wales and Victoria feature higher population densities than are found in most of the nation's arid heartland. Inland Aboriginal populations are widely but thinly scattered across districts such as northern Western Australia and South Australia, as well as in the Northern Territory, accounting for smaller but regionally important centers of settlement.

The population geography of the rest of Oceania shows a broad distribution of peoples, both native and European, who have clustered near favorable resource opportunities. In New Zealand, more than 70 percent of the country's 3.9 million residents live on the North Island, with Auckland (1 million) dominating the metropolitan scene in the north and the capital city of Wellington (165,000) anchoring settlement along the Cook Strait in the south. Settlement on the South Island is mostly located in the somewhat drier lowlands and coastal districts east of the mountains, with Christchurch (320,000) serving as the largest urban center. Elsewhere, rugged and mountainous terrain on both the North and South Islands feature much lower population densities. Such is not the case in Papua New Guinea: Less than 15 percent of the country's population is urban, and many people live in the isolated, interior highlands.

The nation's largest city is the capital of Port Moresby (200,000), located along the narrow coastal lowland in the far southeastern corner of the country. The largest city on the northern margin of Oceania is Honolulu (1 million), on the island of Oahu, where rapid metropolitan growth since World War II has occurred because of U.S. statehood and the scenic attractions of its mid-Pacific setting.

Legacies of Human Occupancy

The historical settlement of the Pacific realm can never be precisely reconstructed, but humans in several major migrations succeeded in occupying the region over time. The region's remoteness from many of the world's early population centers meant that it often lay beyond the dominant migratory paths of earlier peoples. Even so, settlers found their way to the isolated Australian interior and the far reaches of the Pacific. The pace of new in-migrations increased once Europeans identified the region and its resource potential.

Peopling the Pacific   The large islands of New Guinea and Australia, given their nearness to the Asian landmass, were settled much earlier than the more distant islands of the Pacific. Around 40,000 years ago, the ancestors of today's native Australian or **Aborigine** populations were making their way out of Southeast Asia and into Australia (Figure 14.16). The first Australians most likely arrived using some kind of watercraft. However, since such boats were probably not very

FIGURE 14.16 • PEOPLING THE PACIFIC   Ancestors of Australia's aboriginal population may have made their way into the island continent more than 40,000 years ago. Much more recent settlement of Pacific islands by Austronesian peoples from Southeast Asia shaped cultural patterns across the oceanic portions of the realm. Eastward migrations through the Solomon, Fiji, and Cook islands were followed by later movements to the north and south.

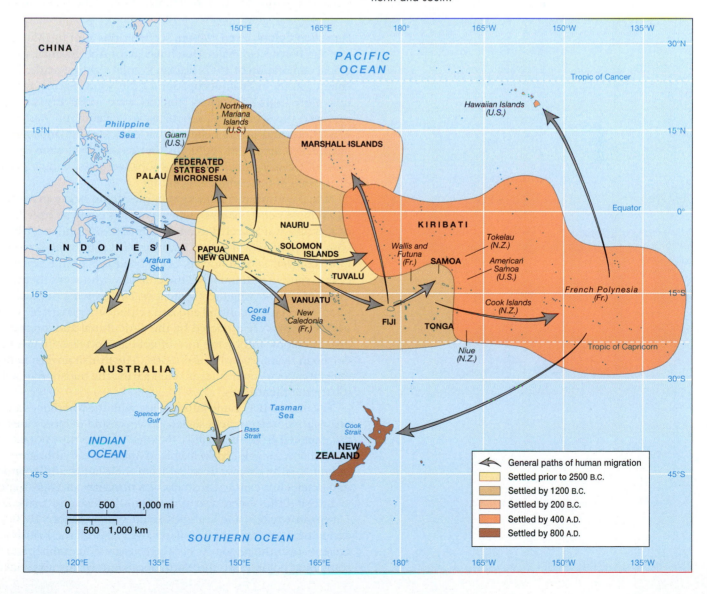

seaworthy, the more distant islands remained inaccessible to humankind for tens of thousands of years. During glacial periods, however, sea levels were much lower than they are now, which would have allowed easier movement to Australia across relatively narrow spans of water. It is not known whether the original Australians arrived in one wave of people or in many, but the available evidence suggests that they soon occupied large portions of the continent, including Tasmania (which was then connected to the mainland by a land bridge).

Eastern Melanesia was settled much later than Australia and New Guinea. By approximately 3,500 years ago, certain Pacific peoples had mastered long-distance sailing and navigation, which eventually opened the entire oceanic realm to human habitation. In that era, people gradually moved east to occupy New Caledonia, the Fiji Islands, and Samoa. From there, later movements took seafaring folk north into Micronesia, with areas such as the Marshall Islands occupied around 2,000 years ago. Continuing movements from Asia further complicated the story of these migrating Melanesians. Some of the migrants mixed culturally and eventually reached western Polynesia, where they formed the core population of the Polynesian people. By A.D. 800, they had reached such distant places as New Zealand, Hawaii, and Easter Island. Debate has centered on whether these Polynesians purposefully set out to colonize new lands or whether they were blown off course during routine voyages, only to end up on new islands. Certainly, population pressures could quickly reach a crisis stage on relatively small islands, encouraging people to make dangerous voyages. Equipped with sturdy outrigger sailing vessels and ample supplies of food, the Polynesians were quickly able to colonize most of the islands they discovered.

European Colonization   About six centuries after the Maori brought New Zealand into the Polynesian realm, Dutch navigator Abel Tasman spotted the islands on his global exploration of 1642. Tasman's initial sighting marked the beginning of a new chapter in the human occupation of the South Pacific. Late in the following century, more lasting European contacts were made. British sea captain James Cook surveyed the shorelines of both New Zealand and Australia between 1768 and 1780. Cook and others believed that these distant lands might be worthy of European development. In addition, other expeditions were exploring the Pacific, and most of Oceania's major island groups were assuming a place on European maps by the end of the eighteenth century.

Actual European colonization of the region began in Australia. The British needed a remote penal colony to which convicts could be exiled. The southeastern coast of Australia was selected as an appropriate site, and in 1788 the First Fleet arrived with 750 prisoners in Botany Bay near what is now modern Sydney. Other fleets and more convicts soon followed, as did boatloads of free settlers. Before long, however, free settlers outnumbered the convicts, who were themselves gradually gaining freedom after serving their sentences. The growing population of English-speaking people soon moved inland and also settled other favorable coastal areas. British and Irish settlers were attracted by the agricultural and stock-raising potential of the distant colony and by the lure of gold and other minerals (a major gold rush occurred in Australia during the 1850s). The British government also encouraged the emigration of its own citizens, often paying the transportation fare of those too poor to afford it themselves.

The new settlers came into conflict with the Aborigines almost immediately after arriving. No treaties were signed, however, and in most cases Aborigines were simply expelled from their lands. In some places, most notably Tasmania, they were hunted down and killed. In mainland Australia, the Aborigines were greatly reduced in numbers by disease, removal from their lands, and economic hardship. By the mid-nineteenth century, Australia was primarily an English-speaking land.

British settlers were also attracted to the lush and fertile lands of New Zealand. European whalers and sealers arrived shortly before 1800, but more permanent agricultural settlement took shape after 1840 as the British formally declared sovereignty over the region. As new arrivals grew in number and the scope of planned settlement colonies on the North and South islands expanded, tensions increased with the native Maori population. Organized in small kingdoms or chiefdoms, the Maori were formidable fighters. In 1845 one native group decided to resist further settlement by Europeans, leading to the widespread Maori wars that engulfed New Zealand until 1870. The British eventually prevailed, however, and the Maori lost most of their land as well as control of their country.

The native Hawaiians also lost control of their lands to more numerous immigrants. Hawaii emerged as a united and powerful kingdom in the early 1800s, and for many years its native rulers limited U.S. and European claims to their islands. Increasing numbers of missionaries and settlers from the United States were allowed in, however, and by the late nineteenth century, control of the Hawaiian economy had largely passed to foreign plantation owners. By 1898, U.S. forces were strong enough to overthrow the Hawaiian monarchy and to annex the islands to the United States.

Modern Settlement Landscapes

The settlement geography of Australia and Oceania offers an interesting mixture of local and global influences. The contemporary cultural landscape still reflects the imprint of indigenous peoples in those settings where native populations remain numerically dominant. Elsewhere, patterns of recent colonization have produced a modern scene mainly shaped by Europeans. The result includes everything from German-owned vineyards in South Australia to houses on New Zealand's South Island that appear to be plucked directly from the British Isles. In addition, processes of economic and cultural globalization during the twentieth century have resulted in urban forms that make cities such as Perth or Auckland look strikingly similar to such places as San Diego or Seattle.

The Urban Transformation Both Australia and New Zealand are highly urbanized, Westernized societies, and thus the vast majority of their populations live in urban and suburban environments (Figure 14.17). As in Europe and North America, much of this urban transformation came during the twentieth century as the rural economy became less labor-intensive and as opportunities for urban manufacturing and service employment grew. As urban landscapes evolved, they took on many of the characteristics of their largely European populations, but they blended these with a strong dose of North American influences, as well as with the unique settings native to each urban place. The result is an urban landscape in which many North Americans are quite comfortable, even though the varied local accents heard on the street and many features of the metropolitan scene are reminders of the strong and lasting attachments to British traditions (Figure 14.18).

The Western-style urban settings in Australia and New Zealand offer a dramatic contrast with the urban landscapes found in less-developed settings in the region. Walk the streets of Port Moresby in Papua New Guinea, and a very different urban landscape is evidence of the large gap between rich and poor within Oceania (Figure 14.19). Rapid growth in Port Moresby, the country's political capital and largest commercial center, has produced many of the classic problems of urban underdevelopment: There is a shortage of adequate housing, the building of roads and schools lags far behind the need, and street crime and alcoholism are on the rise. Elsewhere, urban centers such as Suva (Fiji), Noumea (New Caledonia), and Apia (Samoa) also reflect the economic and cultural tensions generated as local populations are exposed to Western influences. Rapid growth is a common problem in the smaller cities of Oceania because native people from rural areas and nearby islands gravitate toward the job opportunities available. In the past 50 years, the huge global growth of tourism in

FIGURE 14.17 • SYDNEY'S SUBURBS Like North American cities, urban settlements in both New Zealand and Australia now sprawl far beyond their traditional centers as residents seek larger lots, open space, and other amenities of the countryside. Urban planners note how this suburban sprawl complements the outdoor orientation of Australians and New Zealanders. *(Patrick Ward/Corbis)*

FIGURE 14.18 • DOWNTOWN MELBOURNE Metropolitan Melbourne lies along the Yarra River. Capital of the Australian state of Victoria, Melbourne resembles many growing North American cities with its high-rise office buildings, entertainment districts, and downtown urban redevelopment. *(Fritz Prenzel/ Peter Arnold, Inc.)*

FIGURE 14.19 • PORT MORESBY, PAPUA NEW GUINEA
Urban poverty and high crime are common in the city of Port Moresby, the capital of Papua New Guinea. The city's slums, many built out on the water, reflect stresses of recent urban growth as rural residents emigrate from nearby highlands. *(Chris Rainier/Corbis)*

FIGURE 14.20 • CANTERBURY PLAIN   The varied agricultural landscape of South Island's Canterbury Plain offers a mix of grain fields, livestock, orchard crops, and vegetable gardens. The rugged Southern Alps are a dramatic backdrop to this productive region. *(Robert Frerck/Woodfin Camp & Associates)*

places such as Fiji and Samoa has also transformed the urban scene: Nineteenth-century village life has often been replaced by a landscape of souvenir shops, honking taxicabs, and crowded seaside resorts.

The Rural Scene   Rural landscapes across Australia and the Pacific region also express a complex mosaic of cultural and economic influences. In some settings, Australian Aborigines or native Papua New Guinea Highlanders can still be found in their familiar homelands, their traditional lifeways and settlements barely changed from pre-European times. Yet such settlement landscapes are becoming increasingly rare. Global influences penetrate the scene as the cash economy, foreign tourism and investment, and the currents of popular culture work their way from city to countryside.

Much of rural Australia is too dry for farming or serves as only marginally valuable agricultural land. Although modest in size, the area in crops has doubled since 1960 as increased use of fertilizers, more widespread irrigation, and more aggressive rabbit eradication efforts have opened up new areas for development. Much of the remainder of the interior, however, features range-fed livestock, areas beyond the pale of any agricultural potential, and isolated areas where Aboriginal peoples still pursue their traditional forms of hunting and gathering.

Sheep and cattle dominate rural Australia's livestock economy. Many rural landscapes in the interior of New South Wales, Western Australia, and Victoria, for example, are oriented around isolated sheep stations, or ranch operations that move the flocks from one large pasture to the next. Cattle can sometimes be found in these same areas, although many of the more extensive, range-fed cattle operations are concentrated farther north in Queensland. Croplands also vary across the region. Sometimes mingling with the sheep country, a band of commercial wheat farming includes southern Queensland; the moister interiors of New South Wales, Victoria, and South Australia; and a swath of favorable land east and north of Perth. Elsewhere, specialized sugarcane operations thrive along the narrow, warm, and humid coastal strip of Queensland. To the south and west, productive irrigated agriculture has developed in places such as the Murray River Basin, allowing for the production of orchard crops and vegetables. **Viticulture**, or grape cultivation, increasingly shapes the rural scene in places such as South Australia's Barossa Valley, New South Wales' Riverina district, and Western Australia's Swan Valley. Indeed, the area under grape cultivation grew by 50 percent between 1991 and 1998 as the popular Chardonnay, Cabernet Sauvignon, and Shiraz varieties boosted wine production to revenues of more than $540 million per year.

New Zealand's Landscapes   Although much smaller in area, New Zealand's rural settlement landscape includes a variety of agricultural activities. Pastoral activities clearly dominate the New Zealand scene, with the vast majority of agricultural land devoted to livestock production, particularly sheep grazing and dairying. Commercial livestock outnumber people in New Zealand by a ratio of more than 20 to 1, and this is apparent everywhere on the rural scene. Dairy operations are present, mostly in the lowlands of the north, where they sometimes mingle with suburban landscapes in the vicinity of Auckland. One of the largest zones of more specialized cropping spreads across the fertile Canterbury Plain near Christchurch (Figure 14.20). This spectacular South Island setting proved fertile ground for English settlement and continues to feature a varied landscape of pastures, grain fields, orchards, and vegetable gardens, all spread beneath the towering peaks of the Southern Alps.

Rural Oceania   Elsewhere in Oceania, varied influences shape the rural scene. On high islands with more water, denser populations take advantage of more diverse agricultural opportunities than are usually found on the more-barren low islands, where fishing is often more important. Several types of rural settlement can be identified across the island realm. In rural New Guinea, village-centered shifting cultivation dominates: Farmers clear a patch of forest and then, after a few years, shift to another patch, thus practicing a form of land rotation. Subsistence foods such as sweet potatoes, taro (another starchy root crop), coconut palms, bananas, and other garden crops often are found in the same field, and growing numbers of planters also include commercial crops such as coffee. In other parts of Oceania, traditional agricultural patterns are similar (Figure 14.21). Commercial plantation agriculture has also made its mark in many more accessible rural settings. In these places, settlements consist of worker housing near crops that are typically controlled by absentee landowners. For example, copra (coconut), cocoa, and coffee operations have transformed many agricultural settings in places such as the Solomon Islands and Vanuatu. Sugarcane plantations have reshaped other island settings, particularly in Fiji and Hawaii.

Diverse Demographic Paths

A variety of population-related issues face residents of the region today. In Australia and New Zealand, while populations grew rapidly (mostly from natural increases) in the twentieth century, today's low birthrates parallel the pattern in North America. Just as in the United States and Canada, however, significant population shifts within these countries continue to create challenges. For example, the departure of farmers from Australia's wheat-growing and sheep-raising interior mirrors similar processes at work in the rural Midwest of the United States and in the Canadian prairies. Communities see many of their productive young people and professionals leave for the better employment opportunities of the city.

Different demographic challenges grip many less-developed island nations of Oceania. Population growth rates are often above 2 percent per year and are even higher in countries such as Vanuatu and the Solomon Islands. While the larger islands of Melanesia contain some room for settlement expansion, competitive pressures from commercial mining and logging operations limit the amount of new agricultural land that will probably be available in the future. On some of the smaller island groups in Micronesia and Polynesia, population growth is a more pressing problem. Tuvalu (north of Fiji), for example, has just over 10,000 inhabitants, but they are crowded onto a land area of about 10 square miles (26 square kilometers), making it one of the world's more thickly populated countries. In atoll environments, a single island might be considered overpopulated if it supports only 100 people. Thus, even relatively small population centers in the Pacific Islands find it difficult to cope with higher birthrates. Making matters worse, many young people are migrating to already crowded urban centers in these island nations.

FIGURE 14.21 • YAM HARVEST These farmers on the Melanesian island of Vakuta (Papua New Guinea) are harvesting yams. Traditional tropical agriculture features a mix of crops, often grown in the same field. Where possible, fields are periodically rotated to maintain productivity. *(Peter Essick/Aurora & Quanta Productions)*

Cultural Coherence and Diversity: A Global Crossroads

Australia and Oceania offer excellent examples of how cultural geographies are transformed as different groups migrate to a region, interact with one another, and evolve over time. As Europeans and other outsiders arrived in the region, colonization forced native peoples to adjust. Worldwide processes of globalization have also redefined the region's cultural geography, provoking fears of homogenization while at the same time promoting more cultural preservation efforts as native groups attempt to protect their heritage. A contemporary snapshot of the region reveals that all of these processes remain at work today and help account for the diverse cultural landscapes found in Australia, New Zealand, and the other Pacific islands.

Multicultural Australia

Australia's cultural patterns illustrate many of these fundamental processes at work. Today, while still dominated by its colonial European roots, the country's multicultural character is becoming increasingly visible as native peoples assert their cultural identity and as varied immigrant populations play larger roles in society, particularly within major metropolitan areas.

Aboriginal Imprints For thousands of years, Australia's Aborigines dominated the cultural geography of the continent. They never practiced agriculture, opting instead for a hunting-gathering way of life that persisted up to the time of the European conquest. As the population consisted of foragers and hunters living in a relatively dry land, settlement densities remained low; tribal groups were often isolated from one another; and overall populations probably never numbered more than 300,000 inhabitants. To survive, Aborigines developed great adaptive skills, often subsisting in harsh environments that Europeans avoided. Because these people were clustered in many different areas, their language became fragmented. Although precise counts vary, there were probably 250 languages spoken at the time of European contact, and almost 50 indigenous languages can still be found.

Radical cultural and geographical changes accompanied the arrival of Europeans, and Aboriginal populations were decimated in the process. The geographical results of colonization were striking as Aboriginal settlements were relocated to the sparsely settled interior, particularly in northern and central Australia, where fewer Europeans competed for land. In most cases, the European attitude toward the Aboriginal population was even more discriminatory than it was toward native peoples of the Americas.

Today, Aboriginal cultures persevere in Australia, and a growing native peoples movement is similar to activities in the Americas. Indigenous peoples account for approximately 2 percent (or 400,000) of Australia's population but their geographical distribution changed dramatically in the last century. Aborigines account for almost 30 percent of the Northern Territory's population (many of these near Darwin), and other large native reserves are located in northern Queensland and Western Australia. Most native peoples, however, live in the same urban areas that dominate the country's overall population geography. Indeed, more than 70 percent of Aborigines live in cities, and very few of them still practice traditional hunting-and-gathering lifestyles. Processes of cultural assimilation are clearly at work: Urban Aborigines are frequently employed in service occupations, Christianity has often replaced traditional animist religions; and only 13 percent of the native population still speaks a native language (Figure 14.22).

Still, forces of diversity are at work, suggesting a growing Aboriginal interest in preserving traditional cultural values. Particularly in the outback, a handful of Aboriginal languages remain strong and have growing numbers of speakers. In addition, cultural leaders are preserving some aspects of Aboriginal spiritualism, and these religious practices often link local populations to surrounding places and natural features that are considered sacred. In fact, a growing number of these sacred locations are at the center of land-use controversies (such as mining on sacred lands) between Aboriginal populations and Australia's European majority. The future for Aboriginal cultures remains unclear: Pressures for cultural assimilation will be intense as many native peoples move to more Western-oriented urban settlements and lifestyles. At the same time, rapid rates of natural increase (almost twice the national average) and a growing cultural awareness of Aboriginal traditions will work to preserve elements of the country's indigenous cultures.

A Land of Immigrants Most Australians reflect the continent's more recent European-dominated migration history, but even these patterns have become more complex as a rising tide of Asian cultures becomes important. Overall, more than 70 percent of Australia's population continues to reflect a British or Irish cultural heritage. These groups dominated many of the nineteenth- and early twentieth-century migrations into the country, and the close cultural ties to the British Isles remain strong. Large numbers of Italians, Greeks, and Germans have added diversity to the European mix. Some of these migrants became involved in specialized agricultural activities, particularly in southeastern Australia, while others moved to the cities, where they added ethnic variety to the mostly Anglo mix.

A need for laborers along the fertile Queensland coast also caused European plantation owners to import inexpensive workers from the Solomons and New Hebrides. These Pacific Island laborers, known as **kanakas**, were spatially and socially segregated from their Anglo employers but further diversified the cultural mix of Queens-

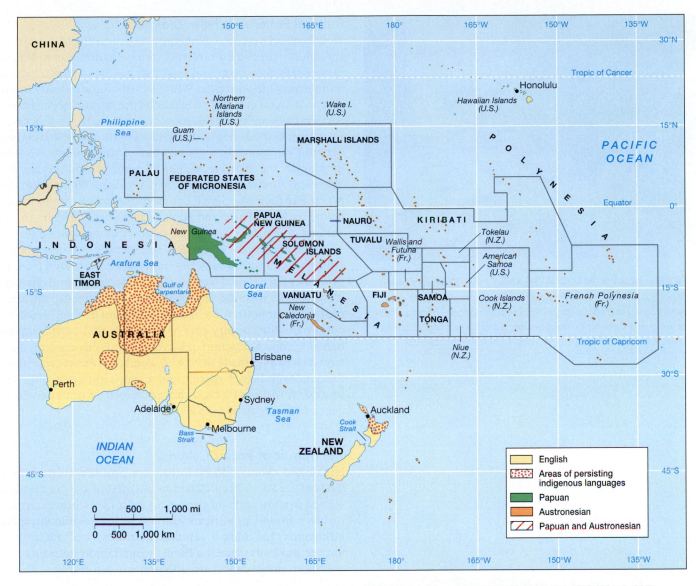

FIGURE 14.22 • LANGUAGE MAP OF AUSTRALIA AND OCEANIA   While English is spoken by most residents, native peoples and their linguistic traditions remain an important cultural and political force in both Australia and New Zealand. Elsewhere, traditional Papuan and Austronesian languages dominate Oceania. The French colonial legacy also persists in select Pacific locations. Great linguistic diversity has shaped the cultural geography of Melanesia, and more than 1,000 languages have been identified in Papua New Guinea.

land's "sugar coast." Historically, however, nonwhite migrations to the country were strictly limited by what is often termed a **White Australia Policy**, in which governmental guidelines promoted European and North American immigration at the expense of other groups. This remained national policy until 1973.

Recent migration trends have reversed this historical bias, and more diverse inflows of new workers and residents are adding to the country's multicultural character. Since the 1970s, the government's Migration Program has been dominated by a variety of people chosen on the basis of their educational background and potential for succeeding economically in Australian society. For example, a growing number of families have come from places such as China, India, Malaysia, and the Philippines. Smaller numbers have qualified as migrants through their New Zealand citizenship, while others have arrived as refugees from troubled parts of the world, such as Vietnam and Yugoslavia. The result is a more diverse foreign-born population. Indeed, 22 percent of Australia's people are now immigrants, reflecting the country's global popularity as a migration destination. In the early twenty-first century, almost 40 percent of the settlers arriving in the country have been from Asia. Major cities offer particularly attractive possibilities: Sydney's Asian population already exceeds 10 percent and is growing rapidly, while Perth's culture and economy are increasingly linked to its Asian neighbors.

FIGURE 14.23 • MAORI ARTISANS New Zealand's native Maori population actively preserves its cultural traditions and has recently increased its political role in national affairs. These artisans are carving decorations for a traditional Maori canoe. *(Arno Gasteiger/Bilderberg/Aurora & Quanta Productions)*

FIGURE 14.24 • TONGA VILLAGE Although the economic and technological effects of globalization have arrived in Tonga, village life remains important in many Polynesian settings. Most village housing in these tropical environments reflects the use of locally available construction materials. *(Ted Streshinsky/Corbis)*

Cultural Patterns in New Zealand

New Zealand's cultural geography broadly reflects the patterns seen in Australia, although the precise cultural mix differs slightly. Native Maori populations are more numerically important and culturally visible in New Zealand than their Aboriginal counterparts in Australia. While British colonization clearly mandated the dominance of Anglo cultural traditions by the late nineteenth century, Maori populations survived, although they lost most of their land in the process. After the initial decline, native populations began rebounding in the twentieth century, and today the Maori account for more than 15 percent of the country's 3.9 million residents. Geographically, the Maori remain most numerous on the North Island, including a sizable concentration in metropolitan Auckland. While urban living is on the rise, many Maori, like their Aboriginal counterparts, are also committed to preserving their religion, traditional arts, and Polynesian lifeways (Figure 14.23). In addition, Maori is now an official language in the country, along with English.

While many New Zealanders still identify with their largely British heritage, the country's cultural identity has increasingly separated from its British roots. Several processes have forged New Zealand's special cultural character. As Britain tightened its own links with the European continent after World War II, New Zealanders increasingly formed a more independent and diverse identity. In many ways, popular culture ties the country ever more closely to Australia, the United States, and continental Europe, a function of increasingly global mass media. For example, the most famous film ever made in New Zealand, *The Piano*, was produced by an Australian, financed with French money, and reached its largest audiences in the United States. Indeed, by the late 1990s, locally produced programming made up less than 20 percent of New Zealand television.

The Mosaic of Pacific Cultures

Native and exotic cultural influences produce a variety of cultures across the islands of the South Pacific. In more isolated places, traditional cultures are largely insulated from outside influences. In most cases, however, modern life in the islands revolves around an intricate cultural and economic interplay of local and Western influences. One thing is certain: The relative cultural insularity of the past is gone forever, and in its place is a Pacific realm rapidly adjusting to powerful forces of colonization, global capitalism, and popular culture.

Language Geography The modern language map reveals some significant cultural patterns that both unite and divide the region (see Figure 14.22). Most of the native languages of Oceania belong to the Austronesian language family, which encompasses wide expanses of the Pacific, much of insular Southeast Asia, and Madagascar. Linguists hypothesize that the first great oceanic mariners spoke Austronesian languages, and thus spread them throughout this vast realm of islands and oceans. Within the broad Austronesian family, the Malayo-Polynesian subfamily includes most of the related languages of Micronesia and Polynesia, suggesting a common cultural and migratory history for these widespread peoples.

Melanesia's language geography is more complex and still incompletely understood by outside experts: While coastal peoples often speak languages brought to the region by the seafaring Austronesians, more isolated highland cultures, particularly on the island of New Guinea, speak varied Papuan languages. Indeed, the linguistic complexity of that island is so complex—more than 1,000 languages have been identified—that many experts question whether they even constitute a unified "Papuan family" of related languages. Some scholars estimate that half of New Guinea's languages are spoken by fewer than 500 persons, suggesting the region's topography plays a strong role in isolating cultural groups. These New Guinea highlands may hold some of the world's few remaining **uncontacted peoples**, cultural groups that have yet to be "discovered" by the Western world.

Village Life   Traditional patterns of social life are as complex and varied as the language map. In many cases, however, life revolves around predictable settings. For example, across much of Melanesia, including Papua New Guinea, most people live in small villages often occupied by a single clan or family group. Many of these traditional villages contain fewer than 500 residents, although some larger communities may house more than 1,000 people. Life often revolves around the gathering and growing of food, annual rituals and festivals, and complex networks of kin-based social interactions

Traditional Polynesian settings also focus on village life (Figure 14.24), although there are often strong class-based relationships between local elites (often religious leaders) and ordinary residents. Polynesian villages are also more likely linked to other islands by wider cultural and political ties. Despite the Western stereotype of depicting Polynesian communities in idyllic and peaceful terms, violent warfare was actually quite common across much of the region prior to European contact.

External Cultural Influences   While traditional culture worlds persist in some settings, most Pacific islands have witnessed tremendous cultural transformations in the past 150 years. Outsiders from Europe, the United States, and Asia brought new settlers, values, and technological innovations that have forever changed Oceania's cultural geography and its place in the larger world. The result is a modern setting where Pidgin English has mostly replaced native languages, Hinduism is practiced on remote Pacific Islands, and traditional fishing peoples now work at resort hotels and golf course complexes.

European colonialism transformed the cultural geography of the Pacific world by introducing new political and economic systems. As well, the region's cultural makeup was changed by new people migrating into the Pacific islands. Hawaii illustrates the pattern. By the mid-nineteenth century, Hawaii's King Kamehameha was already entertaining a varied assortment of whalers, Christian missionaries, traders, and navy officers from Europe and the United States. A small elite group of **haoles**, or light-skinned European and American foreigners, were successfully profiting from commercial sugarcane plantations and Pacific shipping contracts.

Labor shortages on the islands, however, led to the importation of Chinese, Portuguese, and Japanese workers who further complicated the region's cultural geography. By 1900, the Japanese had become a dominant part of the island workforce. The United States formally annexed the islands in 1898. The cultural mix revealed in the Hawaiian census of 1910 suggests the magnitude of change: More than 55 percent of the population was Asian (mostly Japanese and Chinese), native peoples made up another 20 percent, and about 15 percent (mostly imported European workers) were white. By the end of the twentieth century, the Asian population was less dominant but more ethnically varied, about 40 percent of Hawaii's residents were white, and the small number of remaining native Hawaiians had been joined by an increasingly diverse group of other Pacific Islanders. In addition, ethnic mixing has produced a rich mosaic of Hawaiian cultures that offer a unique blend of North American, Asian, Pacific Island, and European influences (Figure 14.25).

Hawaii's story has been played out in many other Pacific island settings. In the Mariana Islands, Guam was absorbed into America's Pacific empire as part of the Spanish-American War in 1898. Thereafter, not only did native peoples feel the effects of Americanization (the island remains a self-governing U.S. territory today), but thousands of Filipinos were moved there to supplement its modest labor force. To the southeast, the British-controlled Fiji Islands offered similar opportunities for redefining Oceania's cultural mix. The same sugar plantation economy that spurred changes in Hawaii prompted the British to import thousands of South Asian laborers to Fiji. The descendents of these Indians (most practice Hinduism) now constitute almost half the island country's population and often come into sharp conflict with the native Fijians (Figure 14.26). In French-controlled portions of the realm, small groups of traders and plantation owners filtered into the Society Islands (Tahiti), but a larger group of French colonial settlers (many originally a part of a penal colony) had a major impact on the cultural makeup of New Caledonia. Still a French

FIGURE 14.25 • MULTICULTURAL HAWAIIANS   Many residents of the Hawaiian Islands represent a blend of Pacific Island, Asian, and European influences. These young women express a mixture of Polynesian and Asian ancestors. *(Porterfield/ Chickering/Photo Researchers, Inc.)*

FIGURE 14.26 • SOUTH ASIANS IN FIJI   British sugar plantation owners imported thousands of South Asian workers to Fiji during the colonial era. Today almost half of Fiji's population is South Asian, including many urban residents. This group is often in conflict with native Fijians. *(Frank Fournier/Woodfin Camp & Associates)*

colony, New Caledonia's population is more than one-third French, and its capital city of Noumea reveals a cultural setting forged from French and Melanesian traditions.

Given the frequency of contact between different island cultures, it is no surprise that people have generated new forms of intercultural communication. For example, several forms of **Pidgin English** (also known simply as *Pijin*) are found in the Solomons, Vanuatu, and New Guinea, where it is the major language used between ethnic groups. In Pijin, a largely English vocabulary is reworked and blended with Melanesian grammar. Pijin's origin is commonly traced to nineteenth-century Chinese sandalwood traders ("pijin" is the Chinese pronunciation of the word for "business"). While of historical origin, Pijin is now becoming a "globalized" language of sorts in Oceania as trade and even political ties develop between different native island groups.

A tidal wave of outside influences since World War II has produced cultural changes as well as growing indigenous responses designed to preserve traditional values. Some groups, particularly in Melanesia, remain more isolated from the outside world, although even there growing demands for natural resources offer an avenue for increasing Western or Asian contacts. In many settings, however, the global growth of tourism has brought Oceania into the relatively easy reach of wealthier Europeans, North Americans, Asians, and Australians. The Hawaiian Islands, Fiji, French Polynesia, and American Samoa are being joined by an increasing number of other island tourist destinations. While offering tremendous economic benefits to certain places, the onrush of tourists and their consumer-driven values has often come into sharp conflict with native cultures.

Geopolitical Framework: A Land of Changing Boundaries

Geopolitical space across Australia and Oceania has been reshaped many times as different cultural groups and political powers have asserted themselves across the region. Present patterns of political organization are merely a snapshot in time and perhaps more likely than in most areas of the world to be redefined on future maps.

More specifically, Pacific geopolitics reflect a complex interplay of local, colonial-era, and global-scale forces. The complexities become apparent in the story of Micronesia's Marshall Islands. This sprinkling of islands and atolls (covering 70 square miles, or 180 square kilometers, of land) historically consisted of many various ethnic groups that made up small political units. In 1914, the Japanese moved into the islands, and the area remained under their control until 1944, when U.S. troops occupied the region. Following World War II, a United Nations trust territory (administered by the United States) was created across a wide swath of Micronesia, including the Marshall group. Demands for local self-government grew during the 1960s and 1970s, resulting in a new constitution and independence for the Marshall Islanders by the early 1990s. Today, still benefiting from U.S. aid, government officials in the modest capital city on Majuro Atoll struggle to unite island populations, protect large maritime sea claims, and resolve a generation of legal and medical problems that grew from U.S. nuclear bomb testing in the region (Figure 14.27). Similar stories are typical across the realm, suggesting a twenty-first-century political geography that is still very much in the making.

Roads to Independence
The modern political states of the region have arrived at independence along many different paths. Other political units remain colonial entities to this day. The newness and fluidity of the political boundaries are remarkable: The region's oldest independent states are Australia and New Zealand, and both were twentieth-century creations that are only now considering whether they want to complete their formal political separation from the British Crown. Elsewhere, political ties between colony and mother country are even closer and more lasting. Even many of the newly independent Pacific **microstates**, with their tiny overall land areas, keep special political and economic ties to countries such as the United States.

FIGURE 14.27 • MARSHALL ISLANDS The political control of Micronesia has shifted numerous times during the last two centuries. While now part of the independent Marshall Islands, these palm-fringed beaches were once claimed by Spanish, German, Japanese, and U.S. interests. *(Douglas Peebles Photography)*

Independent Australia (1901) and New Zealand (1907) gradually created their own political identities, yet both still struggle with the final shape of these identities. Although Australia became a commonwealth in 1901, it still acknowledges the British Crown as the symbolic head of its government (Figure 14.28). A national referendum in 1999 forced Australians to decide whether they would like their country to drop this remaining tie to Britain and instead become a genuine republic with its own president replacing the British queen as head of state. However, a majority of 55 percent voted to retain Australia's ties to the Crown. Australia, like the United States, is a federal country, with each of its six states having significant powers. The Northern Territory, for unique historical reasons, however, remains directly under the authority of the central government. In New Zealand, formal legislative links with Great Britain were not broken until 1947. In 1994 some New Zealand officials began discussing the same formal break with the British Crown being debated by the Australians.

Elsewhere in the Pacific, colonial ties were cut even more slowly, and the process has not yet been completed. In the 1970s, Britain and Australia began giving up their colonial empires in the Pacific. Fiji (Great Britain) gained independence in 1970, followed by Papua New Guinea (Australia) in 1975 and the Solomon Islands (Great Britain) in 1978. The small island nations of Kiribati and Tuvalu (Great Britain) also became independent in the late 1970s.

The United States has recently turned over most of its Micronesian territories to local governments, while still holding a large influence in the area. After gaining these islands from Japan in the 1940s, the U.S. government provided large monetary subsidies to islanders but also utilized a number of islands for military purposes. Bikini Atoll was destroyed by nuclear tests, and the large lagoon of Kwajalein Atoll was used as a giant missile target. A major naval base, moreover, was established in Palau, the westernmost archipelago of Oceania. By the early 1990s, both the Marshall Islands and the Federated States of Micronesia (including the Caroline Islands) had gained independence. Their ties to the United States, however, remain close. A number of other Pacific islands remain under U.S. administration. Palau is a U.S. "trust territory," although the Palauans have some local autonomy. The people of the Northern Marianas chose to become a "self-governing commonwealth in association with the United States," a rather vague political position that allows them to become U.S. citizens. The residents of self-governing Guam and American Samoa are also U.S. citizens.

Other colonial powers were less inclined to give up their oceanic possessions. New Zealand still controls substantial territories in Polynesia, including the Cook Islands, Tokelau, and the island of Niue. France has even more extensive holdings in the region. Its largest maritime possession is French Polynesia, which includes a large expanse of mid-Pacific territory. To the west, France still controls the much smaller territory of Wallis and Futuna in Polynesia and the larger island of New Caledonia in Melanesia.

Persisting Geopolitical Tensions

Cultural diversity, colonial legacy, youthful states, and a rapidly changing political map contribute to ongoing geopolitical tensions within the Pacific world (Figure 14.29). Indeed, some of these conflicts have consequences that extend far beyond the boundaries of the region. Others are more locally based but are still reminders of the difficulties that occur as political space is redefined across varied natural and cultural settings.

Native Rights in Australia and New Zealand
Indigenous peoples in both Australia and New Zealand have used the political process to gain more control over land and resources in their two countries. Indeed, the strategies these native groups have used parallel efforts in the Americas and elsewhere. In Australia, Aboriginal groups are discovering newfound political power from both more effective lobbying efforts by native groups and a more sympathetic federal government. Since land treaties were generally not signed with Aborigines as whites conquered the continent, native

FIGURE 14.28 • BRITAIN'S QUEEN IN AUSTRALIA   By the voters' choice, Australia retains its links to the British Crown as their symbolic leader. Despite governmental action to make Australia a fully independent republic with its own president—instead of the Queen—as head of state, in a national referendum in 1999, voters chose to retain traditional ties to the British Crown. Here, Queen Elizabeth II visits Aboriginal dancers at the Muda Aboriginal Language and Culture Centre in the Australian outback. *(AP/Wide World Photos)*

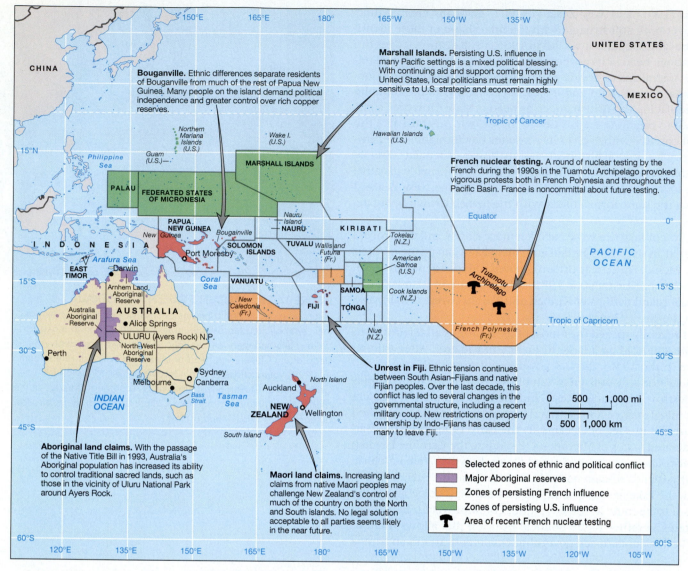

FIGURE 14.29 • GEOPOLITICAL ISSUES IN AUSTRALIA AND OCEANIA Native land claim issues increasingly shape domestic politics in Australia and New Zealand. Elsewhere, ethnic conflicts have raised political tensions in settings such as Fiji and Papua New Guinea. Colonialism's impact continues as well: American and French interests remain particularly visible in the region.

peoples originally had no legal land rights whatsoever. More recently, the Australian government established a number of Aboriginal reserves, particularly in the Northern Territory, and expanded Aboriginal control over sacred national parklands such as Uluru (Ayers Rock; Figure 14.30). Further concessions to indigenous groups were made in 1993 as the government passed the **Native Title Bill**, which compensated Aborigines for lands already given up, gave them the right to gain title to unclaimed lands they still occupied, and provided them with legal standing to deal with mining companies in native-settled areas. However, efforts to expand Aboriginal land rights have met strong opposition. In 1996 an Australian court ruled that pastoral leases (the form of land tenure held by the cattle and sheep ranchers who control most of the outback) do not necessarily negate or replace Aboriginal land rights. Grazing interests were infuriated, which led the government to respond that Aboriginal claims allow the visiting of sacred sites and some hunting and gathering but do not give native peoples complete economic control over the land.

In New Zealand, Maori land claims have generated similar controversies in recent years. The Maori constitute a far larger proportion of the overall population, and the lands that they claim tend to be much more valuable, further complicating the issue. Recent protests include civic disobedience and demonstrations; growing Maori land claims over much of North and South islands; and a call to return the country's name to the indigenous **Aotearoa**, "Land of the Long White Cloud." The government response, complete with a 1995 visit from

Queen Elizabeth, has been to acknowledge increased Maori land and fishing rights as well as to propose a series of financial and land settlements that have yet to be agreed to by the Maoris.

Conflicts in Oceania Other geopolitical issues simmer elsewhere in the Pacific, periodically threatening to further redefine the region's territorial boundaries. As mentioned earlier, ethnic differences in Fiji have threatened to tear apart that small island nation.

Papua New Guinea (PNG) must also contend with ethnic tensions. The country is composed of different cultural groups, many of which have a long history of mutual hostility. Most of these peoples now get along with each other reasonably well, although tribal fighting occasionally breaks out in highland market towns. A much bigger problem for the national government has been the rebellion on Bougainville. This sizable island, which has large reserves of copper and other minerals, is located in the Solomon archipelago but belongs to PNG because Germany colonized it in the late 1800s and it thus became politically attached to the eastern portion of New Guinea. Many of Bougainville's native residents believe that their resources are being exploited by foreign interests and by an unsympathetic national government, and they demand local control. Papua New Guinea has reacted with military force; about 5 percent of the island's entire population has already been killed in the conflict; and recent efforts have failed to create a more stable regional government.

The continued French colonial presence within the Pacific region has also created political uncertainties, both in relations with native peoples and between the French and other independent states in the area. Continued French rule in New Caledonia has provoked much local opposition. This large island has large mineral reserves (especially of nickel) and sizable numbers of French colonists. French settlement and exploitation of mineral resources angered many native peoples. By the 1980s a local independence movement was gaining strength, but in 1987 and 1998 the island's residents (indigenous and French immigrants alike) voted to remain under French rule, at least until 2018. Still, an underground independence movement continues to operate. Troubles have also arisen in French Polynesia. Although the region receives large subsidies from the French and residents have voted to remain a colony, a large minority of the population opposes French control and demands independence. The independence movement was greatly strengthened in 1995 when France decided to resume nuclear testing in the Tuamotu Archipelago. Those activities angered not only the people of French Polynesia (there was antigovernment rioting in Tahiti), but also those of other Pacific countries and territories. Australian leaders called the tests "an act of stupidity," New Zealand recalled its ambassador in protest, and other Pacific nations from Japan to Chile registered their disapproval. Future French nuclear policies in the Pacific will undoubtedly continue to impact geopolitical relations far beyond Polynesia.

A Regional and Global Identity?

Australia and New Zealand have emerged to play key political roles in the South Pacific. Although these two countries sometimes disagree on strategic and military matters, their size, wealth, and collective political influence in the region make them important forces for political stability. Special colonial relationships still connect these nations with present and former Pacific holdings. Australia maintains close political ties to its former colony of Papua New Guinea, and New Zealand's continuing control over Niue, Tokelau, and the Cook Islands in Polynesia suggests that its political influence extends well beyond its borders. When political and ethnic conflicts arise elsewhere in Oceania, Australia and New Zealand are often involved in negotiating peace settlements. Recently, for example, both nations assisted in mediating ongoing disputes on Papua New Guinea's island of Bougainville. In general, the two countries enjoy close political and strategic relations and participate in joint military efforts in the region. Given the other global interests in the region, however, it remains unclear whether these nations can or wish to assert their political dominance across the entire South Pacific.

FIGURE 14.30 • ABORIGINES AT ULURU NATIONAL PARK
Australian Aborigines gathered recently at Uluru National Park to celebrate their increased political control over the region. Since then, the Native Title Bill has promoted numerous land cessions and further legal settlements. *(Michael Jensen/Auscape International Pty. Ltd.)*

Economic and Social Development: A Hard Path to Paradise

Wealth and poverty coexist in the Pacific realm, but regional patterns are complex. Affluent Australia and New Zealand, for example, also contain pockets of pronounced poverty. On the other hand, Oceania offers varied settings that include well-fed subsistence-based populations, relatively prosperous and more commercialized economies, and truly malnourished and impoverished peoples highly dependent on limited government assistance. The twenty-first century poses significant economic challenges for the entire region. Its nations have small populations and domestic markets, thus the realm retains a continuing peripheral position in the global economy.

The Australian and New Zealand Economies

Much of Australia's past economic wealth has been built upon the cheap extraction and export of abundant raw materials. Export-oriented agriculture has long been one of the key supports of Australia's economy. Australian agriculture is highly productive in terms of labor input, and it produces a wide variety of both temperate and tropical crops, as well as huge quantities of beef and wool for world markets. While farm exports are still important to the economy, the mining sector has grown much more rapidly since 1970. Today, Australia is one of the world's mining superpowers. The years since the 1850s gold rush in Victoria have seen a huge expansion in the nation's mineral output, a pattern that accelerated in the 1960s. Among Australia's many assets are coal; rich reserves of iron ore, particularly in Western Australia; and an assortment of other metals, such as bauxite (for aluminum), copper, gold, nickel, lead, and zinc. Indeed, the New South Wales-based Broken Hill Proprietary Company (BHP) is one of the world's largest mining corporations. Given the country's small population, much of that mineral base is consumed elsewhere, making Australia a huge source of raw materials for developed nations, particularly Japan.

Although the nation's mineral wealth served it well in the boom years of the 1970s, recent global economic trends have been less favorable for its economic fortunes. Australia's manufacturing sector continues to be a concern. Many of its industries focus on the simple processing of raw materials, but often these are not highly profitable operations. At the same time, small domestic markets discourage the growth of more specialized industries. What has been lacking is an innovative spirit in the fast-growing high-technology and information-based industries that have powered recent economic expansion in North America, Japan, and Europe. That may be changing: Recent changes in government policy have encouraged more domestic investment, higher savings rates, and more rapid economic growth. Growing numbers of Asian immigrants and economic links with potential Asian markets also offers promise for the future. In addition, an expanding tourism industry is helping to diversify the economy. More than 7 percent of the nation's workforce is now devoted to serving the needs of more than 4 million visitors annually. Popular destinations include Melbourne and Sydney, as well as recreational settings such as Queensland's resort-filled Gold Coast, the Great Barrier Reef, and the vast arid outback. Along the Gold Coast, most luxury hotels are owned by Japanese firms and provide a bilingual resort experience for their Asian clientele.

New Zealand is also a wealthy country, but it is somewhat less well off than Australia. Before 1970, New Zealand relied heavily on exports to Great Britain, mostly agricultural products such as wool and butter. Problems with this strategy occurred, however, once Britain joined the European Union, which then adopted strict agricultural protection policies. Unlike Australia, however, New Zealand lacked a rich base of mineral resources to export to global markets. By the 1980s, the country had slipped into a serious recession. Eventually, the New Zealand government enacted drastic reforms. The country had previously been noted for its lofty taxes, high levels of social welfare, and state ownership of large companies. Suddenly "privatization" (a change from state to private ownership) became the watchword, and most state industries were sold off to private parties. As a result, New Zealand has been transformed into one of the most market-oriented countries of the world. To diversify from its traditional export base, the nation has also encouraged more aggressive development of its timber resources, fisheries, and tourist industry.

Oceania's Economic Diversity

Varied economic activities shape the Pacific island nations. One way of life is oriented around subsistence-based economies, such as shifting cultivation or fishing. In other settings, the commercial extractive economy dominates, with large-scale plantations, mines, and timber activities often competing for land and labor with the traditional subsistence sector. Elsewhere, the huge growth in global tourism has transformed the economic geographies of many island settings, forever changing the way that people make a living. Many island nations also benefit from direct subsidies and economic assistance that come from present and former colonial powers, all designed to promote development and stimulate employment.

Melanesian Economies   Melanesia is the least-developed and poorest part of Oceania. Melanesian countries have benefited less from tourism or from subsidies from wealthy colonial and ex-colonial powers. Most Melanesians live in remote villages that remain somewhat isolated from the modern economy. The Solomon Islands, for example, with few industries other than fish canning and coconut processing, has a per capita GNI of only $750 per year. Papua New Guinea's economy is somewhat more commercially oriented, supporting a per capita GNI of $810. Outbound shipments of coconut products and coffee have increasingly been supplemented with the rapid development of tropical hardwoods in the forest products sector. Gold- and copper-mining have also dramatically transformed the landscape, although political instability has often interfered with mineral production in settings such as Bougainville. In New Guinea's interior highlands, however, much of village life is focused on subsistence activities. Fiji remains the most prosperous Melanesian country with a per capita GNI of almost $2,310 because it is a major sugar producer, and, as well, has developed a tourist economy popular with North Americans and Japanese.

The Economic Impact of Mining   Among the smaller islands of Melanesia and Micronesia, mining economies dominate New Caledonia and Nauru. New Caledonia's nickel reserves, the world's second largest, are both a blessing and a curse. While they currently sustain much of the island's export economy, income from nickel mining will lessen in the near future as the reserves dwindle. Dramatic price fluctuations for the industrial economy also hamper economic planning for the French colony. Other activities include coffee growing, cattle grazing, and tourism. To the north, the tiny, phosphate-rich island of Nauru also depends on mining. The citizens of Nauru, for the most part, live directly off the royalties they receive from the mines. Much of the money gained in mining has been invested in a global trust fund to assure Nauruan citizens of an income even after the phosphate deposits have been exhausted. However, much of this trust fund money recently found its way into the Asian real estate market, and these investments have proved risky. Because of these investments, it now seems possible that the Nauruans will end up with an environmentally devastated island from mining and little financial security.

Micronesia and Polynesian Economies   Elsewhere in Micronesia and Polynesia, conditions depend upon both local subsistence economies or economic linkages to the wider world beyond. Many archipelagos export a few food products, but native populations survive mainly on fish, coconuts, bananas, and yams. Some island groups, however, enjoy large subsidies from either France or the United States, although such support often comes with a political price. Change is also occurring in some of these island settings: In 1999, Japan agreed to build a spaceport for its future shuttlecraft on Micronesia's Christmas Island (Kiribati), and the Marshall Islands are the site of a planned industrial park financed by mainland Chinese.

Other island groups have been completely transformed by tourism. In Hawaii, more than one-third of the state's economy flows directly from tourist dollars. With almost 5 million visitors annually (including more than 1.25 million from Asia), Hawaii represents all of the classic benefits and risks of the tourist economy. While job creation and economic growth have reshaped the island realm, congested highways, high prices, and the unpredictable spending habits of tourists have left the region at risk for future problems. Elsewhere, French Polynesia has long been a favored destination of the international jet set (Figure 14.31). More than 20 percent of French Polynesia's GNI is derived from tourism, making it one of the wealthiest areas of the Pacific. More recently, Guam has emerged as a favorite destination of Japanese and Korean tourists, especially those on honeymoons. Indeed, on a smaller scale, tourism is on the rise across much of the island realm, and many economic planners see it as the avenue to future prosperity. Critics, however, warn that tourist jobs tend to be low-paying; the local quality of life may in fact decline with the presence of tourists; and the natural environments of small islands can quickly be overwhelmed by a high number of demanding visitors.

The Global Economic Setting

Even as Australia and Oceania remain on the margins of global economic activity, their relationships with the world economy will increasingly shape the quality of life and the prospects for development within the region. Several important questions remain. Will future trade patterns in the region shift away from Europe and North America in favor of closer links to Asia? Will there be a move away from the traditional extractive economies that have shaped economic development in the region everywhere from giant Australia to tiny Nauru? Can growing economic linkages within the region make a difference in a part of the world whose total population is less than that of California? None of these questions have ready answers, but they suggest how residents of the region will need to think about their entry into the twenty-first century's global economy.

FIGURE 14.31 • TAHITIAN RESORT Luxury resort settings in Tahiti (near Papeete) exemplify a growing industry in the South Pacific. While bringing important investment capital, this kind of tourism changes the region's social and economic structure as well as its cultural landscape. (Bob & Suzanne Clemenz)

Many international trade flows link the area to the far reaches of the Pacific and beyond. Australia and New Zealand dominate global trade patterns in the region. In the past 30 years, ties to Great Britain, the British Commonwealth, and Europe have weakened in comparison with growing trade links to Japan, East Asia, the Middle East, and the United States. Australia, for example, now imports more manufactured goods from Japan and the United States than it does from Britain and Europe. Similarly, Australian exports, principally raw materials, increasingly flow toward destinations in Asia and North America, although Europe remains an important trading partner. Other global economic ties have come in the form of capital investment in the region. U.S. and Japanese banks and other financial institutions now dot the South Pacific landscape from Sydney to Suva. Both Australia and New Zealand also participate in the **Asia-Pacific Economic Co-operation Group (APEC)**, an organization designed to encourage economic development in Southeast Asia and the Pacific Basin. The region's economic ties to Asia also carry risks, however; the Asian downturn in the late 1990s, for example, slowed the popular Korean tourist trade in New Zealand and lowered Asian demand for a variety of Australian raw material exports.

Enduring Social Challenges

Australians and New Zealanders enjoy high levels of social welfare but face some of the same challenges evident elsewhere in the developed world. Lifespans average more than 75 years in both countries, and rates of child mortality have fallen greatly since 1960. Paralleling patterns in North America and Europe, cancer and heart disease are leading causes of death, and alcoholism is a continuing social problem, particularly in Australia. Unfortunately, Australia's rate of skin cancer is among the world's highest, the result of having a largely fair-skinned, outdoors-oriented population from northwest Europe in a sunny, low-latitude setting. Overall, Australia's Medicare program (initiated in 1984) and New Zealand's system of social services provide high-quality health care to their populations. The position of women is also high in both countries, including participation in the workforce. Women have recently played key political roles in New Zealand, in particular.

Not surprisingly, the social conditions of the Aborigines and Maoris are much less favorable than those of the population overall. Schooling is irregular for many native peoples, and levels of postsecondary education for Aborigines (12 percent) and Maoris (14 percent) remain far below national averages (32 to 34 percent). Many other social measures reflect the pattern, as well. For example, less than one-third of Aboriginal households own their own homes, while more than 70 percent of white Australian households are homeowners. Furthermore, considerable discrimination against native peoples continues in both countries, a situation that has been aggravated and publicized with the recent assertion of indigenous political rights and land claims. As with North American African-American, Hispanic, and Native American populations, simple social policies do not yet exist as solutions to these lasting problems.

Levels of social welfare in Oceania are higher than one might expect, based on the region's economic situation. Many of its countries and colonies have invested heavily in health and education services and have achieved considerable success. For example, the average life expectancy in the Solomon Islands, one of the world's poorer countries as measured by per capita GNI figures, is a respectable 67 years. By other social measures as well, the Solomon Islands and a number of other Oceania states have reached higher levels of human well-being than is the case in most Asian and African countries with similar levels of economic output. This is partly a result of successful policies, but it also reflects the relatively healthy natural environment of Oceania. Many of the tropical diseases that are so troublesome in Africa simply do not exist within the region.

Papua New Guinea (PNG) is the major exception to these relatively high levels of social welfare. Here the average life expectancy is only 56 years, and a recent study suggests that 34 percent of its young people suffer from malnutrition, particularly protein deficiencies. Adult illiteracy (25 percent) is also more prevalent in PNG than elsewhere in the region. PNG has found it difficult to provide even basic educational and health services to its people for two key reasons: The country possesses the largest expanse of land in Melanesia, and much of the population lives in relatively isolated villages in the rugged central highlands.

CONCLUSION

Relative location and small populations contribute to the lasting peripheral position of Australia and Oceania on the global stage. One of the last habitable portions of the planet to be occupied, the region was also a late chapter in the story of Europe's global colonial expansion. Similarly, its contemporary political geography reveals its still changing character as countries struggle to free themselves from colonial ties and assert their own political identities. Globalization complicates the process both culturally and politically: New people, many from Asia, are adding ethnic variety as well as providing flows of investment capital that cut across old colonial relationships.

In addition, the spatial isolation that water and distance once offered is fast disappearing, casting even the most insulated societies of the Australian outback and the Polynesian periphery rapidly into the postindustrial world of the twenty-first century. The transformation has not been easy or predictable: Some native peoples are attracted to modern life, while others resist the cultural, economic, and political currents of the world beyond. Meanwhile, the natural environment, transformed by earlier indigenous peoples and European colonists, has witnessed accelerating changes in the past 50 years as urbanization, extractive economic activities, and tourism reconfigure the landscape. Today, sprawling suburbs, open-pit copper and gold mines, and thatched-hut resorts are the expressions of a once-distant world now brought near by modern communications, improved air travel, and international trade flows.

Australia remains dominant in the greater Pacific realm and should continue to play an important role as a principal economic entry point into the region. Its large land area, resource base, and population are complemented by its increasing emergence as a regional finance center for the South Pacific, as well as nearby portions of Southeast and East Asia. Indeed, the Asian connection seems destined to play a key part in Australia's future economic and even cultural identity. Clearly, the country is no longer simply a distant outpost of Europe. Asian migrants, many of them skilled and wealthy, add a dynamic and creative component to Australia's largely European population. In addition, flows of money, raw materials, and manufactured goods also bind Australia to its Asian neighbors. Will "Asianization" increasingly make Australia a part of that continent, or will political, economic, and cultural resistance to such linkages result in resistance to globalization? These important questions remain to be answered.

If Australia sits at the edge of Asia, New Zealand might be viewed as sitting on the threshold of Polynesia. New Zealand, of course, was once an integral part of Polynesia, but by the early 1900s it seemed to many to have been transformed into a little England exiled to the South Pacific. Today, however, its Maori population is growing quickly, joined by immigrants from other parts of Polynesia. New Zealand has taken an active role in the political affairs of the entire Pacific basin, and it is in many respects Oceania's leading state. Overall, its multicultural identity and its postindustrial economy continue to evolve. Along with the Polynesian links, closer ties to Australia and Asia will also shape its future cultural scene, economic base, and political agenda.

KEY TERMS

Aborigine (page 383)
Aotearoa (page 394)
archipelagos (page 374)
Asia-Pacific Economic Co-operation Group (APEC) (page 399)
atoll (page 381)

haoles (page 391)
high islands (page 381)
hot spot (page 381)
kanakas (page 388)
Maori (page 374)
Melanesia (page 374)

Micronesia (page 375)
microstates (page 392)
Native Title Bill (page 394)
Oceania (page 373)
outback (page 374)
Pidgin English (Pijin) (page 392)

Polynesia (page 374)
tsunamis (page 381)
uncontacted people (page 390)
viticulture (page 386)
White Australia Policy (page 389)

World Data Appendix

On the following pages are population, economic, and social indicator data for the world's countries. The first table presents data for the ten largest countries of the world. Following that, the tables are organized by book chapter for each world region. The different countries of South Asia (India, Bangladesh, Pakistan, and so on), for example, are found under Chapter 12.

The data categories and terms (such as TFR, GNI, and GNP) are discussed in Chapter 1. In several cases, data are separated for males and females. This is done to show that there are significant gender differences in some countries regarding the health and status of women.

Sources for the data are given on page 408.

Country	Population (Millions, 2001)	Population Density, per square mile	Annual Rate of Natural Increase	Total Fertility Rate (TFR)	Percent <15 years of age	Percent >65 years of age	Percent Urban	Total GNI (Millions of $U.S., 1999)
THE WORLD'S TEN LARGEST COUNTRIES								
China	1,273.30	344	0.9	1.8	23	7	36	979,894
India	1,033.00	814	1.7	3.2	36	4	28	441,834
United States	284.5	77	0.6	2.1	21	13	75	8,879,500
Indonesia	206.1	280	1.7	2.7	31	4	39	125,043
Brazil	171.8	52	1.5	2.4	30	5	81	730,424
Pakistan	145	472	2.8	5.6	42	4	33	62,915
Russia	144.4	22	-0.7	1.2	18	13	73	328,995
Bangladesh	133.5	2,401	2	3.3	40	3	21	47,071
Japan	127.1	872	0.2	1.3	15	17	78	4,054,545
Nigeria	126.6	355	2.8	5.8	44	3	36	31,600
CHAPTER 3, NORTH AMERICA								
Canada	31	8	0.3	1.4	19	13	78	614,003
United States	284.5	77	0.6	2.1	21	13	75	8,879,500
CHAPTER 4, LATIN AMERICA								
Argentina	37.5	35	1.1	2.6	28	10	90	276,097
Bolivia	8.5	20	2.4	4.2	40	4	63	8,092
Brazil	171.8	52	1.5	2.4	30	5	81	730,424
Chile	15.4	53	1.3	2.3	28	7	86	69,602
Colombia	43.1	98	1.8	2.6	32	5	71	90,007
Costa Rica	3.7	188	1.8	2.6	32	5	45	12,828
Ecuador	12.9	118	2.2	3.3	34	4	62	16,841
El Salvador	6.4	788	2.3	3.5	36	5	58	11,806
Guatemala	13	309	2.9	4.8	44	3	39	18,625
Honduras	6.7	155	2.8	4.4	43	4	46	4,829
Mexico	99.6	132	1.9	2.8	34	5	74	428,877
Nicaragua	5.2	104	3	4.3	43	3	57	2,012
Panama	2.9	100	2.1	2.6	31	6	56	8,657
Paraguay	5.7	36	2.7	4.3	40	5	52	8,374
Peru	26.1	53	1.8	2.9	34	5	72	53,705
Uruguay	3.4	49	0.7	2.3	24	13	92	20,604
Venezuela	24.6	70	2	2.9	34	5	87	87,313
CHAPTER 5, THE CARIBBEAN								
Anguilla	0.01	—	1	1.8	26	7	—	—
Antigua and Barbuda	0.1	394	1.6	2.4	28	8	37	606
Bahamas	0.3	58	1.5	2.4	31	6	84	—
Barbados	0.3	1620	0.5	1.6	23	9	38	2,294
Belize	0.3	29	1.9	3.2	41	5	49	673
Cayman	0.03	—	0.9	2.1	22	8	—	—
Cuba	11.3	264	0.6	1.6	22	10	75	—
Dominica	0.1	262	0.8	1.8	33	9	71	238
Dominican Republic	8.6	456	2.1	3.1	35	5	61	16,130
French Guiana	0.2	6	2.3	3.4	31	5	79	—
Grenada	0.1	678	1.3	2.4	38	4	34	—
Guadeloupe	0.5	691	1.2	1.9	25	9	48	334
Guyana	0.7	8	1.3	2.5	31	5	36	651
Haiti	7	650	1.7	4.7	43	4	35	3,584
Jamaica	2.6	624	1.5	2.4	31	7	50	6,311
Martinique	0.4	897	0.8	1.8	23	10	93	—
Montserrat	0.01	—	1	1.9	24	12	—	—
Netherlands Antilles	0.2	722	1.1	2.1	26	7	70	—
Puerto Rico	3.9	1139	0.8	1.9	25	10	71	—
St. Kitts and Nevis	0.04	281	0.9	2.5	31	9	43	259
St. Lucia	0.2	656	1.3	2.1	33	6	30	590
St. Vincent and the Grenadines	0.1	757	1.2	2.2	32	6	44	301
Suriname	0.4	7	1.9	3	33	5	69	—
Trinidad and Tobago	1.3	656	0.7	1.7	26	7	72	6,142
Turks and Caicos	0.2	—	2.1	3.3	33	4	—	—

GNI per Capita ($U.S., 1999)	GNI per Capita Purchasing Power Parity (PPP) ($Intl, 1999)	Average Annual Growth % GDP per Capita, 1990–1999	Life Expectancy at Birth		Under Age 5 Mortality (per 1000)		Percent Illiteracy (Ages 15 and over)		Female Labor Force Participation (% of total, 1999)
			Male	Female	1980	1999	Male	Female	
780	3,550	9.5	69	73	65	37	9	25	45
440	2,230	4.1	60	61	177	90	32	56	32
31,910	31,910	2	74	80	15	8	0	0	46
600	2,660	3	65	70	125	52	9	19	41
4,350	6,840	1.5	65	72	81	40	15	15	35
470	1,860	1.3	60	61	161	126	41	70	28
2,250	6,990	-5.9	59	72	—	20	0	1	49
370	1,530	3.1	59	59	211	89	48	71	42
32,030	25,170	1.1	77	84	11	4	0	0	41
260	770	-0.5	52	53	196	151	29	46	36
20,140	25,440	1.7	76	81	13	6	3	3	46
31,910	31,910	2	74	80	15	8	3	3	46
7,550	11,940	3.6	70	77	38	22	3	3	33
990	2,300	1.8	60	64	170	83	8	21	38
4,350	6,840	1.5	65	72	81	40	15	15	35
4,630	8,410	5.6	72	78	35	12	4	5	33
2,170	5,580	1.4	68	74	58	28	9	9	38
3,570	7,880	3	75	79	29	14	5	5	31
1,360	2,820	0	68	73	101	35	7	11	28
1,920	4,260	2.8	67	73	120	36	19	24	36
1,680	3,630	1.5	63	68	121	52	24	40	28
760	2,270	0.3	64	68	103	46	26	26	31
4,440	8,070	1	73	78	74	36	7	11	33
410	2,060	0.4	66	70	143	43	33	30	35
3,080	5,450	2.4	72	76	36	25	8	9	35
1,560	4,380	-0.2	71	76	61	27	6	8	30
2,130	4,480	3.2	66	71	126	48	6	15	31
6,220	8,750	3	70	78	42	17	3	2	42
3,680	5,420	-0.5	70	76	42	23	5	9	34
—	7,900	—	73	79	—	—	5	5	—
8,990	9,870	2.7	68	72	—	—	10	12	—
—	15,500	-0.1	70	75	—	—	2	2	47
8,600	14,010	1.5	70	75	—	—	2	3	46
2,730	4,750	0.7	70	74	—	—	30	30	24
—	24,500	—	76	81	—	—	2	2	—
—	1,700	—	73	77	22	8	3	4	39
3,260	5,040	1.8	70	76	—	—	6	6	—
1,920	5,210	3.9	67	71	92	47	17	17	30
—	6,000	—	72	79	—	—	16	18	—
—	9,000	—	63	66	—	—	2	2	—
3,440	6,330	2.2	73	80	—	—	10	10	—
760	3,330	5.2	62	68	—	—	1	2	—
460	1,470	-3.4	47	51	—	—	52	58	34
2,430	3,390	-0.6	70	73	200	118	49	53	43
—	10,700	—	76	82	39	24	18	10	46
—	—	—	76	80	—	—	3	3	—
—	11,800	—	72	76	—	—	2	1	—
—	9,800	1.9	71	80	—	—	10	12	—
6,330	10,400	4.9	66	71	—	—	7	6	37
3,820	5,200	0.9	70	73	—	—	35	31	—
2,640	4,990	2.6	70	73	—	—	4	4	—
—	3,780	3.3	68	74	—	—	5	9	—
4,750	7,690	2	68	73	—	—	1	3	33
—	7,700	—	71	76	40	20	5	8	34

Country	Population (Millions, 2001)	Population Density, per square mile	Annual Rate of Natural Increase	Total Fertility Rate (TFR)	Percent <15 years of age	Percent >65 years of age	Percent Urban	Total GNI (Millions of $U.S., 1999)
CHAPTER 6, SUB-SAHARAN AFRICA								
Angola	12.3	26	2.4	6.9	48	3	32	3,276
Benin	6.6	152	3	6.3	48	2	39	2,320
Botswana	1.6	7	1	3.9	41	4	49	5,139
Burkina Faso	12.3	116	3	6.8	48	3	15	2,602
Burundi	6.2	579	2.5	6.5	48	3	8	823
Cameroon	15.8	86	2.7	5.2	43	3	48	8,798
Cape Verde	0.4	287	3	4	43	7	53	569
Central African Republic	3.6	15	2	5.1	44	4	39	1,035
Chad	8.7	18	3.3	6.6	48	3	21	1,555
Comoros	0.6	692	3.5	6.8	46	5	29	189
Congo	3.1	24	3	6.3	43	3	41	1,571
Dem. Rep. of Congo	53.6	59	3.1	7	48	3	29	—
Djibouti	0.6	71	2.7	6.1	43	3	83	511
Equatorial Guinea	0.5	43	3.1	5.9	44	4	37	516
Eritrea	4.3	95	3	6	43	3	16	779
Ethiopia	65.4	153	2.9	5.9	44	3	15	6,524
Gabon	1.2	12	1.6	4.3	40	6	73	3,987
Gambia	1.4	323	3	5.9	46	3	37	415
Ghana	19.9	216	2.2	4.3	43	3	37	7,451
Guinea	7.6	80	2.3	5.5	44	3	26	3,556
Guinea-Bissau	1.2	88	2.2	5.8	44	3	22	194
Ivory Coast	16.4	132	2	5.2	42	2	46	10,387
Kenya	29.8	133	2	4.4	44	3	20	10,696
Lesotho	2.2	186	2	4.3	40	5	16	1,158
Liberia	3.2	75	3.1	6.6	43	3	45	—
Madagascar	16.4	71	3	5.8	45	3	22	3,712
Malawi	10.5	231	2.3	6.4	47	3	20	1,961
Mali	11	23	3	7	47	3	26	2,577
Mauritania	2.7	7	2.8	6	44	2	54	1,001
Mauritius	1.2	1,520	1	2	26	6	43	4,157
Mozambique	19.4	63	2.1	5.6	44	3	28	3,804
Namibia	1.8	6	1.9	5	43	4	27	3,211
Niger	10.4	21	2.9	7.5	50	2	17	1,974
Nigeria	126.6	355	2.8	5.8	44	3	36	31,600
Reunion	0.7	744	1.5	2.3	27	7	73	—
Rwanda	7.3	719	1.8	5.8	44	3	5	2,041
São Tomé and Principe	0.2	445	3.5	6.2	48	4	44	40
Senegal	9.7	127	2.8	5.7	44	3	43	4,685
Seychelles	0.1	449	1.1	2	29	8	63	520
Sierra Leone	5.4	196	2.6	6.3	45	3	37	653
Somalia	7.5	30	3	7.3	44	3	28	—
South Africa	43.6	92	1.2	2.9	34	5	54	133,569
Sudan	31.8	33	2.4	4.9	43	3	27	9,435
Swaziland	1.1	165	2	5.9	46	3	25	1,379
Tanzania	36.2	99	2.8	5.6	45	3	22	8,515
Togo	5.2	235	2.9	5.8	47	2	31	1,398
Uganda	24	258	2.9	6.9	51	2	15	6,794
Zambia	9.8	34	2.3	6.1	45	2	38	3,222
Zimbabwe	11.4	75	0.9	4	44	3	32	6,302
CHAPTER 7, SOUTHWEST ASIA AND NORTH AFRICA								
Algeria	31	34	1.9	3.1	39	4	49	46,548
Bahrain	0.7	2,688	1.9	2.8	31	2	88	0.8
Cyprus	0.9	247	0.6	1.8	23	10	66	9,086
Egypt	69.8	181	2.1	3.5	36	4	43	86,544
Gaza and West Bank	3.3	1,365	3.7	5.9	47	4		5,063
Iran	66.1	108	1.2	2.6	36	5	64	113,729
Iraq	23.6	139	2.7	5.3	42	3	68	
Israel	6.4	791	1.6	3	29	10	91	99,574
Jordan	5.2	150	2.2	3.6	40	5	79	7,717
Kuwait	2.3	297	1.8	4.2	26	1	100	
Lebanon	4.3	1,061	1.7	2.5	29	7	88	15,796
Libya	5.2	8	2.4	3.9	37	4	86	

GNI per Capita ($U.S., 1999)	GNI per Capita Purchasing Power Parity (PPP) ($Intl, 1999)	Average Annual Growth % GDP per Capita, 1990–1999	Life Expectancy at Birth Male	Life Expectancy at Birth Female	Under Age 5 Mortality (per 1000) 1980	Under Age 5 Mortality (per 1000) 1999	Percent Illiteracy (Ages 15 and over) Male	Percent Illiteracy (Ages 15 and over) Female	Female Labor Force Participation (% of total, 1999)
270	1,100	-2.8	37	39	261	208	—	—	46
380	920	1.8	49	51	214	145	45	76	48
3,240	6,540	1.8	41	42	94	95	26	21	45
240	960	1.4	47	47	—	210	67	87	47
120	570	-5.0	46	47	193	176	44	61	49
600	1,490	-1.5	55	56	173	154	19	31	38
1,330	4,450	3.2	65	72	—	—	—	—	39
290	1,150	-0.3	43	46	—	151	41	67	—
210	840	-0.9	48	52	235	189	50	68	45
350	1,430	-3.1	54	59	—	—	—	—	42
550	540	-3.3	47	52	125	144	13	27	43
—	—	-8.1	45	50	210	161	28	51	43
790	—	-5.1	44	48	—	—	—	—	—
1,170	3,910	16.3	48	52	—	—	—	—	36
200	1,040	2.2	53	57	—	105	33	61	47
100	620	2.4	51	53	213	180	57	68	41
3,300	5,280	0.6	51	54	194	133	—	—	45
330	1,550	-0.6	51	54	216	110	57	72	45
400	1,850	1.6	56	59	157	109	21	39	51
490	1,870	1.5	43	47	280	167	—	—	47
160	630	-1.9	44	46	290	214	42	82	40
670	1,540	0.6	45	47	—	—	—	—	—
360	1,010	-0.3	48	49	115	118	12	25	46
550	2,350	2.1	52	55	168	141	28	7	37
—	—	—	49	52	—	—	—	—	40
250	790	-1.2	52	56	216	149	27	41	45
180	570	0.9	39	40	44	39	26	55	49
240	740	1.1	45	47	—	223	53	67	46
390	1,550	1.3	49	52	175	142	48	69	44
3,540	8,950	3.9	67	74	40	23	12	19	32
220	810	3.8	69	76	223	203	41	72	48
1,890	5,580	0.8	47	45	114	108	18	20	41
190	740	-1.0	41	41	317	252	77	92	44
260	770	-0.5	52	53	196	151	29	46	36
—	—	—	70	79	—	—	—	—	—
250	880	-3.0	39	40	—	203	27	41	49
270	—	-0.9	63	66	—	—	—	—	—
500	1,400	0.6	51	54	—	124	54	73	43
6,500	—	1.3	67	73	—	—	—	—	—
130	440	-7.0	42	47	336	283	—	—	37
—	—	—	45	48	—	—	—	—	43
3,170	8,710	-0.2	52	54	91	76	14	16	38
330	—	—	55	57	145	109	31	55	29
1,350	4,380	-0.2	40	41	—	—	—	—	38
260	500	-0.1	52	54	176	152	16	34	49
310	1,380	-0.5	53	58	188	143	26	60	40
320	1,160	4	42	43	180	162	23	45	48
330	720	-2.4	37	38	149	187	15	30	45
530	2,690	0.6	41	39	108	118	8	16	44
1,550	4,840	-0.5	68	70	139	39	23	44	27
			70	75	—	—	—	—	20
11,950	19,080	2.8	75	79	—	—	—	—	39
1,380	3,460	2.4	65	68	175	61	34	57	30
1,780		-0.2	70	74	—	26	—	—	—
1,810	5,520	1.9	69	71	126	33	17	31	27
			58	60	95	128	35	55	19
16,310	18,070	2.3	76	80	19	8	2	6	41
1,630	3,880	1.1	69	71	49	31	6	17	24
			72	73	35	13	16	21	31
3,700		5.7	68	73		32	8	20	29
			73	77	80	28	10	33	23

Country	Population (Millions, 2001)	Population Density, per square mile	Annual Rate of Natural Increase	Total Fertility Rate (TFR)	Percent <15 years of age	Percent >65 years of age	Percent Urban	Total GNI (Millions of $U.S., 1999)
Morocco	29.2	169	2	3.4	33	5	55	33,715
Oman	2.4	29	3.5	6.1	41	2	72	
Qatar	0.6	139	2.7	3.9	27	2	91	
Saudi Arabia	21.1	25	2.9	5.7	43	2	83	139,365
Sudan	31.8	33	2.4	4.9	43	3	27	9,435
Syria	17.1	231	2.6	4.1	41	3	50	15,172
Tunisia	9.7	154	1.3	2.3	31	6	62	19,757
Turkey	66.3	221	1.5	2.5	30	6	66	186,490
United Arab Emirates	3.3	103	1.4	3.5	26	1	84	
Western Sahara	0.3	3	2.9	6.8	95			
Yemen	18	88	3.3	7.2	48	3	26	6,088

CHAPTER 8, EUROPE

Western Europe

Country	Population (Millions, 2001)	Population Density, per square mile	Annual Rate of Natural Increase	Total Fertility Rate (TFR)	Percent <15 years of age	Percent >65 years of age	Percent Urban	Total GNI (Millions of $U.S., 1999)
Austria	8.1	251	0	1.3	17	15	65	205,743
Belgium	10.3	872	0.1	1.6	17	17	97	252,051
France	59.2	278	0.4	1.9	19	16	74	1,453,211
Germany	82.2	597	-0.1	1.3	16	16	86	2,103,804
Liechtenstein	0.03	534	0.6	1.4	19	10	23	—
Luxembourg	0.4	446	0.4	1.7	19	14	88	18,545
Netherlands	16	1,018	0.4	1.7	19	14	62	397,384
Switzerland	7.2	453	0.2	1.5	18	15	68	273,856
United Kingdom	60	635	0.1	1.7	19	16	90	1,403,843

Eastern Europe

Country	Population (Millions, 2001)	Population Density, per square mile	Annual Rate of Natural Increase	Total Fertility Rate (TFR)	Percent <15 years of age	Percent >65 years of age	Percent Urban	Total GNI (Millions of $U.S., 1999)
Bulgaria	8.1	190	-0.5	1.2	16	16	68	11,572
Czech Republic	10.3	337	-0.2	1.1	17	14	77	51,623
Hungary	10	278	-0.4	1.3	17	15	64	46,751
Moldova	4.3	328	-0.1	1.4	24	9	46	1,481
Poland	38.6	310	0	1.4	20	12	62	157,429
Romania	22.4	243	-0.1	1.3	18	13	55	33,034
Slovakia	5.4	286	0	1.3	20	11	57	20,318

Southern Europe

Country	Population (Millions, 2001)	Population Density, per square mile	Annual Rate of Natural Increase	Total Fertility Rate (TFR)	Percent <15 years of age	Percent >65 years of age	Percent Urban	Total GNI (Millions of $U.S., 1999)
Albania	3.4	310	1.2	2.8	33	6	75	3,146
Bosnia-Herzegovina	3.4	173	0.4	1.6	20	8	72	4,706
Croatia	4.7	197	-0.2	1.4	20	12	77	20,222
Greece	10.9	214	0	1.3	15	17	81	127,648
Italy	57.8	497	0	1.3	14	18	82	1,162,910
Macedonia	2	205	0.5	1.9	23	10	75	3,348
Malta	0.4	3,157	0.3	1.7	21	12	80	3,492
Portugal	10	282	0.1	1.5	17	15	79	110,175
San Marino	0.03	1,166	0.4	1.3	15	16	89	—
Slovenia	2	256	-0.1	1.2	16	14	50	19,862
Spain	39.8	204	0	1.2	15	17	64	583,082
Yugoslavia (Serbia)	10.7	270	0.1	1.6	21	13	52	—

Northern Europe

Country	Population (Millions, 2001)	Population Density, per square mile	Annual Rate of Natural Increase	Total Fertility Rate (TFR)	Percent <15 years of age	Percent >65 years of age	Percent Urban	Total GNI (Millions of $U.S., 1999)
Denmark	5.4	322	0.2	1.7	18	15	72	170,685
Estonia	1.4	78	-0.4	1.3	18	14	69	4,906
Finland	5.2	40	0.2	1.7	18	15	60	127,764
Iceland	0.3	7	0.8	2	23	12	93	8,197
Ireland	3.8	142	0.6	1.9	22	11	58	80,559
Latvia	2.4	95	-0.6	1.2	18	15	69	5,913
Lithuania	3.7	147	-0.1	1.3	20	13	68	9,751
Norway	4.5	36	0.3	1.8	20	15	74	149,280
Sweden	8.9	51	0	1.5	19	17	84	236,940

CHAPTER 9, THE RUSSIAN DOMAIN

Country	Population (Millions, 2001)	Population Density, per square mile	Annual Rate of Natural Increase	Total Fertility Rate (TFR)	Percent <15 years of age	Percent >65 years of age	Percent Urban	Total GNI (Millions of $U.S., 1999)
Armenia	3.8	330	0.3	1.1	24	9	67	1,878
Belarus	10	125	-0.4	1.3	19	13	70	26,299
Georgia	5.5	203	0	1.2	20	13	56	3,362
Russia	144.4	22	-0.7	1.2	18	13	73	328,995
Ukraine	49.1	211	-0.7	1.1	18	14	68	41,991

CHAPTER 10, CENTRAL ASIA

Country	Population (Millions, 2001)	Population Density, per square mile	Annual Rate of Natural Increase	Total Fertility Rate (TFR)	Percent <15 years of age	Percent >65 years of age	Percent Urban	Total GNI (Millions of $U.S., 1999)
Afghanistan	26.8	106	2.4	6	43	3	22	—
Azerbaijan	8.1	243	0.9	2	32	6	51	3,705
Kazakstan	14.8	14	0.5	1.8	28	7	56	18,732

GNI per Capita ($U.S., 1999)	GNI per Capita Purchasing Power Parity (PPP) ($Intl, 1999)	Average Annual Growth % GDP per Capita, 1990–1999	Life Expectancy at Birth		Under Age 5 Mortality (per 1000)		Percent Illiteracy (Ages 15 and over)		Female Labor Force Participation (% of total, 1999)
			Male	Female	1980	1999	Male	Female	
1,190	3,320	0.4	67	71	152	62	39	65	35
		0.3	69	73	95	24	21	40	16
			69	74	—	—	—	—	15
6,900	11,050	-1.1	66	69	85	25	17	34	15
330			55	57	145	109	31	55	29
970	3,450	2.7	70	70	73	30	12	41	27
2,090	5,700	2.9	70	74	100	30	20	41	31
2,900	6,440	2.2	67	71	133	45	7	24	37
			71	76		9	26	22	14
		-1.6	—	—	—	—	—	—	—
360	730	-0.4	57	61	198	97	33	76	28
25,430	24,600	1.4	75	81	17	5	0	0	40
24,650	25,710	1.4	75	81	15	6	0	0	41
24,170	23,020	1.1	75	83	13	5	0	0	45
25,620	23,510	1	74	81	16	5	0	0	42
—	—	—	—	—	—	—	—	—	—
42,930	41,230	3.8	75	81	—	—	0	0	37
25,140	24,410	2.1	75	81	11	5	0	0	40
38,380	28,760	-0.1	77	83	11	5	0	0	40
23,590	22,220	2.1	75	80	14	6	0	0	44
1,410	5,070	-2.1	68	75	25	17	1	2	48
5,020	12,848	0.9	71	78	19	5	0	0	47
4,640	11,050	1.4	66	75	26	10	1	1	48
410	2,100	-10.8	64	72	—	22	1	2	49
4,070	8,390	4.4	68	77	—	10	0	0	46
1,470	5,970	-0.5	67	74	36	24	1	3	44
3,770	10,430	1.6	69	77	23	10	0	0	48
930	3,240	2.8	69	75	57	—	9	23	41
1,210	—	32.7*	65	72	—	18	0	0	38
4,530	7,260	1	70	77	23	9	1	3	44
12,110	15,800	1.8	76	81	23	7	2	4	38
20,170	22,000	1.2	76	82	17	6	1	2	38
1,660	4,590	-1.5	70	75	69	17	—	—	42
9,210	—	4.2*	74	80	—	—	0	0	28
11,030	15,860	2.3	72	79	31	6	6	11	44
—	—	—	76	83	—	—	0	0	—
10,000	16,050	2.5	72	79	16	6	0	0	46
14,800	17,850	2	74	82	16	6	2	3	37
—	—	—	70	75	—	16	0	0	43
32,050	25,600	2	74	79	10	6	0	0	46
3,400	8,190	-0.3	65	76	25	12	0	0	49
24,730	22,600	2	74	81	9	5	0	0	48
29,540	27,210	1.8	78	81	—	—	0	0	45
21,470	22,460	6.1	74	79	14	7	0	0	34
2,430	6,220	-3.7	65	76	26	18	0	0	50
2,640	6,490	-3.9	67	77	24	12	0	1	48
33,470	28,140	3.2	76	81	11	4	0	0	46
26,750	22,150	1.2	77	82	9	4	0	0	48
490	2,360	-3.9	71	76		18	1	3	48
2,620	6,880	-2.9	62	74		14	0	1	49
620	2,540		69	77		20			47
2,250	6,990	-5.9	59	72		20	0	1	49
840	3,360	-10.3	63	74		17	0	1	48
—	800**	—	46	44	—	—	49**	79**	—
460	2,450	-10.7	68	75	—	21	1**	4**	44
1,250	4,790	-4.9	60	71	—	28	1**	2**	47

Country	Population (Millions, 2001)	Population Density, per square mile	Annual Rate of Natural Increase	Total Fertility Rate (TFR)	Percent <15 years of age	Percent >65 years of age	Percent Urban	Total GNI (Millions of $U.S., 1999)
Kyrgyzstan	5	65	1.3	2.4	35	5	35	1,465
Mongolia	2.4	4	1.4	2.2	34	4	57	927
Tajikistan	6.2	112	1.4	2.4	42	4	27	1,749
Tibet	2.5	4.5	—	—	—	—	—	—
Turkmenistan	5.5	29	1.3	2.2	38	4	44	3,205
Uzbekistan	25.1	145	1.7	2.7	38	4	38	17,613
Xinjiang	17	24	—	—	—	—	—	
CHAPTER 11, EAST ASIA								
China	1,273.30	344	0.9	1.8	23	7	36	979,894
Hong Kong	6.9	16,743	0.3	1	17	11	100	165,122
Japan	127.1	872	0.2	1.3	15	17	78	4,054,545
South Korea	48.8	1,274	0.9	1.5	22	7	79	397,910
North Korea	22.2	472	1.5	2.3	26	6	59	—
Taiwan	22.5	1,608	0.8	1.7	21	9	77	—
CHAPTER 12, SOUTH ASIA								
India	1,033.00	814	1.7	3.2	36	4	28	441,834
Pakistan	145	472	2.8	5.6	42	4	33	62,915
Bangladesh	133.5	2,401	2	3.3	40	3	21	47,071
Nepal	23.5	413	2.4	4.8	41	3	11	5,173
Bhutan	0.9	50	3.1	5.6	42	4	15	399
Sri Lanka	19.5	771	1.2	2.1	28	6	22	15,578
Maldives	0.3	2,495	3.2	5.8	46	3	25	322
CHAPTER 13, SOUTHEAST ASIA								
Burma (Myanmar)	47.8	183	1.6	3.3	33	5	27	—
Brunei	0.3	156	2	2.7	32	3	67	—
Cambodia	13.1	187	1.7	4	43	4	16	3,023
Indonesia	206.1	280	1.7	2.7	31	4	39	125,043
Laos	5.4	59	2.5	5.4	44	4	17	1,476
Malaysia	22.7	178	2	3.2	33	4	57	76,944
Philippines	77.2	666	2.2	3.5	37	4	47	77,967
Singapore	4.1	17,320	0.9	1.6	17	6	100	95,429
Thailand	62.4	315	0.8	1.8	24	6	30	121,051
Vietnam	78.7	623	1.4	2.3	33	6	24	28,733
CHAPTER 14, AUSTRALIA AND OCEANIA								
Australia	19.4	6	0.6	1.7	20	12	85	397,345
Fed. States of Micronesia	0.1	444	2.5	4.6	44	4	27	212
Fiji	0.8	119	1.9	3.3	33	4	46	1,849
French Polynesia	0.2	153	1.6	2.6	30	5	54	3,909
Guam	0.2	744	2.5	4.2	32	5	38	—
Kiribati	0.1	337	2.4	4.5	40	3	37	81
Marshall Islands	0.1	1,007	2.2	6.6	49	3	65	99
Nauru	0.01	1,412	1.4	3.7	43	1	100	—
New Caledonia	0.2	30	1.7	2.6	30	6	71	3,169
New Zealand	3.9	37	0.8	2	23	12	77	53,299
Palau	0.02	107	1	2.5	28	6	71	—
Papua New Guinea	5	28	2.3	4.8	39	4	15	3,834
Samoa	0.2	155	2.4	4.5	32	6	33	181
Solomon Islands	0.5	41	3.4	5.7	43	3	13	320
Tonga	0.1	349	2.1	4.2	41	4	32	172
Tuvalu	0.01	1,100	2.1	3.1	34	3	18	—
Vanuatu	0.2	44	3	4.6	42	3	21	227

* Not a 10-year average

** Source: *CIA World Factbook*, 2002.

Data sources: Population Reference Bureau, *World Population Data Sheet, 2001.* (Columns 1–7, 12–13); *The World Bank Atlas, 2001.* (Columns 9–11, 14–18)

GNI per Capita ($U.S., 1999)	GNI per Capita Purchasing Power Parity (PPP) ($Intl, 1999)	Average Annual Growth % GDP per Capita, 1990–1999	Life Expectancy at Birth Male	Life Expectancy at Birth Female	Under Age 5 Mortality (per 1000) 1980	Under Age 5 Mortality (per 1000) 1999	Percent Illiteracy (Ages 15 and over) Male	Percent Illiteracy (Ages 15 and over) Female	Female Labor Force Participation (% of total, 1999)
300	2,420	-6.4	65	72	—	38	1**	4**	47
390	1,610	-0.6	65**	70**	—	73	2	3	47
280	1,090**	2.0**	66	71	—	34	1	3	45
			—	—	—	—	—	—	—
670	3,340	-9.6	63	70	—	45	1**	3**	46
720	2,230	-3.1	68	73	—	29	1	1	47
780	3,550	9.5	69	73	65	37	10	27	45
24,570	22,570	1.9	77	82	—	5	4	12	37
32,030	25,170	1.1	77	84	11	4	1	1	41
8,490	15,530	4.7	71	78	27	9	1	3	43
—	1,000**	1.0**	67	73	43	93	1**	1**	41
—	16,100**	5.5**	72	78	—	—	7**	21**	—
440	2,230	4.1	60	61	177	90	34	62	32
470	1,860	1.3	60	61	161	126	45	71	28
370	1,530	3.1	59	59	211	89	37	51	42
220	1,280	2.3	58	57	180	109	59	86	40
510	1,260	3.4	66	66	—	—	44**	72**	40
820	3,230	4	70	74	48	19	7	13	36
1,200	1,800**	3.9	60	63	—	—	7**	7**	43
—	1,200**	—	54	57	134	120	11	22	43
—	17,400**	-0.5	71	76	—	—	7**	17**	35
260	1,350	1.9	54	58	330	143	52	78	52
600	2,660	3	65	70	125	52	10	22	41
290	1,430	3.8	51	54	200	143	30	66	—
3,390	7,640	4.7	70	75	42	10	11	22	38
1,050	3,990	0.9	64	70	81	41	5	6	38
24,150	22,310	4.7	76	80	13	4	3	10	39
2,010	5,950	3.8	70	75	58	33	4	8	45
370	1,860	6.2	63	69	105	42	4	9	49
20,950	23,850	2.9	76	82	13	5			40
1,830	—	-1.8	65	67				30	
2,310	4,780	1.2	65	69					
16,930	22,220	-0.1	69	74					
—	—	—	72	77					
910	—	1	59	65					
1950	—	—	63	67					
—	—	—	57	65					
15,160	21,130	-0.8	70	76					
13,990	17,630	1.8	74	80	16	6			45
—	—	—	64	71					
810	2,260	2.3	56	55			20	30	42
1,070	4,070	1.4	65	72					47
750	2,050	0.3	67	68					47
1,730	—	0.7	70	72					34
—	—	—	64	70					
1,180	2,880	-0.8	64	67					

Glossary

Aborigine   An indigenous inhabitant of Australia.

acid rain   Harmful form of precipitation high in sulfur and nitrogen oxides. Caused by industrial and auto emissions, acid rain damages aquatic and forest ecosystems in regions such as eastern North America and Europe.

African diaspora   The forced dispersion of African peoples from their native lands to various parts of the Americas and the Middle East through the slave trade.

agrarian reform   A popular but controversial strategy to redistribute land to peasant farmers. Throughout the twentieth century, various states redistributed land from large estates or granted title from vast public lands in order to reallocate resources to the poor and stimulate development. Agrarian reform occurred in various forms, from awarding individual plots or communally held land to creating state-run collective farms.

agricultural density   The number of farmers per unit of arable land. This figure indicates the number of people who directly depend upon agriculture, and it is an important indicator of population pressure in places where rural subsistence dominates.

alluvial fan   A fan-shaped deposit of sediments dropped by a river or stream flowing out of a mountain range.

Altiplano   The largest intermontane plateau in the Andes, which straddles Peru and Bolivia and ranges in elevation from 10,000 to 13,000 feet (3,000 to 4,000 meters).

altitudinal zonation   The relationship between higher elevations, cooler temperatures, and changes in vegetation that result from the environmental lapse rate (averaging 3.5°F for every 1,000 feet). In Latin America, four general altitudinal zones exist: tierra caliente, tierra templada, tierra fria, and tierra helada.

animism   A wide variety of tribal religions based on the worship of nature's spirits and human ancestors.

anthropogenic   An adjective for human-caused change to a natural system, such as the atmospheric emissions from cars, industry, and agriculture that are causing global warming.

anthropogenic landscape   A landscape heavily transformed by human agency.

Aotearoa   Maori name for New Zealand, meaning "Land of the Long White Cloud."

apartheid   The policy of racial separateness that directed the separate residential and work spaces for white, blacks, coloureds, and Indians in South Africa for nearly 50 years. It was abolished when the African National Congress came to power in 1994.

archipelagos   Island groups, often oriented in an elongated pattern.

areal differentiation   The geographic description and explanation of spatial differences on Earth's surface; this includes physical as well as human patterns.

areal integration   The geographic description and explanation of how places, landscapes, and regions are connected, interactive, and integrated with each other.

Asia-Pacific Economic Co-operation Group (APEC)   An international group of Asian and Pacific Basin nations that fosters coordinated economic development within the region.

Association of Southeast Asia Nations (ASEAN)   A supranational geopolitical group linking together the 10 different states of Southeast Asia.

asymmetrical warfare   Military action between a superpower using strategies dependent on high-technology weapons and the low technology and guerilla tactics used by small insurgent groups.

atoll   Low, sandy islands made from coral, often oriented around a central lagoon.

autonomous areas   Minor political subunits created in the former Soviet Union and designed to recognize the special status of minority groups within existing republics.

autonomous region   In the context of China, provinces that have been granted a certain degree of political and cultural autonomy, or freedom from centralized authority, owing to the fact that they contain large numbers of non-Han Chinese people. Critics contend that they have little true autonomy.

Baikal-Amur Mainline (BAM) Railroad   Key central Siberian railroad connection completed in the Soviet era (1984), which links the Yenisey and Amur rivers and parallels the Trans-Siberian Railroad.

Balfour Declaration   Statement issued by Great Britain in 1917 pledging its support for establishing a home for the Jewish people in Palestine.

balkanization   Geopolitical process of fragmentation of larger states into smaller ones through independence of smaller regions and ethnic groups. The term takes its name from the geopolitical fabric of the Balkan region.

barrios   Urban Hispanic neighborhoods, often associated with low-income groups in North America.

Berlin Conference   The 1884 conference that divided Africa into European colonial territories. The boundaries created in Berlin satisfied European ambition but ignored indigenous cultural affiliations. Many of Africa's civil conflicts can be traced to ill-conceived territorial divisions crafted in 1884.

biofuels   Energy sources derived from plants or animals. Throughout the developing world, wood, charcoal, and dung are primary energy sources for cooking and heating.

biome   Ecologically interactive flora and fauna adapted to a specific environment. Examples are deserts or tropical rain forests.

bioregion   A spatial unit or region of local plants and animals adapted to a specific environment, such as a tropical savanna.

Bolsheviks   A faction within the Russian Communist movement led by Lenin that successfully took control of the country in 1917.

boreal forest   Coniferous forest found in high-latitude or mountainous environments of the Northern Hemisphere.

brain drain   Migration of the best-educated people from developing countries to developed nations where economic opportunities are greater.

British East India Company   Private trade organization that acted as arm of colonial Britain—backed by the British army—in monopolizing trade in South Asia until 1857, when it was abolished and replaced by full governmental control.

bubble economy   A highly inflated economy that cannot be sustained. Bubble economies usually result from rapid influx of international capital into a developing country.

buffer zone   An array of nonaligned or friendly states that "buffer" a larger country from invasion. In Europe, keeping a buffer zone has been a long-term policy of Russia (and also of the former Soviet Union) to protect its western borders from European invasion.

Bumiputra   The name given to native Malay (literally, "sons of the soil"), who are given preference for jobs and schooling by the Malaysian government.

Burakumin   The indigenous outcast group of Japan, a people whose ancestors reputedly worked in leather-craft and other "polluting" industries.

bustees   Settlements of temporary and often illegal housing in Indian cities, caused by rapid urban migration of poorer rural people and the inability of the cities to provide housing for this rapidly expanding population.

capital leakage   The gap between the gross receipts an industry (such as tourism) brings into a developing area and the amount of capital retained.

cargo cult   Quasi-animist Melanesian religious practice, originating with military cargo supply drops during World War II.

Caribbean Community and Common Market (CARICOM)   A regional trade organization established in 1972 that includes former English colonies as its members.

Caribbean diaspora   The out-migration of Caribbean peoples to various destinations in North America and Europe. This livelihood strategy has been widely practiced since the 1950s, when many people began leaving due to the region's limited economic opportunities. Although difficult for migrants and their families, the Caribbean diaspora has become a way of life and explains the region's dependence on foreign remittances.

caste system   Complex division of South Asian society into different hierarchically ranked hereditary groups. Most explicit in Hindu society, but also found in other cultures to a lesser degree.

Central Place theory   A theory used to explain the distribution of cities and the relationships between different cities, based on retail marketing.

centralized economic planning   An economic system in which the state sets production targets and controls the means of production.

centrifugal forces   Those cultural and political forces, such as linguistic minorities, separatists, and fringe groups, that pull away from and weaken an existing nation-state.

centripetal forces   Those cultural and political forces, such as a shared sense of history, a centralized economic structure, and the need for military security, that promote political unity in a nation-state.

chain migration   A pattern of migration in which a sending area becomes linked to a particular destination, such as Dominicans with Queens, New York.

chernozem soils   A Russian term for dark, fertile soil, often associated with grassland settings in southern Russia and Ukraine.

China proper   The eastern half of the country of China where the Han Chinese form the dominant ethnic group. The vast majority of China's population is located in China proper.

Chipko movement   The "tree-hugging" movement of northern India in which women, drawing upon Hindu tradition, attempt to save forests from destruction by embracing the trees as loggers approach.

circular migration   Temporary labor migration in which an individual seeks short-term employment overseas, saves money, and then returns home.

clan   A social unit that is typically smaller than a tribe or ethnic group but larger than a family, based on supposed descent from a common ancestor.

climate region   A region of similar climatic conditions. An example would be the marine west coast climate regions found on the west coasts of North America and Europe.

climograph   Graph of average annual temperature and precipitation data by month and season.

Closer Economic Relationship (CER) Agreement   An agreement signed in 1982 between Australia and New Zealand designed to eliminate all economic and trade barriers between the two countries.

Cold War   The ideological struggle between the United States and the Soviet Union that was conducted between 1946 and 1991.

collective farms   Group-farmed agricultural units organized around state-mandated production goals.

collectivization   The agglomeration of small, privately owned agricultural parcels into larger, state-owned farms. This was a central component of communism in eastern Europe and the Soviet Union.

colonialism   The formal, established (mainly historical) rule over local peoples by a larger imperialist government for the expansion of political and economic empire.

coloureds   A racial category used throughout South Africa to define people of mixed European and African ancestry.

Columbian Exchange   An exchange of people, diseases, plants, and animals between the Americas (New World) and Europe/Africa (Old World) initiated by the arrival of Christopher Columbus in 1492.

command economy   Centrally planned and controlled economies, generally associated with socialist or communist countries, in which all goods and services, along with agricultural and industrial products, are strictly regulated. This was done during the Soviet era in both the Soviet Union and its eastern European satellites.

Commonwealth of Independent States (CIS)   A loose political union of former Soviet republics (without the Baltic states) established in 1992 after the dissolution of the Soviet Union.

concentric zone model   A simplified description of urban land use: A well-defined central business district (CBD) is surrounded by concentric zones of residential activity, with higher-income groups living on the urban periphery.

Confucianism   The philosophical system developed by Confucius in the sixth century B.C.

connectivity   The degree to which different locations are linked with one another through transportation and communication infrastructure.

continental climate   Climate regions in continental interiors, removed from moderating oceanic influences, that are characterized by hot summers and cold winters. At least one month must average below freezing.

convection cells   Large areas of slow-moving molten rock in Earth's interior that are responsible for moving tectonic plates.

copra   Dried coconut meat.

core-periphery model   A conceptualization of the world into two economic spheres. The developed countries of western Europe, North America, and Japan form the dominate core, with less-developed countries making up the periphery. Implicit in this model is that the core gained its wealth at the expense of peripheral countries.

Cossacks   Highly mobile Slavic-speaking Christians of the southern Russian steppe who were pivotal in expanding Russian influence in sixteenth- and seventeenth-century Siberia.

Council of Mutual Economic Assistance (CMEA)   The communist agency that coordinated economic planning and development between the Soviet Union and satellite countries in eastern Europe until 1991.

counterurbanization   The movement of people out of metropolitan areas toward smaller towns and rural areas.

creolization   The blending of African, European, and even some Amerindian cultural elements into the unique sociocultural systems found in the Caribbean.

crony capitalism   A system in which close friends of a political leader are either legally or illegally given business advantages in return for their political support.

cultural assimilation   The process in which immigrants are culturally absorbed into the larger host society.

cultural imperialism   The active promotion of one cultural system over another, such as the implantation of a new language, school system, or bureaucracy. Historically, this has been primarily associated with European colonialism.

cultural landscape Primarily the visible and tangible expression of human settlement (house architecture, street patterns, field form, etc.), but also includes the intangible, value-laden aspects of a particular place and its association with a group of people.

cultural nationalism A process of protecting, either formally (with laws) or informally (with social values), the primacy of a certain cultural system against influences (real or imagined) from another culture.

cultural syncretism The blending of two or more cultures, which produces a synergistic third culture that exhibits traits from all cultural parents. Also called *cultural hybridization*.

culture Learned and shared behavior by a group of people empowering them with a distinct "way of life"; it includes both material (technology, tools, etc.) and immaterial (speech, religion, values, etc.) components.

culture hearth An area of historical cultural innovation.

Cyrillic alphabet Based on the Greek alphabet and used by Slavic languages heavily influenced by the Eastern Orthodox Church. Attributed to the missionary work of St. Cyril in the ninth century.

dacha A Russian country cottage used especially in the summer.

Dalit The currently preferred term used to denote the members of India's most discriminated-against ("lowest") caste groups, those people previously deemed "untouchables."

decolonialization The process of a former colony's gaining (or regaining) independence over its territory and establishing (or reestablishing) an independent government.

demographic transition A four-stage model of population change derived from the historical decline of the natural rate of increase as a population becomes increasingly urbanized through industrialization and economic development.

denuclearization The process whereby nuclear weapons are removed from an area and dismantled or taken elsewhere.

dependency theory A popular theory to explain patterns of economic development in Latin America. Its central premise is that underdevelopment was created by the expansion of European capitalism into the region that served to develop "core" countries in Europe and to impoverish and make dependent peripheral areas such as Latin America.

desertification The spread of desert conditions into semiarid areas owing to improper management of the land.

diaspora The scattering of a particular group of people over a vast geographical area. Originally, the term referred to the migration of Jews out of their original homeland, but now it has been generalized to refer to any ethnic dispersion.

digital divide The uneven and inequitable access to the Internet and to computer technology based on wealth and education.

distributaries The different channels of a river in its delta area.

dollarization An economic strategy in which a country adopts the U.S. dollar as its official currency. A country can be partially dollarized, using U.S. dollars alongside its national currency, or fully dollarized, when the U.S. dollar becomes the only medium of exchange and a country gives up its own national currency. Panama fully dollarized in 1904; more recently, Ecuador fully dollarized in 2000.

domestication The purposeful selection and breeding of wild plants and animals for cultural purposes.

domino theory A U.S. geopolitical policy of the 1970s that stemmed from the assumption that if Vietnam fell to the communists, the rest of Southeast Asia would soon follow.

Dravidian language One of the earliest (from perhaps 4,000 years ago) language families and, unlike Hindi, not Indo-European. Once spoken throughout South Asia, Dravidian languages are now found only in southern India and part of Sri Lanka.

Eastern Orthodox Christianity A loose confederation of self-governing churches in eastern Europe and Russia that are historically linked to Byzantine traditions and to the primacy of the patriarch of Constantinople (Istanbul).

economic convergence The notion that globalization will result in the world's poorer countries gradually catching up with more advanced economies.

economic growth rate The annual rate of expansion for GNI.

El Niño An abnormally large warm current that appears off the coast of Ecuador and Peru in December. During an El Niño year, torrential rains can bring devastating floods along the Pacific coast and drought conditions in the interior continents of the Americas.

encomienda A social and economic system used in the early Spanish colonies where there were large native populations. Groups of Indians would be "commended" to a Spaniard, who would exact tribute from them in the form of labor or products. In return, the Spaniard was obligated to educate the Indians in the Spanish language and the Catholic faith.

entrepôt A city and port that specializes in transshipment of goods.

environmental lapse rate The decline in temperature as one ascends higher in the atmosphere. On average, the temperature declines 3.5 °F for every 1,000 feet ascended or 6.5 °C for every 1,000 meters.

ethnic religion A religion closely identified with a specific ethnic or tribal group, often to the point of assuming the role of the major defining characteristic of that group. Normally, ethnic religions do not actively seek new converts.

ethnicity A shared cultural identity held by a group of people with a common background or history, often as a minority group within a larger society.

ethnographic boundaries State and national boundaries that are drawn to follow distinct differences in cultural traits, such as religion, language, or ethnic identity.

Euroland The 11 states that form the European Monetary Union, with its common currency, the euro. This monetary unit completely replaced national currencies in July 2002.

European Union (EU) The current association of 25 European countries that are joined together in an agenda of economic, political, and cultural integration.

exclave A portion of a country's territory that lies outside of its contiguous land area.

Exclusive Economic Zone An agreement under the International Law of the Sea that provides for a 200-mile zone of exclusive legal offshore development rights that can be utilized or sold by coastal or island nations.

exotic river A river that issues from a humid area and flows into a dry area otherwise lacking streams.

federal state Political system in which a significant amount of power is given to individual states; in India, these states were created upon independence in 1947 and were drawn primarily along linguistic lines, so that today state power is often associated with specific ethnic groups within the nation.

Fertile Crescent An ecologically diverse zone of lands in Southwest Asia that extends from Lebanon eastward to Iraq and that is often associated with early forms of agricultural domestication.

feudalism The formal power relationship that consists of well-defined responsibilities and obligations between a superior (such as an aristocrat) and those lower in social status (such as serfs or vassals).

fjords Flooded, glacially carved valleys; in Europe, found primarily along Norway's western coast.

forward capital A capital city deliberately positioned near the international border of a contested territory, signifying the state's interest—and presence—in this zone of conflict.

fossil water Water supplies that were stored underground during wetter climatic periods.

Free Trade Area of the Americas (FTAA) A proposed hemispheric trade association that will include Latin America, North America, and the

Caribbean (minus Cuba). If the treaty is successfully negotiated, it will integrate some 800 million people in the largest trade association in the world. The FTAA is scheduled to go into effect in 2005.

free trade zone (FTZ)   A duty-free and tax-exempt industrial park created to attract foreign corporations and create industrial jobs.

gentrification   A process of urban revitalization in which higher-income residents displace lower-income residents in central city neighborhoods.

geomancy   The traditional Chinese and Korean practice of designing buildings in accordance with the principles of cosmic harmony and discord that supposedly course through the local topography.

geometric boundaries   Boundaries of convenience drawn along lines of latitude or longitude without consideration for cultural or ethnic differences in an area.

ghettos   Urban ethnic neighborhoods often associated with low-income groups in North America.

glasnost   A policy of greater political openness initiated during the 1980s by then Soviet President Mikhail Gorbachev.

globalization   The increasing interconnectedness of people and places throughout the world through converging processes of economic, political, and cultural change.

Golden Triangle   An area of northern Thailand, Burma, and Laos that is known as a major source region for heroin and is plugged into the global drug trade.

Gondwanaland   The ancient mega-continent that included Africa, South America, Antarctica, Australia, Madagascar, and Saudi Arabia. Some 250 million years ago, it began to split apart due to plate tectonics.

Gran Colombia   A short-lived Latin American state during the early years of independence (1822–1830). A dream of Simon Bolívar, it included Colombia, Venezuela, Ecuador, and Panama.

grassification   The conversion of tropical forest into pasture for cattle ranching. Typically, this process involves introducing species of grasses and cattle, mostly from Africa.

Great Escarpment   A landform that rims southern Africa from Angola to South Africa. It forms where the narrow coastal plains meet the elevated plateaus in an abrupt break in elevation.

Greater Antilles   The four large Caribbean islands of Cuba, Jamaica, Hispaniola, and Puerto Rico.

Green Line   United Nations-patrolled dividing line separating Greek (southern) and Turkish (northern) portions of Cyprus.

Green Revolution   Term applied to the development of agricultural techniques used in developing countries that usually combine new, genetically altered seeds that provide higher yields than native seeds when combined with high inputs of chemical fertilizer, irrigation, and pesticides.

greenhouse effect   The natural process of lower atmosphere heating that results from the trapping of incoming and reradiated solar energy by water moisture, clouds, and other atmospheric gases.

gross domestic product (GDP)—gross national product (GNP) GDP is the total value of goods and services produced within a given country (or other geographical unit) in a single year. GNP is a somewhat broader measure that includes the inflow of money from other countries in the form of the repatriation of profits and other returns on investments, as well as the outflow to other countries for the same purposes.

gross national income (GNI)   The value of all final goods and services produced within a country's borders (gross domestic product, or GDP) plus the net income from abroad (formerly referred to as gross national product, or GNP).

gross national income, per capita   The figure that results from dividing a country's GNI by the total population.

Group of 7 (G-7)   A collection of powerful countries that confers regularly on key global economic and political issues. It includes the United States, Canada, Japan, Great Britain, Germany, France, and Italy; sometimes Russia also takes part.

growth poles   Planned industrial centers developed by states in order to increase manufacturing and stimulate economic growth in underdeveloped areas.

guest workers   Workers from Europe's agricultural periphery—primarily Greece, Turkey, southern Italy, and the former Yugoslavia—solicited to work in Germany, France, Sweden, and Switzerland during chronic labor shortages in Europe's boom years (1950s to 1970s).

Gulag Archipelago   A collection of Soviet-era labor camps for political prisoners, made famous by writer Aleksandr Solzhenitsyn.

Hajj   An Islamic religious pilgrimage to Makkah. One of the five essential pillars of the Muslim creed to be undertaken once in life, if an individual is physically and financially able to do it.

haoles   Light-skinned Europeans or U.S. citizens in the Hawaiian Islands.

hierarchical diffusion   The spread of an idea or cultural trait through adoption by leaders and other elite at the top of the social structure or hierarchy. In the case of religion, it was common for all members of a clan or tribe to convert if the leader adopted a new religion.

high islands   Larger, more elevated islands, often focused around recent volcanic activity.

Hindi   An Indo-European language with more than 480 million speakers, making it the second-largest language group in the world. In India, it is the dominant language of the heavily populated north, specifically the core area of the Ganges Plain.

Hindu nationalism   A contemporary "fundamental" religious and political movement that promotes Hindu values as the essential—and exclusive—fabric of Indian society. As a political movement, it appears to have less tolerance of India's large Muslim minority than other political movements.

hiragana   The main Japanese syllabary, used for writing indigenous words. Each symbol stands for a particular vowel-consonant combination.

homelands   Nominally independent ethnic territories created for blacks under the grand apartheid scheme. Homelands were on marginal land, overcrowded, and poorly serviced. In the post-apartheid era, they were eliminated.

Horn of Africa   The northeastern corner of Sub-Saharan Africa that includes the states of Somalia, Ethiopia, Eritrea, and Djibouti. Drought, famine, and ethnic warfare in the 1980s and 1990s resulted in political turmoil in this area.

hot spot   A supply of magma that produces a chain of mid-ocean volcanoes atop a zone of moving oceanic crust.

houseyard   A rural subsistence property that is matriarchal in organization.

hurricanes   Storm systems with an abnormally low-pressure center sustaining winds of 75 mph or higher. Each year during hurricane season (July–October), a half dozen to a dozen hurricanes form in the warm waters of the Atlantic and Caribbean, bringing destructive winds and heavy rain.

hydropolitics   The interplay of water resource issues and politics.

ideographic writing   A writing system in which each symbol represents not a sound but rather a concept.

import substitution   A development policy that discourages imports (via high tariffs) and encourages substituting domestically produced manufactured goods.

indentured labor   Foreign workers (usually South Asians) contracted to labor on Caribbean agricultural estates for a set period of time, often several years. Usually the contract stipulated paying off the travel debt incurred by the laborers. Similar indentured labor arrangements have existed in most world regions.

Indian diaspora   The historical and contemporary propensity of Indians to migrate to other countries in search of better opportunities. This has led to large Indian populations in South Africa, the Caribbean, and the Pacific islands, along with western Europe and North America.

informal sector A much-debated concept that presupposes a dual economic system consisting of formal and informal sectors. The informal sector includes self-employed, low-wage jobs that are usually unregulated and untaxed. Street vending, shoe shining, artisan manufacturing, and even self-built housing are considered part of the informal sector. Some scholars include illegal activities such as drug smuggling and prostitution in the informal economy.

insolation Incoming solar energy that enters the atmosphere adjacent to Earth.

internally displaced persons Groups and individuals who flee an area due to conflict or famine but still remain in their country of origin. These populations often live in refugee-like conditions but are harder to assist because they technically do not qualify as refugees.

Iron Curtain A term coined by British leader Winston Churchill during the Cold War that defined the western border of Soviet power in Europe. The notorious Berlin Wall was a concrete manifestation of the Iron Curtain.

irredentism A state or national policy of reclaiming lost lands or those inhabited by people of the same ethnicity in another nation-state.

Islamic fundamentalism A movement within both the Shiite and Sunni Muslim traditions to return to a more conservative, religious-based society and state. Often associated with a rejection of Western culture and with a political aim to merge civic and religious authority.

island biogeography The study of the ecology of plants and animals unique to island environments.

isolated proximity A concept that explores the contradictory position of the Caribbean states, which are physically close to North America and economically dependent upon that region. At the same time, Caribbean isolation fosters strong loyalties to locality and limited economic opportunity.

Jainism A religious group in South Asia that emerged as a protest against orthodox Hinduism about the sixth century B.C. Its ethical core is the doctrine of noninjury to all living creatures. Today, Jains are noted for their nonviolence, which prohibits them from taking the life of any animal.

kanakas Melanesian workers imported to Australia, historically often concentrated along Queensland's "sugar coast."

kanji The Chinese characters, or ideographs, used in Japanese writing.

Khmer Rouge Literally, "Red (or communist) Cambodians." The left-wing insurgent group led by French-educated Marxists rebelled against the royal Cambodian government in the early 1960s and again in a peasants' revolt in 1967.

kibbutzes Collective farms in Israel.

kleptocracy A state where corruption is so institutionalized that politicians and bureaucrats siphon off a huge percentage of a country's wealth.

laissez-faire An economic system in which the state has minimal involvement and in which market forces largely guide economic activity.

latifundia A large estate or landholding.

Lesser Antilles The arc of small Caribbean islands from St. Maarten to Trinidad.

Levant The eastern Mediterranean region.

lingua franca An agreed-upon common language to facilitate communication on specific topics such as international business, politics, sports, or entertainment.

linguistic nationalism The promotion of one language over others that is, in turn, linked to shared notions of nationalism. In India, some Hindu nationalists promote Hindi as the national language, yet this is resisted by many other groups in which that language is either not spoken or does not have the same central cultural role, as in the Ganges Valley. The lack of a national language in India remains problematic.

location factors The various influences that explain why an economic activity takes place where it does.

loess A fine, wind-deposited sediment that makes fertile soil but is very vulnerable to water erosion.

low islands Low, small, sandy islands formed from eroding coral reefs. Generally less fertile and populated than high islands.

machismo A stereotypical cultural trait of male dominance.

Maghreb A region in northwestern Africa, including portions of Morocco, Algeria, and Tunisia.

maharaja Regional Hindu royalty, usually a king or prince, who ruled specific areas of South Asia before independence, but who was usually subject to overrule by British colonial advisers.

mallee A tough and scrubby eucalyptus woodland of limited economic value that is common across portions of interior Australia.

Mandarin A member of the high-level bureaucracy of Imperial China (before 1911). Mandarin Chinese is the official spoken language of the country and is the native tongue of the vast majority of people living in north, central, and southwestern China.

Maori Indigenous Polynesian people of New Zealand.

maquiladora Assembly plants on the Mexican border built by foreign capital. Most of their products are exported to the United States.

marianismo An idealized model for women that stresses the virtues of patience, deference, and working in the home.

marine west coast climate Moderate climate with cool summers and mild winters that is heavily influenced by maritime conditions. Such climates are usually found on the west coasts of continents between latitudes of 45 to 50 degrees.

maritime climate Climate moderated by proximity to oceans or large seas. It is usually cool, cloudy, and wet and lacks the temperature extremes of continental climates.

maroons Runaway slaves who established communities rich in African traditions throughout the Caribbean and Brazil.

Marxism The philosophy developed by Karl Marx, the most important historical proponent of communism. Marxism, which has many variants, presumes the desirability and, indeed, the necessity of a socialist economic system run through a central planning agency.

medieval landscape Urban landscapes from A.D. 900 to 1500 characterized by narrow, winding streets, three- or four-story structures (usually in stone, but sometimes wooden), with little open space except for the market square. These landscapes are still found in the centers of many European cities.

medina The original urban core of a traditional Islamic city.

Mediterranean climate A unique climate, found in only five locations in the world, that is characterized by hot, dry summers with very little rainfall. These climates are located on the west side of continents, between 30 and 40 degrees latitude.

mega-city Urban conglomerations of more than 10 million people.

megalopolis A large urban region formed as multiple cities grow and merge with one another. The term is often applied to the string of cities in eastern North America that includes Washington, D.C.; Baltimore; Philadelphia; New York City; and Boston.

Melanesia Pacific Ocean region that includes the culturally complex, generally darker-skinned peoples of New Guinea, the Solomon Islands, Vanuatu, New Caledonia, and Fiji.

Mercosur The Southern Common Market established in 1991 that calls for free trade among member states and common external tariffs for non-member states. Argentina, Paraguay, Brazil, and Uruguay are members; Chile is an associate member.

mestizo A person of mixed European and Indian ancestry.

Micronesia Pacific Ocean region that includes the culturally diverse, generally small islands north of Melanesia. Micronesia includes the Mariana Islands, Marshall Islands, and Federated States of Micronesia.

microstates Usually independent states that are small in both area and population.

mikrorayons Large, state-constructed urban housing projects built during the Soviet period in the 1970s and 1980s.

minifundia A small landholding farmed by peasants or tenants who produce food for subsistence and the market.

mono-crop production Agriculture based upon a single crop.

monotheism A religious belief in a single God.

Monroe Doctrine A proclamation issued by U.S. President James Monroe in 1823 that the United States would not tolerate European military action in the Western Hemisphere. Focused on the Caribbean as a strategic area, the doctrine was repeatedly invoked to justify U.S. political and military intervention in the region.

monsoon The seasonal pattern of changes in winds, heat, and moisture in South Asia and other regions of the world that is a product of larger meteorological forces of land and water heating, the resultant pressure gradients, and jet-stream dynamics. The monsoon produces distinct wet and dry seasons.

moraines Hilly topographic features that mark the path of Pleistocene glaciers. They are composed of material eroded and carried by glaciers and ice sheets.

Mughal Empire (also spelled Mogul) The preeminent Islamic period of rule that covered most of South Asia during the early sixteenth to late seventeenth centuries and attempted to unify both Muslims and Hindus into a large South Asian state. The capital of this empire was Lahore, in what is now Pakistan. The last vestiges of the Mughal dynasty were dissolved by the British following the uprisings of 1857.

multinational corporation A corporation that produces goods and services in a wide range of different countries. The classical multinational corporation, in contrast to the truly transnational corporation, remains solidly based in a single country.

nation-state A relatively homogeneous cultural group (a nation) with its own political territory (the state).

Native Title Bill Australian legislation signed in 1993 that provides Aborigines with enhanced legal rights over land and resources within the country.

neocolonialism Economic and political strategies by which powerful states indirectly (and sometimes directly) extend their influence over other, weaker states.

neoliberal policies Economic policies widely adopted in the 1990s that stress privatization, export production, and few restrictions on imports.

neotropics Tropical ecosystems of the Americas that evolved in relative isolation and support diverse and unique flora and fauna.

North America Free Trade Agreement (NAFTA) An agreement made in 1994 between Canada, the United States, and Mexico that established a 15-year plan for reducing all barriers to trade among the three countries.

Oceania A major world subregion that usually includes New Zealand and the major island regions of Melanesia, Micronesia, and Polynesia.

offshore banking Islands or microstates that offer financial services that are typically confidential and tax-exempt. As part of a global financial system, offshore banks have developed a unique niche, offering their services to individual and corporate clients for set fees. The Bahamas and Cayman Islands are leaders in this sector.

Organization of African Unity (OAU) Founded in 1963, the organization grew to include all the states of the continent except South Africa, which finally was asked to join in 1994. It is mostly a political body that has tried to resolve regional conflicts.

Organization of American States (OAS) Founded in 1948 and headquartered in Washington, D.C., the organization advocates hemispheric cooperation and dialog. Most states in the Americas belong except Cuba.

Organization of Petroleum Exporting Countries (OPEC) An international organization of 12 oil-producing nations (formed in 1960) that attempts to influence global prices and supplies of oil. Algeria, Gabon, Indonesia, Iran, Iraq, Kuwait, Libya, Nigeria, Qatar, Saudi Arabia, UAE, and Venezuela are members.

orographic rainfall Enhanced precipitation over uplands that results from lifting (and cooling) of air masses as they are forced over mountains.

Ottoman Empire A large, Turkish-based empire (named for Osman, one of its founders) that dominated large portions of southeastern Europe, North Africa, and Southwest Asia between the sixteenth and nineteenth centuries.

outback Australia's large, generally dry, and thinly settled interior.

overurbanization A process in which the rapid growth of a city, most often because of in-migration, exceeds the city's ability to provide jobs, housing, water, sewers, and transportation.

oxisols Reddish or yellowish soils found in the tropical shields of Latin America. They are formed over long periods of time and accumulate rusted or oxidized iron. When these fine soils are disturbed or compacted, a hard-pan surface results that is virtually impossible to farm.

Palestinian Authority (PA) A quasi-governmental body that represents Palestinian interests in the West Bank and Gaza.

Pan-African Movement Founded in 1900 by U.S. intellectuals W. E. B. Du Bois and Marcus Garvey, this movement's slogan was "Africa for Africans" and its influence extended across the Atlantic.

particularism A mode of thought that emphasizes the uniqueness of different places and different phenomena. Particularism is opposed to universalism, which emphasizes locality-transcending commonalties.

pastoral nomadism A traditional subsistence agricultural system in which practitioners depend on the seasonal movements of livestock within marginal natural environments.

pastoralism A way of life and form of livelihood centered around raising large animals, usually cattle, sheep, goats, and horses, but also sometimes including camels, yaks, and other species.

pastoralists Nomadic and sedentary peoples who rely upon livestock (especially cattle, camels, sheep, and goats) for sustenance and livelihood.

perestroika A program of partially implemented, planned economic reforms (or restructuring) undertaken during the Gorbachev years in the Soviet Union designed to make the Soviet economy more efficient and responsive to consumer needs.

permafrost A cold-climate condition in which the ground remains permanently frozen.

physiological densities A population statistic that relates the number of people in a country to the amount of arable land.

Pidgin English A version of English that also incorporates elements of other local languages, often utilized to foster trade and basic communication between different culture groups.

plantation America A cultural region that extends from midway up the coast of Brazil, through the Guianas and the Caribbean, and into the southeastern United States. In this coastal zone, European-owned plantations, worked by African laborers, produced agricultural products for export.

plate tectonics The theory that explains the gradual movement of large geological platforms (or plates) along Earth's surface.

podzol soils A Russian term for an acidic soil of limited fertility, typically found in northern forest environments.

polders Reclaimed agricultural areas along the Dutch coast that have been diked and drained. Many are at or below sea level.

pollution exporting The process of exporting industrial pollution and other waste material to other countries. Pollution exporting can be direct, as when waste is simply shipped abroad for disposal, or indirect, as when highly polluting factories are constructed abroad.

Polynesia Pacific Ocean region, broadly unified by language and cultural traditions, that includes the Hawaiian Islands, Marquesas Islands, Society

Islands, Tuamotu Archipelago, Cook Islands, American Samoa, Samoa, Tonga, and Kiribati.

postindustrial economy   An economy in which the tertiary and quaternary sectors dominate employment and expansion.

prairie   An extensive area of grassland in North America. In the more humid eastern portions, grasses are usually longer than in the drier western areas, which are in the rain shadow of the Rocky Mountain range.

primate city   The largest urban settlement in a country that dominates all other urban places, economically and politically. Often—yet not always—the primate city is also the country's capital.

privatization   The process of moving formerly state-owned firms into the contemporary capitalist private sector.

protectorate   During the period of global Western imperialism, a state or other political entity that remained autonomous but sacrificed its foreign affairs to an imperial power in exchange for "protection" from other imperial powers.

purchasing power parity (PPP)   A method of reducing the influence of inflated currency rates by adjusting a local currency to a composite baseline of one U.S. dollar based upon its ability to purchase a standardized "market basket" of goods.

qanat system   A traditional system of gravity-fed irrigation that uses gently sloping tunnels to capture groundwater and direct it to needed fields.

Quran   (also spelled Koran) A book of divine revelations received by the prophet Muhammad that serves as a holy text in the religion of Islam.

rain-shadow effect   A weather phenomenon in which mountains block moisture, producing an area of lower precipitation on the leeward side of the uplift.

rate of natural increase (RNI)   The standard statistic used to express natural population growth per year for a country, region, or the world based upon the difference between birth and death rates. RNI does not consider population change from migration. Though most often a positive figure (such as 1.7 percent), RNI can also be expressed as a negative (-.08) for no-growth countries.

refugee   A person who flees his or her country because of a well-founded fear of persecution based on race, ethnicity, religion, ideology, or political affiliation.

remittances   Money sent by immigrants to their country of origin to support family members left behind. For many countries, remittances are a principal source of foreign exchange.

Renaissance–Baroque landscape   Urban landscapes generally constructed during the period from 1500 to 1800 that are characterized by wide, ceremonial boulevards, large monumental structures (palaces, public squares, churches), and ostentatious housing for the urban elite. A common landscape feature in European cities.

rift valley   A surface landscape feature formed where two tectonic plates are diverging or moving apart. Usually, this forms a depression or large valley.

rimland   The mainland coastal zone of the Caribbean, beginning with Belize and extending along the coast of Central America to northern South America.

rural-to-urban migration   The flow of internal migrants from rural areas to cities that began in the 1950s and intensified in the 1960s and 1970s.

Russification   A policy of the Soviet Union designed to spread Russian settlers and influences to non-Russian areas of the country.

rust belt   Regions of heavy industry that experience marked economic decline after their factories cease to be competitive.

Sahel   The semidesert region at the southern fringe of the Sahara, and the countries that fall within this region, which extends from Senegal to Sudan. Droughts in the 1970s and early 1980s caused widespread famine and dislocation of population.

salinization   The accumulation of salts in the upper layers of soil, often causing a reduction in crop yields, resulting from irrigation with water of high natural salt content and/or irrigation of soils that contain a high level of mineral salts.

samurai   The warrior class of traditional Japan. After 1600, the military role of the samurai declined as they assumed administrative positions, but their military ethos remained alive until the class was abolished in 1868.

Sanskrit   The original Indo-European language of South Asia, introduced into northwestern India perhaps 4,000 years ago, from which modern Indo-Aryan languages evolved. Over the centuries, it has become the classical literary language of the Hindus and is widely used as a scholarly second language, much like Latin in medieval Europe.

scheduled castes   Part of the contemporary reform movement in India to lessen the social stigma of lower castes, particularly of the dalits (formerly untouchables), by reserving a certain number of places in universities, governmental offices, and seats in national and state legislatures for them.

Schengen Agreement   The 1985 agreement between some—but not all—European Union member countries to reduce border formalities in order to facilitate free movement of citizens between member countries of this new "Schengenland." For example, today there are no border controls between France and Germany, or between France and Italy.

sectoral transformation   The evolution of a labor force from being highly dependent on the primary sector to being oriented around more employment in the secondary, tertiary, and quaternary sectors.

secularization   The widespread movement in western Europe away from regular participation and engagement with traditional organized religions such as Protestantism or Catholicism.

sediment load   The amount of sand, silt, and clay carried by a river.

shatterbelt   A geopolitically unstable area where the superpowers vie for power. Eastern Europe was the classic shatterbelt until World War II.

shield landscape   Barren, mostly flat lands of southern Scandinavia that were heavily eroded by Pleistocene ice sheets. In many places, this landscape is characterized by large expanses of bedrock with little or no soil that resulted from glacial erosion.

shields   Large upland areas of very old exposed rocks that range in elevation from 600 to 5,000 feet (200 to 1,500 meters). The three major shields in South America are the Guiana, Brazilian, and Patagonian.

shifted cultivators   Migrants, with or without agricultural experience, who are transplanted by government relocation schemes.

shifting cultivation   An agricultural system in which plots of land are farmed and then abandoned for a number of years until fertility is restored, at which point they are again brought under cultivation. See also *swidden*.

Shiites   Muslims who practice one of the two main branches of Islam; especially dominant in Iran and nearby southern Iraq.

Shogun, Shogunate   The true ruler of Japan before 1868, as opposed to the emperor, whose power was merely symbolic.

Sikhism   An Indian religion combining Islamic and Hindu elements, founded in the Punjab region in the late fifteenth century. A long tradition of militarism continues today, and a large proportion of Sikh men are in the Indian armed forces.

Slavic peoples   A group of peoples in eastern Europe and Russia who speak Slavic languages, a distinctive branch of the Indo-European language family.

social and regional differentiation   "Social differentiation" refers to a process by which certain classes of people grow richer when others grow poorer; "regional differentiation" refers to a process by which certain places grow more prosperous while others become less prosperous.

socialist realism   An artistic style once popular in the Soviet Union that was associated with realistic depictions of workers in their patriotic struggles against capitalism.

Special Economic Zones (SEZ)   Relatively small districts in China that have been fully opened to global capitalism.

spheres of influence   In countries not formally colonized in the nineteenth and early twentieth centuries (particularly China and Iran), limited areas called "spheres of influence" were gained by particular European countries for trade purposes and more generally for economic exploitation and political manipulation.

squatter settlements   Makeshift housing on land not legally owned or rented by urban migrants, usually in unoccupied open spaces within or on the outskirts of a rapidly growing city.

steppe   Semiarid grasslands found in many parts of the world. Grasses are usually shorter and less dense in steppes than in prairies.

structural adjustment programs   Controversial yet widely implemented programs used to reduce government spending, encourage the private sector, and refinance foreign debt. Typically, these IMF and World Bank policies trigger drastic cutbacks in government-supported services and food subsidies, which disproportionately affect the poor.

subcontinent   A large segment of land separated from the main landmass on which it sits by lofty mountains or other geographical barriers. South Asia, separated from the rest of Eurasia by the Himalayas, is often called the "Indian subcontinent."

subduction zones   Areas where two tectonic plates are converging or colliding. In these areas, one plate usually sinks below another. They are characterized by earthquakes, volcanoes, and deep oceanic trenches.

subnational organizations   Groups that form along ethnic, ideological, or territorial lines that can induce serious internal divisions within a state.

subsistence agriculture   Farming that produces only enough crops or animal products to support a farm family's needs. Usually, little is sold at local or regional markets.

Suez Canal   Pivotal waterway connecting the Red Sea and the Mediterranean opened by the British in 1869.

Sunda Shelf   An extension of the continental shelf from the Southeast Asia mainland to the outlying islands. Because of the shelf, the overlying sea is generally shallow (less than 200 feet, or 61 meters, deep).

Sunnis   Muslims who practice the dominant branch of Islam.

superconurbation   A massive urban agglomeration that results from the coalescing of two or more formerly separate metropolitan areas.

supranational organizations   Governing bodies that include several states, such as trade organizations, and often involve a loss of some state powers to achieve the organization's goals.

sustainable development   A vision of economic change and growth seeking a balance with environmental protection and social equity so that the short-term needs of contemporary society do not compromise needs of future generations. The operational scale of sustainable development is local rather than global.

sweatshops   Crude factories in developing countries in which workers perform labor-intensive tasks for extremely low wages.

swidden agriculture   Also called *slash-and-burn agriculture*. A form of cultivation in which forested or brushy plots are cleared of vegetation, burned, and then planted to crops, only to be abandoned a few years later as soil fertility declines.

syncretic religions   The blending of different belief systems. In Latin America, many animist practices were folded into Christian worship.

taiga   The vast coniferous forest of Russia that stretches from the Urals to the Pacific Ocean. The main forest species are fir, spruce, and larch.

tectonic plates   The basic building blocks of Earth's crust; large blocks of solid rock that very slowly move over the underlying semimolten material.

terraces   Flat areas carved across the face of slopes, usually in a steplike fashion.

theocratic state   A political state led by religious authorities. Also called a *theocracy*.

tonal language   Language in which the same set of phonemes (or basic sounds) may have very different meanings, depending on the pitch in which they are uttered.

total fertility rate (TFR)   The average number of children who will be borne by women of a hypothetical, yet statistically valid, population, such as that of a specific cultural group or within a particular country. Demographers consider TFR a more reliable indicator of population change than the crude birthrate.

township   Racially segregated neighborhoods created for nonwhite groups under apartheid in South Africa. They are usually found on the outskirts of cities and classified as black, coloured, or South Asian.

transhumance   A form of pastoralism in which animals are taken to high-altitude pastures during the summer months and returned to low-altitude pastures during the winter.

transmigration   The planned, government-sponsored relocation of people from one area to another within a state territory.

transnational firm   Firms and corporations that, although they may be chartered and have headquarters in one specific country, do international business through an array of global subsidiaries.

transnationalism   Complex social and economic linkages that form between home and host countries through international migration. Unlike earlier generations of migrants, early twenty-first-century immigrants can maintain more enduring and complex ties to their home countries as a result of technological advances.

Trans-Siberian Railroad   Key southern Siberian railroad connection completed during the Russian empire (1904) that links European Russia with the Russian Far East terminus of Vladivostok.

Treaty of Tordesillas   A treaty signed in 1494 between Spain and Portugal that drew a north-south line some 300 leagues west of the Azores and Cape Verde islands. Spain received the land to the west of the line and Portugal the land to the east.

tribal peoples   Peoples who were traditionally organized at the village or clan level, without broader-scale political organization.

tribalism   Allegiance to a particular tribe or ethnic group rather than to the nation-state. Tribalism is often blamed for internal conflict within Sub-Saharan states.

tribe   A group of families or clans with a common kinship, language, and definable territory but not an organized state.

tsars   A Russian term (also spelled *czar*) for "Caesar," or ruler; the authoritarian rulers of the Russian empire before its collapse in the 1917 revolution.

tsetse fly   A fly that is a vector for a parasite that causes sleeping sickness (typanosomiasis), a disease that especially affects humans and livestock. Livestock is rarely found in those areas of Sub-Saharan Africa where the tsetse fly is common.

tsunamis   Very large sea waves induced by earthquakes.

tundra   Arctic region with a short growing season in which vegetation is limited to low shrubs, grasses, and flowering herbs.

Turkestan   That portion of Central Asia populated primarily by Turkish-speaking peoples; "eastern Turkestan" comprises those areas presently controlled by China, and "western Turkestan" comprises areas formerly held by the Soviet Union.

typhoons   Large tropical storms, similar to hurricanes, that form in the western Pacific Ocean in tropical latitudes and cause widespread damage to the Philippines and coastal Southeast and East Asia.

uncontacted peoples   Cultures that have yet to be contacted and influenced by the Western world.

unitary state   A political system in which power is centralized at the national level.

United Provinces of Central America   Formed in 1823 to avoid annexation by Mexico, this union collapsed in the 1830s, yielding the independent states of Guatemala, Honduras, El Salvador, Nicaragua, and Costa Rica.

universalizing religion   A religion, usually with an active missionary program, that appeals to a large group of people regardless of local culture and conditions. Christianity and Islam both have strong universalizing components. This contrasts with ethnic religions.

urban decentralization   The process in which cities spread out over a larger geographical area.

urban form   The physical arrangement or landscape of the city, made up of building architecture and style, street patterns, open spaces, housing types, etc.

urban primacy   A state in which a disproportionately large city, such as London, New York, or Bangkok, dominates the urban system and is the center of economic, political, and cultural life.

urban realms model   A simplified description of urban land use, especially descriptive of the modern North American city. It features a number of dispersed, peripheral centers of dynamic commercial and industrial activity linked by sophisticated urban transportation networks.

urban structure   The distribution and pattern of land use, such as commercial, residential, or manufacturing, within the city. Often, commonalties give rise to models of urban structure characteristic of the cities of a certain region or of a shared history, such as cities shaped by European colonialism.

urbanized population   That percentage of a country's population living in settlements characterized as cities. Usually, high rates of urbanization are associated with higher levels of industrialization and economic development, since these activities are usually found in and around cities. Conversely, lower urbanized populations (less than 50 percent) are characteristic of developing countries.

Urdu   Although Urdu originated in the region of northern India and arose from a similar colloquial base as Hindi, it is one of the official languages of Pakistan today because of its long association with Muslim culture. Urdu borrows heavily from Persian vocabulary and Arabic grammatical construction and, further, is written in a modified form of the Persian Arabic alphabet.

viticulture   Grape cultivation.

Wallace's Line   A line in Southeast Asia delineating the abrupt difference in flora and fauna from that found on the Asian mainland to plants and animals more common to Australia.

water stress   An environmental planning tool used to predict areas that have—or will have—serious water problems based upon the per capita demand and supply of freshwater.

White Australia Policy   Before 1975, a set of stringent Australian limitations on nonwhite immigration to the country. Largely replaced by a more flexible policy today.

World Trade Organization (WTO)   Formed as an outgrowth of the General Agreement on Tariffs and Trade (GATT) in 1995, a large collection of member states dedicated to reducing global barriers to trade.

Index

419

Rowntree, Lewis, Price, Wyckoff
Globalization and Diversity: Geography of a Changing World
Author Field Trip CD
0-13-147985-7
© 2005 by Pearson Education, Inc.
Upper Saddle River, NJ 07458
Pearson Prentice Hall. All rights Reserved
Pearson Prentice Hall™ is a trademark of Pearson Education, Inc.
Pearson® is a registered trademark of Pearson plc
Prentice Hall® is a registered trademark of Pearson Education, Inc.

YOU SHOULD CAREFULLY READ THE TERMS AND CONDITIONS BEFORE USING THE CD-ROM PACKAGE. USING THIS CD-ROM PACKAGE INDICATES YOUR ACCEPTANCE OF THESE TERMS AND CONDITIONS.

Pearson Education, Inc. provides this program and licenses its use. You assume responsibility for the selection of the program to achieve your intended results, and for the installation, use, and results obtained from the program. This license extends only to use of the program in the United States or countries in which the program is marketed by authorized distributors.

LICENSE GRANT

You hereby accept a nonexclusive, nontransferable, permanent license to install and use the program ON A SINGLE COMPUTER at any given time. You may copy the program solely for backup or archival purposes in support of your use of the program on the single computer. You may not modify, translate, disassemble, decompile, or reverse engineer the program, in whole or in part.

TERM

The License is effective until terminated. Pearson Education, Inc. reserves the right to terminate this License automatically if any provision of the License is violated. You may terminate the License at any time. To terminate this License, you must return the program, including documentation, along with a written warranty stating that all copies in your possession have been returned or destroyed.

LIMITED WARRANTY

THE PROGRAM IS PROVIDED "AS IS" WITHOUT WARRANTY OF ANY KIND, EITHER EXPRESSED OR IMPLIED, INCLUDING, BUT NOT LIMITED TO, THE IMPLIED WARRANTIES OR MERCHANTABILITY AND FITNESS FOR A PARTICULAR PURPOSE. THE ENTIRE RISK AS TO THE QUALITY AND PERFORMANCE OF THE PROGRAM IS WITH YOU. SHOULD THE PROGRAM PROVE DEFECTIVE, YOU (AND NOT PEARSON EDUCATION, INC. OR ANY AUTHORIZED DEALER) ASSUME THE ENTIRE COST OF ALL NECESSARY SERVICING, REPAIR, OR CORRECTION. NO ORAL OR WRITTEN INFORMATION OR ADVICE GIVEN BY PEARSON EDUCATION, INC., ITS DEALERS, DISTRIBUTORS, OR AGENTS SHALL CREATE A WARRANTY OR INCREASE THE SCOPE OF THIS WARRANTY. SOME STATES DO NOT ALLOW THE EXCLUSION OF IMPLIED WARRANTIES, SO THE ABOVE EXCLUSION MAY NOT APPLY TO YOU. THIS WARRANTY GIVES YOU SPECIFIC LEGAL RIGHTS AND YOU MAY ALSO HAVE OTHER LEGAL RIGHTS THAT VARY FROM STATE TO STATE. Pearson Education, Inc. does not warrant that the functions contained in the program will meet your requirements or that the operation of the program will be uninterrupted or error-free. However, Pearson Education, Inc. warrants the CD-ROM(s) on which the program is furnished to be free from defects in material and workmanship under normal use for a period of ninety (90) days from the date of delivery to you as evidenced by a copy of your receipt. The program should not be relied on as the sole basis to solve a problem whose incorrect solution could result in injury to person or property. If the program is employed in such a manner, it is at the user's own risk and Pearson Education, Inc. explicitly disclaims all liability for such misuse.

LIMITATION OF REMEDIES

Pearson Education, Inc.'s entire liability and your exclusive remedy shall be:
1. the replacement of any CD-ROM not meeting Pearson Education, Inc.'s "LIMITED WARRANTY" and that is returned to Pearson Education, or 2. if Pearson Education is unable to deliver a replacement CD-ROM that is free of defects in materials or workmanship, you may terminate this agreement by returning the program. IN NO EVENT WILL PEARSON EDUCATION, INC. BE LIABLE TO YOU FOR ANY DAMAGES, INCLUDING ANY LOST PROFITS, LOST SAVINGS, OR OTHER INCIDENTAL OR CONSEQUENTIAL DAMAGES ARISING OUT OF THE USE OR INABILITY TO USE SUCH PROGRAM EVEN IF PEARSON EDUCATION, INC. OR AN AUTHORIZED DISTRIBUTOR HAS BEEN ADVISED OF THE POSSIBILITY OF SUCH DAMAGES, OR FOR ANY CLAIM BY ANY OTHER PARTY. SOME STATES DO NOT ALLOW FOR THE LIMITATION OR EXCLUSION OF LIABILITY FOR INCIDENTAL OR CONSEQUENTIAL DAMAGES, SO THE ABOVE LIMITATION OR EXCLUSION MAY NOT APPLY TO YOU.

GENERAL

You may not sublicense, assign, or transfer the license of the program. Any attempt to sublicense, assign or transfer any of the rights, duties, or obligations hereunder is void. This Agreement will be governed by the laws of the State of New York. Should you have any questions concerning this Agreement, you may contact Pearson Education, Inc. by writing to:
ESM Media Development
Higher Education Division
Pearson Education, Inc.
1 Lake Street
Upper Saddle River, NJ 07458

Should you have any questions concerning technical support, you may write to:
New Media Production
Higher Education Division
Pearson Education, Inc.
1 Lake Street
Upper Saddle River, NJ 07458

YOU ACKNOWLEDGE THAT YOU HAVE READ THIS AGREEMENT, UNDERSTAND IT, AND AGREE TO BE BOUND BY ITS TERMS AND CONDITIONS. YOU FURTHER AGREE THAT IT IS THE COMPLETE AND EXCLUSIVE STATEMENT OF THE AGREEMENT BETWEEN US THAT SUPERSEDES ANY PROPOSAL OR PRIOR AGREEMENT, ORAL OR WRITTEN, AND ANY OTHER COMMUNICATIONS BETWEEN US RELATING TO THE SUBJECT MATTER OF THIS AGREEMENT.

PROGRAM INSTRUCTIONS

Windows:
When the CD is inserted it should run at startup. If not, double click on the START.exe icon to launch the program.

Macintosh:
When the CD is inserted it should run at startup. If not double click on the MACSTART icon to launch the program.

SYSTEM REQUIREMENTS

Windows:
Processor
    K6-2 300mhz or Intel Celeron 300mhz processor
RAM
    32 megs of RAM for Windows 98/ME (Windows system minimum)
    64 megs of RAM for Windows NT (Windows system minimum)
    64 megs of RAM for Windows 2000 (Windows system minimum)
    128 megs or RAM for Windows XP (Windows system minimum)
Operating System
    Windows 98, NT, 2000 , XP
Monitor Resolution
    800x600
Software Required
    Adobe PDF reader is required to view and print the maps.
Required Hardware
    8X CD-ROM
    Graphic card capable of 800x600 resolution and 8bit color
    Sound Card
    Printer
Internet Connection
    Not Required

Macintosh:
Processor
    PowerMac 9500 series processor or greater
RAM
    64 megs of RAM
Operating System
    OS 9 - X.3
Monitor Resolution
    800x600
Software Required
    Adobe PDF reader is required to view and print the maps.
Required Hardware
    8X CD-ROM
    Graphic card capable of 800x600 resolution and 8bit color
    Sound Card
    Printer required to print maps
Internet Connection
    Not Required

Adobe Acrobat Reader 5 is required to print the exercises from the CD. Acrobat for both Mac and PC systems are included on the CD.

Technical Support

If you are having problems with this software, call (800) 677-6337 between 8:00 a.m. and 5:00 p.m. CST, Monday through Friday. You can also get support by filling out the web form located at:

http://247.prenhall.com/mediaform.html